THIRD EDITION

Introduction to PCM Telemetering Systems

THIRD EDITION

Introduction to PCM Telemetering Systems

Stephen Horan

CRC Press is an imprint of the
Taylor & Francis Group, an **informa** business

CRC Press
Taylor & Francis Group
6000 Broken Sound Parkway NW, Suite 300
Boca Raton, FL 33487-2742

Printed on acid-free paper
Version Date: 20170605

International Standard Book Number-13: 978-1-138-74693-0 (Paperback)

Visit the Taylor & Francis Web site at
http://www.taylorandfrancis.com

and the CRC Press Web site at
http://www.crcpress.com

Contents

List of Figures

List of Tables

Preface

Telemetering systems provide the opportunity for scientists, engineers, and users across a wide spectrum of practitioners to acquire and make sense of data describing the world around them, no matter where they are and where the data originates. In this environment, the practitioners are often called upon to cross the traditional discipline boundaries to communicate with people in other disciplines to explain how they do their jobs and understand how the work of others enables the practitioners to do their jobs. It is not unusual for a computer scientist to need to communicate with a physicist and radio engineer about some aspect of data transmission and interpretation. For systems engineers, this type of cross-training is typical. However, for those newly graduated from traditional academic disciplines, the confusing array of terms in a new jargon can be bewildering. It is my hope that this text may help to narrow the differences between the fields and show that many similar concepts have a variety of names. I based the text, in large part, on concepts that I wish I had known when I first went to work in the telemetering industry after graduation.

This specific text grew from the development of the telemetering systems class at New Mexico State University as part of the master's degree program for electrical engineers. The selection of topics and level of depth were chosen for a one-semester class that was suitable for students from a variety of backgrounds. The references are accessible, often historical, and good for further in-depth study of the topics. Our class enrollment has included students at the advanced undergraduate level and the entering graduate student level from engineering, physical science and computer science backgrounds. The students have also come from local industry and government facilities in the state.

One question that students and colleagues have posed to me over the years is why there is not more concentration on specific hardware designs in the text. The short answer to that question involves the rapid pace of technology development. Designs that are valid today will not be relevant in two or three years. In the years since the first edition was finished, the Internet has revolutionized concepts for distributed data transmission both over landlines and wireless links. These trends will only continue with new devices supporting network-based standards, software-defined technologies, and the Internet of Things (IoT). Bottom line: the industry's design details will change but the fundamental concepts and standards upon which these technologies are based are reasonably stable and they are presented here. Technology textbooks do not have a long shelf life if the text is too dependent upon current design details.

I wish to acknowledge the help and encouragement of the faculty and students at New Mexico State University who have pushed me to make improvements since the first edition. I also wish to thank NASA and the Air Force for the research support over the academic portion of my career. Colleagues at the International Telemetering Conference, especially the students who took my short courses there, have provided good examples to incorporate in the text. Colleagues at NASA's Langley Research Center and other NASA Centers have also helped expand the content of this text. Finally, I wish to thank Dr. Sheila Horan for her critical reading of the manuscript and her support during this process.

The contents of this text are intended for general educational use. Neither the National Aeronautics and Space Administration nor any of its directorates at Langley Research Center are responsible for the content of this text or any materials referenced in this text. Reference herein to any specific commercial products, processes, or services by trade name, trademark, manufacturer, or otherwise, does not constitute or imply its endorsement or recommendation by the United States Government. NASA does not control or guarantee the accuracy, relevance, timeliness, or completeness of information in this presentation. This presentation represents the work of the author alone.

Stephen Horan
Poquoson, Virginia

Cover Photo The cover photo illustrates astronaut Mark Lee on an EVA with the Lidar In Space Technology Experiment (LiTE) payload in the cargo bay of *Discovery* during mission STS-64. Astronaut Lee is free flying around the shuttle as his own individual spacecraft using the Simplified Aid for EVA Rescue (SAFER) backback. During this ten-day mission in September 1994, LiTE provided forty-three hours of high-rate data on cloud structures, storm systems, dust clouds, pollutants, forest burning and surface reflectance based on gathering the return Lidar pulses. This photo was chosen because of the telemetry systems used by the entities in the photograph and because the LiTE payload was developed at NASA's Langley Research Center where this author works. Photograph courtesy of NASA.

Author

Dr. Stephen Horan, Ph.D., has over 30 years of engineering experience in the disciplines of communications and telemetry systems, small spacecraft design, and software within academic and government facilities. Presently, he is an electronics engineer with the National Aeronautics and Space Administration's Langley Research Center (LaRC). Dr. Horan joined LaRC in 2009 as a lead spacecraft communications engineer developing satellite and hosted payload concepts and their associated ground systems. He has also served as the branch head for the Remote Sensing Flight Systems Branch. Starting in August 2013, he became the Principal Technologist for Avionics in NASA's Space Technology Mission Directorate. In 2017, he also assumed the duties of the LaRC Spectrum Manager.

Dr. Horan continues as a professor emeritus at New Mexico State University (NMSU) where he held the Frank Carden endowed chair in Telemetering and Telecommunications from 1996 to 2009 when he retired as a professor and department head in the Klipsch School of Electrical and Computer Engineering. At NMSU, he developed the graduate telemetering systems course at the Klipsch School and the industry short course upon which this text is based. He taught other courses in communications systems and was the principal investigator on several telemetry and telecommunications projects at NMSU.

Dr. Horan earned his A.B. in physics from Franklin and Marshall College, and his M.S. in astronomy, M.S.E.E., and Ph.D. in electrical engineering from New Mexico State University.

Dr. Horan's research focus was in the area of space communications under the support of NASA and the Air Force, including leading the development of the communications system for the *3 Corner Satellite* and the *NMSUSat* small satellite projects. Prior to joining the NMSU faculty, Dr. Horan was with Space Communications Company (now part of General Dynamics) working in the areas of satellite telemetry and telecommand systems, operator interfaces, and systems engineering at NASA's White Sands Ground Terminal.

Dr. Horan is a senior member of the IEEE and AIAA. He also holds amateur radio license NM4SH. He has been a NASA/ASEE Summer Faculty Fellow at Johnson Space Center and Goddard Space Flight Center.

CHAPTER 1

INTRODUCTION

1.1 SYSTEM CONTEXT

The modern world runs on data. While much of this is e-mail, video, and audio data, there is an increasing volume of data and supporting applications dedicated to monitoring environments such as cars, buildings, and even people. With the coming Internet of Things (IoT), these data exchanges will become more prevalent in all areas of our lives and on our communications devices. While this text started with an emphasis on government-related telemetry systems, these same techniques are now finding their way into many aspects of our everyday life to support all of these modern data applications on our computers and smart phones. Telemetry systems, as we will discuss in this text, are part of our overall telecommunications infrastructure — even to becoming an enabling technology for the IoT as the next major utility [Stan14]. Before discussing the characteristics of the telemetry system, we will first start with the basics of data communications.

The field of telecommunications had traditionally been divided into two major segments: analog communications and digital communications. In the modern telecommunications world, we view nearly all communications as some form of digital data transmission. We often find it convenient, from regulatory and operational points of view, to make the distinction between pure data communications and analog signals that an electronic signal-processing device has converted to a digital format. In digital data communications, we have file exchanges, message exchanges, and measurement exchanges. The first example corresponds to using the File Transfer Protocol (ftp) to move a file from a server across the Internet to a destination computer, while the second example corresponds to an electronic mail message with an attachment. The measurement

Figure 1.1: Astronaut, Space Shuttle, and LiTE atmospheric science instrument in Earth orbit during STS-64. Photograph courtesy of NASA.

exchange category is, in large part, the concern of this text.

In this text, we look at techniques used to measure phenomena present in the natural world and communicating the results. While practitioners give these techniques different names in different disciplines, the users commonly refer to them as *telemetry systems*. Many of us first heard the term *telemetry* in connection with space flight, as Figure 1.1 exemplifies. All of the human-made items in the figure utilize telemetering systems: the environmental system for the astronaut, the system performance and positioning of the Space Shuttle, and the *LiTE* atmospheric science instrument's measurements and its operational state. The non-aerospace world is also filled with telemetry systems. There are further examples in the power system industry for control of the electrical grid and generators, medical electronics for remote patient monitoring, weather instrumentation for automatic data gathering and aggregation, and the automotive fields where cars have both wireless tire pressure gauges and data links back to the factory for vehicle health monitoring, to name a few. Future telemetry systems are expected to include elements that are part of our clothing and even embedded in our persons [Amen14; Stan14].

Figure 1.2 provides a high level overview of a generic telemetry system that is common in any application area. All telemetry systems have three major pieces: (1) the *payload segment* (the place where the measurements are made) containing the necessary measurement electronics, (2) the *user*

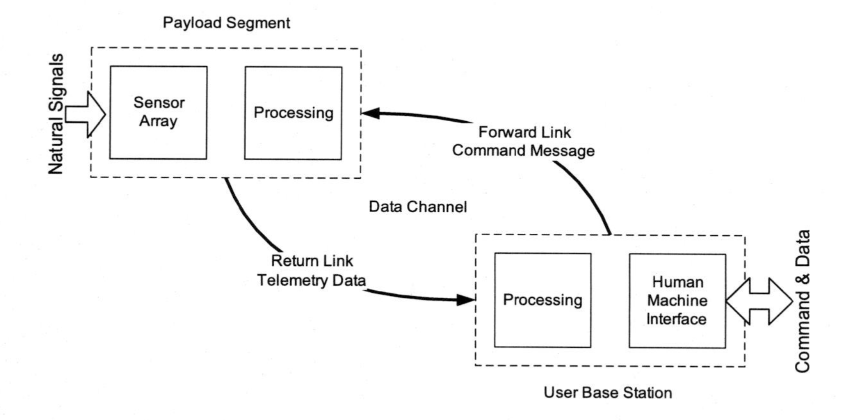

Figure 1.2: Data flow and processing modules in a telemetry and telecommand system.

base station for collecting the data and interacting with the operator, and (3) the *data channel* between the two. We refine the major pieces further into the following sub-components that Figure 1.2 shows:

Payload Segment using

- Natural signals that the system measures, for example atmospheric properties, engine properties, bodily functions, and manufacturing processes
- Sensors that make electronic signals from the natural signals
- Electronic components to process the sensor signals

Data Channel using

- Telecommunications links, such as electronic cables, radio links, telephone lines, and the Internet, over which data are sent to the user and user commands to control the process are sent to the payload

User Base Station using

- Processing electronics to prepare the data for the operator and accepts the operator's commands for the system to transmit to the payload

- User interfaces that display the data to the operator and accept the commands from the user

Before examining the details of each of these parts of the overall system, let us start with the basic telecommunications definitions that apply to the telemetry system and then set the context for how we will organize the overall system and the characteristics of each segment.

1.1.1 Definition of Telemetry and Telecommand

The standard *telemetry* definition in the telecommunications industry is [ATIS]

1. The use of telecommunication for automatically indicating or recording measurements at a distance from the measuring instrument.

2. The transmission of non-voice signals for the purpose of automatically indicating or recording measurements at a distance from the measuring instrument.

From this formal definition, we say that telemetering systems include the payload electronics to gather the data in a remote location and then transport the data to an operator/recorder in the user base station location. We refer to the transported information as *telemetry.* We call the process of the payload gathering the data as *telemetering.* For the moment, we defer any discussion of the data's nature or the means by which they are transported from the source to the destination other than to note that some form of telecommunications is used. For this and other definitions (see, e.g., [IEEE100]), the user acquires the measurements through an electronic data acquisition and sampling device. Based on that criterion, a person making an observation and then telephoning the result to another party is not a telemetry system. Alternatively, having your car send periodic performance measurements to the manufacturer over a cell phone link is a telemetering system.

The measurement process often has a complementary control activity whereby the operator sends commands to the measurement device. The command activity happens remotely so designers usually call it a telecommand process. The standard definition of *telecommand* is the

> use of telecommunication for the transmission of signals to initiate, modify or terminate functions of equipment at a distance [ATIS].

We use the terms *telecommand* and *command* interchangeably. The commanding activity configures the measurement system and controls the functions it should perform.

1.1.2 Link Definitions

As we saw above, a telemetry system needs a telecommunications link. We classify the links as operating in one of three modes [IEEE100]

Half Duplex Link: a channel that transfers data in two directions but only one direction at a time

Full Duplex Link: a channel that transfers data in two directions simultaneously

Simplex Link: a channel that transfers data in one direction only

A telemetry system without a commanding function only requires a simplex link. A telemetry and telecommand system requires one of the duplex options. In the system illustrated in Figure 1.2, the telemetry data link between the sensors and the user is called the *return link*. The telecommand data link between the user and the sensors is called the *forward link*. The forward and return data channels in Figure 1.2 are shown separately because they often are two different physical links; that is, they are two separate simplex links forming a duplex link. In other applications such as sending telemetry and telecommand data across the Internet, the data channel is one physical link that the forward and return links share.

1.1.3 Pulse Code Modulation Definition

Unless the system is making a very simple set of measurements, the system designer must find a way to transmit several sensor channels simultaneously from the data acquisition system to the operator's base station. The designer has two options when transmitting the system's measurements: as either analog data or digital data. The telemetry community has a long history of sending data in the original analog format. With the advent of computer-based data processing, system designers and end users generally consider digital data more useful for signal processing (although at a slight loss of the total information content in the system).

We call the process of sampling and converting an analog signal to a

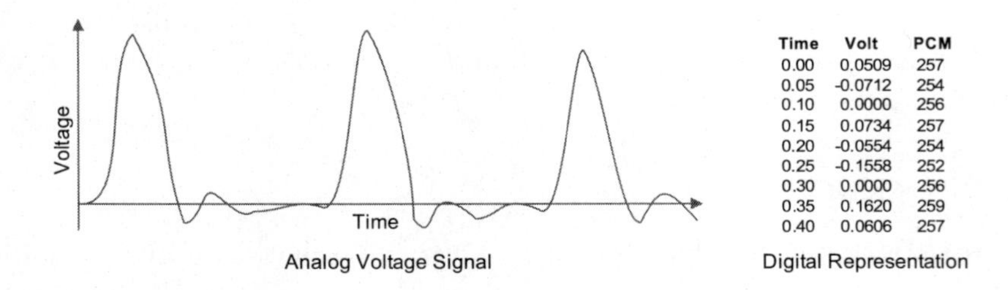

Time	Volt	PCM
0.00	0.0509	257
0.05	-0.0712	254
0.10	0.0000	256
0.15	0.0734	257
0.20	-0.0554	254
0.25	-0.1558	252
0.30	0.0000	256
0.35	0.1620	259
0.40	0.0606	257

Figure 1.3: Synthetic ECG representation: analog voltage representation of original signal, digital representation of sampled voltage, and PCM converted values.

digital data representation and then transmitting the digital data *Pulse Code Modulation (PCM)* [Oliv48]. To make a PCM system, the designer must have

1. An input signal with a fixed maximum bandwidth
2. A means to sample the signal at a rate at least twice the maximum bandwidth
3. A means to encode the samples into discrete values
4. A means to transmit and receive the encoded values
5. A means to regenerate the signal from the encoded values and then low-pass filter the values with filter bandwidth equal to the maximum signal bandwidth

For example, Figure 1.3 illustrates a synthesized Electrocardiograph (ECG) signal. The analog signal has a continuous-amplitude voltage versus continuous time, which is a pattern that the heart should produce. The sampled function is a table of voltages at the sample times and the associated converted values for the first several sample times. With a PCM transmission, the payload sends a coded value, and not the actual measured voltage value. We will discuss the PCM process in detail in Chapter 5.

When the system designer needs to transmit multiple sensor measurements, the designer has three methods for resource sharing or multiplexing to provide a "fair" allocation for each sensor or data source. These options are for data sharing in the

Time Domain where data are assigned to a unique time slot for transmission

Frequency Domain where data are assigned to a unique frequency for transmission

Code Domain where data are assigned to a unique, orthogonal mathematical code for transmission

In each of these domains, only one data source can occupy a time, frequency, or code, respectively, at a given instant. Figure 1.4 illustrates these three ways of sharing resources. The details of accomplishing this resource division are the subject of Chapter 6. If time-domain multiplexing is used, then the system is referred to as a *Time Division Multiplexing (TDM)* system where each sensor is given a specified time slice of the available transmission time as shown in the first panel in the figure. Likewise, if the system uses frequency-domain multiplexing, then the system is a *Frequency Division Multiplexing (FDM)* system where each channel is transmitted over a unique frequency as shown in the middle panel in Figure 1.4. Finally, if code division multiplexing is used, as in the last panel of Figure 1.4, the system is referred to as a *Code Division Multiple Access (CDMA)* system where the data sources share a common transmission frequency band and transmit at the same time but use the unique, orthogonal code properties to differentiate the data sources.

All strategies have advantages and disadvantages. TDM and CDMA systems are used in digital transmission environments, while FDM systems are used in analog transmission environments. Within the CDMA system, a TDM process is frequently used to allow multiple sensors from the same data source to share the code space. In this text, we concentrate on the TDM and PCM methods; however, many of the topics have FDM counterparts that work in much the same way, while the CDMA systems build on the TDM techniques.

1.2 SYSTEM COMPONENTS

Figure 1.2 showed the overall structure of a telemetry and telecommand system. To give readers a better idea of the issues we discuss in the remainder of the text, we expand Figure 1.2 to show the major details in each segment. Figure 1.5 illustrates this expanded telemetering and telecommand system. This block diagram has much in common with generalized communications systems. This commonality should be no

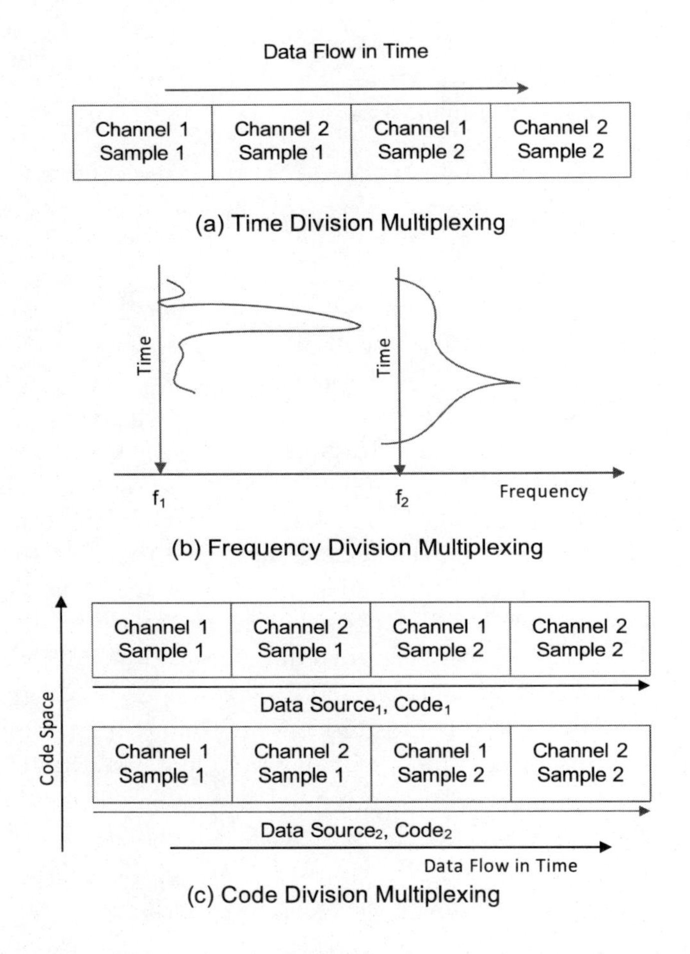

Figure 1.4: Methods for transmitting sensor information via (a) time division, (b) frequency division, and (c) code division multiplexing communications techniques. Time division allocates time slots, frequency division allocates carrier frequencies, and code division allocates orthogonal mathematical codes.

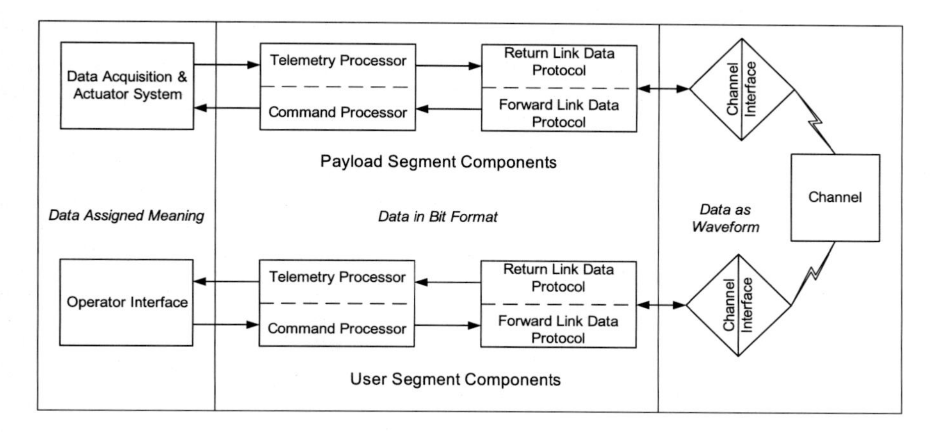

Figure 1.5: Major elements of the overall telemetry and telecommand system.

surprise because a telemetering and telecommand system is really a specialized communications system. The system structure presented in Figure 1.5 is the icon around which we organize the telemetry-system component discussion throughout the text.

Because the user may often need to configure the payload electronics, Figure 1.5 shows a telecommand data path along with the telemetry data path. We must keep in mind that the two-way nature of many payloads influences the data that are gathered. For the purposes of this text, we assume that all telemetering systems are really bidirectional systems even if we only discuss one side. An implied symmetry usually exists between the telecommand and telemetering sides. Commands tend to be intermittent, short data groups while the telemetry tends to be either continuous or at least considerably longer data groups. However, from a functional point of view, what is present on one link usually is present on the other.

As we saw earlier, the command and telemetry system's components are divided into two major blocks: the user segment and the payload segment. This is a designation to indicate the relative relationship between the components. The payload might be an instrumentation rack on one side of a room with the operator sitting at a desktop computer on the other side of the room, and a coaxial cable serving as a channel between them. The paragraphs below describe the functions of each major system.

The *user segment* primarily deals with the operator's interface and data processing functions. The user segment has as its major subsystems the operator interface, the baseband data processing, the signal processing hardware for data transmission, and the interface to the transmission channel in the user base station. For the purposes of this text, we assume that the channel is a radio link, although this is by no means the only way to transmit data. We make the assumption because most telemetry systems use radio links. The functions of each of the subsystems shown in Figure 1.5 are as follows:

Operator Interface provides data displays for the users and the means for the operator to enter commands to configure the interface, the payload, or support entities.

User Segment Telemetry Processor/Command Processor processes raw inputs into a format understood by either the operator or the payload.

Return Data Link Protocol/Forward Data Link Protocol processes the data for the transmission protocol format; these components also apply any encryption/decryption algorithms.

User Segment Channel Interface provides an interface with the channel for data transmission or reception; this may be an antenna, fiber optic, wire cable, etc.

The *channel* is the transmission medium between each of these components in the payload segment and their counterparts in the user segment. Engineers characterize the channel by how it affects the transmitted signals. Some characteristic parameters are

- Noise characteristics (white noise or narrow-band noise)
- Fading characteristics (multipath or atmospheric effects)
- Jamming or interference (jamming is intentional interference while unspecified interference is assumed to be unintentional)
- Power loss effects due to rain loss, atmospheric absorption, etc.

The *payload segment* deals with data acquisition from sensors that measure the environment and the payload operating characteristics. Usually, the payload has only limited data processing functions — although this is changing rapidly as embedded computer systems become feasible and cost effective. The payload segment has as its subsystems the sensor system, the signal conditioning electronics, the baseband data collection, the signal processing hardware for data transmission, and the interface to the transmission channel. The functions of each of these subsystems are as follows:

Data Acquisition and Actuator System is composed of various devices for producing electronic signals that are proportional to a desired measurement; this includes both the *science* sensors and the *health and welfare* sensors in the system. It also includes the electronics for configuring the measurement system and interacting with the external environment via the actuators.

Payload Telemetry and Command Processor provides any necessary in-payload telemetry data processing as well as processing commands sent to the payload that cause the payload to take some action.

Return and Forward Link Data Protocol Processor performs the necessary transmission protocol data wrapping/unwrapping specified in the protocol and applies any encryption/decryption algorithms.

Payload Channel Interface provides an interface with the channel for data transmission or reception; this may be an antenna, fiber optic, wire cable, etc.

Naturally, the degree of complexity of these components is a strong function of the complexity of the overall system. Systems with several hundred sensors, multiple operator consoles, and high-speed transmission are more complicated than equipment that measures the output of a handful of sensors across the room. However complicated its purpose, we still break the system into these functions.

1.3 ORGANIZATION OF THE TEXT

We group the text into three major parts: (1) System Elements, (2) Data Transport, Timing, and Synchronization, and (3) Data Transmission Techniques. Our goal in System Elements is to understand how the system designers generate the measurement data, configure the two endpoints of the overall command and telemetry system, and represent that data at the endpoints. This part includes discussions of the sensors that perform the measurements and the operator interface. We also present topics related to data sampling, signal processing, and real-time computer systems. These topics are vital to understanding how the telemetry system generates the data and how the users display and analyze the data. If the system designer executes the design correctly, major changes in the data transport and hardware systems can be made and the same data generations and user interfaces remain at the payload and user nodes.

Our goal in Data Transport, Timing, and Synchronization is to understand the issues the system designer faces in formatting and transmitting the data from the sensors in the payload to the operators in the base station. The components in the subsystems found here do not really place any interpretation on the sensor data — all the bits look the same to the subsystems! The only data interpreted here are those relative to maintaining the synchronization between the payload and the user segment. In this section, we examine data transport from the traditional frame format used in space and military systems as well as the trends in packet techniques. Packet techniques are common in those telemetry systems that act more like standard computer-to-computer communications. In this section, we make a distinction between telemetry subsystems and command subsystems with the realization that from a communications perspective, they are two halves of the same link; from the transmission hardware's perspective, the link direction is not important.

With the final part, Data Transmission Techniques, our goal is to understand the techniques designers use in selecting components to sustain the data transmission links. It has a certain bias toward traditional radio telemetry systems because we find radio-based systems in many applications. However, fiber optic and laser systems are very important emerging technologies. In addition, as computer communications become more integrated into telemetry systems, standard communications protocols and services are used on the links. In that case, the command and telemetry system becomes just another user on a shared, commercial link that may be carrying voice, video, and data.

With this organization, we have broken the problem of data acquisition and transport along natural partitions matching the ways practitioners typically work. The reader should view these partitions, at least from a pedagogical point of view, as soft partitions. To understand fully the command and telemetry system, one must know how the parts interact and how the results of one subsystem influence the actions of another. If it were possible, it would be best to study all of the chapters in parallel to see the interaction. Given that many of us process information serially, keep asking yourself how this chapter interacts with the material presented in the surrounding chapters as you progress from one topic to the next.

1.4 REFERENCES

[Amen14] S.A. Amendola et al. "RFID Technology for IoT-Based Personal Healthcare in Smart Spaces." In: *Internet of Things Journal* 1.2 (Apr. 2014), pp. 144 –152. DOI: 10.1109/JIOT.2014.2313981.

[ATIS] Alliance for Telecommunications Industry Solutions. *ATIS Telecom Glossary 2011.* 2011. URL: http://www.atis.org/glossary/.

[IEEE100] Institute of Electrical and Electronics Engineers. *IEEE 100 The Authoritative Dictionary of IEEE Standards Terms Seventh Edition.* IEEE-STD-100. IEEE Press. Piscataway, NJ, 2000. DOI: 10.1109/IEEESTD.2000.322230.

[Oliv48] B. M Oliver, J. R. Pierce, and C. E. Shannon. "The Philosophy of PCM." In: *Proc. IRE* 36 (11 Nov. 1948), pp. 1324 –1331. DOI: 10.1109/JRPROC.1948.231941.

[Stan14] J.A. Stankonic. "Research Directions for the Internet of Things." In: *Internet of Things Journal* 1.1 (Feb. 2014), pp. 3 –9. DOI: 10.1109/JIOT.2014.2312291.

I

SYSTEM ELEMENTS

A telemetry and telecommand system forms a specific type of telecommunications network. The payload and the operator interface are communications peers that exchange messages as illustrated with the direct arrows in the graphic below. The Transmission Channel pipes connecting the Payload and the User Interface represent the actual system that formats and manages the message flow between the endpoints. This first part of the text is concerned with the system's data and the payload and operator interface segments representing that data. The balance of the text after this part deals with how these two segments communicate.

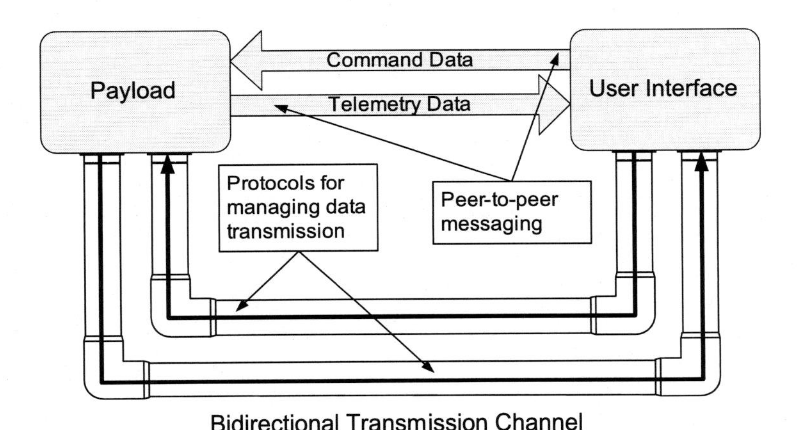

The telemetering and telecommand system pipe diagram illustrates the payload and user as communications peers with the communications protocols supporting the data flow.

In Part I, we investigate the following system elements that are instrumental in generating and representing the data:

Sensor Technology: An overview of example sensor technologies for telemetry measurements

Modeling and Calibration: Mathematical modeling and calibration of sensor data to make proper measurements, fit the data, and express the uncertainty in the measurements

Telemetry and Command Computer Systems: Computer data and user interfaces for controlling the telecommand and telemetry processes

Signal Processing: Sampling, filtering, and signal processing of sensor data

The sensor technology overview covers the concepts behind several sensor families and their outputs. The modeling and calibration discussion reviews the ways in which experimenters use mathematical models, such as least squares analysis, to represent the mapping between inputs and outputs. The computer system discussion includes user interface functions, data input and output methods, and supporting software concepts. The signal processing discussion includes data sampling, analog-to-digital conversion, filtering, and limiting signal bandwidth.

CHAPTER 2

MEASUREMENT TECHNOLOGY

2.1 INTRODUCTION

The telemetering system's function is to make one or more measurements for transmission. The International Vocabulary of Metrology (VIM) [VIM12] defines a *measurement* as the "process of experimentally obtaining one or more quantity values that can reasonably be attributed to a quantity." The system designer needs a *measurement instrument*, which is a "device used for making measurements, alone or in conjunction with one or more supplementary devices," to make each measurement [VIM12]. To make the ensemble of measurements for the telemetering system, the designer needs a *measuring system* composed of a

> set of one or more measuring instruments and often other devices ... assembled and adopted to give information used to generate measured quantity values within specified intervals for quantities of specified kinds [VIM12].

The heart of any measurement instrument is its *sensor*, which is the

> element of a measuring system that is directly affected by a phenomenon, body, or substance carrying a quantity to be measured [VIM12].

The electronics industry has produced many different sensors for the practitioner to use in measurement applications. Quite often, the manufacturers' data sheets and catalogs are the best source for applications information that explains how to use a given sensor to produce meaningful and repeatable results. In this chapter, we examine the characteristics of several representative sensors to see how the process works and the types of outputs generated. Our intention is not to examine all of the available sensors and the corresponding ways in which they are used. Rather, the purpose is to demonstrate the characteristics found in representative measurements and to become familiar with the typical concerns found in all systems.

When we discuss a sensor, we are referring to an electronic device that is composed of two major parts: a detector and support electronics. A *detector* is a

> device or substance that indicates the presence of a phenomenon, body, or substance when a threshold value of an associated quantity is exceeded [VIM12].

In our context, a detector responds to a natural signal, such as temperature, to produce an electrical signal, such as current or voltage. The sensor contains at least one detector responding to the natural signals plus other associated electronics necessary to produce a stable output signal such as voltage stabilization electronics to keep the sensor's output steady. Figure 2.1 illustrates an overall arrangement for a measurement instrument showing the sensor, signal processing electronics, and supplemental electronics to make the instrument workable in the overall measurement system. The sensor converts the input signal into a voltage that the Analog-to-Digital Converter (ADC) measures. The ADC bits are the instrument's output signal. The support electronics control the output interface and regulate the power supplied to the components. These signals enter or exit the device through the Input/Output (I/O) port.

Figure 2.2 looks at the overall telemetry and telecommand system from a signal processing perspective where the sensor is part of the data acquisition system highlighted in the Figure. In Section 3.3.1, we will look at sensor calibration techniques that make them function as reliable measurement instruments. In Chapter 5, we will see how signal processing electronics are used to make the sensor output into a Pulse Code Modulation (PCM) data stream.

2.2 OBJECTIVES

Our objectives for this chapter are to understand the models describing the performance of representative sensors. We use these models in sample

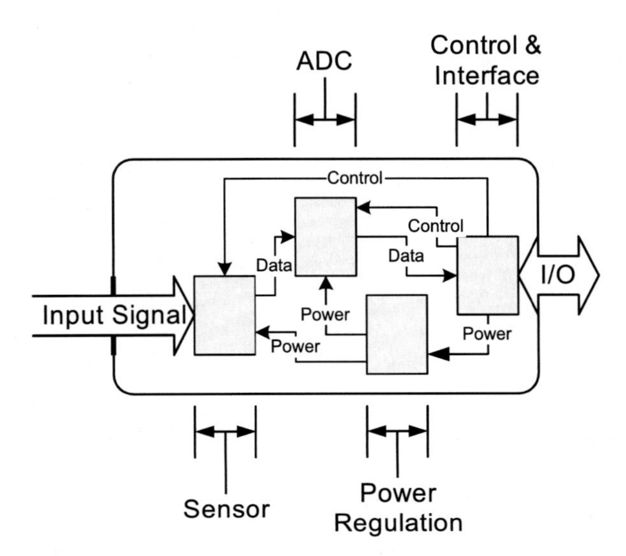

Figure 2.1: Major electrical and signal processing components found in a typical measurement instrument. The detector is inside the sensor component.

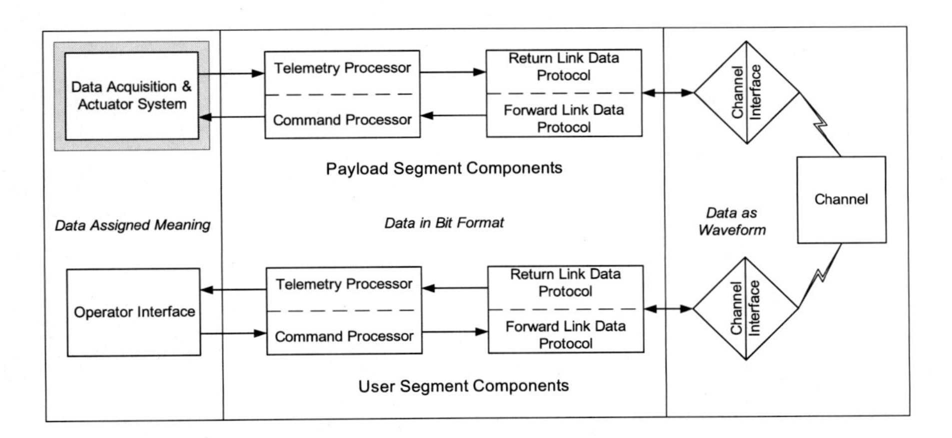

Figure 2.2: The highlighted region illustrates the placement of the sensors that are part of the data acquisition system within the overall telemetry and telecommand system.

circuits to compute parameters such as voltage and resistance as a function of the input stimulus. This chapter does not go into the full details of all sensor classes. Rather, we discuss the common elements of sensors found in telemetering systems. At the conclusion of this chapter, the reader will be able to

- Characterize sensor attributes such as precision and accuracy
- Describe the sensor input-output relationship
- Classify sensors by the output produced (voltage, current, or numeric)
- Configure a simple measurement circuit using a Wheatstone bridge
- Explain the operating characteristics of several common sensors

Once the reader attains familiarity with the sensors and their usage, the reader should be able to recognize other sensor examples using many of the same techniques and generalize the results to other applications.

2.3 MEASUREMENT ENVIRONMENT

2.3.1 General Components

Figure 2.3 shows the sensor environment within the data acquisition system. The *applied stimulus* is the environmental signal that the telemetering system is trying to measure, i.e., the payload "science" information. The sensor, which is composed of the detector and the supporting signal conditioning electronics, converts the applied stimulus into an electrical signal that a device such as an ADC measures. The ADC output is then processed into a format suitable for transmission across the channel. Typical environmental measurements are

- Temperature or humidity
- Intensity or color of light
- Force or pressure exerted on a surface or body
- Flow rate of a fluid
- Position or distance, velocity, and acceleration
- pH of a substance

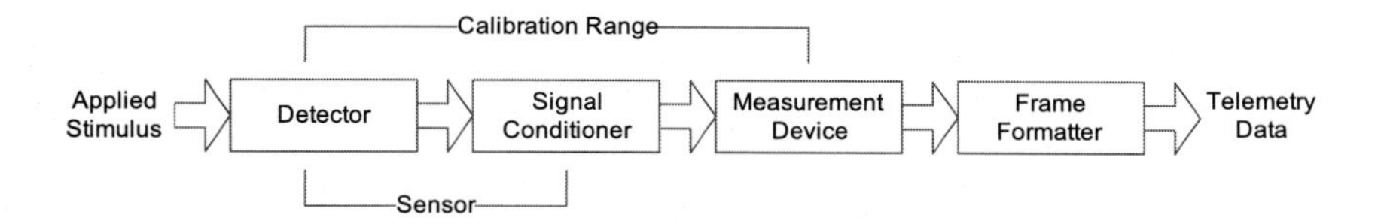

Figure 2.3: Overall measurement architecture. This shows the signal flow from the source to the output the system is readying for transmission to the user.

- Biomedical variables, e.g., blood pressure or heartbeat

The sensors may make these measurements as a function of time of day or location. The data analyst then processes these measurements to determine some environmental parameter, for example, sampling temperature and humidity to construct a model of an aspect of the Earth's atmosphere. This latter stage is beyond the discussion of telemetering systems; however, the data analyst who is constructing the detailed model needs to know the technology and limitations of the data acquisition methods.

Telemetering systems typically include measurements both of the payload science and the payload's state. The measurement system's purpose, for example atmospheric measurements, biomedical measurements, etc., define *payload science* measurements. The measurements about the payload's state are often called *health and welfare* or *housekeeping* measurements and they report parameters such as battery voltage, temperatures in the payload, switch settings, fault settings, etc. Operators and analysts use these measurements to verify that the payload is operating correctly and to know how to interpret the payload science measurements properly. The telemetry system transmits both measurement sets and the users process both sets with the same types of mathematical measurement models.

The environmental stimulus produces an electrical output, such as a voltage or a current, in the sensor's detector; voltage output is the most common. The sensor, with its associated electronics, produces a signal proportional to the analog signal present at the sensor's input terminal. The sensor's response function and the associated electronics modify this proportional signal. For example, an environmental variable has a value of 0.995 in normalized units. The overall sensor package has a response sensitivity (input signal gain) of 0.1 for an input of that magnitude. The sensor package output is then 0.0995 and the data analyst must know what input setting produced this output value. The development of this mapping is the *calibration process* that we discuss in Section 3.3.1. Some measurements in the measurement set are in terms of time and not electrical measurements; for example, time-of-flight measurements between two pulses that correspond with measuring a distance.

As a practical note, the designer must match the sensor to the conditions under which the sensor makes the measurement. As expected, acceleration meters must be able to withstand the full range of accelerations experienced. However, if the sensor is subject to unusual temperature extremes while measuring the acceleration, then the designer must select the acceleration sensor with both the temperature range and the acceleration levels in mind.

Many sensors, for example, thermocouples, produce outputs that are static electrical voltages lacking the ability to source sufficient current to drive the measurement devices and produce an accurate reading. This implies the need for signal conditioning electronics to provide the necessary electrical drive and hence the need for the signal conditioner that is shown in Figure 2.3. Signal conditioners may also provide filtering circuitry to eliminate noise generated by the environment.

The next block in Figure 2.3 is the measurement device that estimates the sensor's voltage or current value as generated by the signal conditioning electronics. The measurement device's output is a quantized representation of the sensor's voltage or current. The quantization converts the continuous analog signal into a discrete digital signal. Next, the measurement device presents its output to the frame formatter. The *frame formatter* is the component in the telemetry system that packages the information from all of the sources in the payload. The formatter gives each measurement its assigned time slot for transmission to the user base station.

Table 2.1 provides an example of this process where we examine a set of pressure measurements to deduce height for a rocket payload starting from the raw data. The first column in the table contains the raw data that the payload sent to represent the actual height values. The second column removes a formatting artifact of this particular data acquisition system. In this case, the measurement device produced a 10-bit output, but the payload memory could only accommodate eight bits so the designer deleted the two least significant bits with a subsequent slight loss of data resolution. To recover the data, we multiply each raw datum by four. At this point, we have an estimate of the measurement device's output. The third column converts the digital value back to an approximation of

Table 2.1: Recovering Height Measurements from Recorded Data

Recorded Measurand Raw Count	Actual ADC Output Unitless	Sensor Output Volts	Pressure Millibar	Height Kilometers
175	700	3.418	865.104	1.278
176	704	3.438	869.444	1.238
175	700	3.418	865.104	1.278
174	696	3.398	860.764	1.319
172	688	3.359	852.083	1.401
165	660	3.223	821.701	1.695
160	640	3.125	800.000	1.912

the sensor's original analog voltage value. The fourth column shows how we use the manufacturer's sensor calibration to convert from measured voltage back to atmospheric pressure in millibar. The fifth column shows how we use a model atmosphere to convert the pressure to an estimate of the payload's altitude which is the result we seek.

The entire system should undergo periodic calibrations either at its usage site or in a special calibration laboratory to verify that measurements are accurate. The calibration process ensures a reliable mapping between the measurement device's output and the applied stimulus. The calibration should include the mapping between the applied stimulus and the output plus any changes in the output that may happen due to environmental variables such as temperature, humidity, and aging.

In the next section, we review standard definitions related to measurements, sensors, and the quality of the results. Then we examine several sensor classes used in telemetering systems.

2.3.2 Measurement Definitions

This section covers standard definitions for the measurement process that apply to all the sensors that we examine later. The goal is to provide a common framework for the electrical characterization of the sensors and to provide some background for the statistical discussions in Chapter 3.

2.3.2.1 Measurement

We start by exploring the measurement process in more detail. A slightly different definition to the one given earlier is that a measurement

> is the determination of the magnitude or amount of a quantity by comparison (direct or indirect) with the prototype standards of the system of units employed [IEEE100].

Notice that (a) measurement involves an experimental approach and (b) it is not enough to capture the signal; we also must be able to relate the output to an agreed-upon system of units to make the measurement meaningful. In this process, the physical quantity, property, or condition to be measured is called the *measurand* [IEEE100; VIM12]. The measurand is the knowledge that we are seeking, while the measurement is a sample representation of the knowledge we seek in the proper units.

The detector in the sensor is usually a transducer. A *transducer* is a

> device to receive energy from one system and supply energy (of either the same or of a difference kind) to another system, in such a manner that the desired characteristics of the energy input appear at the output [IEEE100].

A transducer has two possible realizations: a measuring transducer and an actuator. A *measuring transducer* is a "device, used in measurement, that provides an output quantity having a specified relation to the input quantity" [VIM12]. An *actuator* works in the reverse direction in that it takes an input signal generated in the payload and produces an output that works on the environment. The measuring transducers create the telemetry data and the actuators respond to the telecommand data. Figure 2.4 illustrates the relationship between transducers, sensors, actuators, and the operator. The sensor supplies information about the environment to the user. The actuator changes something in the environment at the operator's direction. The hybrid allows both actions in one device.

2.3.2.2 Input-Output Relationship

Figure 2.5 illustrates an example mathematical input-output relationship that characterizes electronic devices. This mapping provides the input excitation, I, as the independent variable and the sensor output (voltage or current), S, as the dependent variable. Analysts expect that the

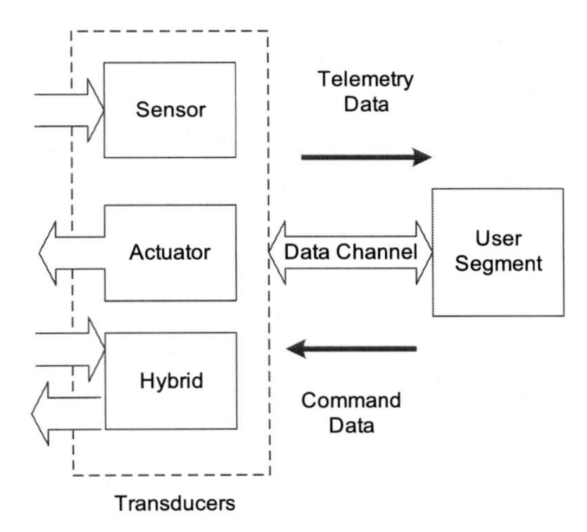

Figure 2.4: Transducer, sensor, and actuator interactions from the data flow point of view.

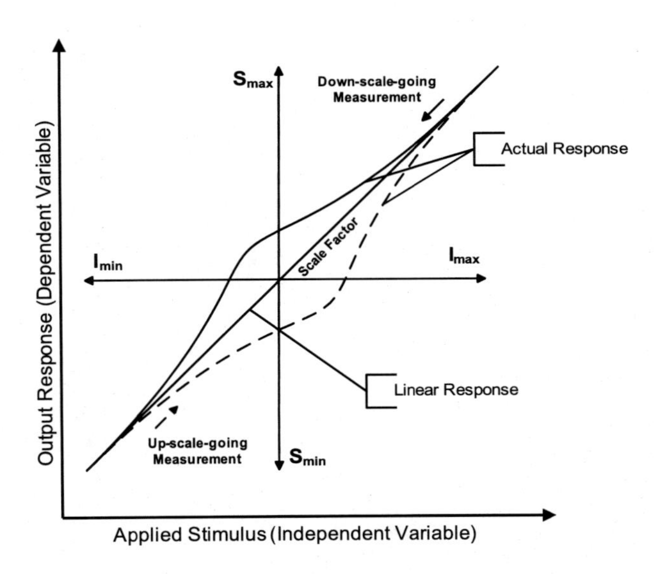

Figure 2.5: Representative sensor input-output mapping. The output variables are typically either voltages or currents. The actual and ideal linear responses are illustrated.

input/output relationship is approximately linear over the region of interest. This relationship is often determined in two directions: one plot with the independent variable changing from low to high and a second plot with the independent variable changing from high to low. A naive expectation is that both curves are exactly aligned but, in reality, they frequently are not due to hysteresis effects (see Section 3.4.4 for the formal definition of hysteresis).

The sensor designer associates defined limits for the sensor with the input-output mapping. The sensor is limited both in terms of the valid independent input variable and dependent output variable. If the experimenter exceeds these ranges, the sensor may experience problems. The input full range on the sensor is $\Delta_I = I_{max} - I_{min}$ while the sensor's output span is $\Delta_S = S_{max} - S_{min}$. Experimenters define the *scale factor*, SF, also known as the *sensitivity*, for the sensor as the ratio of Δ_S to Δ_I or [Frad97; IEEE100; Klaa96; VIM12]

$$SF = \frac{\Delta_S}{\Delta_I} \tag{2.1}$$

As may be expected, realistic sensors cannot correctly respond to every possible measurand level that nature generates nor do they provide differentiation between two arbitrarily closely spaced measurand values. The *sensor resolution* is the "smallest change in a quantity being measured that causes a perceptible change in the corresponding indication" [VIM12]. The complement to the resolution is the *discrimination threshold* which is the "largest change in a value of a quantity being measured that causes no detectable change in the corresponding indication" [VIM12]. In practice, the sensor does not reach its steady-state value output immediately, but will take a finite to to do so. This is rated with the *step response time* or the "duration between the instant when an input quantity value of a measuring instrument or measuring system is subjected to an abrupt change between two specified constant quantity values and the instant when a corresponding indication settles within specified limits around its final steady value" [VIM12].

Often, we wish to know the magnitude of the smallest and largest signals that we measure relative to the sensor's resolution. At the low end, the *sensor threshold* is a "value of voltage or other measure that a signal must exceed in order to be detected or retained for further processing" [IEEE100]. This is similar to the *detection limit* which is the "measured quantity value, obtained by a given measurement procedure, for which the probability of falsely claiming the absence of a component in a material is α, given a probability β of falsely claiming its presence" [VIM12]. The VIM suggests $\alpha = \beta = 0.05$ and recommends not using sensitivity for detection limit. The detection limit level is a function of noise level in the measurement, low-level distortion, interference, and the sensor resolution level. The *sensor dynamic range* is the "difference, in decibels, between the overload level and the minimum acceptable signal level in a system or transducer" [IEEE100]. The maximum input level is also known as the *overload level.*

2.3.2.3 Precision, Accuracy, and Reproducibility

An experimenter making measurements at a specific time cannot guarantee that the measurement system produces exactly the same results on a different day. Nor does the experimenter guarantee that a different person obtains exactly the same results, even if nothing in the system appears to have changed. Precision and accuracy give measures of how well behaved the sensor system is over time or with different users. *Measurement precision* is the

> closeness of agreement between indications or measured quantity values obtained by replicate measurements on the same or similar objects under specified conditions [VIM12].

Experimenters frequently express precision by using the standard deviation in the measurements [IEEE100]. *Measurement accuracy* is the

> closeness of agreement between a measured quantity value and a true quantity value of a measurand [VIM12].

Figure 2.6 schematically illustrates precision and accuracy where we see three representative clusters of measurements relative to the true value. A related characteristic is the *reproducibility* of the sensor or its ability to produce the same value or result under the same input conditions and operating in the same environment [IEEE100]. A good measurement set must be accurate and precise. A valuable measurement system must also provide reproducible results.

Occasionally, someone asks whether it is better to have high precision or high accuracy. The experimenter needs high precision to determine whether the measurements have high accuracy, so the experimenter should design the measurement process to produce both. However, given the choice between low precision and low accuracy, the experimenter generally prefers lower accuracy as long as the process is repeatable. Lower accuracy

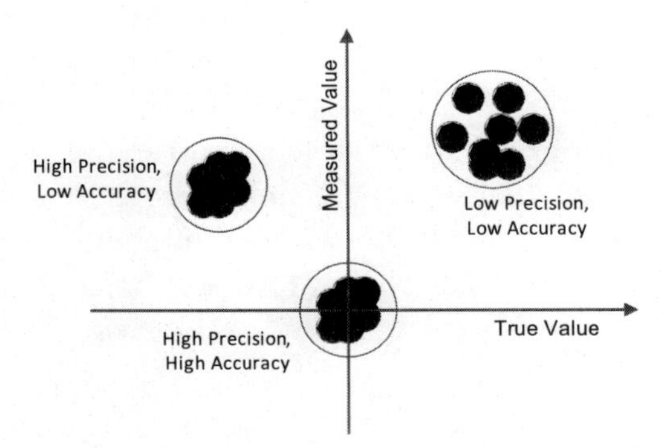

Figure 2.6: The relationship between precision and accuracy relative to the true value. Only the high precision, high accuracy case has the measurements tightly clustered near the true value.

with high precision offers the experimenter the potential to calibrate for the difference and treat it as a systematic bias. Low precision leaves the experimenter guessing about the nature of the measurements.

Experimenters use mathematical definitions to quantify precision and accuracy [IEEE100]. Let x_n be the nth measurement in a series that has an average value of μ and an estimated standard deviation of σ. Also, let X be the "true" value the measurements are attempting to replicate with a "true" standard deviation of σ_X for an infinitely large number of measurements. We define the accuracy of the nth measurement, A_n, as

$$A_n = \frac{x_n - X}{X} \times 100\% \approx \frac{x_n - \mu}{\mu} \times 100\% \tag{2.2}$$

In a similar manner, we define the precision of the nth measurement, P_n, as

$$P_n = \frac{x_n - X}{\sigma_X} \times 100\% \approx \frac{x_n - \mu}{\sigma} \times 100\% \tag{2.3}$$

Notice accuracy in this definition is related to the "true" or standard value, while precision is related to the measurement set's standard deviation. Also notice that the larger the value for each, the worse the quality of the measurement. In a sense, we are computing the measurement's *inaccuracy* and the *imprecision*. Since we do not know the exact values, we approximate the exact values with the estimated values found from the measurement set. Table 2.2 provides an example of these concepts using a measurement set with a designed true value of 1.125. We compute

Table 2.2: Example Accuracy and Precision Computation for a Sample Data Set with Random Measurement Errors

	$X = 1.125$	$\mu = 1.109$	$\sigma = 0.091$
$\mathbf{x_n}$	**True $\mathbf{A_n}$**	**Est. $\mathbf{A_n}$**	$\mathbf{P_n}$
1.165	3.56%	5.05%	61.7%
1.205	7.11%	8.66%	106%
1.085	-3.56%	-2.16%	-26.2%
1.125	0.00%	1.44%	17.6%
1.165	3.56%	5.05%	61.7%
0.885	-21.3%	-20.2%	-246%
0.125	0.00%	1.44%	17.62%
1.125	0.00%	1.44%	17.62%
1.045	-7.11%	-5.77%	-70.5%
1.165	3.56%	5.05%	61.7%

the mean, A_n, and P_n from the sample measurements using Equations (2.2) and (2.3). The table lists both the actual and the estimated accuracy values and the estimated precision.

2.3.2.4 *Absolute Measurement and Differential Measurement*

The measurement designer may configure the sensors to operate in either absolute or differential measurement modes. In *absolute measurement* mode, the sensor manufacturer references the sensor's output to an absolute system of units. For example, a pressure sensor may be set to make measurements relative to either a vacuum or one atmosphere. In *differential measurement* mode, the output is proportional to the difference between two input values. The measurement designer uses this mode, for example, in pressure measurements to determine the relative pressure on either side of a barrier.

2.4 REPRESENTATIVE SENSOR TECHNOLOGY

The sensor's function is to translate some physical parameter into an electrical analog, which is the information that the payload transmits to the base station and/or records on a physical medium for later analysis. The telemetry system transmits a modified and sampled version of that analog signal. We will investigate the sampling process more in Chapter

5. Here, we use the different sensor types to illustrate points of concern and not attempt to describe all possible variations. In this section, we examine

1. Sensors based on generalized resistance properties
2. Sensors based on capacitive properties
3. Sensors based on fundamental physical effects
4. Sensors based on semiconductor technology
5. Sensors based on time measurement
6. Sensors based on technology hybrids

These sensors are used in a variety of applications in addition to the typical applications we mention here.

2.4.1 Resistive Sensors

Sensor designers make a large number of measurements by causing a resistive device to change in response to the measurand. The measurement designer changes the process in this case into making a precise voltage measurement. An added benefit or bane, depending upon the application, is that resistive devices are also temperature sensitive. This is an advantage to the designer making a thermometer but it causes problems if the designer is making different measurements in an uncontrolled environment. In this section, we examine some of the ways resistive devices are used and how the measurement designer compensates for temperature effects. In particular, we examine

1. Pressure, strain, or force measurements
2. Temperature measurements
3. Light measurements
4. Position measurements

The designer bases each measurement on some change in resistance in response to the measurand. Each of these becomes an exercise in voltage measurement. The reader can then generalize these voltage measurement techniques to other devices that operate in the same way.

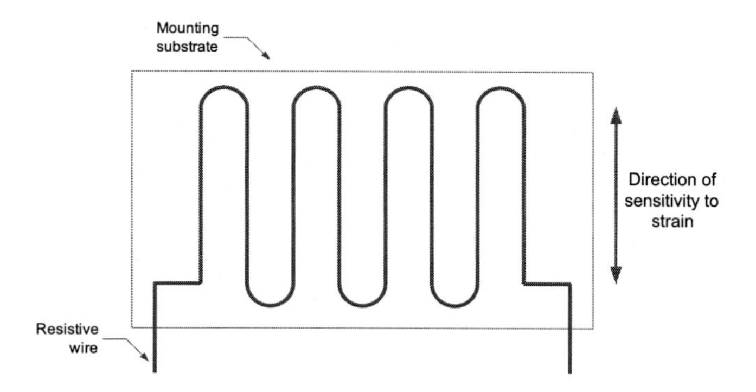

Figure 2.7: Components of a single-resistor strain gauge.

2.4.1.1 Pressure, Strain, or Force Measurements

One technique for measuring pressure, strain, and force is to use a *strain gauge* or a resistor that changes its resistance in response to a deformation imposed by a force that acts upon it. Manufacturers package strain gauge varieties in either single-gauge or multigauge configurations [Frad97; Klaa96; Norn89; Norp97]. Figure 2.7 illustrates the basic parts of a single-gauge device. The operating principle is to deform a wire that has a well-defined resistance by the application of an external stimulus. The external force causes a change in the resistance that the sensor measures, usually by sensing a voltage drop across the gauge. Designers use a *Wheatstone bridge circuit* to measure the change in resistance precisely through a voltage change across the circuit.

We see how this sensor works by examining the properties of electrical resistance. For a given wire in the gauge, we obtain the *gauge resistance*, R_g, by

$$R_g = \frac{\rho}{L} A \tag{2.4}$$

In Equation (2.4), ρ is the wire's intrinsic resistivity, L is the wire's length, and A is the wire's cross-sectional area. The gauge's resistance changes as the applied force deforms the gauge in the direction of sensitivity. If the designer uses the gauge as a resistor with a constant voltage supply, then the changing resistance appears as a voltage change across the gauge.

Figure 2.8 shows a Wheatstone bridge circuit for making measurements with a strain gauge. The active gauge is the resistor labeled $R_{straingauge}$ and it is used to measure the applied strain. The purpose

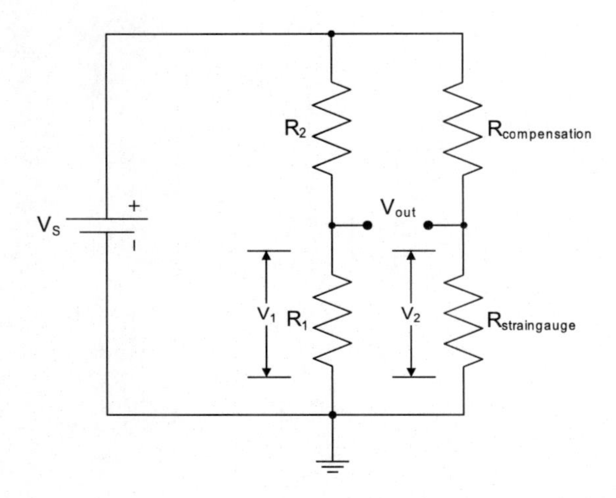

Figure 2.8: Wheatstone bridge circuit for making strain gauge measurements.

of the compensating gauge, $R_{compensation}$, which is mounted perpendicular to the expected strain direction and therefore does not undergo a resistance change, is to provide temperature compensation (see Problem 3). The designer must provide temperature compensation in the configuration because most resistors change their values with a change in temperature. The voltage measurement, V_{out}, is made across the center of the two voltage dividers and it is proportional to the resistance change. Manufacturers commonly supply four-gauge devices in a Wheatstone bridge configuration so that the measurement designer easily configures the bridge. The voltage, relative to ground, in the middle of the first voltage divider formed by R_1 and R_2 is given in terms of the battery voltage, V_s, by

$$V_1 = V_s \frac{R_1}{R_1 + R_2} \tag{2.5}$$

Similarly, the voltage, relative to ground, in the middle of the second voltage divider formed by $R_{compensation}$ and $R_{straingauge}$ is given by

$$V_2 = V_s \frac{R_{straingauge}}{R_{compensation} + R_{straingauge}} \tag{2.6}$$

We obtain the measured output voltage by computing

$$V_{out} = V_1 - V_2 \tag{2.7}$$

We can relate the measurements of pressure and force to strain if we know the dimensions and some of the physical properties of the materials under test. To assist in this process, we use a unitless variable called the *gauge factor*, ϵ, that is defined as

$$\epsilon = \frac{\Delta R_g / R_g}{\Delta L / L} = \frac{\Delta R_g / R_g}{G} \tag{2.8}$$

In Equation (2.8), G is the *unit strain* (dimensionless). We may think of this as the relative change in resistance, per unit resistance, to the relative change in length, per unit length, for the strain gauge. Pressure, P, causes a deformation in objects and a strain gauge measures the deformation by noting that for an applied force, F, and a cross-sectional area, A, on the object (not the cross-section of the gauge wire!), that pressure is defined as the force per unit area or

$$P = \frac{F}{A} \tag{2.9}$$

We relate this to Young's modulus for the material, E, by using

$$E = \frac{P}{G} \tag{2.10}$$

where G is the unit strain given above. From these relationships, the analyst determines the applied forces, pressures, and displacements.

2.4.1.2 Temperature Measurements

Designers know well that resistors are temperature dependent. If the sensor designer controls this dependence, then the designer can make the resistor into a temperature sensor with an appropriate circuit. This section discusses two different temperature sensors: the resistance thermometer and the thermistor.

Resistance Thermometers Manufacturers base temperature measurement devices on the principal that the device's resistance changes with temperature. Many experimenters use a Resistance Thermometer (ReT), also called a Resistance Temperature Detector (RTD), for this purpose [Frad97; Klaa96; Norn89]. A ReT is a linear device compared with a thermistor or a thermocouple. The ReT typically has an operating range of $\approx$ 900°C (−182.96°C through 630.74°C), for example). Manufacturers take care to match components precisely so that thermal expansion does not cause device failure. Some ReTs are not useful in high-vibration environments.

As with many types of resistive devices, designers use an equation for the temperature-voltage relationship for a ReT of the form

$$R_T = R_0 \left(1 + \alpha T\right) \tag{2.11}$$

In Equation (2.11), R_T is the resistance at a temperature, T, relative to the reference temperature in the measurement environment, R_0 is the resistance at a specified reference temperature (usually 0°C), and α is the coefficient of resistance change in units of $\Omega/\Omega/°\text{C}$. The most common ReT material at noncryogenic temperatures is platinum. In that case, designers use a more detailed equation, the *Callendar-Van Dusen* equation, for the resistance-voltage characteristic. This equation for this ReT resistance-to-temperature conversion is

$$R_T = R_0 + R_0\alpha \left[T - \delta\left(\frac{T}{100}\right)\left(\frac{T}{100} - 1\right) - \beta\left(\frac{T}{100}\right)^3\left(\frac{T}{100} - 1\right)\right] \tag{2.12}$$

For platinum ReTs, the constants α, β, and δ are as follows [Norn89]:

1. $\alpha = 0.003\,92\,\Omega/\Omega/°\text{C}$
2. $\beta = 0$ for $T > 0$ or $\beta = 0.11$ for $T < 0$, and
3. $\delta = 1.49$

The National Institute of Standards and Technology (NIST) uses a more exact series equation to provided a higher-accuracy calibration for the platinum ReT [Stro08].

Figure 2.9 shows a Wheatstone bridge circuit that designers use with a ReT to make a temperature measurement. The voltage divider equations describing the output are the same as those given in Equations (2.5), (2.6), and (2.7) with the resistor designations changed appropriately in Equation (2.6).

Thermistors The *thermistor* is a semiconductor-based device whose name is a contraction of *thermal* and *resistor*. Figure 2.10 shows thermistors frequently exhibit a negative temperature coefficient [Frad97; Norp97; Norn89]. Designers use the following equation to obtain ±0.02°C resolution over a 100°C temperature range:

$$\frac{1}{T} = A + B\ln\left(R\right) + C\ln\left(R\right)^3 \tag{2.13}$$

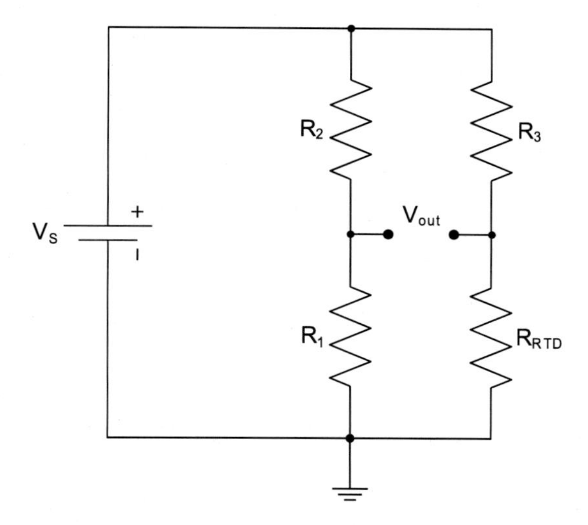

Figure 2.9: Wheatstone bridge circuit for making RT temperature measurements.

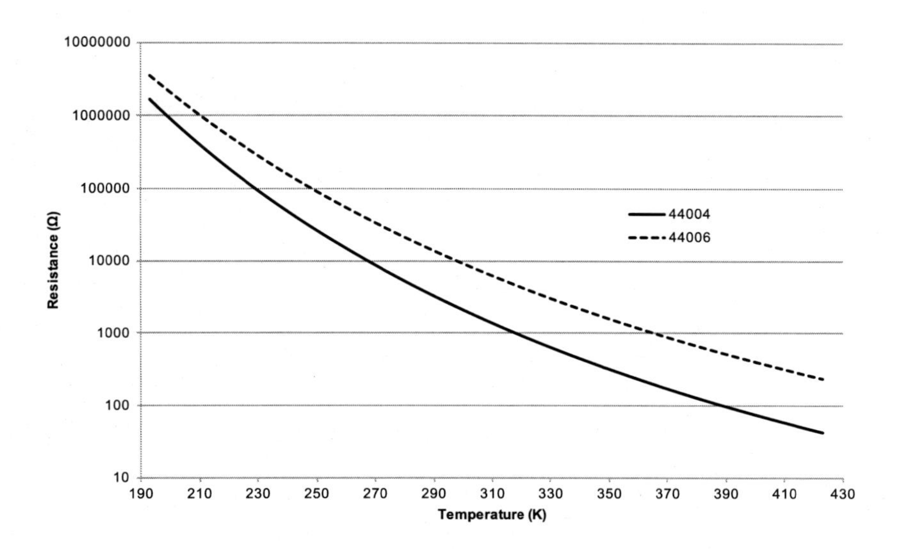

Figure 2.10: Resistance versus temperature for two example thermistor models.

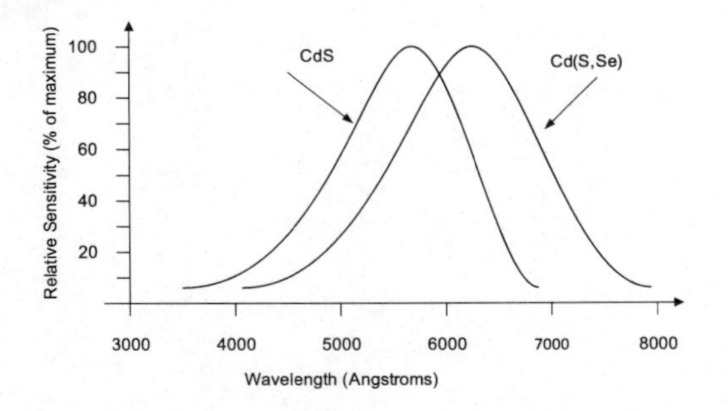

(a) Photocell response as a function of incident light wavelength for two different photocell detector technologies.

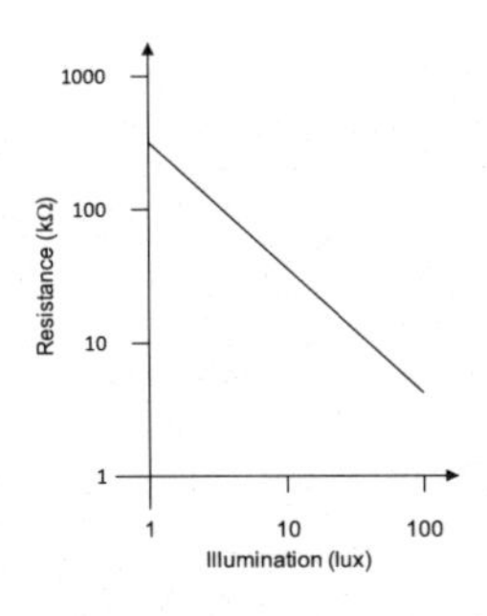

(b) Photocell resistance as a function of incident light intensity.

Figure 2.11: Example photocell response function and resistivity changes due to incident light.

A, B, and C are constants specific to the device, while T is the temperature in Kelvins. For example, the type 44004 device shown in Figure 2.10 has an A of 0.001 471, a B of 0.000 238, and a C of 1.0363×10^{-7}. One consequence of this temperature-resistance relationship is that a 1°C change in temperature produces different voltage-change amplitudes at different temperatures.

The thermistor is a semiconductor device that the designer keeps below the maximum operating temperature and does not subject to much mechanical stress. The designer can make thermistor temperature measurements using the ReT's Wheatstone bridge circuit shown in Figure 2.9 by substituting the thermistor for the ReT.

2.4.1.3 Light Measurements

One property of *photocells*, or *photoresistors*, is that light exposure changes the resistance of the cell material in proportion to the light's intensity and wavelength. The first part of Figure 2.11 illustrates the broad spectral responses, while the second part of Figure 2.11 illustrates the characteristic resistance change. Again, an experimenter can measure the resistance changes using a Wheatstone bridge (Figure 2.9) and replacing the ReT with a photocell.

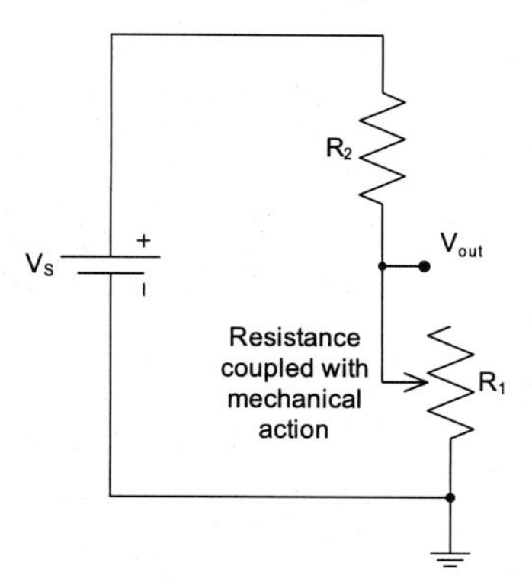

Figure 2.12: Example of using a potentiometer to measure displacement.

2.4.1.4 *Position Measurements*

Resistive position measurement devices are inexpensive and are rather inexact. Here, the designer makes the measurement by mechanically coupling a potentiometer to the displacement. Figure 2.12 shows a circuit for this measurement. As the displacement arm moves in this configuration, the potentiometer's resistance changes in proportion to the displacement. The measured voltage change is proportional to the changing resistance. This method has the disadvantage that the resistance in the potentiometer is subject to temperature changes and other non-linearities in the system.

2.4.2 Capacitive Sensors

While designers do not generally regard capacitors as precision devices, capacitors have the ability to hold a charge and produce a voltage proportional to a desired measurement. Temperature drift in the capacitor is a major problem if the designer does not properly configure the circuit. We now address two ways that we can use capacitors for sensors: rain gauge and time measurement.

2.4.2.1 *Capacitive Rain Gauge*

Designers use capacitors to generate a response in proportion to a measurand by changing the capacitance in the sensor in response to the input. This is the basis behind a Capacitive Rain Gauge (CRG). The CRG uses the internal water level to set the sensor's capacitance. The

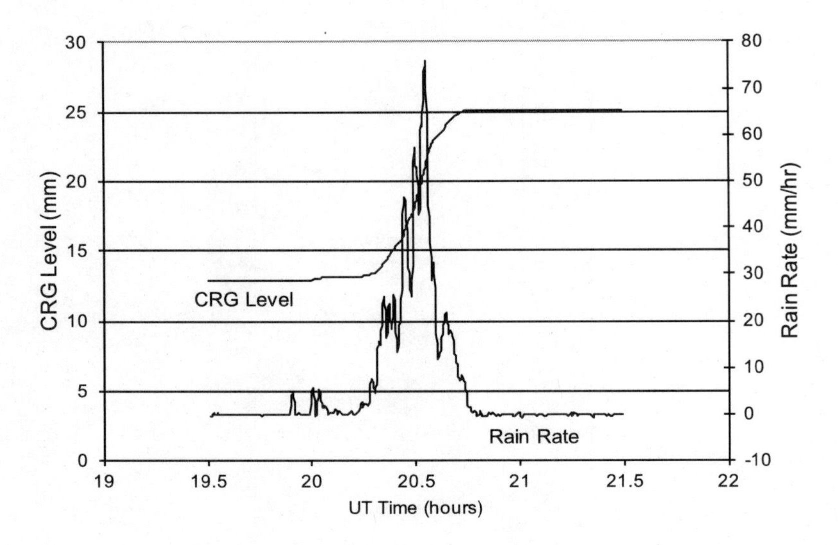

Figure 2.13: Capacitive rain gauge measurement showing both the gauge level (left) and the rate of change in the level (right).

gauge uses a secondary circuit to convert the physical water level into a voltage measurement. Figure 2.13 illustrates a capacitive rain gauge's operation by showing the measured water level in millimeters in the gauge during a rainstorm as a function of time on the left-hand axis. For rain measurement, the rate of change in the level is important. Figure 2.13 illustrates the rainfall rate in millimeters per hour on the figure's the right-hand axis. This rate is the time rate of change in the gauge volume measurement.

2.4.2.2 Time Measurement

Figure 2.14 shows a time measurement made by using a capacitor to fashion a *time-to-amplitude converter* [Hora84]. This converter produces a voltage across a capacitor proportional to the time interval, $\Delta\tau$, between the reception of the start and stop pulses. The timer depends upon having a constant-current source, i, to charge the capacitor over the time interval. We determine the voltage, V_{out}, across the capacitor, C, by using

$$V_{out} = \frac{i\Delta\tau}{C} = k\Delta\tau \tag{2.14}$$

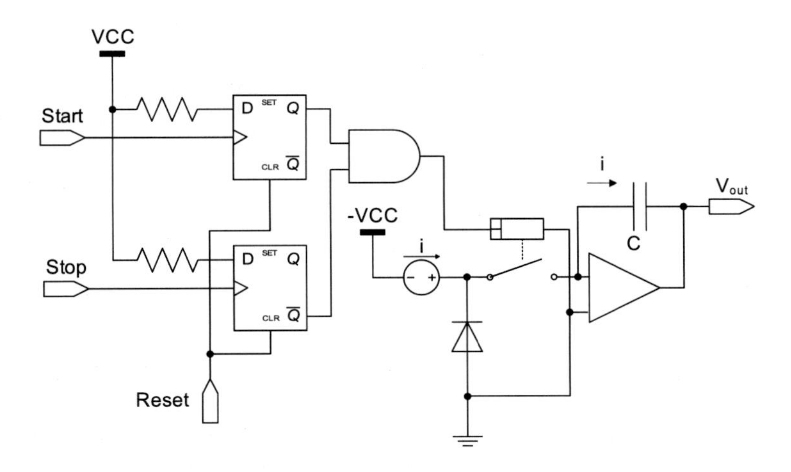

Figure 2.14: Capacitive time-to-amplitude converter circuit. The Start and Stop signals control the timing interval. Reset clears the timer.

The experimenter can calibrate the circuit to produce a voltage versus time relationship. The converter is very straightforward but depends upon the stability of the constant current source and the capacitor.

2.4.3 Physical Effect Sensors

Sensor designers use elementary physical effects or properties of materials to make a sensor for detecting a measurand. Two examples of physical effects are the Seebeck effect and the piezoelectric effect. Due to the nature of physical materials, a designer basing a sensor on either of these physical effects requires that the design include both the sensor and additional support circuitry to make a valid measurement.

2.4.3.1 Seebeck Effect

The *thermocouple* is one of the most commonly used devices for measuring temperature. The operating principal behind the thermocouple is the *Seebeck effect*, i.e., every junction of two different metals produces a potential voltage across the junction [Frad97; Klaa96; Norn89; Norp97]. Table 2.3 lists typical metals used in *thermocouple junctions* [ITS90]. The table lists the thermocouples by their common names (E, J, K, etc.) and the associated metal compositions. The table lists trade names for some types along with manufacturer's specifications for the alloy

Table 2.3: Thermocouple Junction Metals

TC Type	+ Metal	– Metal	Useful Range (°C)
B	70% Platinum and 30% Rhodium	94% Platinum and 6% Rhodium	0 to 1820
E	90% Nickel and 10% Chromium	55% Copper and 45% Nickel	-270 to 1000
J	Iron	55% Copper and 45% Nickel	-210 to 1200
K	90% Nickel and 10% Chromium	95% Nickel, 2% Aluminum, 2% Manganese, 1% Silicon	-270 to 1372
R	Platinum and 13% Rhodium	Platinum	-50 to 1767
S	Platinum and 10% Rhodium	Platinum	-50 to 1767
T	Copper	55% Copper and 45% Nickel	-270 to 400
	CONSTANTAN: 55% Cu and 45% Ni	CHROMETAL: 90% Ni and 10% Cr	ALUMEL: 95% Ni, 2% Mn, and 2% Al

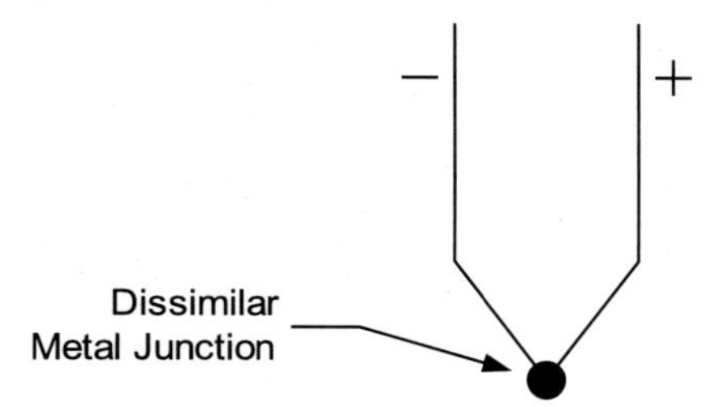

Figure 2.15: Thermocouple temperature sensor. The manufacturer forms the thermocouple by joining two different metals or alloys at the bead.

concentrations. Figure 2.15 shows how a sensor is fabricated by joining two different metals or alloys into a bead and implanting sensor leads. The joined metals produce a voltage across the junction that is a function of the materials and the temperature.

While the Seebeck effect is useful and makes for an obvious way to measure temperature, it also implies that **every** dissimilar metallic junction in the measurement system produces a voltage at that junction. This means that all probes, inputs, and sockets have a voltage offset if the designer does not configure system carefully. Figure 2.16 illustrates a thermocouple measurement configuration to control these voltage offsets [Frad97; Norp97]. The designer uses the isothermal block to control the junction temperatures produced by the dissimilar metals used in the leads and wires coming from the thermocouple to the voltage measurement point. Notice that the output wiring is copper for both leads to avoid a voltage offset at that point.

The thermocouple has advantages over the ReT in the areas of

- Greater resistance to damage
- Ease of manufacture and placement within the system
- Larger operating range

Thermocouples have a disadvantage in that they have nonlinear operating characteristics. The NIST has extensively documented the relationship between the thermocouple junction temperatures and the measured voltages across the thermocouple [ITS90]. NIST uses mathematical power series relationships for each thermocouple type. These equations are a limited series, typically five to eight terms, with unique coefficients for

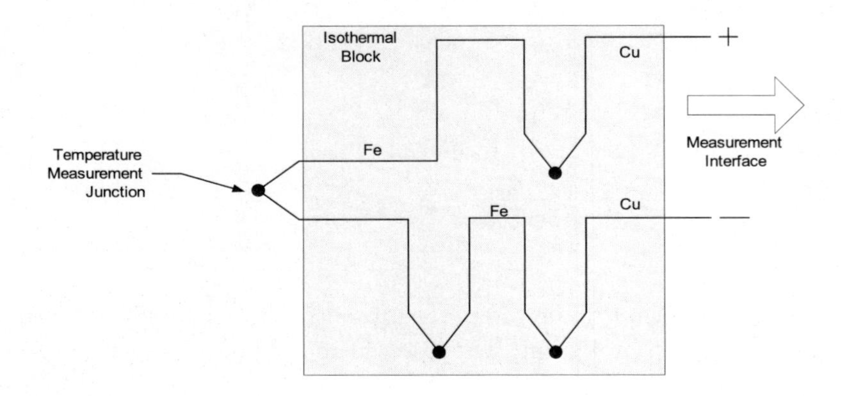

Figure 2.16: Typical thermocouple measurement configuration. The isothermal block is to control voltage offsets due to dissimilar metals.

each device type. For all but the K-type thermocouples, NIST expresses the *temperature-voltage* equation as

$$V = \sum_{i=0}^{M} a_i T^i \tag{2.15}$$

where V is the output voltage, T is the junction temperature, and M is the maximum order coefficient used to fit the relationship. For the K-type thermocouples, NIST uses the following temperature-voltage relationship:

$$V = \sum_{i=0}^{M} a_i T^i + 0.118598 \exp\left[-0.000118343\,(T - 0.0126969)^2\right] \tag{2.16}$$

Table 2.4 lists the coefficients for polynomial fits for selected thermocouple types and the valid ranges of the fits. The output voltage is in mV and temperature is in °C. Figure 2.17 gives the resulting calibration curves and Figure 2.18 illustrates the slopes of these curves, which gives the *Seebeck coefficient* (μV/°C) for the junction type [ITS90].

From the point of view of calibration linearity, it would be handy if the Seebeck coefficient were flat (constant slope) over the range of temperature measurement. As we see from the figures, the junction types do not have linear curves over their full operating range. However, the K-type material is close to being linear over a wide temperature range. In all cases, the thermocouple output is typically a few microvolts to a few millivolts. The designer must reliably read the voltages to a few

Table 2.4: Thermocouple Calibration Coefficients

Type E 0 to 1000°C	**Type J** -210 to 760°C	**Type K** 0 to 1372°C	**Type R** -50 to 1064°C	**Type S** -50 to 1064°C	**Type T** 0 to 400°C	**a$_i$**
0.000 00	0.000 00	$-1.760\,04 \times 10^{-2}$	0.000 00	0.000 00	0.000 00	a_0
$5.866\,55 \times 10^{-2}$	$5.038\,12 \times 10^{-2}$	$3.892\,12 \times 10^{-2}$	$5.289\,62 \times 10^{-3}$	$5.403\,13 \times 10^{-3}$	$3.874\,81 \times 10^{-2}$	a_1
$4.503\,23 \times 10^{-5}$	$3.047\,58 \times 10^{-5}$	$1.855\,88 \times 10^{-5}$	$1.391\,67 \times 10^{-5}$	$1.259\,34 \times 10^{-5}$	$3.329\,22 \times 10^{-5}$	a_2
$2.890\,84 \times 10^{-8}$	$-8.568\,11 \times 10^{-8}$	$-9.945\,76 \times 10^{-8}$	$-2.388\,56 \times 10^{-8}$	$-2.324\,78 \times 10^{-8}$	$2.061\,82 \times 10^{-7}$	a_3
$-3.305\,69 \times 10^{-10}$	$1.322\,82 \times 10^{-10}$	$3.184\,09 \times 10^{-10}$	$3.569\,16 \times 10^{-11}$	$3.220\,29 \times 10^{-11}$	$-2.188\,23 \times 10^{-9}$	a_4
$6.502\,44 \times 10^{-13}$	$-1.705\,30 \times 10^{-13}$	$-5.607\,28 \times 10^{-13}$	$-4.623\,48 \times 10^{-14}$	$-3.314\,65 \times 10^{-14}$	$1.099\,69 \times 10^{-11}$	a_5
$-1.919\,75 \times 10^{-16}$	$2.094\,81 \times 10^{-16}$	$5.607\,51 \times 10^{-16}$	$5.007\,77 \times 10^{-17}$	$2.557\,44 \times 10^{-17}$	$-3.081\,58 \times 10^{-14}$	a_6
$-1.253\,66 \times 10^{-18}$	$-1.253\,84 \times 10^{-19}$	$-3.202\,07 \times 10^{-19}$	$-3.731\,06 \times 10^{-20}$	$-1.250\,69 \times 10^{-20}$	$4.547\,91 \times 10^{-17}$	a_7
$2.148\,92 \times 10^{-21}$	$1.563\,17 \times 10^{-23}$	$9.715\,11 \times 10^{-23}$	$1.577\,16 \times 10^{-23}$	$2.714\,43 \times 10^{-24}$	$-2.751\,29 \times 10^{-20}$	a_8
$-1.438\,80 \times 10^{-24}$	—	$-1.210\,47 \times 10^{-26}$	$-2.810\,39 \times 10^{-27}$	—	—	a_9
$3.596\,09 \times 10^{-28}$	—	—	—	—	—	a_{10}

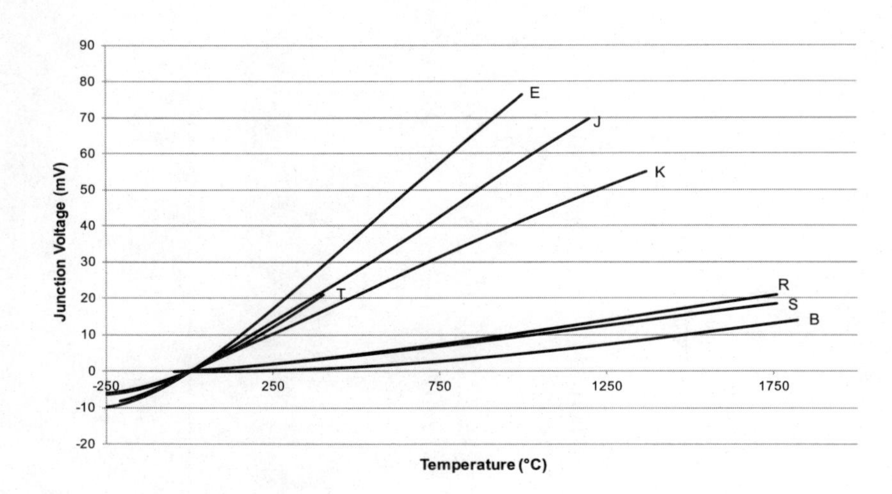

Figure 2.17: Thermocouple calibration curves for common thermocouple types based on the data in [ITS90].

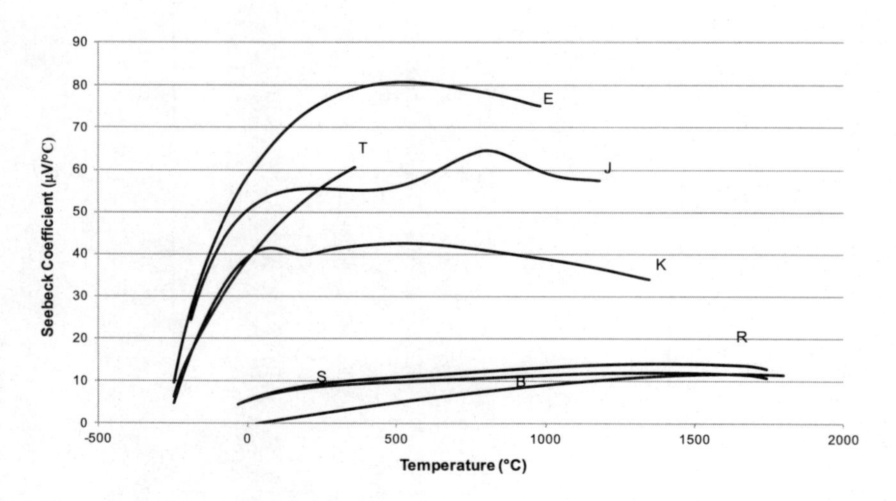

Figure 2.18: Seebeck coefficients for various thermocouple types. Curves based on data in [ITS90].

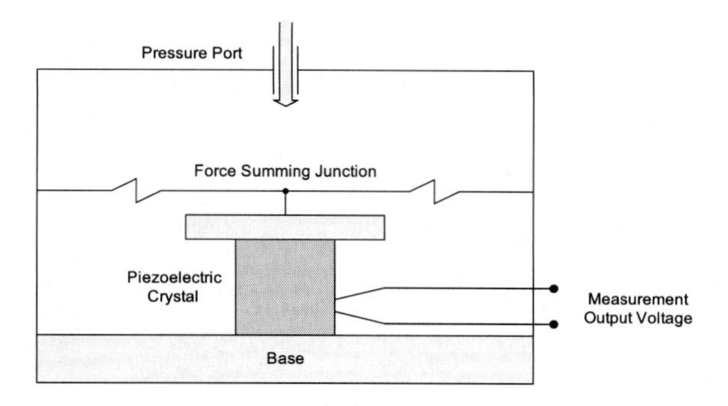

Figure 2.19: Piezoelectric pressure gauge schematic.

microvolts to achieve good resolution! This requires the experimenter to have careful setup procedures, proper signal conditioning, and careful placement of the measurement leads.

2.4.3.2 Piezoelectric Effect

A *piezoelectric gauge* reacts to dynamic changes in the crystalline detector. Figure 2.19 shows an example piezoelectric gauge configuration. When the crystal is compressed, it produces a voltage output, V, that we model using

$$V = h\epsilon t \tag{2.17}$$

where h is the strain coefficient (V/m), ϵ is the strain (m/m), and t is the crystal thickness. We treat this voltage as an instantaneous quantity because it dissipates if left static. The advantage of this transducer is that crystal gauges are linear over a large temperature range. A disadvantage is that the crystal is not capable of supplying electrical current like a battery and does not have the capability to provide the current to drive a low-impedance load on the input side of a measurement device. Therefore, the experimenter needs to buffer the inputs to the measurement device properly so that the measurement device's signal represents the compression.

Figure 2.20 is an example circuit diagram for a modern pressure gauge. The user needs to provide power and ground to the sensor and the sensor has necessary signal conditioning electronics built into the sensor. The sensor output is a 0 to 5 V level proportional to the atmospheric pressure. Figure 2.21 shows how manufacturers produce this sensor family in both absolute and differential gauge modes. The experimenter typically

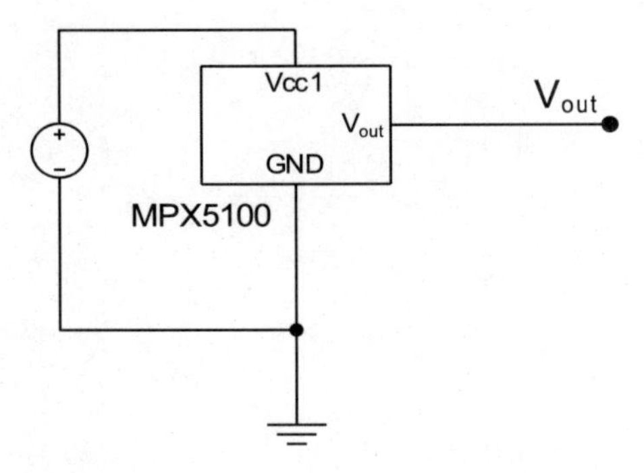

Figure 2.20: Circuit diagram for an integrated pressure measurement circuit.

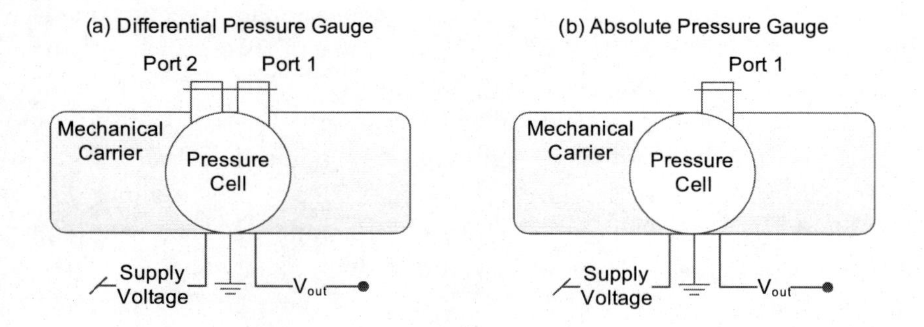

Figure 2.21: Differential and absolute pressure gauges.

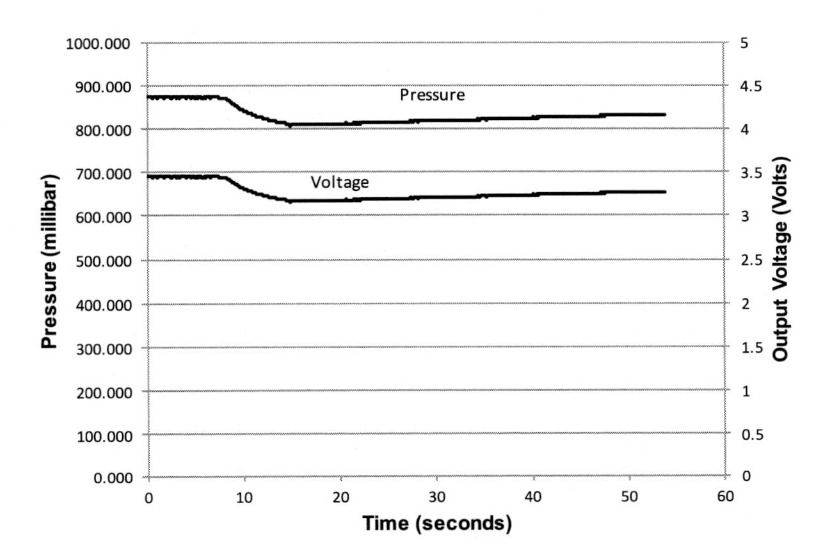

Figure 2.22: Example pressure output for a high-power model rocket flight. Left-hand y-axis is the sensor voltage output, while the right-hand y-axis is the voltage converted to atmospheric pressure.

uses the absolute mode to measure air pressure relative to one standard atmosphere. Similarly, the experimenter uses the differential mode to measure the pressure difference between two environments such as two separate rooms. The manufacturer of the MPX5100 gives a voltage versus atmospheric pressure relationship, for the pressure, P, in millibar, over its linear region for the absolute-mode sensor as [MP5100]

$$V = 5.0\,(0.0009P + 0.04) \tag{2.18}$$

Figure 2.22 provides an example use of this sensor as an altimeter for a high-power model rocket. We inverted Equation (2.18) to convert the sensor output voltage to atmospheric pressure as a function of time. We used a model atmosphere to convert atmospheric pressure to an altitude estimate and the result was given in Table 2.1 on page 25.

2.4.4 Semiconductor Sensors

Device designers use the p-n junction in semiconductor devices to form transistors and diodes. However, that junction is also sensitive to incident

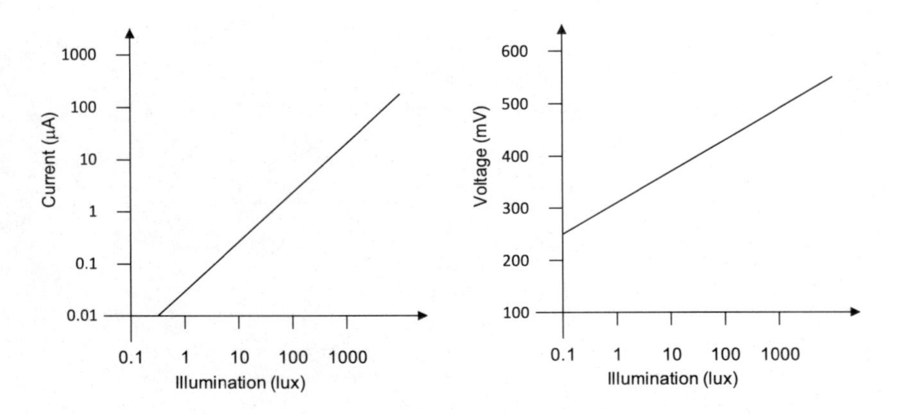

Figure 2.23: Typical photodiode response curves for the current and voltage as a function of input intensity.

light and changes in temperature. Sensor designers exploit this sensitivity to produce sensors for light detection and temperature measurement.

2.4.4.1 Photodetectors

Phototransistors and photodiodes are discrete devices for measuring the incident light [Frad97; Norn89]. Phototransistors and photodiodes allow their junction to change its electrical state by the application of light. For phototransistors, the light becomes a substitute for the base current that circuit designers normally apply when they, for example, design a semiconductor-based device as an amplifier. Figure 2.23 illustrates a typical characteristic curve for a photodiode. Phototransistors generally have greater sensitivity than photodiodes. Photodiodes generally have faster responses than do phototransistors.

2.4.4.2 Temperature Sensors

Sensor designers also use semiconductor devices as temperature sensors. A semiconductor junction is temperature sensitive so the sensor designer uses this property to fabricate a thermometer. With some additional circuitry included in the sensor, the designer adjusts the output to measure the temperature on either the Kelvin or the Celsius temperature scales. Figure 2.24 shows an example measurement circuit with a semiconductor temperature sensor. The LM35 produces an output accurate to 0.5°C with a voltage scale factor of 10 mV/°C [LM35].

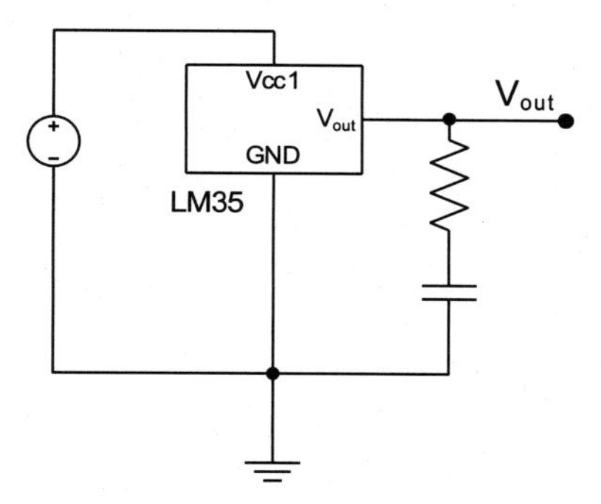

Figure 2.24: Circuit for an integrated circuit semiconductor temperature sensor.

2.4.5 Digital Time Measurement

Sensor designers employ direct digital techniques to produce sensor outputs without the need to convert the measurement from an analog voltage first. One example of this technique is with time measurements. For example, the capacitive technique mentioned earlier requires a constant current source. By using digital clocks, the designer avoids using analog sources.

Experimenters use a digital clock in *time-of-flight measurements* that measure elapsed time or distance in, for example, laser range finders. Figure 2.25 shows the generation of a time-of-flight measurement by measuring the elapsed time between the reception of the *START* and *STOP* pulses that measurement device uses to control the timing. The range finder generates the *START* pulse when it emits the laser pulse and the *STOP* pulse when it receives the return pulse. The sensor determines the distance by measuring the elapsed time, multiplying it by the speed of light, and dividing the result by two. This technique has a dynamic range in excess of 1000. Often, the pulse width, and not the clocking circuitry, dominates the resolution error. The heart of these time-of-flight converters is a high-accuracy clock circuit that produces a result proportional to the flight time. Figure 2.25 illustrates that the counter accumulates the number of clock cycles during the *Counting Enable Pulse* between receiving the *START* and *STOP* gating pulses.

For accurate timing, designers typically use a crystal-based oscillator to drive the gated oscillator in Figure 2.25. This timer has the disadvantage of lacking knowledge of the arrival phase for the *START* and

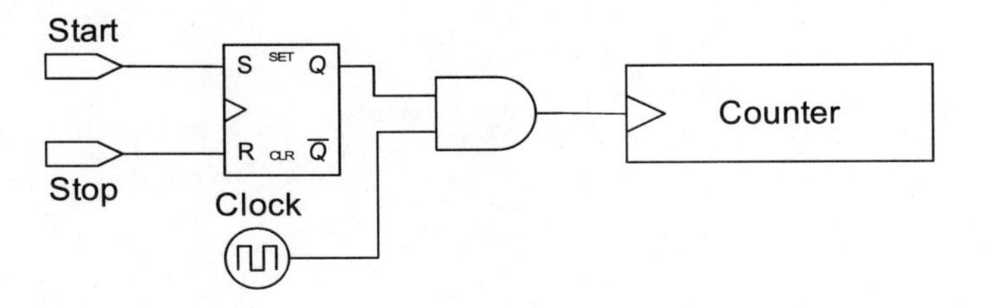

(a) Example gated oscillator timing circuit.

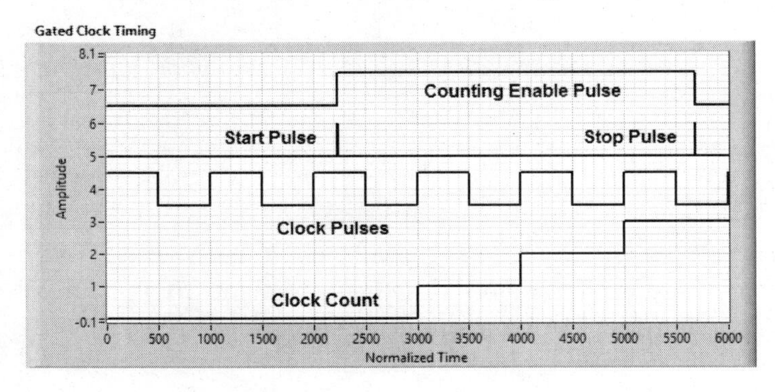

(b) Gated oscillator timing process.

Figure 2.25: Elapsed time measurement using a gated oscillator.

STOP gating signals relative to the free-running crystal oscillator, as Figure 2.26 illustrates. Hence, the timing accuracy is typically no better than ± 1 count. The inherent frequency stability and low drift for crystal-oscillator-based counters allow designers to generate configurations that are essentially as long as the designer cares to make them. The elapsed time duration that the timer is capable of measuring frequently offsets the limited ± 1 count resolution uncertainty.

One way the designer can mitigate some of the counting uncertainty of the crystal oscillator is to use an oscillator that is capable of starting in phase with the arrival of the *START* pulse. Figure 2.27 illustrates such an oscillator that uses a logic gate and a delay line to form a clock. When the *START* pulse arrives as a step function, the logic gate changes state and the delay line prevents the NAND gate from immediately seeing this result. This sets up an oscillation as Figure 2.27 shows. This type of gated oscillator has the advantage of starting in phase with the *START* pulse, but it still has the ± 1 count resolution problem at the end of the

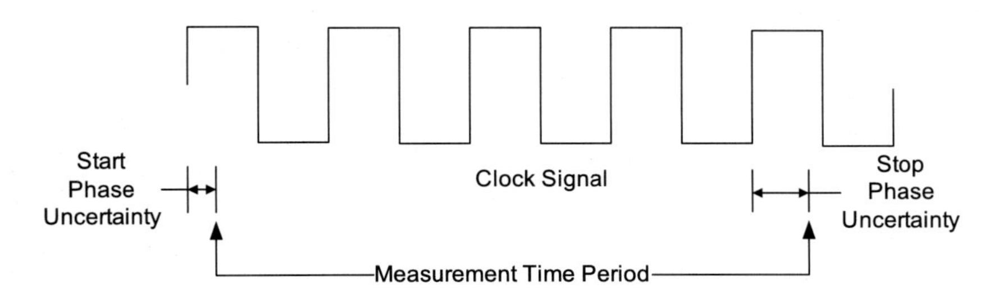

Figure 2.26: Measuring the time between start and stop pulses through digital clock timing.

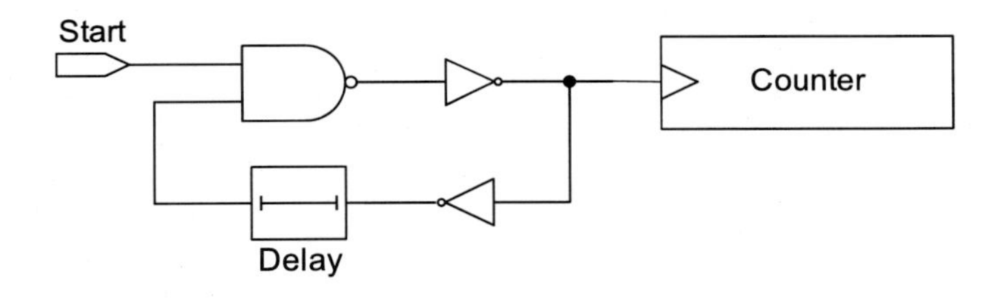

(a) Example logic-delay oscillator timing circuit.

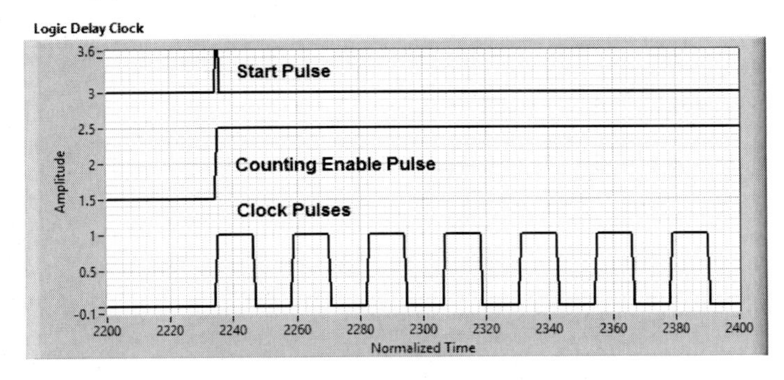

(b) Logic-delay clock timing process.

Figure 2.27: Elapsed time measurement using a logic-delay oscillator.

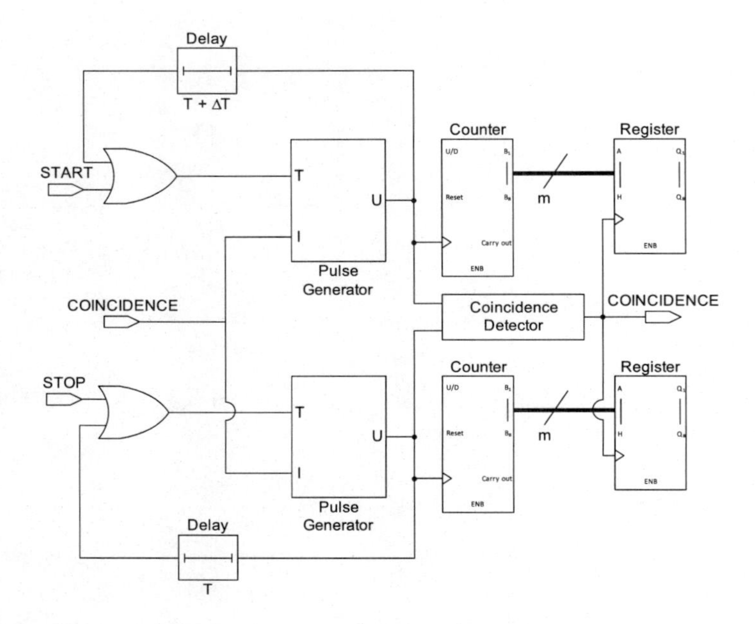

Figure 2.28: Vernier timer circuit diagram.

Counting Enable Pulse timing gate. Designers use this type of clock for a relatively short duration because the delay line is not as stable as the crystal oscillator.

By using the logic-delay clock, the designer is able to remove some of the clocking uncertainty. To further improve the situation, the designer needs a means to capture the clock's phase for both the $START$ and $STOP$ gating signals. *Vernier timing* is a technique to do this by effectively stretching time by a factor k to permit better resolution. A Vernier timer uses two clocks of slightly different periods to generate the timing information (Figure 2.28) [Hora84]. The first clock has the longer period and the timer starts it upon the arrival of the start signal, i.e., the timer generates the signal when it initiates the timing pulse. When the timer receives the reflected timing pulse, it both generates a stop signal and starts the shorter-period second clock. The coincidence detector looks for both clocks to align in phase, at which point the timer quenches both clocks and it reads the contents of the counters (also known as *scalers*). If we represent the time from start pulse until the coincidence of the two clock pulses by the quantity $k\tau$ and τ is the time between the start and stop pulses, then we relate $k\tau$ to the contents of the two counters via the equation

$$k\tau = (N_{START} - 1)(T + \Delta T) = t + (N_{STOP} - 1)T \tag{2.19}$$

where N_{START} is the count stored in the start counter and N_{STOP} is the count stored in the stop counter (see Problems 8 and 9). With T as the period of the stop clock circuit and $(T + \Delta T)$ as the period of the start clock circuit, the time difference, τ, between the start and stop signals is

$$\tau = (N_{START} - N_{STOP})T + (N_{START} - 1)\Delta T \tag{2.20}$$

As we see in Equation (2.20), the timing resolution is $\pm\Delta T$. The timer's counter size and the stability of its clock period limits the time duration and, therefore, the maximum distance the device measures. One disadvantage is that the system cannot make another timing measurement prior to the time kt when the two clocks are in coincidence. The advantage is that the results are generated in a digital form and do not need a separate analog-to-digital conversion stage.

2.4.6 Hybrid Sensors

One of the frontiers of sensor technology is the integration of mechanical and electrical devices into a single package. One example of this type of technology is the single-chip *accelerometer* composed of signal processing electronics, a machined beam, and a capacitor array to enable the measurement of accelerations from $\pm 1\,\text{g}$ through over $\pm 100\,\text{g}$, depending on sensitivity and sensor scale factor. The construction of the accelerometer's internal components provides for sensitivity to acceleration in one preferred direction. Modern chip manufacturers allow the designer to choose between single-axis, dual-axis, or tri-axis configurations in a single chip package.

We express the acceleration in terms of the output voltage, V_{out}, and the accelerometer scale factor, *SF*, by using the equation

$$Accel = \frac{V_{out} - 2.5V}{SF} \tag{2.21}$$

The 2.5 V output level is the nominal 0 g acceleration level and the measured accelerations produce voltages symmetrically around this level. Designers use gain resistors with a noninverting amplifier attached to the accelerometer's output to set the scale factor over a wider range.

The designer configures the accelerometer to measure accelerations in

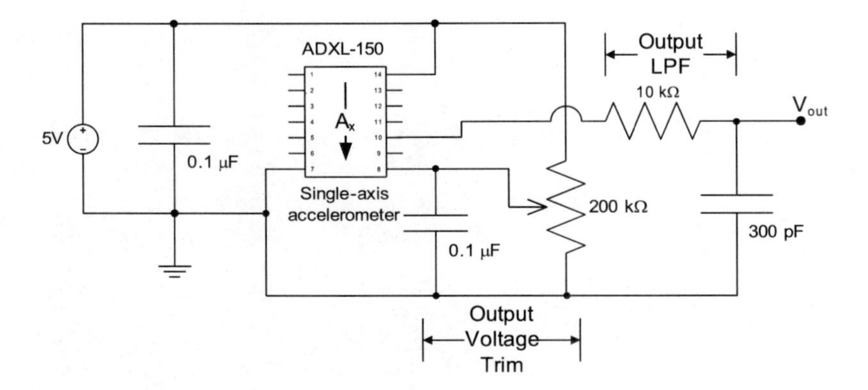

Figure 2.29: Accelerometer absolute mode circuit. The potentiometer provides offset adjustment. The output has a low-pass filter to restrict the measurement bandwidth based on device characteristics. The arrow on the accelerometer chip indicates the direction of sensitivity.

either *absolute mode* or *differential mode* [ADX150]. When the designer needs to measure both the acceleration due to gravity as well as motion-based acceleration, then absolute mode is used. Figure 2.29 provides an example of an accelerometer configured in absolute mode. Figure 2.30 provides an example with the output from an absolute mode accelerometer circuit during a high-power model rocket flight. The first few seconds show the acceleration due to gravity at 0 g prior to engine ignition. We then add the remaining acceleration to the acceleration due to gravity to obtain the total acceleration felt by the rocket payload.

Designers often use the differential mode to measure shock and vibration; designers also call this mode the Alternating Current (AC) coupled mode. Figure 2.31 provides an example circuit for differential mode. Designers replace the output trim potentiometer from the absolute mode with a single capacitor for the AC coupling. The capacitor on the output blocks any Direct Current (DC) component; the designer may attach this to the input of a noninverting amplifier to obtain a stronger signal. Figure 2.32 illustrates an example of the shock and vibration for the high-power model rocket launch. We made the absolute measurement example in the direction of the rocket's motion. Similarly, we made the differential measurements in the x-y axis perpendicular to the direction of motion.

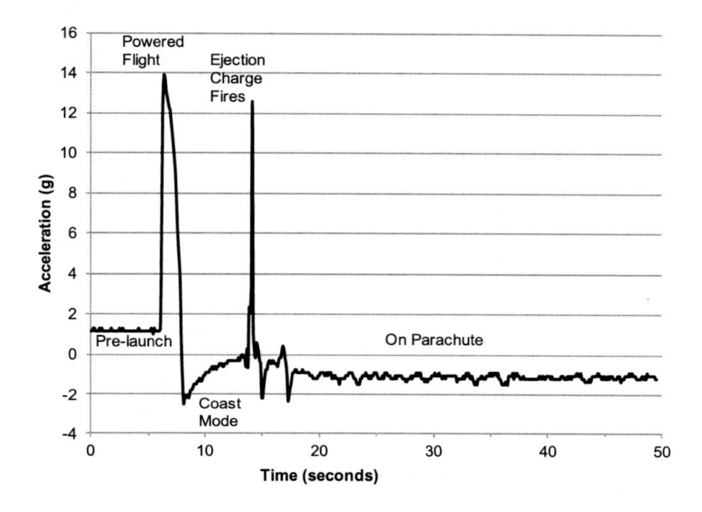

Figure 2.30: Absolute acceleration measurement example for a high-power model rocket. Flight activities include pre-launch, power flight with the engine burning, coast mode after engine burnout, ejection charge firing to deploy the parachute, and parachute float time.

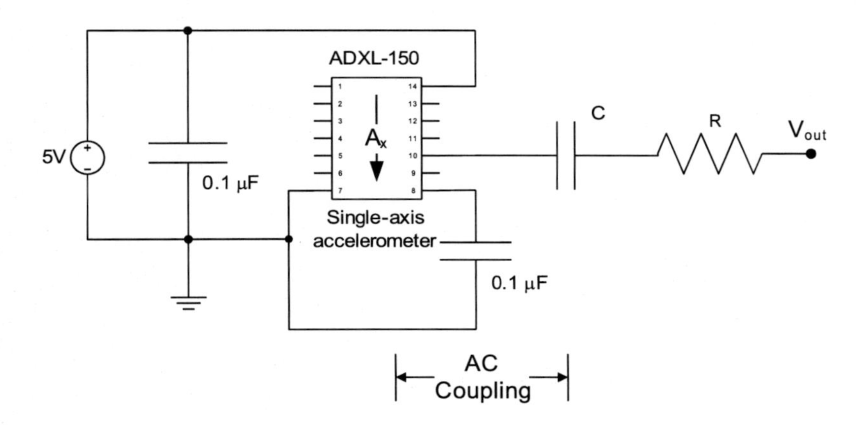

Figure 2.31: Differential model accelerometer circuit.

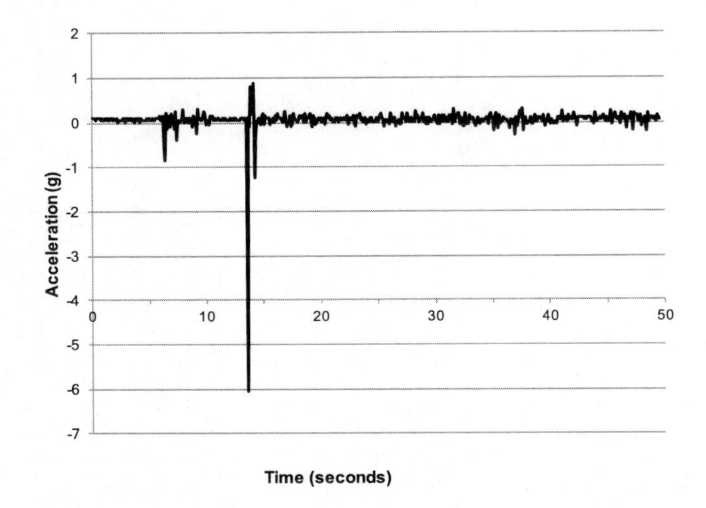

Figure 2.32: An example differential acceleration measurement for a high-power model rocket. This plot corresponds to the same events as the absolute acceleration measurement.

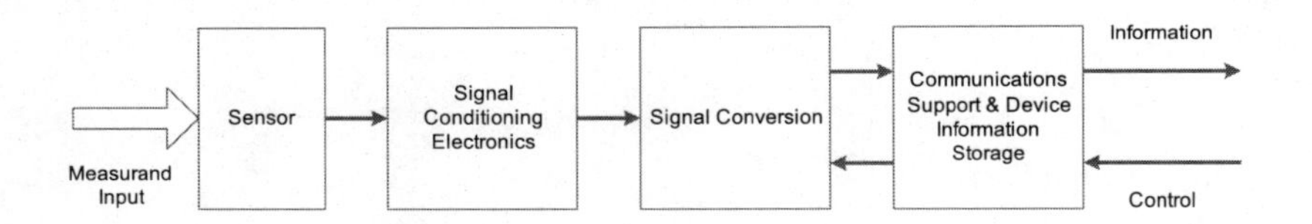

Figure 2.33: Block diagram for a generic smart sensor configuration.

2.5 SMART SENSORS AND RFID SENSORS

With the advent of miniaturization provided by modern integrated circuits, sensor designers provide additional functionality beyond simple voltage output. In this section, we will examine two examples of these new sensor technologies: smart sensors and Radio Frequency Identification (RFID) sensors.

The enhanced sensor and actuator devices are often called *smart sensors* and *smart actuators* [Lee01]. Figure 2.33 is a block diagram for a generic smart sensor. The intent of smart sensor technology development is to enhance the available information flow by providing the user with advantages such as:

- Self-identification of sensor and/or data sets

- Internal storage of calibration and other related data to assist in data analysis
- Self-testing of sensor functionality
- Autonomous communications with other, related sensors in the system
- Improved communications with the host system

This additional functionality requires defined standards for communications, data formatting, and interfacing support. The hope is that system designers are able to advance technology beyond simply transmitting *data* and transmit *information* in the future.

The RFID sensors are an outgrowth of the RFID tag technology used to identify and track objects based on a fixed number embedded with the tag. The industry identifies three classes of tags: [Ruha08]

Active Tag: a tag with some form of power such as a battery or energy harvesting circuit to power sensors, memory elements, and even processing circuits as well as the transmission back to the reader.

Passive Tag: a tag without a power source and the tag is energized by the Radio Frequency (RF) energy from an external reader.

Semi-passive Tag: a tag that uses a battery or energy harvesting circuit to power a sensor and support electronics but only responds to activation signals from the reader using the reader's transmitted energy.

The RFID sensors have several distinct advantages such as:

- The reduction in system weight due to the elimination of many wires
- The ability to place sensors in hard-to-reach locations using a "lick and stick" approach or place sensors in dangerous locations such as inside an engine
- The ability for the system designer to add new sensors to an existing system and having them easily integrate with other sensors

The RFID sensors also have several distinct disadvantages such as:

- The relatively low data rates for data transmission

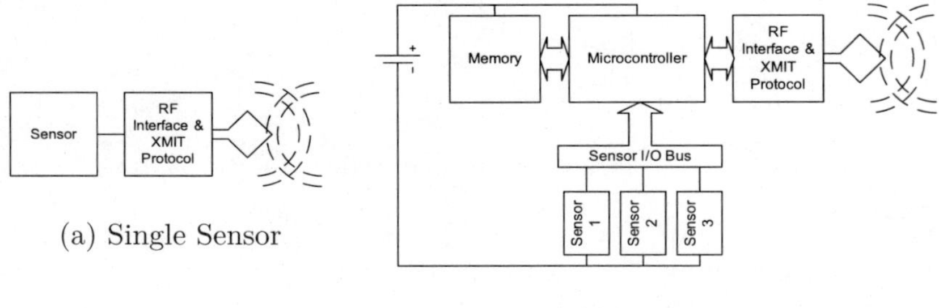

(a) Single Sensor

(b) Sensor Array

Figure 2.34: Example RFID sensor configurations. Panel (a) illustrates a single sensor/tag configuration. Panel (b) illustrates a multiple array system including a microcontroller and memory that are supported by a battery.

- The complicated channel reflections that interfere with high-quality data transmissions
- The relatively short transmission distances in many applications

Figure 2.34 illustrates two approaches to RFID sensor applications. The first panel illustrates a single sensor such as a pressure sensor or a temperature sensor with its tag. The sensor only reports the immediate measurement when interrogated. The second panel illustrates a sensor system servicing several sensors such with a microcontroller to acquire the data from each sensor. A memory is used to store data between interrogations and a battery powers the electronics. The incoming RFID signal powers the transmission modules when the system is interrogated.

The next extension to this smart sensor and smart actuator technology is the anticipated Internet of Things (IoT) where the ubiquity of these types of devices will enable new products and services. One frequent model for the IoT is built around smart, wireless sensors frequently utilizing RFID technologies. The sensors utilize batteries or energy-harvesting power modes. Data will be seamlessly stored, and even processed, utilizing Cloud-based techniques. Sensors will have the ability to self-integrate into communications networks and discover cooperating sensors in their immediate environment and across the Internet [Amen14; Bija16; Mahb16; Want15].

The IoT will also raise many concerns, including [Stan14; Webe10]:

Data Privacy Are data values properly hidden from disclosure to unauthorized parties?

Data Robustness Do the sensors still have the correct position, time base, and relationship to allied sensors and actuators preserved over time?

Data Security Is there proper access control to sensors and actuators, do sensors and actuators have resilience against cyber attacks?

Data Trust Is the data actually as it purports to be, i.e., does it come from the expected source without corruption, was it collected at the correct time, is the sensor calibration still valid?

Network Architecture Will the architecture permit operations and enhanced data services over a variety of ranges of numbers of devices and physical operational areas while also permitting upgrades and enhancements?

Network Openness Are sensors and actuators utilizing protocols and techniques that permit cooperation and enhanced operations?

The resolution of these issues will be evolving with the technology over the next years.

2.6 REFERENCES

[ADX150] *ADXL150/ADXL250* ± 5 g *to* ± 50 g *Low Noise, Low Power Single/Dual Axis iMEMS Accelerometers.* Norwood, MA: Analog Devices, Inc., 1998. 16 pp.

[Amen14] S.A. Amendola et al. "RFID Technology for IoT-Based Personal Healthcare in Smart Spaces." In: *Internet of Things Journal* 1.2 (Apr. 2014), pp. 144 –152. DOI: 10.1109/JIOT.2014.2313981.

[Bija16] F.H. Bijarbooneh et al. "Cloud-Assisted Data Fusion and Sensor Selection for Internet of Things." In: *Internet of Things Journal* 3.3 (June 2016), pp. 257 –268. DOI: 10.1109/JIOT.2015.2502182.

[Frad97] J. Fraden. *Handbook of Modern Sensors Physics, Designs, and Applications.* 2nd Edition. New York, NY: American Institute of Physics, 1997.

[Hora84] S. Horan. "High Speed Time to Digital Conversion." PhD thesis. Las Cruces: New Mexico State Univ., May 1984.

[IEEE100] Institute of Electrical and Electronics Engineers. *IEEE 100 The Authoritative Dictionary of IEEE Standards Terms Seventh Edition*. IEEE-STD-100. IEEE Press. Piscataway, NJ, 2000. DOI: 10.1109/IEEESTD.2000.322230.

[ITS90] National Institute of Standards and Technology. *NIST Standard Reference Database 60, Version 2.0 (Web Version), NIST ITS-90 Thermocouple Database*. U.S. Commerce Department. July 2000. URL: http://srdata.nist.gov/its90/main/.

[Klaa96] K. B. Klaassen. *Electronic Measurement and Instrumentation*. Trans. by S. M. Gee. Cambridge: Cambridge University Press, 1996. ISBN: 0-521-47157-5.

[Lee01] K. Lee. "Sensor Networking and Interface Standardization." In: *Proc. 18th IEEE Instrumentation and Measurement Technology Conference, 2001*. Vol. 1. Budapest: IEEE, May 2001, pp. 147 –152. DOI: 10.1109/IMTC.2001.928803.

[LM35] *LM35 Precision Centigrade Temperature Sensors*. Dallas, TX: Texas Instruments Inc., 2016. 31 pp. URL: http://www.ti.com/lit/ds/symlink/lm35.pdf.

[Mahb16] H. Mahboubi, A.G. Aghdam, and K. Sayrafian-Pour. "Toward Autonomous Mobile Sensor Networks Technology." In: *Transactiona on Industrial Informatics* 12.4 (Apr. 2016), pp. 576 –586. DOI: 10.1109/TII.2016.2521710.

[MP5100] *MPX5100 Integrated Silicon Pressure Sensor On-Chip Signal Conditioned, Temperature Compensated and Calibrated*. Tempe, AZ: Freescale Semiconductor, Inc., 2010. 17 pp. URL: http://cache.freescale.com/files/sensors/doc/data_sheet/MPX5100.pdf?pspll=1.

[Norn89] H. N. Norton. *Handbook of Transducers*. Englewood Cliffs, NJ: Prentice Hall, 1989. ISBN: 0-13-382599-X.

[Norp97] R. B. Northrop. *Introduction to Instrumentation and Measurements*. Boca Raton, FL: CRC Press, 1997. ISBN: 0-8493-7898-2.

[Ruha08] A. Ruhanen et al. *Sensor-enabled RFID tag handbook*. IST-2005-033546. BRIDGE, Jan. 2008. 47 pp. URL: http://bridge-project.eu/data/File/BRIDGE_WP01_RFID_tag_handbook.pdf.

[Stan14] J.A. Stankonic. "Research Directions for the Internet of Things." In: *Internet of Things Journal* 1.1 (Feb. 2014), pp. 3 –9. DOI: 10.1109/JIOT.2014.2312291.

[Stro08] G. F. Strouse. *Standard Platinum Resistance Thermometer Calibrations from the Ar TP to the Ag FP*. NIST Special Publication 250-81. National Institute of Standards and Technology. U.S. Department of Commerce, Jan. 2008. 79 pp. URL: `http://www.nist.gov/calibrations/upload/sp250-81.pdf`.

[VIM12] Joint Committee for Guides in Metrology Working Group 2. *International vocabulary of metrology - Basic and general concepts and associated terms (VIM)*. JCGM 200:2012. Bureau International des Poids et Mesures. 2012. URL: `http://www.bipm.org/utils/common/documents/jcgm/JCGM_200_2012.pdf`.

[Want15] R. Want, B.N. Schilit, and S. Jenson. "Enabling the Internet of Things." In: *Computer* 48.1 (Jan. 2015), pp. 28 –35. DOI: `10.1109/MC.2015.12`.

[Webe10] R.H. Weber. "Internet of Things – New security and privacy challenges." In: *Computer Law & Security Review* 26 (1 Jan. 2010), pp. 23 –30. DOI: `10.1016/clsr.2009.11.008`.

2.7 PROBLEMS

1. Given a vector of measurements $X = \{5.4952, 5.4645, 5.4778, 5.4172, 4.9878, 5.2662, 5.0600, 5.4231, 5.2034, 5.0878\}$ with a known standard value of 5.2489, determine an estimate for the precision and the accuracy of each measurement. Do you believe that this is an accurate and precise set of measurements? Explain your reasoning.

2. For the measurements in Problem 1, would you consider the measurements to be valid if the sign on the accuracy was all positive or all negative? What would you expect and why?

3. You are given a strain gauge and a compensating transducer in a Wheatstone bridge configuration:

   (a) Determine the output voltage as a function of the resistance values.

   (b) If the two transducers have a linear variation with temperature, show by analysis whether or not the compensating transducer really does eliminate the voltage change due to temperature.

   Assume that the temperature/resistance relationship for the strain gauge is given by $R = R_0(1 + \alpha\Delta T)$ where R_0 is the resistance value at the reference temperature and ΔT is the temperature difference from the reference temperature.

4. Using a type 44004 thermistor, approximately what resistor values would be needed to have a voltage difference of 0 V in the bridge at room temperature? Using those resistor values, what would be the expected voltage reading if the thermistor were at 0°C and 100°C? Assume we use a 12 V battery to drive the circuit. What would be the change in voltage from the 0°C and 100°C readings if the temperature changed by 1°C at each point? Hint: Use the coefficients given to fit the temperature as a function of the thermistor resistance as given in the text.

5. Repeat Problem 4 with a platinum ReT.

6. Suppose that atmospheric pressure, in millibar, varies with height, h, in kilometers according to the model:

$$P(h) = 1034.827\,\mathrm{e}^{[-0.12807h]}$$

Determine the expected output voltage from a pressure sensor when the altitude is (a) sea level, (b) one kilometer, (c) two kilometers, and (d) the altitude at your location.

7. For the delay-line based gated oscillator shown in Figure 2.27, how would you use a stop pulse to extinguish the clock pulses within the circuit and reset the clock so that a new oscillation begins with the arrival of a new start pulse? Show by a timing diagram that your change works.

8. Draw the timing diagram for the Vernier time-to-digital converter for the output from both the start and stop pulse generators under the condition that the time you are measuring is shorter than the period of the start oscillator. Verify that the contents on each counter can be related to the total time to coincidence by $kt = (N_{START} - 1)(T + \Delta T) = t + (N_{STOP} - 1)T$ and that $N_{START} = N_{START}$.

9. Draw the timing diagram for the Vernier time-to-digital converter for the output from both the start and stop pulse generators under the condition that the time you are measuring is longer than the period of the start oscillator. Verify that the contents on each

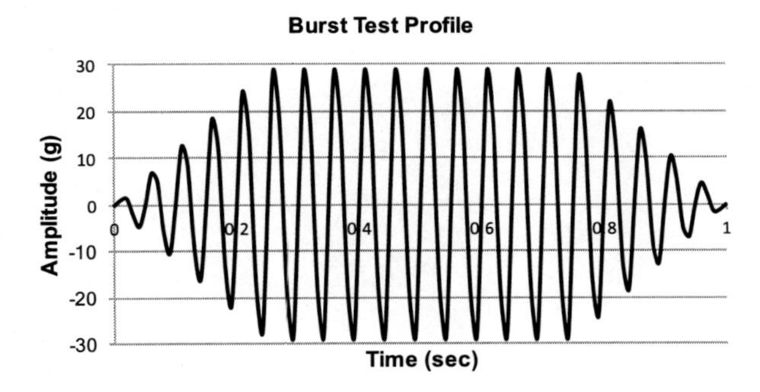

Figure P-1. Sine burst acceleration profile.

counter can be related to the total time to coincidence by $kt = (N_{START} - 1)(T + \Delta T) = t + (N_{STOP} - 1)T$.

10. Suppose you are performing a vibration test on an electronics package by exciting the package with a profile like that shown in Figure P-1. The full-scale oscillation is at 20 Hz with amplitude of 30 g. There is a five-cycle ramp-up and ramp-down time in addition to the 10 cycles at full amplitude. First determine the necessary scale factor to have the accelerometer measure ± 25 g about a nominal 0 g set point. The 0 g set point corresponds to a 2.5 V output reading. Full scale is 0 V for -25 g and 5 V for $+25$ g. Plot the expected output voltage for the test duration.

CHAPTER 3

MODELING AND CALIBRATION

3.1 BACKGROUND

Whenever experimenters make measurements, the result is not normally in a user-friendly form, which efficiently expresses the knowledge that the measurements represent. The experimenter wants measurements in quantities with appropriate units attached to represent the knowledge contained in the measurement. Additionally, they usually desire a mathematical model to encapsulate the entire measurement set so that a function or set of equations describe the data rather than a collection of individual points. In both tasks, analysts use mathematical techniques to give the best match possible between measured data and the mathematical relationships representing the data.

Recently, Ferraro [Ferr15] outlined three pillars of metrology: uncertainty, calibration, and traceability. These three concepts let us judge the quality of the measurement instrument's output and provide an indication of how close we may be to the true value of the measurement. We also need to look at *measurement uncertainty* because we can never know exactly what the true values should be and what the measurement instrument is producing. The *Guide to the expression of Uncertainty in Measurement (GUM)* classifies two types of measurement uncertainty [GUM08]:

Type A: The uncertainty of the result of a measurement is expressed as a standard deviation.

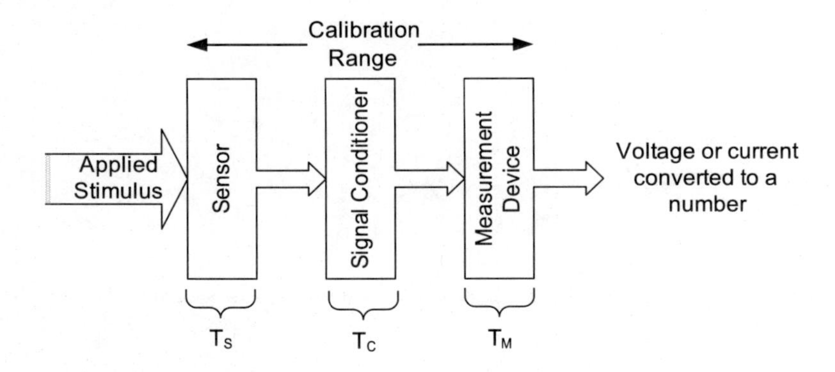

Figure 3.1: Measurement devices from Section 2.3 included in the calibration process.

Type B: The method for the evaluation of uncertainty is by means other than the statistical analysis of series of observations.

In this chapter, we will use the Type A approach. See Salicone [Sali14a] for an example of a Type B approach.

As we indicated in Chapter 2, the experimenter must calibrate the data acquisition instrumentation against a traceable standard. The analyst's first task is to convert the raw measurement to a quantity with appropriate units that represents the measurement. To do this, the analyst must develop a mathematical mapping that relates the measured quantity to the data collected. Figure 3.1 shows that this calibration includes the sensor included in Section 2.3 responding to the measured quantity and the signal processing devices used prior to sampling the resulting voltage or current. The mathematical representation of the entire data set is also an analytical function the analyst develops to map the acquired data to a theoretical model. This mapping may also need to smooth over inaccuracies and noise in the measured data.

Both tasks use the same mathematical tools to achieve their desired goals. This chapter explores the concepts analysts use to develop this mathematical mapping and discusses the method of least squares for modeling the results. These tools allow us to model data and obtain a "best" representation. The initial discussion deals with calibration, but we keep in mind the other uses for the tools as well.

3.2 OBJECTIVES

Our objectives for this chapter are to understand the topics related to modeling the data for calibration and analysis purposes. At the end of this discussion, the reader will be able to

- Understand the need for and general strategy for instrument calibration
- Perform a least squares fit of a polynomial to a data set
- Justify the order of the model chosen to fit the data
- Estimate the statistical confidence region of a data point

We use standard statistical tools such as those found in Microsoft Excel®, MATLAB®, or other packages for making the computations. The reader is encouraged to use these packages whenever possible rather than making the computations by hand.

3.3 BASICS

Before beginning the mathematical techniques, we first look at issues related to calibration and data modeling. The two uses have different approaches and needs, even if the mathematical tools are the same.

3.3.1 Calibration

Calibration has several official definitions. The *International Vocabulary of Metrology (VIM)* defines *calibration* as an

> operation that, under specified conditions, in a first step, establishes a relation between the quantity values with measurement uncertainties provided by measurement standards and corresponding indications with associated measurement uncertainties and, in a second step, uses this information to establish a relation for obtaining a measurement result from an indication [VIM12].

A more operational definition for calibration is the

> process of determining the numerical relationship, within an overall stated uncertainty, between the observed output of a measurement system and the value, based on standard sources, of the physical quality being measured [IEEE100].

From this, we see two approaches that the experimenter should follow: certifying that the measurement is within some pre-approved standard range for the measurement, and determining the amount of deviation from the desired standard and then documenting that quantity numerically. In the first case, the experimenter adjusts the instrument to correct improper readings. In the second case, adjustment may be possible but it may also be adequate to leave the instrument alone and apply the numerical corrections to the measurements to give the correct result. Both techniques require that the experimenter apply known standard inputs to the instrument and the resulting output observed. In this chapter, we concentrate on the analysis methods. The result of the process is to provide some form of a mapping, i.e., a mathematical function or curve, diagram, or table, to express the result for users.

There are two strategies that the analyst may take for developing the calibration equation. The first is to model the transfer functions analytically for each stage in the data collection process (T_S, T_C, and T_M in Figure 3.1) and develop a *process transfer function*, also known as the *response function* [IEEE100], $T_{process}$, such as

$$T_{process} = T_S T_C T_M \tag{3.1}$$

The analyst then fits any free parameters such as gain, sensitivity, and bandwidth to the data set. The second approach is for the analyst to develop an overall equation describing the mapping using a polynomial equation of the form

$$T_{process} = \sum_i a_i V^i \tag{3.2}$$

and then fit the coefficients without trying to model the underlying device characteristics.

The first strategy results in a predefined equation with a certain, small number of unknown parameters that must be empirically determined. The analyst may need to determine parameters individually for each device to account for device-to-device variations. The second strategy results in a polynomial or power law model that the analyst develops to produce a reasonable mapping to the data. The mapping may not have any physical significance as in the first strategy. For expediency, most analysts prefer the second technique. With this method, we do not try to fully understand and account for all of the potential effects at the individual component level; rather, we try to model the overall trend, thereby treating the measurement devices as a system.

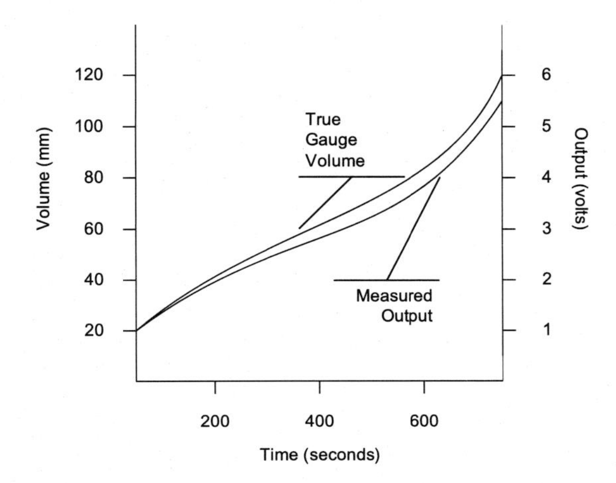

Figure 3.2: Simulated output of a capacitive rain gauge during a rainstorm.

3.3.1.1 Capacitive Rain Gauge Calibration Example

Let us begin the calibration discussion by considering the Capacitive Rain Gauge (CRG) that we saw in Section 2.4.2.1 as the sensor needing calibration. The CRG works by using a captured volume of water to make a capacitor. The output voltage across the CRG is proportional to the stored water volume. As rain falls during a storm, the gauge fills and the output voltage gives an integrated rainfall measurement. Figure 3.2 shows an example of the readings from the CRG during a simulated rainstorm where the figure gives the actual volume, the measured volume, and the output voltage level. In the figure, we see a slight difference between the actual volume and the measured volume of water in the gauge.

We know from experience that this difference is a function of air temperature so we repeated our calibrations on a regular basis to minimize temperature drift. To understand how the CRG is behaving at any time, we periodically add known quantities of water to the gauge and measure the output. Table 3.1 gives a calibration measurement set and Figure 3.3 illustrates the input-output relationship. Because the relationship appears linear, two related questions arise: is this truly linear and, if so, what are the slope and intercept values? We use mathematical and statistical techniques to answer these questions so that we can apply the results to make accurate measurements.

Table 3.1: Rain Gauge Data

Exact Input (mm)	Measured Output (mm)
0	0.07
5	4.37
10	10.31
15	16.38
20	22.19
25	27.70
30	33.07
35	38.27
40	43.56
45	48.76
50	54.04

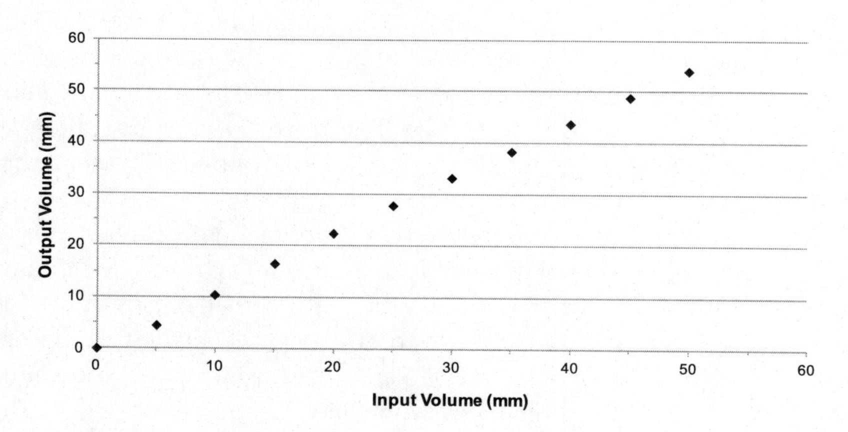

Figure 3.3: Actual versus measured output values for the CRG.

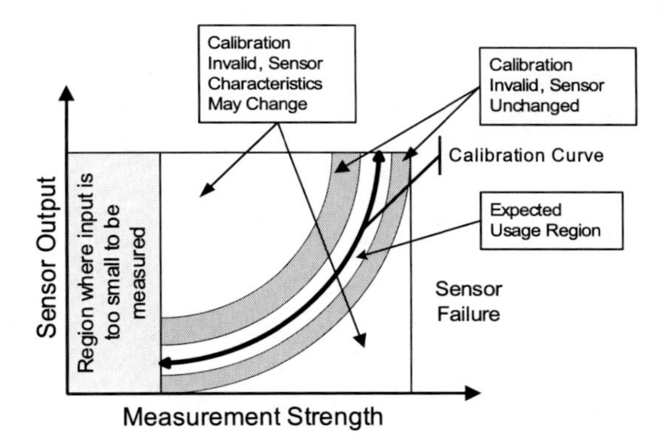

Figure 3.4: Sensor calibration and sensitivity regions. The normal operational region and the out-of-bounds regions are indicated.

3.3.1.2 Calibration Range

Figure 3.4 shows that sensors have a natural range of valid input and output levels. First, there is a minimum input signal level necessary to excite the sensor. There is also a maximum input signal level, which causes sensor failure. Between them is a normal sensor response region where the experimenter desires to make the measurements. This region has the calibration function defined to match the sensor. On either side of that boundary, signal conditions may exceed the sensor's specifications and the calibration is no longer valid.

These specifications define the usable signal level and environmental variables. For example, a sensor may have temperature, humidity, and vibration specifications that the measurement environment must not exceed for the sensor to operate properly. If the sensor exceeds these specifications, then varying results occur. A device that is slightly out of specification usually gives inaccurate readings but the environment does not permanently damage the device. The device continues to operate when the experiment environment returns to normal specifications. However, it may need to be re-calibrated to achieve accurate readings again. If the device is subject to extreme conditions, these conditions may damage the sensor permanently and the experimenter needs to replace it.

3.3.1.3 Measurement Calibration Process

Calibration is the process whereby the experimenter applies specific inputs to the measurement system and measures the corresponding outputs. The input is the known quantity or the independent variable, while the measurement output is the unknown quantity or dependent variable. The input values themselves must relate to a known and reliable standard. This standard may be a local laboratory standard and it provides a reasonably close approximation to the exact definition for the quantity. Alternatively, the laboratory may calibrate its standard against another standard, which the laboratory has calibrated against a primary standard as defined by a national standards laboratory. In all cases, we wish to "truth" the measurement.

The experimenter should calibrate regularly. Manufacturers often suggest the instrument's *calibration interval*, which is defined as the

> maximum length of time between calibration services during which each standard and test and measuring equipment is expected to remain within specific performance levels under normal conditions of handling and use [IEEE100].

Experimenters should also perform the calibration if the measurement system has been subject to large stresses. Some experimenters make it a standard procedure to calibrate the sensors at the start of each experiment or major use to prevent subsequent problems. This is especially important for uses involving human health and safety. Calibration is not a random process. Experimenters calibrate using a method such as the following sequence of steps:

1. Start the standard input in the middle of the input-output range.

2. Record input and output values at the desired step interval until the input reaches its maximum level.

3. Start to decrease the input level and record input and output values at the desired step interval until the input reaches its minimum level.

4. Start to increase the input level and record input and output values at the desired step interval until the input reaches its midpoint level again.

In all cases, the calibration process should not exceed the maximum and minimum input levels [ISA95]. This results in a calibration curve with which the experimenter determines hysteresis error and other sensitivity changes.

3.3.1.4 Calibration Curve Variables

The purpose of calibration and modeling is to relate the measured quantity to the measurand via some form of mathematical relationship. The mathematical mapping relates the measurand to a physical output variable such as voltage or current. As in the CRG example, this mapping is also a function of temperature, and to a certain extent, age, and usage history. In the most general form, this functional mapping is

$$measurand = \mathscr{F}\{outputvariables; inputvariables\} \tag{3.3}$$

The mapping, $\mathscr{F}\{*\}$, that the analyst develops is either linear or nonlinear, depending on the assumed system. Here, we examine a technique to perform the analysis regardless of the variables chosen. For example, suppose the system's output variable is voltage, V, and the environmental variables are the temperature, T, and the humidity, H. Suppose we know from physical analysis that a second-order relationship between the measurand and the system output voltage exists. Then the mapping function is

$$measurand = a_0 + a_1V + a_2V^2 + a_3T + a_4H \tag{3.4}$$

This allows us to fit the expected functional variation plus the environmental influence. There are two approaches that analysts use in practice when performing the calibration: full parameter space and interpolations.

For the analyst to calibrate the instrumentation over the full parameter space, the calibration process must make a great number of measurements over all combinations of the variables. This effort is justified if the instrumentation should make very precise measurements or if the experimenter must use the instrumentation over a wide range of conditions.

The simpler approach that the analyst can take is to perform several calibrations at key variable locations and then linearly interpolate between the curves. This approach can result in the analyst developing simpler, linear equations that need to be sampled at many fewer points. The user then interpolates between the simpler curves in practice.

One alternative to probing the whole environmental variable space is for the experimenter to control the operations space. If the experimenter operates the instrumentation under limited, well-controlled conditions, the experimenter may be able to ignore environmental effects and produce a simpler calibration variable space.

3.3.1.5 Difference between Calibration and Usage

During calibration, the experimenter replaces the measurand by a known standard and treats the standard as the independent variable. The measurement output is the unknown and the experimenter treats it as the dependent variable. In actual usage, the experimenter reverses the roles. In usage mode, the measurement process output is the known quantity and the measurand is the unknown. At this point, the experimenter reverses the calibration process to deduce the correction required to make a measurement. Depending upon how the experimenter performs the modeling, this difference implies that the experimenter may need to invert the modeling equation to use it with the measurement data.

3.3.2 Data Modeling

In many aspects, data modeling and developing the calibration mapping are similar processes. Next, we discuss the differences and extensions to the basic concepts.

3.3.2.1 Difference between Calibration and Data Modeling

In the calibration process, the mathematical equations used to model the process work on the specifics of the measurement system's input-output process. The next step is to perform a mathematical modeling of the entire data set. There may be an underlying physical theory that unites the data. The data analyst can also apply the least squares techniques discussed later in Section 3.6 to the entire data set to fit the data to the model. The goal is the same: the analyst wishes to develop a numerical technique that represents the entirety of the data set and allow the user to work with the equation and not with individual data points. Generally, the analyst uses the calibrated measurand values in the modeling process instead of raw uncalibrated measurements. While the analyst could, in principle, combine the two steps, usually they are not. Keeping them separate allows easier manipulation of different model classes.

3.3.2.2 Modeling as Filtering

The least squares technique discussed later in Section 3.6 also applies in the filtering and signal processing methods developed in Section 5.8.2.2. In a sense, the least squares technique acts as a filter to smooth the data through noise if it is present in the measurement process. This is important because the real-world measurement always includes a certain amount of uncertainty and noise. Therefore, these techniques are more important than simply modeling the underlying data process.

3.4 ERROR TYPES

A *measurement error* is the "measured quantity value minus a reference quantity value" [VIM12]. The reference value is assumed to be the true value the experimenter seeks. The accuracy and precision of the measurement instrument contributes to these measurement errors. This section covers the measurement errors that the experimenter may find in the measurements. The next section explores statistical ways of describing random errors. The data analyst can quantify other error types and compensate for them with proper calibration.

3.4.1 Systematic Errors

A *systematic measurement error* is the

> component of measurement error that in replicate measurements remains constant or varies in a predictable manner [VIM12].

A related concept is the *measurement trueness* or the

> closeness of agreement between the average of an infinite number of replicate measured quantity values and a reference quantity value [VIM12].

Measurement trueness is inversely related to the systematic error. By the definition, an infinite number of measurements is needed to find the systematic error so we often use the *measurement bias* as an "estimate of a systematic measurement error" [VIM12]. Figure 3.5 illustrates an offset between the true value and the measured value if there is bias in the measurement equipment or observer making the measurement. The line on the right side of the figure represents the true value that the measurement process is attempting to discover. The systematic error acts

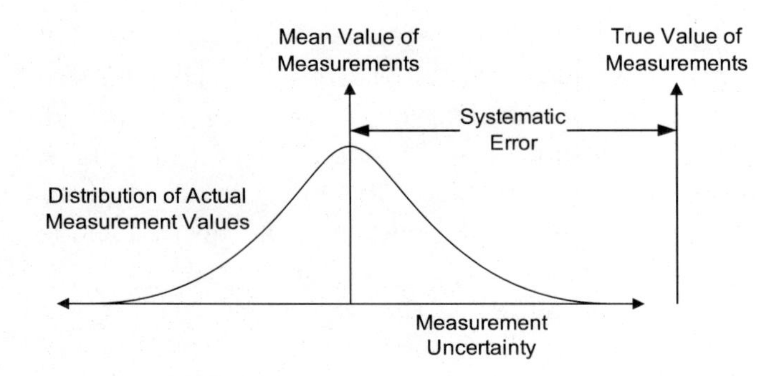

Figure 3.5: Relationship between random and systematic errors with respect to the true value.

to displace the measurements away from the true value to the new one. The systematic error acts like a Direct Current (DC) voltage applied to the measurement process. In this case, the systematic error displaces the measurements towards the value in the middle of the figure. By proper calibration, the data analyst estimates this displacement and then applies the appropriate compensation.

3.4.2 Random Errors

A *random measurement error* is the "component of measurement error that in replicate measurements varies in an unpredictable manner" [VIM12]. Random errors in the measurement process, which Figure 3.5 also shows as the distribution of values about the mean, cause the second effect. Random errors come from a number of sources, such as electrical noise, and they tend to obscure the central value, even when the experimenter makes a large number of measurements. The best that we can do is estimate the underlying value. From a voltage point of view, random errors do not have DC values as systematic errors have. Rather, we consider them as zero-mean Alternating Current (AC) processes. The result is that random errors provide a spread in the measurement values. However, the random error is not apparent to the analyst based on only one value. The analyst needs many measurement samples to see the extent of the random error. The mathematical fitting procedure tends to smooth the measurements and represent the estimates for the underlying values.

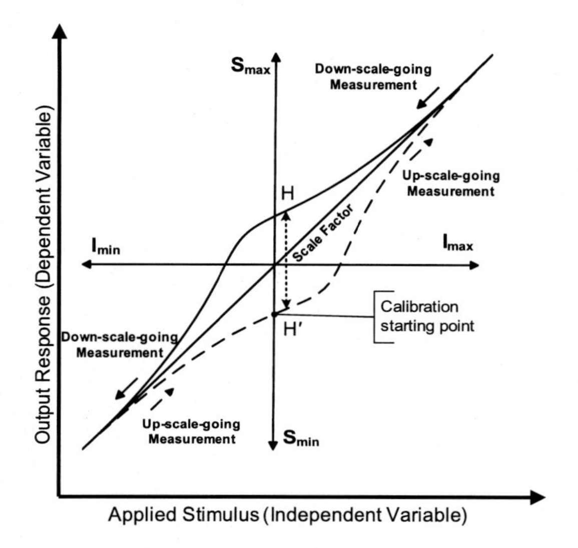

Figure 3.6: Hysteresis error in the measurement process relative to a linear relationship.

3.4.3 Interference

Sometimes a signal that is not directly part of the measurement process seeps into the electronics and causes *interference*. This stray signal biases the result and it is difficult to remove via calibration alone because the interfering signal is not part of the system, and usually it is not present during the calibration process. Additionally, the interfering signal is not always similar to a random error because it may have both DC and AC components. The best solution for interference is to shield the measurement process from it so that the interference does not corrupt the measurement.

3.4.4 Hysteresis Error

Figure 3.6 illustrates how the experimenter determines whether a hysteresis error exists in the measurement system by starting the calibration at the middle of the measurand range and scanning from there to the highest input level. Then the experimenter draws the input down to the minimum level. Finally, the experimenter takes the input back up to the midpoint level. Because the measurement system may react differently as the measurand increases from a low to a high value than it does as the measurand changes in the opposite sense, we define *hysteresis error* as the

> maximum separation due to hysteresis between up-scale-going and downscale-going indications of the measured variable (during a full-range traverse, unless otherwise specified) after transients have decayed [IEEE100]. (see also [Frad97; ISA95])

We define the hysteresis error, e_H, in terms of the maximum voltage separation, $H - H\prime$, and the scale factor, SF, using

$$e_H = \frac{H - H\prime}{SF} \tag{3.5}$$

SF is the scale factor defined in Equation (2.1) and it represents the slope of the line defining the linear representation of the up-scale and downscale measurements.

3.4.5 Dead Band Error

Experimenters sometimes find that their measurement system is insensitive to changes in the measurand over certain input values. These insensitivity regions produce a *dead band* error, defined as the

> maximum interval through which a value of a quantity being measured can be changed in both directions without producing a detectable change in the corresponding indication [VIM12]. (See also [Frad97; ISA95])

Figure 3.7 illustrates this concept where a small flat spot in the middle of the graph represents the dead band that causes measurement inaccuracy.

3.5 STATISTICAL CONCEPTS

To properly characterize the noise effects and better quantify the data modeling results, we use basic statistical measures that deal with both single measurements and whole measurement sets. The Gaussian statistical assumption is used to model random noise because it really serves as the "normal" assumption and describes most of the effects that we are concerned with.

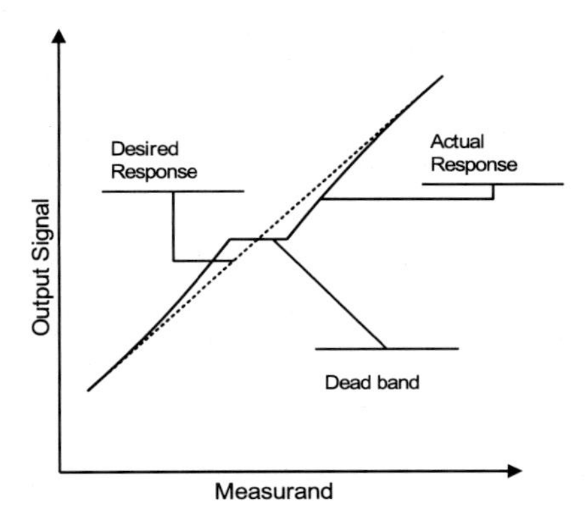

Figure 3.7: Dead band error in the measurement process. The dead band is the flat portion of the actual response.

3.5.1 Measurement Mathematical Model

Analysts assume that the actual measurement is the linear sum of a true value plus noise (superposition model). If we make a number of measurements of the same point, X_0, we write the resulting *measurement vector*, $\vec{X}$, as

$$\vec{X} = X_0 + \vec{N} \tag{3.6}$$

The *noise vector*, $\vec{N}$, associated with the measurement process, modifies the measurement of the true value, X_0. In principle, if we know the value of the noise vector elements, we can invert the equation and uniquely recover the true value. In practice, we *estimate* the noise and then arrive at our best estimate for the true value. Figure 3.8 illustrates this process where we made a series of measurements at each point along the x-axis. We have enlarged one of these points to show the distribution of the measurements contained in that point. In the next section, we describe the probability concepts used to describe the noise process.

3.5.2 Probability Concepts

Analysts use probability measures to describe the noise processes. The important functions that analysts use to describe the probability that a noise takes on certain values are the probability density and the probability distribution. We also concentrate on Gaussian probability functions because they cover many practical situations. This is a brief discussion; for

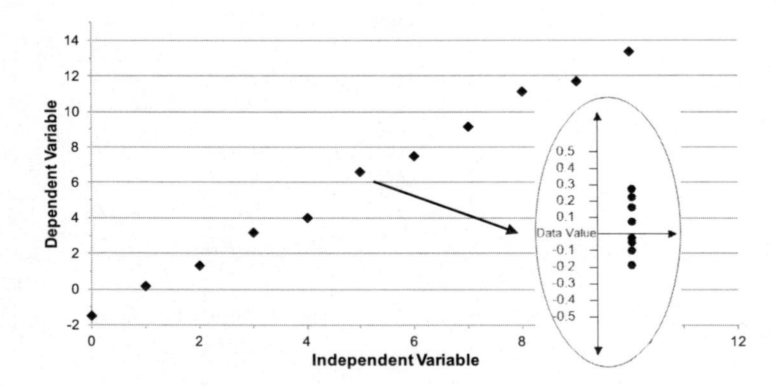

Figure 3.8: A series of measurements with one measurement selected to show the distribution of measurements comprising one of the data points.

more information, consult standard references such as [Leon94; Dave87; Papo65; Woze90].

3.5.2.1 Relative Frequency

One of the most intuitive methods for representing the concept of the probability that an event occurs is by using the relative frequency probability definition. For example, suppose A is an event such as winning at *21*, losing a baseball game, or having a meteor hitting your car. We relate the probability of event A, $p(A)$, out of a total sample of N events, to the number of times the event A occurs, $n(A)$, by the equation

$$p(A) = \lim_{N \to \infty} \frac{n(A)}{N} \tag{3.7}$$

From this equation, we see that we really need a large sample for the concept of a probability to be meaningful. A single measurement is not enough. We need to make "reasonable" approximations since we do not have the time to make an infinite number of measurements.

3.5.2.2 Probability Density

Generally, we wish to know the probability of events at many sampling instances. This gives rise to the concept of a *Probability Density Function (PDF)*. This function can be either continuous or discrete. Noise tends to be a continuous quantity so we concentrate on the continuous form.

Let us start by extending the relative frequency concept and make a

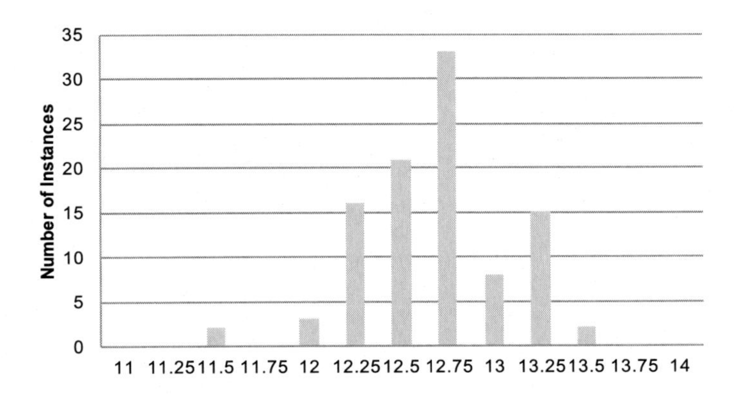

Figure 3.9: The relative frequency of measurements with noise added.

histogram of the relative frequency of a number of measurements. For example, suppose we design an experiment where the "true" measurement value is known to be 12.5 in magnitude and each measurement has a little noise added. Let us make 100 measurements and plot the relative frequency of each value obtained as shown in Figure 3.9. Since we know the "correct" value is 12.5, the plot gives a relative frequency of the noise values. The histogram intervals are chosen by computing the bin width, Δ, knowing the total number of points, N, and the range of the points by using the relationship [Klaa96]

$$\Delta = \frac{(x_{max} - x_{min})}{\sqrt{N}} \tag{3.8}$$

As we can see from the figure, not all of the histogram bins are well occupied which is a result of the limited number of measurements. If we extend this notion to a case where an "infinite" number of noise measurements are made and then plot the histogram, we arrive at the PDF, $p(x)$, for the noise. This function describes the spread of the random variable that we are measuring. The PDF for a random variable, x, has an *average value* or *mean value*, μ, given by

$$\mu = \int_{-\infty}^{\infty} x p(x) \mathrm{d}x \tag{3.9}$$

Statisticians also call this the *first moment* of the variable x. We compute the *variance* of the random variable, σ^2, by using the PDF in the equation

$$\sigma^2 = \int_{-\infty}^{\infty} (x-\mu)^2 \, p(x)\mathrm{d}x \tag{3.10}$$

The variance is also called the *second moment* of x and it measures the dispersion around the mean value. The *standard deviation*, σ, is related to the variance by $\sigma = \sqrt{\sigma^2}$. The standard deviation is also known as the *standard measurement uncertainty* [VIM12].

The PDF has two properties that distinguish its function from an arbitrary function. The PDF is positive; that is, a probability is always a positive number. The integral of the PDF is unity. We use the PDF to compute event probabilities once we know the functional form. How do we determine the correct PDF for a set of measured data since we do not have an infinite number of measurements? Commercial software packages exist that allow users to enter the measured data and they rank order the best estimates for the mathematical PDF to represent the data. Lacking that assistance, that analyst often makes an engineering "best guess" based on experience.

Once we have the PDF, we compute the probability that the continuous variable x lies between the limits a and b by using the relationship

$$p\,(a < x \leqslant b) = \int_{a}^{b} p(x)\mathrm{d}x \tag{3.11}$$

As the limits a and b approach each other, the probability of finding x between them becomes zero; that is, the probability that a continuous variable is found at exactly one point is 0. Also, as the limits a and b approach infinity, the probability approaches unity; that is, if one covers the number line, the probability is 1 that x is found somewhere.

For example, is the function $h(x) = {}^1\!/\!_2[u(x) - u(x-2)]$ illustrated in Figure 3.10(a) a valid PDF for a random variable x? The function $u(x)$ is the unit step function. The area of $h(x)$ computes to one and it is strictly positive so it has the correct mathematical properties. We interpret it as a uniform probability that x is between 0 and 2. Using this PDF, what is the probability that $0.125 < x \leqslant 1$? Applying Equation (3.11), we obtain $p(0.125 < x \leqslant 1) = 0.438$.

3.5.2.3 Cumulative Distribution Function

Another way to represent the probability information is with the *Cumulative Distribution Function (CDF)*. It does not contain any different

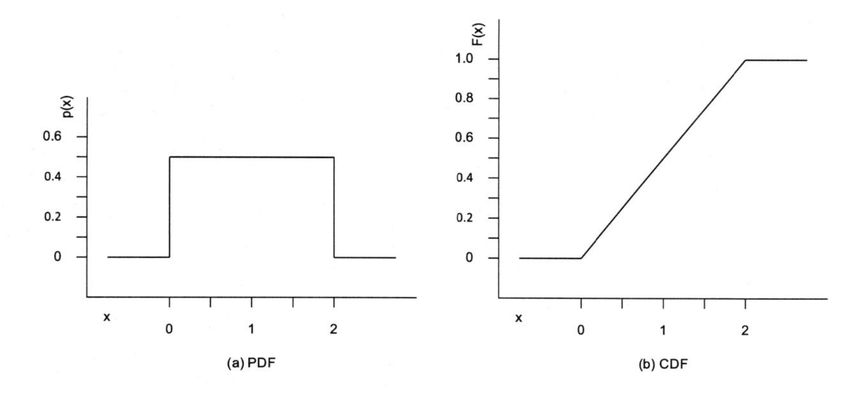

Figure 3.10: An example of a uniform PDF and its associated CDF for a random variable.

information from the PDF; it merely represents it in a different way. The PDF and CDF are related by the definition of the CDF. The CDF, $F(a)$, for a random variable with a probability density, $p(x)$, evaluated at point a is

$$F(a) \equiv \int_{-\infty}^{a} p(x)\mathrm{d}x = p(x \leqslant a) \tag{3.12}$$

That is, the CDF measures the probability that x is less than the point a. We extend this to computing the probability that x is between two limits, a and b, as

$$p(a < x \leqslant b) = F(b) - F(a) \tag{3.13}$$

We see the relationship between the PDF and the CDF for the above example in Figure 3.10(b). For any CDF, the minimum value is 0 and the maximum value is 1. The CDF is also a nondecreasing function. Using the previous example, the CDF for the two end points are $F(0.125) = 0.063$ and $F(1) = 0.5$. The probability that x is between a and b computes to 0.438 as it did above.

3.5.2.4 Gaussian Probability Density Function and Noise Model

Engineers commonly use the Gaussian PDF to model noise processes. This is a reasonable approach because many noise processes are actually the accumulation of multitudes of individual interactions. By the Central

Limit Theorem, these accumulated interactions give the resulting noise a Gaussian PDF, which is written as

$$p(x) = \frac{1}{\sigma\sqrt{2\pi}} \exp\left[\frac{-(x-\mu)^2}{2\sigma^2}\right] \tag{3.14}$$

where μ is the mean value of the random variable x and σ is its standard deviation or standard uncertainty.

We use Equations (3.11), (3.12), and (3.13) to make computations. However, the Gaussian PDF integrals are generally not computable in closed form. Instead, we use the well-known related functions given below to make the Gaussian PDFs computation easier.

The erf Function The first related function is the *Error Function (*erf*)*. There are generally two similar definitions of the erf integral [Papo65]. Here, we use the following integral equation form:

$$\text{erf}(a) = \frac{2}{\sqrt{\pi}} \int_0^a e^{-u^2} du \tag{3.15}$$

This result is the integral of a normalized Gaussian PDF from 0 to the value a. It is not the same Gaussian as in Equation (3.14) but is a transformation of variables as Figure 3.11 shows. **Note:** some references use slightly different variables with these definitions so the reader is warned to check the exact form the author is using.

The error function has the following properties:

$$\text{erf}(-x) = -\,\text{erf}(x)$$

$$\text{erf}(\infty) = 1$$

The erf function is commonly available in computing packages such as Mathcad®, MATLAB®, and Excel® along with other statistical functions. Appendix Table B.1 tabulates the erf function.

The error function has a related function known as the *Complementary Error Function (*erfc*)*. We compute the erfc from the erf by using

$$\text{erfc}(x) = 1 - \text{erf}(x) \tag{3.16}$$

For a Gaussian variable x between the limits a and b, we compute

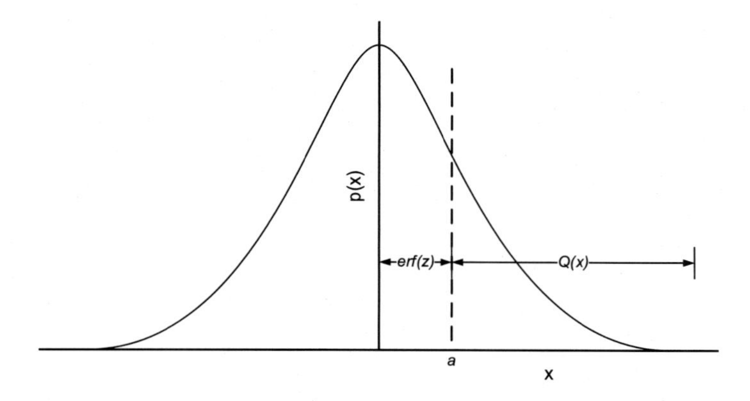

Figure 3.11: A Gaussian PDF with the erf and Q functions indicated ($\mu = 0$ and $\sigma = 1$).

the probability that x is between the limits by using the erf and erfc functions as follows:

$$\begin{aligned} p(a < x \leqslant b) &= \frac{1}{2}\,\mathrm{erf}\left(\frac{b-\mu}{\sigma\sqrt{2}}\right) - \frac{1}{2}\,\mathrm{erf}\left(\frac{a-\mu}{\sigma\sqrt{2}}\right) \\ &= \frac{1}{2}\,\mathrm{erfc}\left(\frac{\mu-b}{\sigma\sqrt{2}}\right) - \frac{1}{2}\,\mathrm{erfc}\left(\frac{\mu-a}{\sigma\sqrt{2}}\right) \end{aligned} \tag{3.17}$$

For example, let $\mu = 2.5$ and $\sigma^2 = 0.5$. What is the probability that the measurement x is between 1.5 and 3.5? Using the erf function

$$\begin{aligned} p(1.5 < x \leqslant 3.5) &= \frac{1}{2}\,\mathrm{erf}\left(\frac{3.5-2.5}{\sqrt{(2)(0.5)}}\right) - \frac{1}{2}\,\mathrm{erf}\left(\frac{1.5-2.5}{\sqrt{(2)(0.5)}}\right) \\ &= 0.8427 \end{aligned} \tag{3.18}$$

What is the probability that the measurement is greater than 10? Using the erf function, we compute

$$\begin{aligned} p(x > 10) = p(10 < x \leqslant \infty) &= \frac{1}{2} - \frac{1}{2}\,\mathrm{erf}\left(\frac{10-2.5}{\sqrt{(2)(0.5)}}\right) \\ &\approx 0 \end{aligned} \tag{3.19}$$

The answer is not exactly zero but the computation is close enough.

The Q Function Another function related to erf and erfc is the Q function, which is defined as

$$Q(a) = \frac{1}{\sqrt{2\pi}} \int_a^\infty e^{-\frac{y^2}{2}} \mathrm{d}y \tag{3.20}$$

If the error function is the integral of the Gaussian PDF from 0 to the point a, then the Q function is related to the integral of the PDF from a to infinity as Figure 3.11 shows. We relate the Q function to the erf and erfc by the following equations:

$$Q(x) = \frac{1}{2} \operatorname{erfc}\left(\frac{x}{\sqrt{2}}\right) \tag{3.21}$$

$$Q(x) = \frac{1}{2} \left[1 - \operatorname{erf}\left(\frac{x}{\sqrt{2}}\right)\right] \tag{3.22}$$

Similar to Equation (3.17), we compute probabilities with the Q function

$$p(a < x \leqslant b) = Q\left(\frac{a-\mu}{\sigma}\right) - Q\left(\frac{b-\mu}{\sigma}\right) \tag{3.23}$$

Appendix Table B.1 also tabulates the Q function. We can also compute the Q function using Equations (3.21) and (3.22), and the erf and erfc functions available in computer analysis packages.

3.5.2.5 Electronic Noise

As we have noted, experimenters find electronic noise in measurement systems and they typically assume that a Gaussian process models the noise. The measurement system has a natural bandwidth, B, measured in Hertz. Random electronic noise is a zero-mean process; that is, the noise does not have a DC offset. We parameterize the noise by an equivalent system temperature, T_{sys}, [Dave87]. This is not a physical temperature; it is a measure of the total noise produced by the object. For example, engineers say that a device, which produces as much noise as a black body source of a specific temperature, has a noise temperature of that same specific temperature regardless of the device's physical temperature. The system temperature is used to compute the noise spectral density, N_o, which describes the noise process in the frequency domain. We compute the spectral density, in W/Hz, from the system temperature and Boltzmann's constant, k, by using

$$N_o = kT_{sys} \tag{3.24}$$

where k is $1.380\,648\,52 \times 10^{-23}$ J/K. Engineers frequently express this fundamental quantity in dB units as -228.6 dBW/K – Hz. From this, we compute the variance of the Gaussian process using the spectral density and the bandwidth

$$\sigma^2 = N_o B \tag{3.25}$$

We will see these relations again in Section 11.5.3 where we will examine how the system noise is related to the data transmission process.

3.5.2.6 Mean, Variance, and Standard Deviation Estimates

In Equations (3.9) and (3.10), we saw the definitions of the mean, μ, the variance, σ^2, and the standard deviation, σ. Statisticians base these definitions on having the full probability density function and, essentially, an infinite number of measurements. Of course, this is impossible to attain in practice. This section covers practical estimates for these parameters. In this section, we will see how well we know these results.

Parameter Estimation In real systems, we need to estimate the mean and the variance based upon a finite number of measurements. We may not even know the underlying PDF. A typical estimate for the mean, $\langle\mu\rangle$ or $\bar{x}$, for a set of N measurements, $\{x_i\}$, assuming that each measurement is equally probable, is to use the customary equation for computing an average

$$\bar{x} = \langle\mu\rangle = \frac{1}{N}\sum_{i=1}^{N} x_i \tag{3.26}$$

In a similar manner, we compute the estimated variance, $\langle\sigma^2\rangle$, of the data set by using

$$\langle\sigma^2\rangle = \frac{1}{N-1}\sum_{i=1}^{N}(x_i - \bar{x})^2 \tag{3.27}$$

We use the data in Table 3.2 to compute the estimated mean and the estimated variance. Applying Equations (3.26) and (3.27), respectively, we arrive at the estimated mean of 12.4024 and the estimated variance of 0.3999. The question now is how good are the estimates? Analysts use the error in the mean, the uncertainty in the mean, and the confidence interval to answer this question.

Error in the Mean The previous example involved a finite number of samples and a single estimate for the mean and variance. Suppose we ran

Table 3.2: Sample Data for Computing Mean and Variance

12.7112	12.2444	12.9988	13.0394	11.6551
11.8167	12.4733	12.6544	13.2575	12.4193
12.9993	12.5057	12.2144	13.0580	12.0337
13.5600	12.8323	11.9057	12.9222	11.8778
12.2843	11.2862	12.6215	13.5281	11.5461
12.2451	13.0092	12.8680	11.3763	12.0059
11.9353	12.6319	12.2218	12.8240	11.7719
13.3503	12.7698	12.6444	12.9008	11.5540
11.2474	11.5000	13.2573	13.2929	12.2232
12.2079	11.1374	11.9733	12.1102	12.6152

the experiment again. It is not hard to believe that we obtain slightly different measures for the mean and variance. Figure 3.12 shows that if we perform the same experiment multiple times, we build up a distribution of mean values. They cluster around the "mean of the means" which is a better estimate for the true value than is any individual sample mean. Statisticians define the variance in the distribution of the mean values as the *error in the mean*. Analysts compute the error in the mean in terms of the estimated variance, $\langle\sigma^2\rangle$, and the number of measurements, N, in the data set using [Rabi95; Youn62]

$$\sigma_\mu^2 = \frac{\langle\sigma^2\rangle}{N} \tag{3.28}$$

The error in the mean for the data set in Table 3.2 is 0.0080. As we see from Equation (3.28), the number of data points in the measurement set influences the error in the mean. The greater the number of points in the data set, the smaller the error in the mean.

Uncertainty in the Mean Just as the standard deviation is the square root of the variance, the *uncertainty in the mean* is the square root of the error in the mean, or [Rabi95; Youn62]

$$\sigma_\mu = \sqrt{\frac{\langle\sigma^2\rangle}{N}} \tag{3.29}$$

For the example data set in Table 3.2, the uncertainty in the mean is 0.0894.

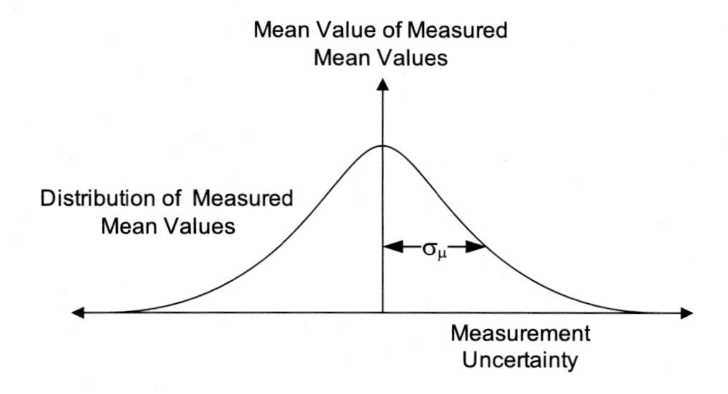

Figure 3.12: The distribution of mean values about the "mean of the means."

3.5.3 Measurement Uncertainty

As we saw at the beginning of this chapter, measurement uncertainty is one of the three "pillars of metrology." While the statistical estimates of the error in the mean and the uncertainty in the mean are quantities that are easy to compute, they do not really tell the user how close they might be to the true value, except in the most general way. The measurement system has other unknowns than just those associated with the measurement instrument itself. Given total environment, the experimenter must consider any measurement as uncertain to some degree. In this section, we will examine ways for the experimenter to express the measurement uncertainty.

3.5.3.1 Uncertainty Definition

We need to think in terms of overall uncertainty because, as the GUM states, the

> uncertainty of the result of a measurement reflects the lack of exact knowledge of the value of the measurand. The result of a measurement after correction for recognized systematic effects is still only an estimate of the value of the measurand because of the uncertainty arising from random effects and from imperfect correction of the result for systematic effects [GUM08].

The *measurement uncertainty* is

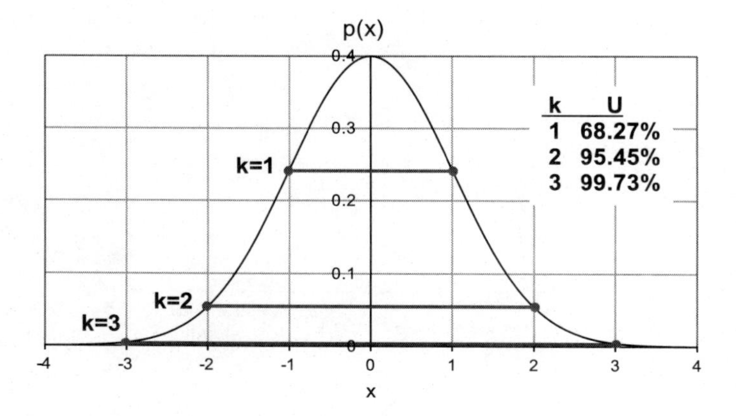

Figure 3.13: Definition of uncertainty regions and k factor for a normal distribution.

> non-negative parameter characterizing the dispersion of the quantity values being attributed to a measurand, based on the information used [VIM12].

For Type A uses, the standard deviation is the parameter used to quantify the uncertainty [GUM08].

In order for the experimenter to give an estimate of total percentage of the uncertainty range being considered by the Type A statistics, the concept of an *expanded uncertainty* is used [GUM08]. The expanded uncertainty, U, is defined as the "product of a combined standard measurement uncertainty and a factor larger than the number one" [VIM12]. The *coverage factor*, k, quantifies the expanded uncertainty in terms of the standard uncertainty, σ, using

$$U = k\sigma \tag{3.30}$$

When the random errors are described by a Gaussian PDF, the percentage of coverage of the whole region is easy to compute. Figure 3.13 illustrates the normal PDF with the corresponding k factors and percentage of coverage. A $k = 1$ uncertainty factor covers 68.3% of the region, while a $k = 3$ covers 99.7% of the region. Typically, k is in the range of 2 to 3, depending upon the coverage needs [GUM08].

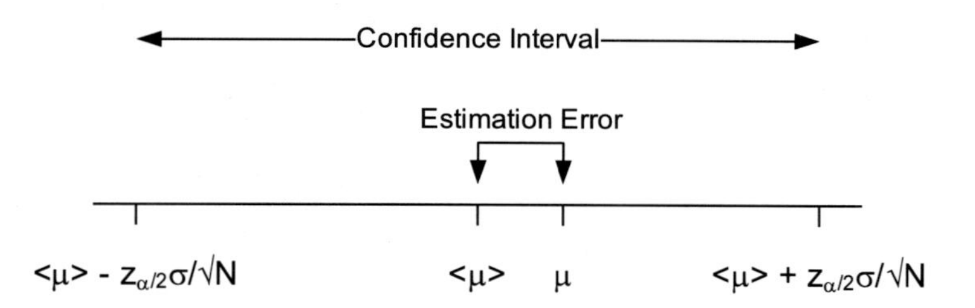

Figure 3.14: The confidence interval relative to the true mean and the estimate of the mean.

3.5.3.2 Confidence Intervals

If the experimenter is dealing with only Type A approaches to measurement uncertainty, then the statistical *confidence interval* technique is a way to express an expanded uncertainty U [GUM08]. Given that the experimenter may have only a finite number of measurements, the means and variances are less than perfect estimates. The confidence interval permits the experimenter to quantify the uncertainty region. Figure 3.14 illustrates the confidence interval for the estimate. The true value for the mean, μ, and the estimated mean, $\langle\mu\rangle$, are shown and their difference is an estimation error. Based on the N samples and the statistical uncertainty in the mean, we define a confidence interval around the estimated mean [Kueh00; Rabi95]. This confidence interval is intended to represent the region in which we expect to find the true mean value if we make an infinite number of measurements.

The confidence interval's width is a function of the statistical uncertainty in the mean and the level of confidence we want in the result. If we wish to be very confident of the result, then the interval needs to be broader than if we are willing to accept a lower confidence level. We define α as the percent uncertainty level in the estimate giving a $(100 - \alpha)$ percent confidence in the estimate. For example, a 5% uncertainty corresponds to a 95% confidence level. Typical values for the confidence interval are 95% or 99%. The uncertainty is mapped to the outer wings of the PDF with $\alpha/2$ % on the far left and $\alpha/2$ % on the far right.

If $\langle\mu\rangle$ is our estimate for the mean of the measurements and σ_μ is the uncertainty in the mean, then we estimate that the true value, μ, is found with $(100 - \alpha)\%$ confidence in a region bounded by

$$\langle\mu\rangle - z_{\alpha/2}\sigma_\mu \leqslant \mu \leqslant \langle\mu\rangle + z_{\alpha/2}\sigma_\mu \tag{3.31}$$

where $z_{\alpha/2}$ is a parameter based on the confidence level chosen. We interpret $z_{\alpha/2}$ as measuring the number of standard deviations about the mean in which we expect to find the true value. We base using $z_{\alpha/2}$ on the assumption that more than 30 measurements are in the data set and that we have a Gaussian process. Appendix Table B.2 gives values for $z_{\alpha/2}$. In the table, P is the Gaussian probability or the confidence level we desire to use. The $z_{\alpha/2}$ is read from the column with the desired confidence level. Continuing the example from Table 3.2, the 95% confidence interval is 0.175 around the estimated mean 12.402. That is, with 95% confidence, we believe the true value lies between 12.227 and 12.578.

Suppose only a smaller number of measurements are available or that we do not know the variance in the data, or that we are not sure Gaussian statistics still hold, is there a way to compute the confidence interval? If the analyst has fewer than 30 measurements or the Gaussian assumption does not hold, the analyst uses a t distribution. In that case, the analyst uses values from the t distribution shown in Table B.3. The value for $z_{\alpha/2}$ is replaced with $t_{\alpha/2}$ which is a function both of the confidence level and the *number of degrees of freedom*, ν, in the measurement set. The number of degrees of freedom is given by $\nu = N - 1$ and this is one of the parameters in Table B.3 for selecting the correct value of t . The confidence interval bounds equation is

$$\langle\mu\rangle - t_{\alpha/2}\sigma_\mu \leqslant \mu \leqslant \langle\mu\rangle + t_{\alpha/2}\sigma_\mu \tag{3.32}$$

3.5.3.3 Number of Measurements Required

The number of measurements, N, is a critical parameter and the confidence of the results directly scales with N. The main question is how many measurements are sufficient? Certainly, we need more than one. Are 10, 100, or 1000 measurements sufficient and how do we know? One approach is to use the confidence interval. If we know the confidence interval we want and the approximate variance in the data, we can invert the equation and solve for the minimum number of measurements needed.

3.5.3.4 Combined Uncertainty and Uncertainty Budget

Figure 3.14 showed the statistical confidence interval associated with a "true" value. When performing the measurement instrument's calibration, we have been assuming that the experimenter knows that "true" value to any desired degree of precision. However, the uncertainty approach dictates that the experimenter consider that the "true value" is itself an estimate with an associated standard uncertainty value.

In the previous sections, we saw both the random or Type A measurement errors and the systematic or bias types of errors. To generate the total measurement uncertainty properly, the analyst needs to combine these various errors. First, we must ask which effects we should consider in the process. The recommendation in [RCC122] suggests that the analyst consider the following list as a starting point:

- Measurement bias
- Random or repeatability error
- Resolution error
- Digital sampling error
- Computation error
- Operator bias
- Environmental factors error
- Stress response error

The analyst should develop a specific list for each measurement or calibration activity.

For example, suppose that S represents the standard deviation in the mean value of a set of measurements that the analyst made due to random error, while B represents the systematic error at the 95% confidence level caused by a DC bias in the instrument. The analyst can estimate the *combined uncertainty* at the 95% confidence level, U_{95}, by computing [RCC122]

$$U_{95} = \mp 2\sqrt{\left(\frac{B}{2}\right)^2 + \left(\frac{S}{\sqrt{N}}\right)^2} \tag{3.33}$$

The factor of 2 in the equation comes from assuming a student's t-statistic value of 2 for the 95% confidence interval. This is a specific application of the more general way to compute the combined uncertainty from

$$U_{95} = \sqrt{U_1^2 + U_2^2 + 2\rho_{1,2}U_1U_2} \tag{3.34}$$

Here, $\rho_{1,2}$ is the correlation coefficient between the two variables. For Equation (3.33), we are assuming that the correlation coefficient between the Type A error and the bias is 0. As pointed out in [RCC122], the analyst can typically assume that the following variables are independent and have $\rho_{1,2} = 0$:

- Random error and measurement bias
- Random error and operator bias
- Measurement bias and operator bias
- Measurement bias and resolution error
- Operator bias and environmental factors error
- Resolution error and environmental factors error
- Digital resolution error and operator bias

Not all of these pairings are found in all operational environments so the analyst must consider which are relevant.

To take all of these uncertainty sources into account, the analyst generates an *uncertainty budget* which is

> statement of a measurement uncertainty, of the components of that measurement uncertainty, and of their calculation and combination [VIM12].

This overall uncertainty budget, builds on Equation (3.34) to find the overall uncertainty based on each of the relevant uncertainty factors, U_i by applying [RCC122]

$$U_{95} = \sqrt{\sum_{i=1}^{n} U_i^2 + 2\sum_{i=1}^{n}\sum_{j>1}^{n} \rho_{i,j} U_i U_j} \tag{3.35}$$

The analyst must check the list of independent effects to determine which $\rho_{i,j}$ values are 0. For more information in applying uncertainty analysis to telemetry systems, the reader should consult [RCC122].

3.6 LEAST SQUARES FITTING

The *least squares method* is the usual procedure analysts employ to fit a model to a data set, especially when measurement noise makes the exact nature of the equation a bit uncertain or when the analyst does not have an *a priori* model for the system. The analyst uses this method to model both the calibration of the instrumentation system and the underlying process producing the data. This section concentrates on fitting polynomials to the data. Once the analyst understands this procedure, the analyst uses other functions that are appropriate to the application (trigonometric, orthogonal polynomials, etc.) and determines the fitting procedure. Since we use the procedure both for calibration and for data analysis, we do not make a distinction between them during the method's development. As a practical matter, we developed the necessary equations here but they are standard functions in analysis packages such as Excel® and MATLAB®.

3.6.1 Least Squares Definition

The least squares process provides the best estimate for the parameters that the analyst chooses to specify a model. The method does not tell the analyst which model to use. A number of methods can determine which of several models is relatively better, but no one method absolutely determines which is the correct model. Statisticians base the least squares method on minimizing the *Mean Square Error (mse)* between the data and the chosen model [Papo65; Rabi95]. The computations involve a set of data points consisting of an independent variable $\{x_i\}$ and a dependent variable $\{y_i\}$. The model produces a set of dependent variable estimates, $\{\hat{y}_i\}$, based upon the independent variable. We define the *mse* between the data and its estimate as

$$mse = \frac{1}{N} \sum_{i=1}^{N} (\hat{y}_i - y_i)^2 \tag{3.36}$$

The *mse* is a function of the data's quality and the model selected. If the analyst chooses an inappropriate model, the *mse* may be large, even if the data are relatively noise free. For an analyst to fit the model, the analyst needs a sufficient number of data points. This number must be greater than the number of parameters determined in the fitting procedure. We can also weight the fit in Equation (3.36) by dividing by the variance in each point. Unless the variances of each point are significantly different, this does not greatly affect the results so we do not use it here.

3.6.2 Linear Least Squares – Mean Square Error Basis

To see how the least squares equations are developed, let us first consider a linear fit to data. In this case, the model becomes

$$\hat{y}_i = a_0 + a_1 x_i \tag{3.37}$$

We next apply the model to the *mse* Equation (3.36). This gives a model-specific *mse* equation of

$$mse = \frac{1}{N}\sum_{i=1}^{N}\left(a_0 + a_1 x_i - y_i\right)^2 \tag{3.38}$$

To minimize the *mse*, we take the partial derivative of the *mse* in Equation (3.38) with respect to each of the fit parameters, a_0 and a_1, giving two equations of the form

$$\frac{\partial}{\partial a_j}\sum_{i=1}^{N}\left(a_0 + a_1 x_i - y_i\right)^2 = 0 \tag{3.39}$$

We then solve the system of equations for the fit parameters. The system of equations we need to solve is

$$\begin{aligned} \sum_{i=1}^{N}\left(a_0 + a_1 x_i - y_i\right) &= 0 \\ \sum_{i=1}^{N}\left(a_0 + a_1 x_i - y_i\right) x_i &= 0 \end{aligned} \tag{3.40}$$

We reorganize the individual equations of Equation (3.40) as follows:

$$\begin{aligned} N a_0 + \sum_{i=1}^{N} a_1 x_i &= \sum_{i=1}^{N} y_i \\ \sum_{i=1}^{N} a_0 x_i + \sum_{i=1}^{N} a_1 x_i^2 &= \sum_{i=1}^{N} x_i y_i \end{aligned} \tag{3.41}$$

It is often easier to manipulate the equations if we write them in matrix form, as follows:

$$\begin{pmatrix} N & \sum_{i=1}^{N} x_i \\ \sum_{i=1}^{N} x_i & \sum_{i=1}^{N} x_i^2 \end{pmatrix}\begin{pmatrix} a_0 \\ a_1 \end{pmatrix} = \begin{pmatrix} \sum_{i=1}^{N} y_i \\ \sum_{i=1}^{N} x_i y_i \end{pmatrix} \tag{3.42}$$

Table 3.3: Example Least Squares Computation

	x	y	x^2	xy
	0	0.07	0	0
	5	4.37	25	21.85
	10	10.31	100	103.1
	15	16.38	225	245.7
	20	22.19	400	443.8
	25	27.7	625	692.5
	30	33.07	900	992.1
	35	38.27	1225	1339.45
	40	43.56	1600	1742.4
	45	48.76	2025	2194.2
	50	54.04	2500	2702
Totals	275	298.72	9625	10477.1

Continuing the example, we next apply this to the data in Table 3.1. Table 3.3 shows the sums needed for using Equation (3.42). Solving the matrix equation for the coefficients gives $a_0 = -0.1991$ and $a_1 = 1.094$. Figure 3.15 illustrates the sample data points and the fit to the data.

3.6.3 Linear Least Squares – Statistical Basis

The matrix formulation of the least squares fit is not the only way to structure the solution. We use the summations listed below to determine the coefficients for the linear fit [Kueh00]. They also help us assess the quality of the fit so that analysts use the summations beyond solving for the coefficients. The necessary summations are

1. the average value of the independent variables

$$\bar{x} = \frac{1}{N} \sum_{i=1}^{N} x_i \tag{3.43}$$

2. the average value of the dependent variables

$$\bar{y} = \frac{1}{N} \sum_{i=1}^{N} y_i \tag{3.44}$$

3. the number of degrees of freedom for the N data points given we

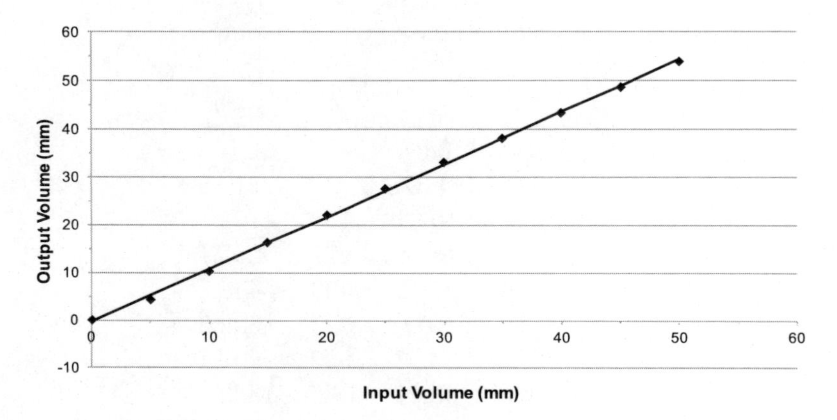

Figure 3.15: Sample data from Table 3.1 and the least squares fit to the data.

have already estimated the mean value from the data points

$$\nu = N - 2 \tag{3.45}$$

4. the spread of the x_i around their mean (Sum Square x)

$$SS_{XX} = \sum_{i=1}^{N} (x_i - \bar{x})^2 \tag{3.46}$$

5. the spread of the y_i around their mean (Sum Square y)

$$SS_{YY} = \sum_{i=1}^{N} (y_i - \bar{y})^2 \tag{3.47}$$

6. the cross-product of the $\{x_i\}$ with the $\{y_i\}$ (Sum Square $x - y$)

$$SS_{XY} = \sum_{i=1}^{N} (x_i - \bar{x})(y_i - \bar{y}) \tag{3.48}$$

With these summations, the linear fit coefficients are

$$\begin{aligned} a_1 &= \frac{SS_{XY}}{SS_{XX}} \\ a_0 &= \bar{y} - a_1 \bar{x} \end{aligned} \tag{3.49}$$

Table 3.4 shows the application of these summations to the sample data from Table 3.1. Using Equation (3.49), we obtain the coefficients $a_0 = -0.1991$ and $a_1 = 1.094$, which are the same values as those found by the matrix method.

Table 3.4: Summations for Least Squares Fit

	x	y	$(x_i - \bar{x})^2$	$(x_i - \bar{x})(y_i - \bar{y})$
	0	0.07	625	677.159
	5	4.37	400	455.727
	10	10.31	225	252.696
	15	16.38	100	107.764
	20	22.19	25	24.832
	25	27.7	0	0.000
	30	33.07	25	29.568
	35	38.27	100	111.136
	40	43.56	225	246.055
	45	48.76	400	432.073
	50	54.04	625	672.091
Mean	25	27.156		
		$\mathbf{SS_{XX}}$	2750	
			$\mathbf{SS_{XY}}$	3009.1

3.6.4 Quality of the Fit

The fit coefficients are only one factor. We need to answer two questions: are the coefficients well determined and is this model correct? This section examines both issues. Figure 3.16 shows one of the first tests experimenters use to determine whether a fit is reasonable by plotting the *residuals* across the data set. The residuals, ρ_i, are the differences between the measured dependent variable and the model output for each point

$$\rho_i = y_i - \hat{y}_i \tag{3.50}$$

Ideally, the residuals' plot should look like a plot of random points, as Figure 3.17 shows in part (a). There should be about as many points with positive residuals as there are with negative residuals and there should not be any obvious structure in the residuals indicating that the model might not be appropriate. Figure 3.16 shows a few more positive residuals than negative residuals; however, with only eleven points, it is difficult to make a definitive determination if a systematic problem exists.

Parts (b) and (c) of Figure 3.17 show nonrandom error cases. The

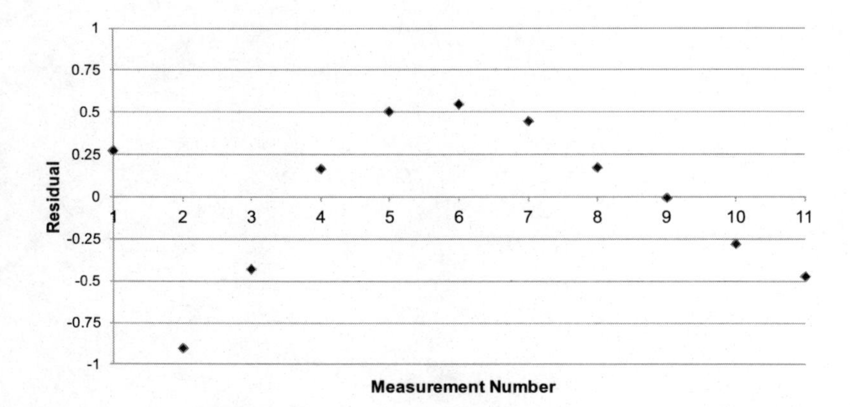

Figure 3.16: Residual plot showing the error between the data and the model as a function of data point number.

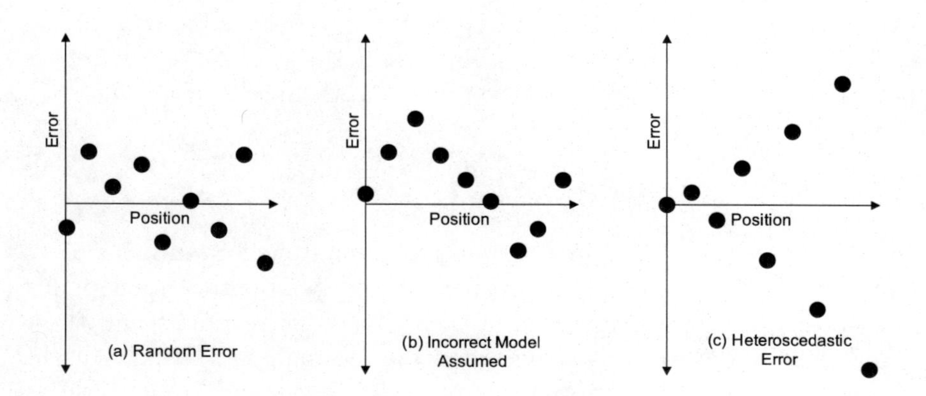

Figure 3.17: Examples of random and nonrandom error residuals.

structure of part (b) indicates that the analyst may have chosen the wrong model. The *heteroscedastic error* in part (c) indicates the sample variance is not uniform in the data set and so data may be sampled with different noise distributions.

In addition to residual plots, we also use numeric indicators to indicate the quality of the modeling process. Analysts use the correlation coefficient to determine the quality of the fit and they use the f statistic to indicate the appropriateness of the model.

3.6.5 Correlation Coefficients

To determine the quality of the fit, we typically use the *Analysis of Variance (ANOVA)* technique to provide indicators of how well the coefficients are determined. Many standard statistical computer software packages and spreadsheet programs such as Excel® contain this analysis. To start the analysis, we define several more summations that we need in addition to the earlier equations. The summations this time require the model for the dependent variable to produce the estimated points, $\hat{y}_i$ given by Equation (3.37). We compute the regression variability or *Sum Square Regression (SSR)*

$$SSR = \sum_{i=1}^{N} (\hat{y}_i - \bar{y})^2 \tag{3.51}$$

We compute the error in the fit as the difference between the measured dependent variable and the model output. With this, we compute the error variability or *Sum Square Error (SSE)*

$$SSE = \sum_{i=1}^{N} (y_i - \hat{y}_i)^2 \tag{3.52}$$

We interpret Equation (3.51) as the total variability in the regression. The quantities SS_{YY}, SSR, and SSE are related by $SS_{YY} = SSR + SSE$. The ratio SSE/SS_{YY} is the proportion of the fit error to the total spread in the data. Table 3.5 gives the computations using Equations (3.47), (3.51), and (3.52) for the data set in Table 3.1.

The *squared correlation*, R^2, measures the proportion of the total variability in the dependent variable that is accounted for by the fit. R^2 is always less than 1 but the closer to 1, the better the linear fit. We compute the correlation by using

$$R^2 = \frac{SSR}{SS_{YY}} = 1 - \frac{SSE}{SS_{YY}} \tag{3.53}$$

Table 3.5: Data for Determining the Error of the Fit

x	y	$\hat{y}$	$(y_i - \bar{y})^2$	$(\hat{y}_i - \bar{y})^2$	$(y_i - \hat{y}_i)^2$
0	0.07	-0.20	733.671	748.321	0.072
5	4.37	5.27	519.218	478.925	0.814
10	10.31	10.74	283.800	269.396	0.188
15	16.38	16.21	116.130	119.731	0.027
20	22.19	21.69	24.665	29.933	0.255
25	27.7	27.16	0.296	0.000	0.296
30	33.07	32.63	34.971	29.933	0.196
35	38.27	38.10	123.513	119.731	0.029
40	43.56	43.57	269.079	269.396	0.000
45	48.76	49.04	466.717	478.925	0.079
50	54.04	54.51	722.730	748.321	0.223
		SS$_{YY}$	3294.79		
			SSR	3292.61	
				SSE	2.178

A related parameter is the *correlation coefficient*, r, that measures the tightness of the fit and whether it is a positive correlation or an anticorrelation. We compute the correlation coefficient by using

$$r = sgn(a_1)\sqrt{R^2} \tag{3.54}$$

The $sgn(x)$ is the function that takes the sign of the argument. Using the data in Table 3.5, $R^2 = 0.9993$ and $r = 0.9997$. This implies the model fits the data nearly perfectly. Once we determine the model's coefficients, we next determine the confidence intervals on each coefficient. To do this, we first determine the Standard Error (SE), s, in the overall model using

$$s = \sqrt{\frac{SS_{YY} - a_1 SS_{XY}}{\nu}} \tag{3.55}$$

and the standard error in each coefficient from

$$\begin{aligned} SE(a_1) &= \frac{s}{\sqrt{SS_{XX}}} \\ SE(a_0) &= s\sqrt{\frac{1}{N} + \frac{\bar{x}^2}{SS_{XX}}} \end{aligned} \tag{3.56}$$

To compute the 95% confidence interval, we use the t distribution with $\nu = N - 2$ degrees of freedom. The two intervals are

$$\begin{aligned} i_1 &= a_1 \pm t_{0.025} SE(a_1) \\ i_0 &= a_0 \pm t_{0.025} SE(a_0) \end{aligned} \tag{3.57}$$

In the example we have been examining, $s = 0.4919$, $SE(a_1) = 0.009381$, and $SE(a_0) = 0.2775$. For a 95% confidence interval, we use $t_{0.025}$ with $\nu = 9$ or 2.262 from Table B.3. The confidence intervals are $i_1 = 1.094 \pm 0.021$ and $i_0 = -0.199 \pm 0.628$.

3.6.5.1 f Statistic

Another parameter that is frequently computed with the model fitting procedure is the f statistic. This computation does not tell the user whether the model is correct in an absolute sense. We use this to indicate whether a model is adequate to explain the data, and we use it to determine if one model fits the data better, worse, or about the same as another model.

To perform the computation, we need the *SSR* and *SSE* results from Equations (3.51) and (3.52). We also define k as the number of parameters in the fitting procedure. For a linear fit, $k = 2$ and we compute the f statistic from those results and the number of measurements, N, using

$$f = \frac{SSR/k}{SSE/(N-k+1)} \tag{3.58}$$

In our example, $f = 7558$. Is this a good result? To make a decision, we need an F-distribution table such Tables B.4 through B.6 in the appendix. These tables list the critical values $f_\alpha(m,n)$ where α gives the desired uncertainty level for the result. The decision rule is that if

$$f > f_\alpha(k, N-k+1) \tag{3.59}$$

the model adequately describes the data at the $1 - \alpha$ confidence level. In our example, if we wanted to have a 95% confidence that the model adequately fits the data, then $\alpha = 0.05$. In the example, $k = 2$ and $N - k + 1 = 10$. The critical value table lists $f_\alpha(2, 10) = 4.1$ so this model does a very good job of explaining the data at a confidence level even better than 99%. Note: as we saw in Equation (3.58), the f statistic's value is proportional to the number of data points so the user's ability to discriminate between candidate fitting models scales with the sample size. Therefore, if we have a relatively short data set over a limited range of the independent variable, then the f values for different models may be similar. The analyst needs to be careful in these situations.

3.6.6 Nonlinear Fits

So far, in this chapter, we have examined linear fits. In this section, we extend the fitting procedure to nonlinear models. First, we look at parametric models and then nonlinear polynomial models. We use the f statistic from the previous section to determine the quality of the model in both cases.

3.6.6.1 Parametric Models

We can apply the least squares method to more than linear equations. Suppose we had a model for a data set in the form

$$\hat{y} = a_0 + a_1 x + a_2 T + a_3 H + a_4 P \tag{3.60}$$

The x is the position variable, while T is for temperature, H is for humidity, and P is for pressure. The mean square error equation is then

$$mse = \frac{1}{N}\sum_{i=1}^{N}(a_0 + a_1 x + a_2 T + a_3 H + a_4 P - y_i)^2 \tag{3.61}$$

We then take the partial derivative of Equation (3.61) with respect to each variable as we did earlier. This leads to a system of equations that we solve for the coefficients a_j as was done in the linear equation case but this time with more coefficients.

3.6.6.2 Power Series Models

We also apply the derivation used for minimizing the mean square error to higher order polynomial models. The general form for the model for each output value, $\hat{y}_i$, at each input point, x_i, is

$$\hat{y}_i = \sum_{j=0}^{M} a_j x_i^j \tag{3.62}$$

As with the linear model, we take the partial derivative of the mse with respect to each of the fit parameters, a_j. This yields a system of

Table 3.6: Data for the Polynomial Fit

x	y
0	3.033
1	3.6805
2	5.5885
3	4.6192
4	3.6122
5	0.8997
6	-2.9238
7	-7.3846
8	-13.3494
9	-20.5442
10	-27.4637

linear equations that we solve for the coefficients. The general form for the matrix that we must solve is

$$\begin{pmatrix} N & \sum_{i=1}^{N} x_i & \sum_{i=1}^{N} x_i^2 & \sum_{i=1}^{N} x_i^3 & \cdots \\ \sum_{i=1}^{N} x_i & \sum_{i=1}^{N} x_i^2 & \sum_{i=1}^{N} x_i^3 & \sum_{i=1}^{N} x_i^4 & \cdots \\ \sum_{i=1}^{N} x_i^2 & \sum_{i=1}^{N} x_i^3 & \sum_{i=1}^{N} x_i^4 & \sum_{i=1}^{N} x_i^5 & \cdots \\ \sum_{i=1}^{N} x_i^3 & \sum_{i=1}^{N} x_i^4 & \sum_{i=1}^{N} x_i^5 & \sum_{i=1}^{N} x_i^6 & \cdots \\ \cdots & \cdots & \cdots & \cdots & \cdots \end{pmatrix} \begin{pmatrix} a_0 \\ a_1 \\ a_2 \\ a_3 \\ \cdots \end{pmatrix} = \begin{pmatrix} \sum_{i=1}^{N} y_i \\ \sum_{i=1}^{N} x_i y_i \\ \sum_{i=1}^{N} x_i^2 y_i \\ \sum_{i=1}^{N} x_i^3 y_i \\ \cdots \end{pmatrix} \tag{3.63}$$

For example, let us fit the data in Table 3.6. Figure 3.18 shows both the data and the fit. The data appear to follow a second-order equation. The necessary set of equations, in matrix form, is

$$\begin{pmatrix} N & \sum_{i=1}^{N} x_i & \sum_{i=1}^{N} x_i^2 \\ \sum_{i=1}^{N} x_i & \sum_{i=1}^{N} x_i^2 & \sum_{i=1}^{N} x_i^3 \\ \sum_{i=1}^{N} x_i^2 & \sum_{i=1}^{N} x_i^3 & \sum_{i=1}^{N} x_i^4 \end{pmatrix} \begin{pmatrix} a_0 \\ a_1 \\ a_2 \end{pmatrix} = \begin{pmatrix} \sum_{i=1}^{N} y_i \\ \sum_{i=1}^{N} x_i y_i \\ \sum_{i=1}^{N} x_i^2 y_i \end{pmatrix} \tag{3.64}$$

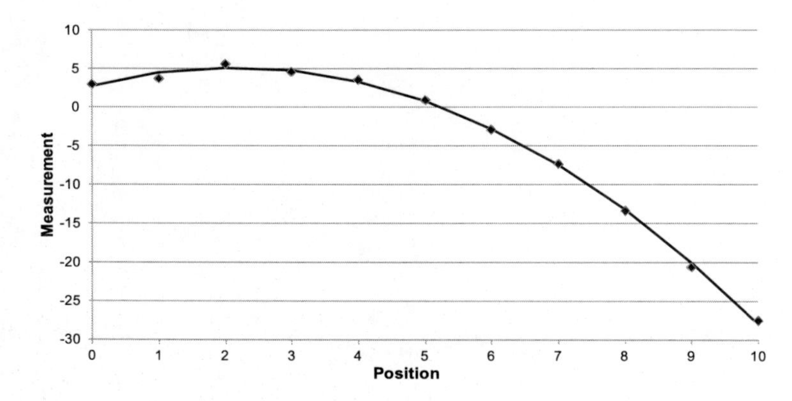

Figure 3.18: Data (markers) and second-order Least Squares fit (line) to the data.

We find the coefficients by solving the matrix equation

$$\begin{pmatrix} a_0 \\ a_1 \\ a_2 \end{pmatrix} = \begin{pmatrix} 11 & 55 & 385 \\ 55 & 385 & 3025 \\ 385 & 3025 & 2533 \end{pmatrix}^{-1} \begin{pmatrix} -50.2327 \\ -587.9025 \\ -5584.0176 \end{pmatrix} \tag{3.65}$$

This yields the following model for the data set, which we also used to generate the line in Figure 3.18:

$$\hat{y} = 2.7243 + 2.2824x - 0.5344x^2 \tag{3.66}$$

Using this model and Equations (3.51) and (3.47), the value for SSR is 1275.8 and SS_{YY} is 1277.4, while R^2 is 0.999. Using Equation (3.52) we find the SSE is 1.540 and using Equation (3.58), we find that f is 2486. This value greatly exceeds the values in the table so we accept this as a good model to fit the data.

3.6.7 Cautions with Least Squares

Like any other mathematical technique, the least squares method cannot be applied blindly and expect good results to follow. This section discusses basic cautions to take with this technique.

3.6.7.1 Model Selection

The least squares method, and its associated statistical analysis, does not tell the analyst whether one model is correct to the exclusion of all

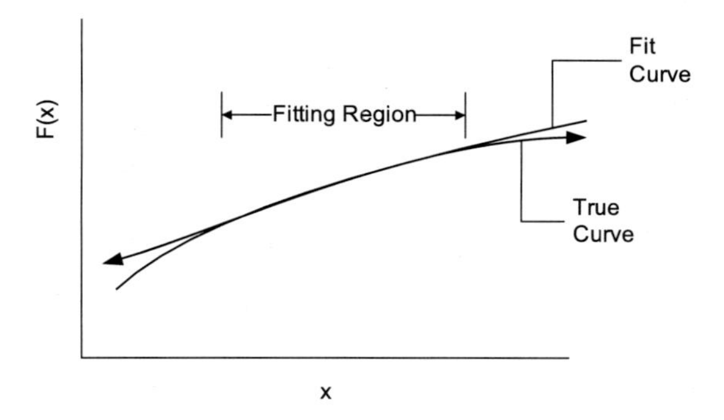

Figure 3.19: Attempting to extend the Least Squares fit beyond the data range. Outside the fit range, the "true" curve may significantly diverge from the "fit" curve.

others. The analyst's best hope is for a statistical indicator that provides some level of statistical confidence whether a model is consistent with the data. If the user picks several incorrect models to investigate, it is possible that none of them really fits the data. The method and the statistics do not "fix" this problem.

The model's range also needs to be correct. The model is not valid outside of the data range. Therefore, greatly extending the model's range could lead to a divergence between the model and where the underlying process is headed, as Figure 3.19 illustrates.

3.6.7.2 Outlying Points

Suppose that the data set has a single point that a large amount of noise corrupts. The effect of the bad point differs depending upon where the point is located in the data set. Figure 3.20 illustrates the effect of a bad point at the upper edge of the data set and at the middle of the data set. The bad point at the upper edge pulls the fit towards the data point. However, when the bad point is located in the middle, it pulls the line by a much smaller distance. This occurs because the linear fit always passes through the point $(\bar{x}, \bar{y})$. This indicates that least squares analysis is sensitive to outlying data points and that outliers at the edges of the data set have the greatest effect on the fit.

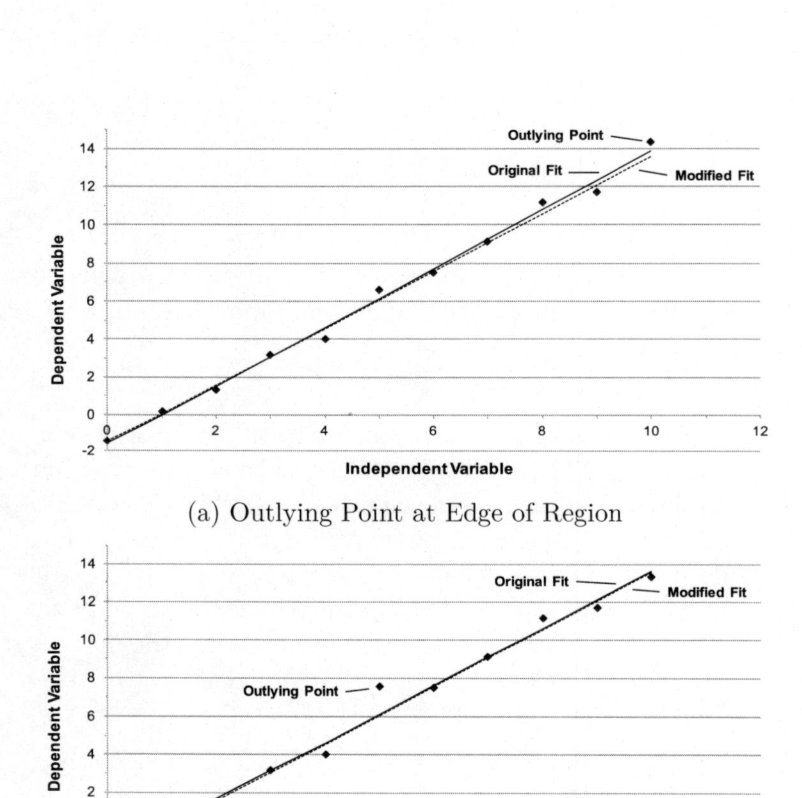

(a) Outlying Point at Edge of Region

(b) Outlying Point at Middle of Region

Figure 3.20: Effects of outlying points on the Least Squares fit when the outlier occurs (a) at the edge and (b) in the middle of the fitting region.

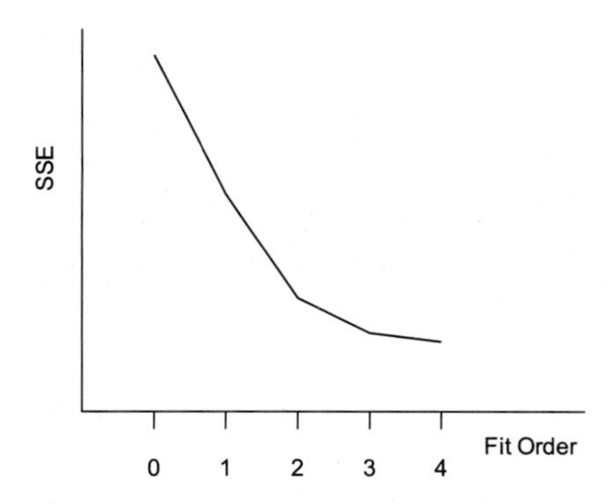

Figure 3.21: The change in SSE with fit order. As the fit improves with higher order terms, the SEE decreases.

3.6.7.3 Over-fitting the Model

Mathematical techniques allow us to fit N data points up to a $N-1$ degree polynomial. However, if the analyst does not know of an underlying physical model, then the analyst is tempted to try to fit the data with a higher order model than justified. We get a sense of this by looking at the *SSE* computation. Typically, the *SSE* decreases with increasing model order until it reaches the point where it begins to flatten out as Figure 3.21 illustrates. Using the f statistic and comparing the *SSE*s between models tells the user that increasing the order of the fit does not substantially improve the quality of the fit. Once the statistics indicate to the user that the model accounts for the data, it is probably best to stop increasing the order of the model unless there is an underlying physical reason for including more terms.

3.7 REFERENCES

[Dave87] W. B. Davenport and W. L. Root. *An Introduction to the Theory of Random Signals and Noise.* New York, NY: IEEE Press, 1987. ISBN: 0-87942-235-1.

[Ferr15] A. Ferrero. "The Pillars of Metrology." In: *IEEE Instrum. Meas. Mag.* 18.6 (Dec. 2015), pp. 7 –11. DOI: 10.1109/MIM.2015.7335771.

[Frad97] J. Fraden. *Handbook of Modern Sensors Physics, Designs, and Applications.* 2nd Edition. New York, NY: American Institute of Physics, 1997.

[GUM08] Joint Committee for Guides in Metrology Working Group 1. *Evaluation of measurement data – Guide to the expression of uncertainty in measurement.* JCGM 100:2008. Bureau International des Poids et Mesures. Sept. 2008. URL: http://www.bipm.org/utils/common/documents/jcgm/JCGM_100_2008_E.pdf.

[IEEE100] Institute of Electrical and Electronics Engineers. *IEEE 100 The Authoritative Dictionary of IEEE Standards Terms Seventh Edition.* IEEE-STD-100. IEEE Press. Piscataway, NJ, 2000. DOI: 10.1109/IEEESTD.2000.322230.

[ISA95] Instrument Society of America. "Instrument Engineers' Handbook." In: *Process Measurement and Analysis. Instrument Terminology and Performance.* Ed. by B.G. Lipták. 3rd Edition. Radnor, PA: Chilton Book Company, 1995. ISBN: 0-8019-8197-2.

[Klaa96] K. B. Klaassen. *Electronic Measurement and Instrumentation.* Trans. by S. M. Gee. Cambridge: Cambridge University Press, 1996. ISBN: 0-521-47157-5.

[Kueh00] R. O. Kuehl. *Design of Experiments: Statistical Principles of Research Design and Analysis.* 2nd Edition. Pacific Grove, CA: Duxbury/
Thomson Learning, 2000.

[Leon94] A. Leon-Garcia. *Probability and Random Processes for Electrical Engineering.* 2nd Edition. Reading, MA: Addison-Wesley, 1994. ISBN: 0-201-50037-X.

[Papo65] A. Papoulis. *Probability, Random Variables, and Stochastic Processes.* New York, NY: McGraw-Hill, 1965.

[Rabi95] S. Rabinovich. *Measurement Errors: Theory and Practice.* New York, NY: American Institute of Physics, 1995.

[RCC122] Telemetry Group. *Uncertainty Analysis Principles and Methods.* 122-07. Secretariat, Range Commanders Council. White Sands, NM, Sept. 2007. URL: http://www.wsmr.army.mil/RCCsite/Pages/Publications.aspx.

[Sali14a] S. Salicone. "The Mathematical Theory of Evidence and Measurement Uncertainty." In: *IEEE Instrum. Meas. Mag.* 17.4 (Aug. 2014), pp. 39 –46. DOI: 10.1109/MIM.2014.6873731.

[VIM12] Joint Committee for Guides in Metrology Working Group 2. *International vocabulary of metrology - Basic and general concepts and associated terms (VIM).* JCGM 200:2012. Bureau International des Poids et Mesures. 2012. URL: http://www.bipm.org/utils/common/documents/jcgm/JCGM_200_2012.pdf.

[Woze90] J. M. Wozencraft and I. M. Jacobs. *Principles of Communication Engineering.* Prospect Heights, IL: Waveland Press, Inc., 1990. ISBN: 0-88133-554-1.

[Youn62] H. D. Young. *Statistical Treatment of Experimental Data.* New York, NY: McGraw-Hill, 1962.

3.8 PROBLEMS

1. For each of the following error sources, comment on whether they primarily affect precision, accuracy or both about equally: systematic errors, random errors, hysteresis error, and interference.

2. Make a histogram of data with noise by using a standard analysis package such as Excel®, Mathcad®, or MATLAB®. Select a "true" value such as 10 and then add random noise to it. Do this 100 times and plot the result in a histogram. If we add Gaussian noise, does the result look like a Gaussian distribution? Experiment with different noise types that are available in your computer package.

3. You are given a measurement set corrupted with Gaussian noise where the measurements have a mean value of 10 and a variance of 2; estimate the following probabilities:

   (a) The measurement is between 9 and 11

   (b) The measurement is between 8 and 12

   (c) The measurement is between 5 and 15

   (d) The measurement is greater than 16 or less than 4

4. Given a Gaussian distribution with a mean of 0 and a variance of 1, what it the probability that a point is found in the range of (a) $\pm 1\sigma$, (b) $\pm 2\sigma$, (c) $\pm 3\sigma$, and (d) $> 3\sigma$ from the mean.

5. Consider measuring a constant 5 V power supply with a noisy voltmeter. The meter adds zero-mean Gaussian noise with a variance of $0.1\,\mathrm{V}^2$ to each measurement. What are the upper and the lower measurement limits such that the probability of finding the measurement outside of these limits is no more than 0.01?

6. For the following set of measurements, estimate the mean, the variance, the estimated error in the mean, and the estimated uncertainty in the mean. Also, find the 95% and 99% confidence intervals

on the mean.

7.894826	5.883171	7.659018	6.045662	5.455724
6.240702	8.185223	6.77516	7.191845	5.811363
7.293361	5.947835	8.317885	7.295075	7.297515
7.283844	7.902557	7.343442	6.675203	7.52899

7. Consider a measurement with an error distribution that is described by a uniform distribution over the interval $(-\alpha, \alpha)$. What are the k-factor intervals for $k = 1$, 2, and 3 based on the percentage of the total PDF area? Use the 68.3%, 95.5%, and 99.7% area values.

8. Repeat Problem 7 with an error distribution described by a cosine distribution $p(x) = \beta \left[1 + \cos\left(\frac{\pi x}{\alpha}\right)\right] [\mathrm{u}(x+\alpha) - \mathrm{u}(x-\alpha)]$ where $|\alpha| \leq \pi/2$. Here, β is a normalization constant to make the PDF valid.

9. Consider a measurement set where we estimate the variance to within a factor of 2. How do you estimate the required number of measurements you must take so that you can estimate its mean to within a 95% confidence interval?

10. Find the combined uncertainty for the case of a bias error described by the uniform distribution of Problem 7 with a $q = 0.05$ and a Type A error given by the cosine distribution of Problem 8 with $\alpha = 3.0$.

11. If a system has a noise temperature of 290 K and a bandwidth of 100 kHz, determine the noise spectral density and the variance of the process.

12. Determine the necessary equations to perform a Least Squares fit to the model $y(t) = a\cos(2\pi f_0 t) + b\sin(2\pi f_0 t)$. Assume that the frequency f_0 is known and the constants a and b are to be found by the fit. How do you modify the procedure if the frequency were also unknown?

13. Determine a model to fit the graph given in Figure P-1. Assume that the noise in the measurements is effectively zero at each point.

14. In an experimental data set with zero-mean Gaussian noise added to each measurement, we have two questionable points that appear

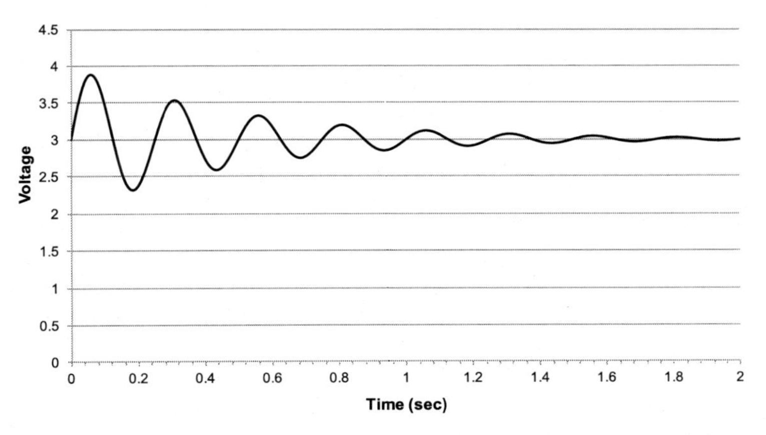

Figure P-1. Data plot for finding the unknown model.

to lie far from the rest of the data. We know that the noise has a variance of 0.1. For the first of the two points, the distance between the mean of the data and the point is 0.3; for the second point, the distance is 0.6. Would you discard either, both, or neither of these points and why did you make that determination?

15. Consider taking data associated with Figure P-2. You only took data for points along the x-axis from -10 through -3 although the full range is larger as the figure illustrates. Answer each of the following questions and give the reasoning behind your answers.

    (a) Would you expect a linear fit to work for your data points?

    (b) Would you expect the f-statistic for a higher-order fit to be better than the f-statistic for your data points?

16. You are given a set of N data points $\{x_i\}$ collected at a single location.

    (a) What is the appropriate model for this set of data collected at a single point?

    (b) What is the equation for the mean square error?

    (c) Derive the equation for the least squares fit to the data given your model.

    (d) How does this equation for the fit compare with the normal equation used to compute an average?

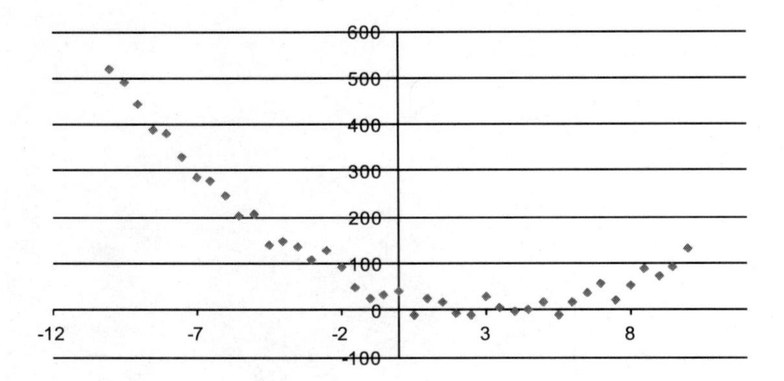

Figure P-2. Data plot for limited fitting range.

17. In the text, we developed a case to adequately explain the data in Table 3.1 as a linear fit. Suppose we wished to use a third-order fit to the data just to see what happens. Compute the necessary coefficients a_0, a_1, a_2, and a_3. Compute the f value as well. Does this model adequately explain the data? How does it compare with the linear fit?

18. We presented a second-order fit for the data in Table 3.6. Try the fit with a linear polynomial. Does the linear model fit the data better than the second-order model? Explain your reasoning.

19. Try to fit the data in Table 3.6 with a third-order polynomial. Is the third-order fit a better fit than the second-order fit? Explain your reasoning.

20. A measurement set includes the following data. The experimenter collected the data at ten settings and gathered ten points at each setting. Each measurement has noise added but we assume the noise variance is the same at each setting. Perform each of the following computations:

    (a) Find the mean, variance, error in the mean, uncertainty in the mean, and 95% confidence interval for the mean value at each of the ten settings.

    (b) Using the mean values, perform a first-, second-, third-, and fourth-order least squares fit to the data. Determine which fit is preferred using the f statistic.

(c) Plot your choice for the best fit to the data. Plot the residuals between the mean values and the model you have chosen.

Use a computer-based analysis package to make the computations easier.

Setting X_1 = 0.5; Voltage values: 0.626, 0.083, 0.139, 0.506, −0.580, 0.071, 0.033, 0.124, 0.257, 0.309

Setting X_2 = 1.025; Voltage values: 0.831, 0.168, 0.007, 0.249, 0.929, 0.822, 0.588, 0.611, 0.444, 0.173

Setting X_3 = 1.875; Voltage values: 0.911, 0.971, 0.820, 1.417, 0.413, 1.332, 1.045, 1.253, −0.037, 0.309

Setting X_4 = 3.425; Voltage values: 2.003, 1.823, 1.272, 1.431, 2.152, 1.751, 1.259, 1.186, 1.487, 1.492

Setting X_5 = 6.050; Voltage values: 2.221, 2.048, 1.330, 1.422, 2.338, 2.110, 2.508, 1.595, 1.843, 2.163

Setting X_6 = 10.125; Voltage values: 2.577, 2.591, 2.885, 3.078, 2.993, 1.990, 2.151, 3.550, 2.320, 2.042

Setting X_7 = 16.025; Voltage values: 2.867, 3.368, 2.911, 3.421, 3.958, 3.204, 3.478, 2.892, 3.063, 2.559

Setting X_8 = 24.125; Voltage values: 4.264, 3.892, 3.525, 3.316, 3.539, 2.869, 3.257, 3.183, 3.643, 3.270

Setting X_9 = 34.800; Voltage values: 3.655, 3.287, 4.684, 4.686, 4.749, 4.724, 3.703, 4.253, 4.016, 3.170

Setting X_{10} = 48.425; Voltage values: 4.779, 4.189, 4.110, 3.834, 5.024, 4.723, 4.444, 3.997, 4.475, 4.133

21. Three different models are used as candidates to fit a data set. There are 15 points in the data set and each model has 3 parameters that are fit using Least Squares. For each of the following cases, explain your reasoning as you answer the question.

    (a) Model 1 has a computed $f = 5$ for a 95% confidence level. Does this model explain the data well?

    (b) Model 2 has a computed $f = 75$ for a 95% confidence level. Does this model explain the data better than Model 1?

    (c) Model 3 has a computed $f = 80$ for a 95% confidence level. Do you choose Model 3 over Model 2?

CHAPTER 4

COMPUTING SYSTEM ELEMENTS

4.1 INTRODUCTION

The operator is an integral component of many telemetry and telecommand systems. For example, the operator of a satellite's control station is an expert who assesses the telemetry in real-time and sends commands to configure both the satellite and the control station for operations. To make the system operate efficiently, the operator must receive system information in a timely fashion. Additionally, the operator needs a quick-acting command entry system. To support this activity, the operator needs a computer system for the telemetry and telecommand processing tasks supporting payload and station data. The highlighted blocks in Figure 4.1 illustrate the locations of the processing components in the overall system.

A generation ago, manufacturers explicitly designed computer systems for these real-time operational needs. Today, conventional desktop computer systems, embedded microcontrollers, industrial-control computers, and even smart phones and tablet computers all find homes in the designs of modern telecommand and telemetry systems. In this chapter, we look at some of the issues involved with the computer systems, both in the payload and in the base station. The operator's hardware and software systems are more complicated than the payload system's since they must deal with the operator interface requirements for both telemetry display output and command data input. If the overall system has a multiuser support configuration, then the computer system also has an interconnection between computer components, typically using a

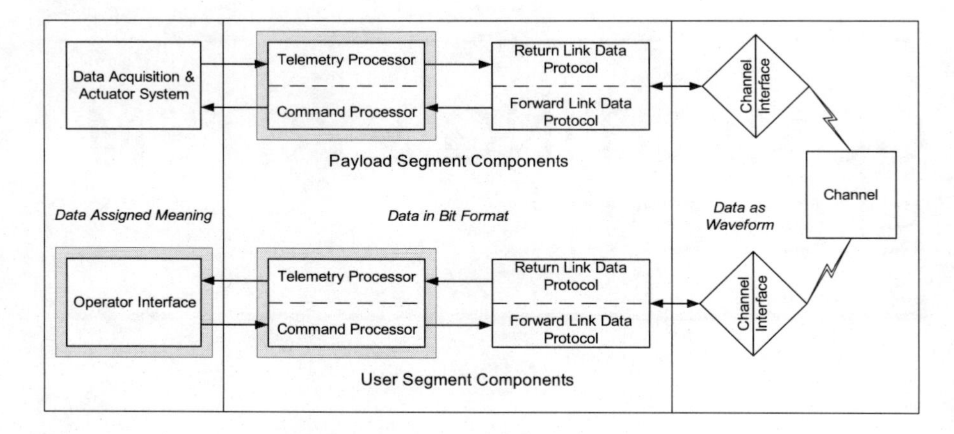

Figure 4.1: The highlighted blocks indicate the locations of the computer hardware and software components for telecommand and telemetry processing, and operator support in the overall telecommand and telemetry system.

Local Area Network (LAN).

On the payload side, many modern systems are "smart," in that they include embedded processors and memory to perform various normal and emergency-mode functions. Here, we examine some of the characteristics of payload software systems to support the necessary control and data processing functions.

4.2 OBJECTIVES

Our objectives for this chapter are to examine the telemetering and telecommand computer system's hardware and software needs. At the conclusion of this chapter, the reader will be familiar with the design issues related to operator interface concepts and the associated computer hardware and software organization. After this chapter, the reader will be able to discuss the following topics:

- Processing states used in the payload and user interface processing software.
- The definition of real-time systems based on interrupt response and other operating system parameters.

- How partitioning of information by subsystem and operator function aid in information organization.
- The use of window display areas to segregate and manage information.
- The use of color to indicate telemetry parameter status and other devices to give operators special warnings.
- The relative strengths and weaknesses of menu, mnemonic and non-mnemonic command line input strategies.
- How designers structure command dictionaries.
- Methods used for command data entry and validation.
- How operators use command files to ease their real-time entry load.

These topics will be illustrated through the background discussions and their application in a balloon payload project.

4.3 COMPUTER SYSTEMS

Designers of computer systems to support both the operator interface and the payload processing functions need certain common characteristics to sustain both the real-time processing algorithms and data movement. In this section, we examine the characteristics of these computer systems and data interfaces.

4.3.1 Real-Time Computing Definition

With the speeds and capabilities of current engineering workstations, supporting application software such as LabVIEW®, and commercially available base station applications, many desktop computer systems and workstations are being adapted for data acquisition and processing. One class of telemetry and telecommand data processing system needs to go beyond mere speed and memory. A computer system's ability to perform real-time data acquisition and processing requires more than just quick clock speeds. In this section, we look at the requirements for real-time systems from hardware and software points of view. Computer systems that need to keep pace with continuous, rapid-input data streams or need to guarantee completion of algorithm-processing operations on a fixed time cycle are the ones most needing these characteristics. A system designer need not incur the extra cost of a true real time system if it is not required to support the design requirements.

A real-time computer system is often defined as "a system that operates faster than the data becomes available." While this definition appears to be functional to the casual user, more is involved from an engineering viewpoint. One standard definition is that a *real-time system* sustains the

> performance of a computation during the actual time of related physical processing in order that results of the computation can be used in guiding the physical process [IEEE100].

Associated with this is *real-time software*

> in which computation is performed during the actual time that an external process occurs, in order that the computation results can be used to control, monitor, or respond in a timely manner to the external process [IEEE100].

Therefore, a more realistic definition for a real-time computer system is one that responds to service calls (hardware generated interrupts or switching between active processing tasks) in a stable and predictable time increment and completes the necessary processing before the next data processing cycle starts.

These characteristics are important because most telemetry computer systems rely on interrupt-driven input and output interfaces to the external world. Speed in processing an isolated segment of data is usually less important than the ability to respond in predictable and time-bounded ways to these interrupt requests. If the timing is predictable and time-bounded, then the system designer knows that when the computer system reads in a data segment for processing, the processing always completes before the next data segment arrives. Usually, the computer's operating system (and not the actual hardware) is the determining limitation on the response to interrupts. Let us look at both of these issues in more detail.

4.3.1.1 Interrupt Characteristics

An interrupt is a signal from a device connected to the computer that indicates when the computer needs to perform a service for the device or software application. The service is either sending data to, or receiving

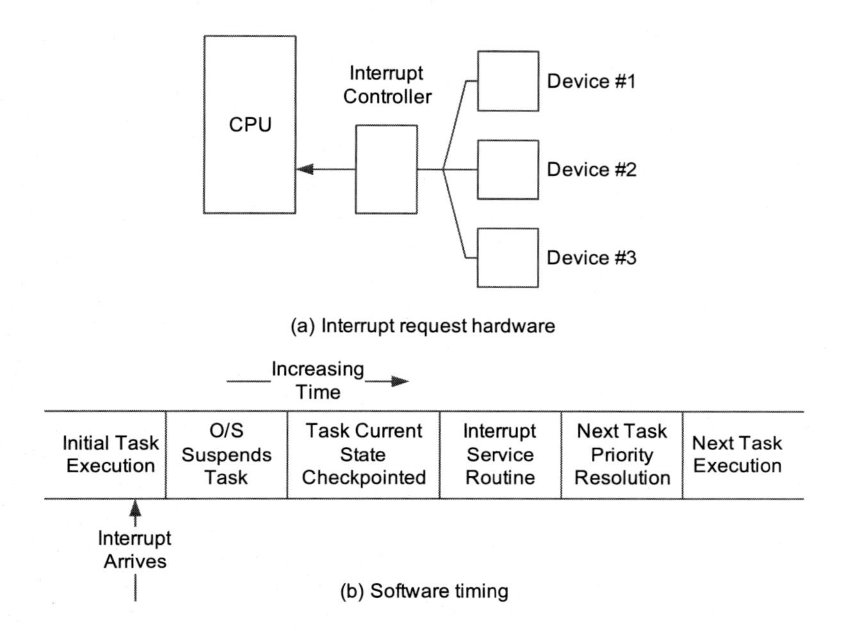

Figure 4.2: The hardware and software involved with interrupt servicing.

data from, the device or application. Both may also involve control information. System designers configure the computer system to service many interrupts from hardware and software sources. This configuration must account for the computer not knowing when the service request comes or what the nature of the request is because the external environment does not synchronize the interrupt signal with the computer's main controlling hardware and software. Figure 4.2 illustrates the computer's interrupt servicing process. For example, the interrupt request arrives from an external device based on some predefined condition such as a data buffer becoming available. If several devices cause interrupts to the system, an interrupt controller may be included to prioritize the service requests. Finally, the controller passes the service request to the computer. Its operating system software processes the request and takes the appropriate action.

The important feature of the system is that the computer services the interrupt request before an external device generates its next interrupt service request for the same type of processing. For example, telemetry data buffers may come on a periodic basis due to the way that the payload packages the data for transmission. It is important that the data processor in the base station processes the first data group before the next data group arrives from the payload. If not, data loss may occur.

4.3.1.2 Software Characteristics

As Figure 4.2 illustrates, the software is also an important part of the interrupt servicing process. Once the computer's Central Processing Unit (CPU) accepts the interrupt, it executes a sequence of procedures in order to service the interrupt request. First, the currently executing application software needs either to complete its task or suspend the task. The Operating System (O/S) has a finite time to accomplish this process. In a process known as *checkpointing*, the O/S saves the state of the application software if the O/S suspends execution. Next, the O/S executes the interrupt service routine. After the servicing is complete, the O/S needs to decide which application to begin next — the application that was suspended or another, higher priority application. Once the O/S decides which application is to be started, that process begins.

The telemetry and telecommand system software has a number of specific software applications tailored to a specific function in the system. The collection of applications initially is resident on the computer's hard disk and started upon system initialization. The computer initiates data processing for each application when its required data are available. The usual description of this computer system is a multitasking operating system. Important O/S characteristics found in many real time telemetry computers include the following abilities [Cutl83; Furh90]:

1. Fix the locations of software applications in the computer's memory because fixing the location reduces the time to initialize the application than recalling the application from disk and then starting it.

2. Fix the priority of the software applications so that the system designers can fine-tune the application order of execution for best overall system performance.

3. Switch between applications to respond to higher priority processing requests at any time in the processing cycle rather than only at application initiation, application completion, or application input/output operation execution.

4. Set a response priority to input and output operations.

5. Reserve a segment of system memory to hold shared data between the applications. This feature is useful and allows for rapid communication between applications.

6. Have reliable timers available to permit elapsed time measurement and allow the application scheduler to generate wake-up calls.

7. Set and respond to application level software processing interrupts to coordinate inter-application process sequencing and to enable processing start and wait cycles.

These attributes enable the system designer to tune and to optimize the system components for maximum throughput.

4.3.2 Computer Input-Output Interfaces

The system designer needs to have a computer system with Input/Output (I/O) capabilities in order to send or receive command data and telemetry data in the payload and base station. These I/O functions are sent over the computer's I/O ports. The computer manufacturer configures these ports according to standards that define the port's electrical, mechanical, and data formats. Here, we look at serial interfaces, MIL-STD-1553, and network standards that the reader may find in telemetry and telecommand systems.

4.3.2.1 Serial Interfaces

The system designer has a choice of serial interfaces for use in the payload and the base station computing systems. Most have standard drivers for the operating system to support the interfaces. Designers can use a Web search on each protocol to find application notes and examples of their use. The following paragraphs include information on common serial data interfaces.

Serial Peripheral Interface The Serial Peripheral Interface (SPI) bus is intended to facilitate connecting single sensors to central computing device and it is described in [SPI04]. The SPI bus has two major electronic units: the master device for controlling the bus and one or more slave devices each connected to a sensor. The designer uses control registers in the master to specify bus handshakes and its transmission baud rate. Data transmission is configured around 8-bit blocks and the data baud rate is selectable from 0.012 21 MHz to 12.5 MHz. Figure 4.3 illustrates

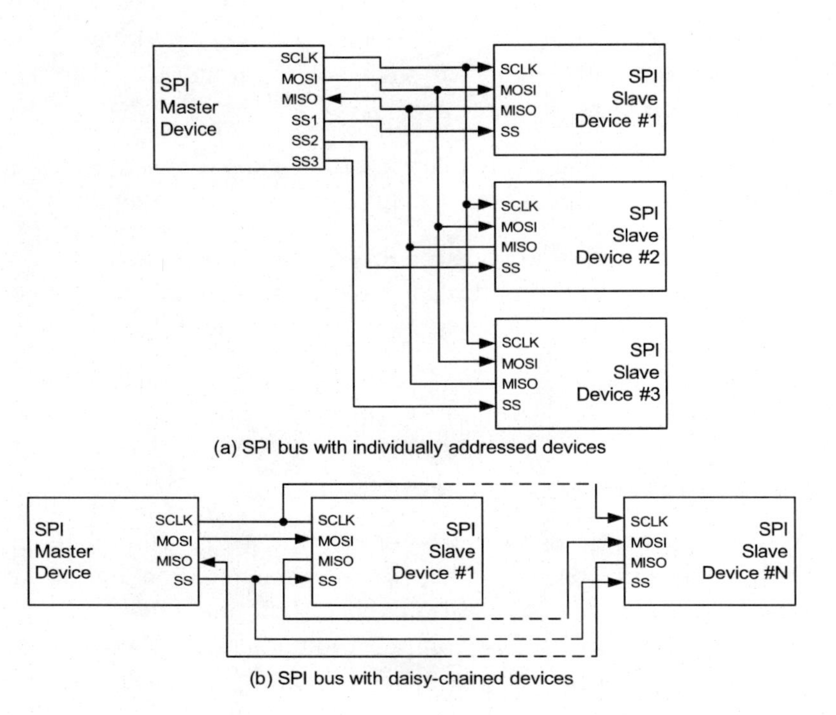

Figure 4.3: Serial Peripheral Interface connections between the bus master and slave devices. The first configuration uses specific device select lines and the second configuration uses a daisy-chain configuration.

possible bus configurations. Figure 4.3(a) shows the master and slave devices configured so that the master uniquely addresses each slave device as needed. Figure 4.3(b) shows a daisy-chain connection that permits a simpler master device to flow the data through the internal shift registers of all of the slaves, in turn, back to the master. In either case, only the master device can initiate a data transfer.

The SPI bus signals are as follows:

SCK — the common Slave Clock for bus timing

MOSI — Master Out, Slave In to transfer data into a slave device

MISO — Master In, Slave Out to transfer data from a slave device

SS — Slave Select signal (active low logic)

The SPI bus is most commonly used in embedded system designs. The

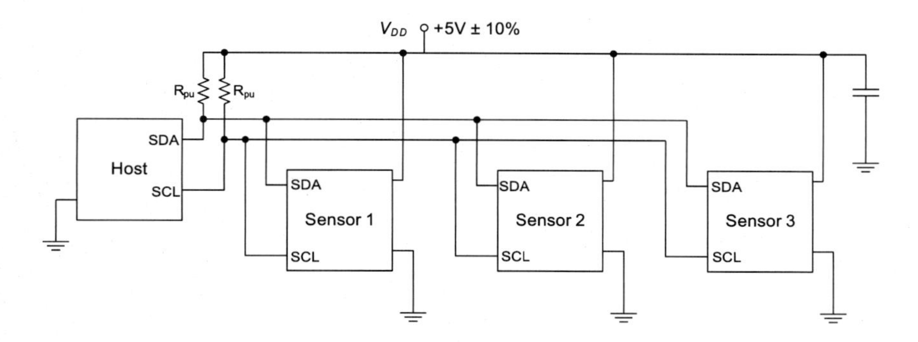

Figure 4.4: Sensors attached to an I^{2}C bus. The bus uses a 5V V_{DD} and pull-up resistors, R_p.

bus does not specify the means for generating the digital sensor data so the user will need to provide that part of the system design.

I^{2}C The Inter-Integrated Circuit (I^{2}C) bus specification permits easy integration of peripherals with a host device (usually a microprocessor or small computer) with minimal wiring. The peripheral devices can be individual sensors, other controllers, memory devices, display devices, or bus extenders. Each peripheral is classified as read only, write only, or read/write, depending upon the function. Each device has a unique identifier based on the manufacturer's unique code, the device model number, and the device revision number.

The two serial signals in I^{2}C are Serial Data (SDA) and Serial Clock (SCL). Additionally, the peripheral devices need the power line, V_{DD}, pull up resistors on the SDA and SCL lines, and the electrical ground. The pull-up resistors are required because the SDA and SCL lines use an open-drain configuration to allow multiple devices to attach to the lines. Figure 4.4 illustrates an example bus configuration with a host and three slave devices.

The I^{2}C bus must have at least one master for data transfers to happen but the bus also supports multiple masters with only one being active at a time. The bus master is responsible for initiating the need for a data transfer, the timing signals for the transfer, including clock generation, and bus access arbitration resolution if two or more masters request access at the same time. Every other device on the I^{2}C bus is a slave device during the time the master has control of the bus. The bus master uses the following steps during the general data transfer process:

1. Asserting the START signal SDA and SCL lines.
2. Resolving any master conflicts of bus access.
3. Sending the slave device address for the data transfer.
4. Indicating the data transfer direction (read or write).
5. Receiving the slave acknowledgment.
6. Sending or receiving the data and associated acknowledgments.
7. Asserting a STOP signal on the SDA and SCL lines.

The slave device needs to supply an acknowledgment bit after the address reception. The device receiving the data must acknowledge every byte with an acknowledgment bit. The master may send multiple bytes to a peripheral or change peripherals after concluding the transaction with the first peripheral as long as the master retains bus control.

The bus protocol uses a 7- or 10-bit address for each slave. The designer sets the addresses using configuration switches on the peripheral. The designer must program the master with the designated addresses before the master accesses the peripheral. Some designers form 8-bit addresses by using the read/write bit in the slave address and using one register for the read and one for the write. Generally, this is outside the bus protocol specification.

The I^2C bus uses a nominal V_{DD} of $5\,V \pm 10\%$, although lower values are supported. The logic levels are referenced to the V_{DD} voltage with the logic 0 threshold at 30% of V_{DD} and the logic 1 threshold at 70% of V_{DD}. The pull-up resistors, R_{pu}, are used to allow common connection to the SDA and SCL lines. The designer computes the resistor values based on the overall bus capacitance and the bus data rate.

While the I^2C bus does not require using a twisted-pair wiring for the SDA and SCL lines, the standard encourages designers to use wiring layouts to reduce noise. Designers should use techniques that minimize crosstalk and unmatched capacitive loads by using configurations such as: SDA, V_{DD}, GND, SCL. The exact configuration depends on if this is in a shielded cable or a circuit board run.

There are several different I^2C bus modes and most are downward compatible. The modes are

Standard mode for bidirectional data transfers up to 100 kbps

Fast mode for bidirectional data transfers up to 400 kbps

Fast-mode Plus for bidirectional data transfers up to 1 Mbps

High-speed mode for bidirectional data transfers up to 3.4 Mbps

Ultra Fast for unidirectional data transfers up to 5 Mbps

More information on the I^2C standard is available in its specification document [I2C14].

TIA-232 The TIA-232-F serial communications standard replaces the earlier RS-232-C standard. In practice, many engineers still refer to this as the "RS-232" standard although the sponsoring organization and revision have been updated. The standard is an electro-mechanical standard specifying voltage levels, connectors, and data handshakes. Users frequently flow ACSII-coded data over the interface; however, the standard does not restrict the data to text only and the user may transmit byte-oriented data over the interface. Horowitz and Hill [Horo89] provide a good discussion of the 232 standard.

The TIA-232 standard uses two voltage regions to specify logic values encoding the data:

Logic 0 is a voltage between +3 V and +15 V

Logic 1 is a voltage between −3 V and −15 V

Voltage levels between −3 V and +3 V are invalid. Designers need to be aware that some devices, claiming to be 232-compatible, use standard 0 to 5 V TTL voltage levels on their interface and therefore the designer may need to add a level converter circuit to be compatible with a device using the standard levels. Control signals use the opposite data polarities for asserted and de-asserted levels.

The original standard limited the maximum data rate to 20 kbps. However, the true limitation on the rate is the connecting cable's capacitance. For short data runs of one or two meters, the cable sustains data rates in excess of 100 kbps with no problems.

The TIA-232 standard defines two actors in the data interface: the Data Terminal Equipment (DTE) and the Data Communications Equipment (DCE). Generally, the DTE is the host device such as a computer and the DCE is the peripheral device. The standard defines a number of

Table 4.1: Five-Wire TIA-232 Interface Control Signal Pin Numbers on DE-9 Connectors

Name	Function	Null Modem DTE	Null Modem DCE	Straight Through DTE	Straight Through DCE
TD	Transmitted Data	3	2	3	3
RD	Received Data	2	3	2	2
SG	Signal Ground	5	5	5	5
RTS	Request To Send	7	8	7	7
CTS	Clear To Send	8	7	8	8

control circuits to provide the handshakes between the actors. However, for most interactions between a host computer and a peripheral, a simple five-wire set of lines suffices. To save space in the system, designers generally opt to use DE-9 connectors rather than DB-25 connectors. In either case, the DTE typically has the male connector and the peripheral device has the female connector, but this is not required in the standard. Table 4.1 provides the pin assignments for the five-wire interface. In the table, we show two variants of the signal connection: the null modem for full-duplex communications and the straight through for half-duplex communications. The designer needs to know which type of cable is required for the peripheral. Generally, manufacturers use the same types of plugs for both cable types although the cables cannot be mixed (this author always labels them with a permanent marker upon taking them from the packaging before putting them into use).

The designer may be able to use a three-wire variant of the null modem connection by cross-strapping the Request To Send (RTS) and Clear To Send (CTS) signals on each side of the interface. Designers generally have options in the computer device driver to set the *flow control* options as well. These options are with control signals or American Standard Code for Information Interchange (ASCII) character codes as follows:

Hardware use the RTS/CTS control lines

None no flow control on the link

X-On/X-Off software flow control using the special control characters. X-On is ASCII DC1 or $0x11$ or CTL+Q. X-Off is ASCII DC3 or $0x13$ or CTL+S

Table 4.2: Recommended Maximum Data Rates for TIA-422 Cables

Cable Length (m)	Data Rate (kbps)
12	10,000
50	2,500
100	1,200
500	230
1,000	115

Frequently, links can use the no flow control option for data transfers with modern equipment over short cable runs.

TIA/EIA-422 The TIA/EIA-422-B standard is a more modern standard for serial communication that permits higher data rates over longer distances than the TIA-232 standard [RS422]. The standard covers the physical layer electrical characteristics over twisted-pair wiring and not the cable connectors or the handshake protocols for actual transmission. There are other standards to cover that aspect; in particular, designers use the 422 standard to extend the 232 standard cable lengths and transmission speeds.

The TIA-422 electrical standard uses differential transmission techniques to improve the noise immunity on the signal lines. The line voltages range from $\pm 2\,\text{V}$ to $\pm 10\,\text{V}$. The standard specifies transmitting logic 0 with a negative voltage and logic 1 with a positive voltage. While the TIA-232 standard was a point-to-point transmission between two devices, the TIA/EIA-422 specification permits the designer to use a single driver to connect up to ten receivers on a single circuit with only one driver per circuit.

The TIA-422 standard transmission speed depends on the length of the cable. Table 4.2 gives recommended maximum data rates as a function of cable length based upon the standard. For those cables up to 12 m in length, data transmission rates up to 10 Mbps are common. As the cable length increases, the transmission rate drops to approximately 100 kbps at a 1-km distance. Depending upon cable quality and electronics, a user may find the maximum cable length is sustainable to 1500 m at the lower data rate.

TIA-485-A The TIA-485-A standard is another physical-layer standard that permits more flexibility than the TIA/EIA-422-B standard. The standard uses a balanced, twisted pair line like the 422 standard. However, the standard removed the restriction of having a single driver on each circuit and permits 32 drivers and 32 receivers on a circuit. This allows the designer to configure the data communications as a shared bus with fewer wires than direct point-to-point communications among all devices.

The data transmission speeds and cable lengths are the same as with the 422 standard and the 422 standard is a subset of the 485 standard. Commercially available serial communications interface boards frequently allow the designer to use the card to support 232/422/485 communications by proper configuration settings.

TIA/EIA-644/Low Voltage Differential Signaling The TIA/EIA-644 or Low Voltage Differential Signaling (LVDS) is a more recent physical-layer standard using differential, twisted-pair wiring standard that host systems use to communicate with peripheral devices. The standard uses low voltages and currents to encode the logic signal levels. The driver on the circuit uses the direction of a constant 3.5-mA current to specify the logic level. The receiver terminates the circuit with a 100-Ω resistor and uses the polarity of the voltage drop across the resistor to indicate the received logic level.

The specifications for the LVDS standard sustain cable lengths up to 10 m. For short, high-quality cables up to 1 m in length, the signaling electronics sustain data rates of 1 Gbps. For 10-m cables, the data rate drops to approximately 200 Mbps. The quality of the cabling and the noise environment factor in the maximum data rate determination.

There is a related standard TIA/EIA-899, or Multipoint Low Voltage Differential Signaling (M-LVDS), to extend LVDS for multidrop signaling. The M-LVDS permits cable runs in excess of 25 m but at a lower data rate.

Universal Serial Bus The Universal Serial Bus (USB) is a serial communication bus that is found in many applications. The balloon payload discussed later in Section 4.5.4 used a USB-based Analog-to-Digital Converter (ADC) for sensor data acquisition and a USB-based camera for photography. In addition to supplying the data to the host computer, the USB connection also supplied power to the ADC and the camera. The ability to power peripherals from the data cable makes this standard very attractive to many designers.

The USB standard is a four-wire standard up through version 2. The D+ and D- data lines are a differential, twisted pair and they are in the cable with +5 V power (called V_{BUS}) and ground lines. Manufacturers shield these cables to improve noise immunity for high-speed applications. The USB standard permits cables lengths up to several meters. In addition to direct point-to-point data transfer, the standard supports multiple hubs to all a single host port to connect to multiple end peripherals. The standard supports three data transfer ranges:

Low Speed — 10 kbps to 100 kbps

Full Speed — 500 kbps to 10 Mbps

High Speed — 25 Mbps to 400 Mbps

The USB specification permits the user to "hot swap" devices without powering them down before making or breaking a connection. The host is responsible for automatically recognizing device presence changes.

The standard specifies the protocols for data encoding, data transfer, and bus management, including how applications are to interface with the bus. The standard also specifies the physical connectors used by devices accessing the bus. More information is found in the USB standard [USB00].

The USB connector is a slip insert connector without locking screws as are often found in DE-9 and DB-25 connectors. This can cause connection problems in high-vibration or acceleration environments. This author often replaces the standard USB connector with a screw-locking DE-9 connector in these types of environments.

4.3.2.2 MIL-STD-1553

The Department of Defense defined the MIL-STD-1553 standard to support a local area network for data acquisition within the payload. The standard originally covered aircraft avionics, but designers also use the standard in non-aircraft situations such as satellites. The standard covers control, electrical, data, and mechanical specifications. The standard allows the designer a great deal of latitude in configuring the system. In this section, we look at the bus connections and handshakes. In Section 6.8 we will look at the packet data structure.

The MIL-STD-1553 standard is a bus-based method for gathering data from modules as well as controlling the modules [MS1553]. Many

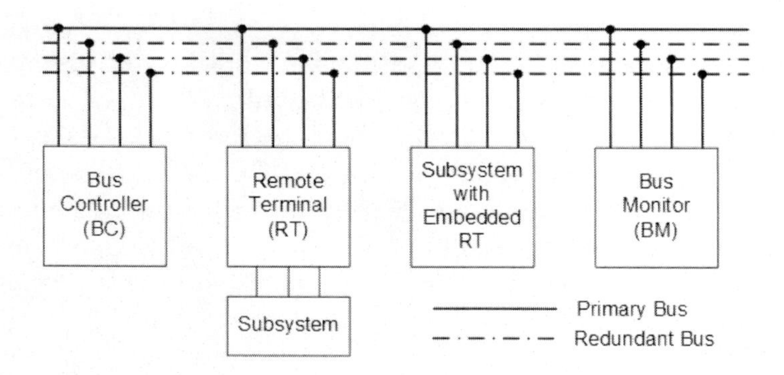

Figure 4.5: MIL-STD-1553 bus structure [MS1553].

vendors supply specialized chips and even complete data acquisition systems to support the standard. The 1553 standard utilizes a master bus controller to coordinate the activities of the instrumentation interfaces. However, the standard does not specify the frequency at which the bus master is to connect with the individual modules to gather data.

Figure 4.5 shows the 1553 primary and backup data bus configuration. The data bus forms a LAN between the individual devices and the LAN operates in a half-duplex mode. Each bus contains a maximum of 31 unique devices. The standard refers to these devices as terminals that fall into three categories: Remote Terminal (RT), Bus Controller (BC), and Bus Monitor (BM).

Designers connect most instrumentation to the bus via a RT. The RT and the instrumentation subsystem can be contained in the same unit (embedded mode) or the RT may act as an intermediate interface between the instrumentation and the data bus. Additionally, the standard includes one or more BCs to allow redundancy, and the BM handles data gathering and recording. The bus structure has one main or primary bus and up to three redundant or backup buses giving a reliable bus architecture which allows the system to detect bus faults and switch to one of the backup buses.

The MIL-STD-1553 bus sustains a maximum bit rate of 1 Mbps. The bus controller manages the data interchanges between entities on the LAN with a command/response handshake protocol. The transmit/receive bit in the 1553 command packet tells the remote terminal the data flow direction. Logic 0 indicates that the RT is to receive data and a logic 1 indicates that the RT is to transmit data. Designers use the command

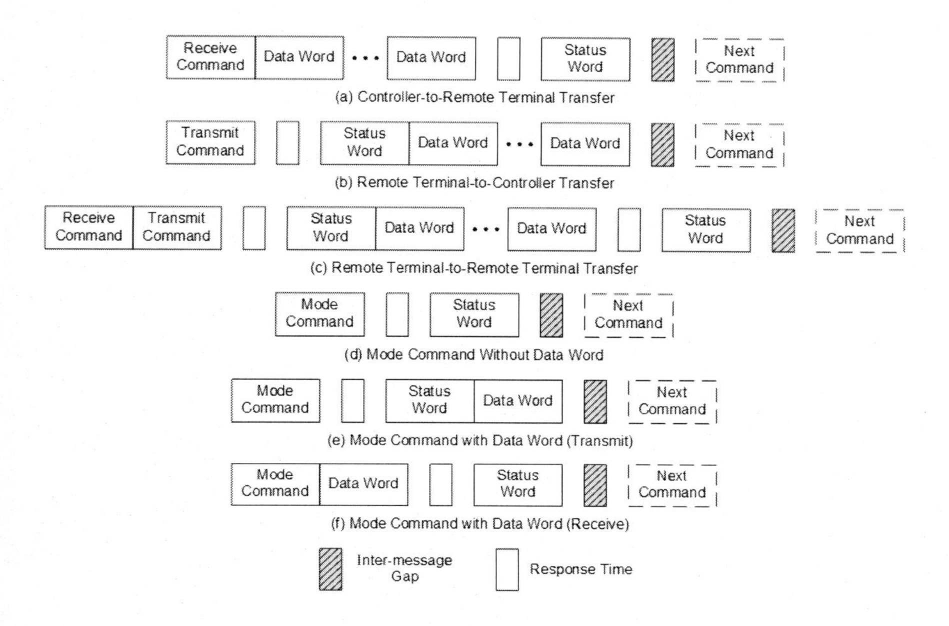

Figure 4.6: The MIL-STD-1553 control and response exchanges for various data transfer modes [MS1553].

word's mode bits to control terminal functions. Between each message is an inter-message gap of at least 4 µs. When a RT is required to transmit a response, it must provide the response word within 4 to 12 µs.

Figure 4.6 illustrates several possible data transfers. In the first example, the BC sends a data block to a RT, while in the second example, the RT sends the data to the BC. The third example shows the data flow between remote terminals. This only occurs after the BC has designated a source and a destination RT. The receive command and the transmit command are issued by the BC, and not the RT. The final example shows the commands to set RT modes with and without data transfers. The standard allows for broadcast commands whereby every RT on the LAN receives a data or mode message.

4.3.2.3 Networks

Designers have a number of networking technologies suitable for transmitting data as well as for connecting peripherals to the computing system. These network protocols utilize both cabled connections and wireless connections. Some of these protocols started out as independent technologies but most are now grouped as part of the IEEE 802 family. While they may have such a designation, that does not imply that they are mutually interoperable. Additionally, the networking protocol does not include any specification on how the user generates the data that is transmitted over the network.

IEEE 802.x System designers use the IEEE 802 family of networking protocols to configure both wired and wireless networking. There are several protocols in the family that designers commonly use for data transmission in telemetry data-acquisition systems. These protocols include:

IEEE 802.3 the Ethernet protocol that is the basis for hard-wired LAN configurations

IEEE 802.11 the Wi-Fi wireless LAN standard on the 2-GHz and 5-GHz Industrial, Scientific, and Medical (ISM) bands

IEEE 802.15.4 the Low-Rate Wireless Personal Area Network (LR-WPAN) low-layer wireless standard that can be used to support upper protocol layers such as ZigBee on the 2-GHz ISM band

Many computing systems come with hardware interfaces for both the Ethernet and Wi-Fi protocols. There are many software applications, such as LabVIEW®, to support data acquisition with these protocols. The designer can configure the applications as either connection-oriented protocols or connectionless protocols. With connection-oriented data exchanges, both nodes in the transaction establish a specific handshake before and during the data transfer. With connectionless data exchanges, the sender does not involve the receiver with a specific data-exchange handshake but sends the data and hopes the data arrives correctly. Designers specify connection-oriented transfers on reliable networks, while they specify connectionless exchanges on unreliable networks.

Many peripheral devices support at least one of the 802.x protocols. In particular, devices intended for application in the Internet of Things (IoT) tend to use the 802.11 and 802.15 protocol families.

Bluetooth The Bluetooth® standard is a low power, short-range protocol for data exchange between a master and peripheral devices. The Bluetooth designers originally conceived the standard as replacing wired standards such as TIA-232. Bluetooth operates in the same radio band as the

Wi-Fi protocol but the two are not interchangeable. Bluetooth was part of the IEEE 802 family as IEEE 802.15.1, but that protocol standard is not maintained. The main advantage of a Bluetooth network is the self-discovery property for integrating peripherals into the network. Many computers currently come with Bluetooth support; system designers can add the protocol by using a USB-based dongle for those without intrinsic support.

ZigBee The ZigBee standard is another low-power, short-range protocol for data exchange. ZigBee utilizes the IEEE 802.15.4 LR-WPAN layers to provide services. While ZigBee is a low-power, short-range link, the protocol permits the devices to use a mesh network topology that enables a wider transmission range. Advantages to ZigBee include a 128-bit encryption for data security, data rates up to 250 kbps with low power consumption. Typical ZigBee applications are in low-data rate environments such as smart homes, energy management, and lighting management.

4.4 USER INTERFACE SYSTEMS

The interface design to support operator data entry and efficient operations is very important in every system. Such systems are also referred to as a *Man-Machine Interface (MMI)* or *Human-Machine Interface (HMI)*. If the interface is realized on a graphical computer, it sometimes is referred to by the generic *Graphical User Interface (GUI)* moniker even though that term is not specifically for telecommand and telemetry interfaces. Typically, the interface software comprises the bulk of the software executing on the operator's computer system. The software architect may distribute the software over several desktop systems or on one large central processing computer/data server. In future systems, the user interface software and data storage may even migrate to cloud-based storage domains and hand-held display devices will download the application and associated data upon session initialization.

The user interface has two major components: the operator's display generation and telemetry database software, and the telecommand interface software. This section describes display and data entry design for telemetry and telecommand operations. The reader is encouraged to look at computer-based systems for data display and entry in other contexts to see which designs are truly useful, and which designs hinder the operator's productivity. This exercise is directly transferable to telemetry

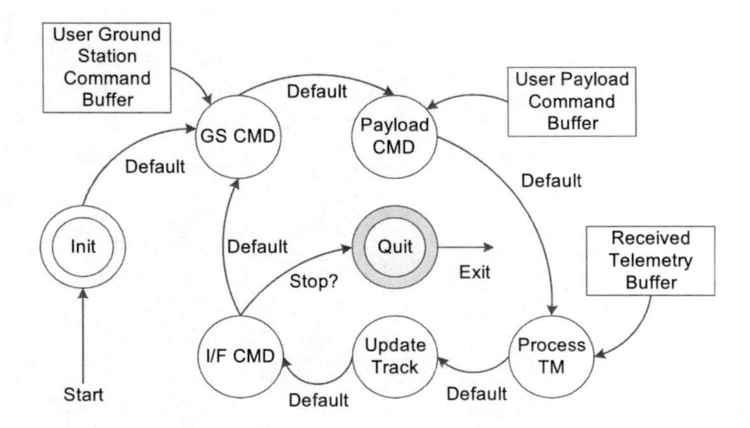

Figure 4.7: Example user base station processing states for the user interface, telecommand, and telemetry processing software.

and telecommand interface design as well. The reader can find further background information in references such as Shneiderman [Shne87].

4.4.1 Processing State Diagram

Figure 4.7 illustrates an example of a *processing state diagram* that a user-interface computer system designer used to describe the control software in the user base station. States define explicit processing modes with defined conditions for leaving the state and deciding the next processing state. The states in Figure 4.7 are as follows:

Initialization (Init) to properly start operations within the user interface control computer and devices within the ground station.

Ground Station Command (GS CMD) to configure ground station hardware based on an operator command input.

Payload Command (Payload CMD) to package a valid command for transmission for the payload based on the operator's input.

Process Telemetry (Process TM) to process the incoming telemetry data.

Update Tracking (Update Track) to update current payload positional information for the operator's display.

Interface Command (I/F CMD) to process user commands to configure the operator's Interface (I/F).

Quit to terminate processing via an orderly system shutdown.

This particular example assumes that the user interface system resides on a single computer without multiple processing threads. Complicated systems may have these functions distributed over multiple processing threads, multiple computers, or even multiple processing sites so the overall state diagram grows more complicated in that case.

4.4.2 Telemetry Database

The *telemetry database* is the structure that holds the incoming data for current display as well as for further processing to support data analysis. We think of this database as a time-variable matrix containing all of the gathered information. However, since that matrix is time-variable, users also need a means to search for past values to support analysis functions. In this section, we see how the database is organized and used.

4.4.2.1 Database Architecture

Once the database designer has the measurement list for each data type, the designer proceeds to configure the database. The top section of Figure 4.8 illustrates an example telemetry database that contains more information than raw numbers generated by the sensors. The bottom section of Figure 4.8 illustrates how a normal range, a caution range, and an alarm range are associated with sensors in the top section. These ranges signal warnings to the operator for sensor readings that indicate problems. Defining the bands requires the operational experience of the system's designers and users. In general, the telemetry database structure may have fields for managing both processing and the actual data values. The following fields are typical:

- The parameter index number that is the database entry number for the parameter, the mnemonic for the parameter or both.
- The parameter type of analog (An), which is for Pulse Code Modulation (PCM) parameters, digital (Di), which is for counters or for other naturally occurring digital representations, and bi-level (Bi), which is for switch or logic settings.

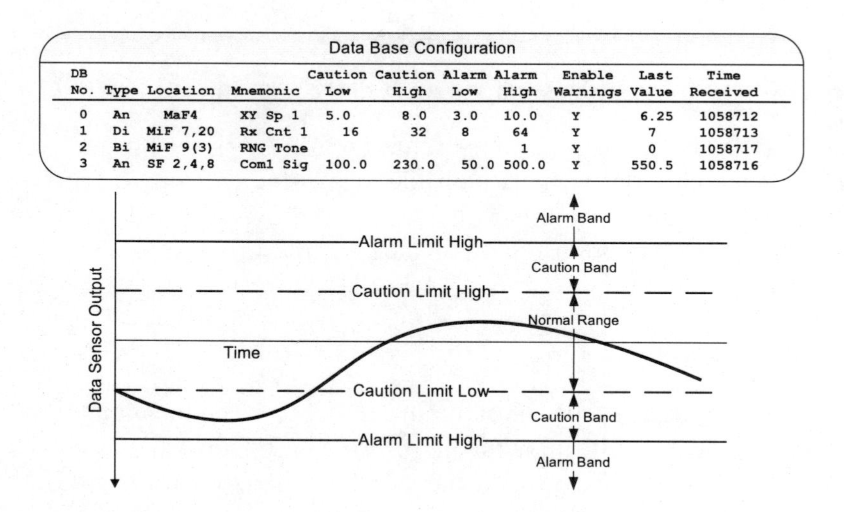

Data Base Configuration

DB No.	Type	Location	Mnemonic	Caution Low	Caution High	Alarm Low	Alarm High	Enable Warnings	Last Value	Time Received
0	An	MaF4	XY Sp 1	5.0	8.0	3.0	10.0	Y	6.25	1058712
1	Di	MiF 7,20	Rx Cnt 1	16	32	8	64	Y	7	1058713
2	Bi	MiF 9(3)	RNG Tone				1	Y	0	1058717
3	An	SF 2,4,8	Com1 Sig	100.0	230.0	50.0	500.0	Y	550.5	1058716

Figure 4.8: The general structure of a telemetry database and the meaning of normal, caution, and alarm bands for the sensors.

- The parameter's location relative to standard synchronization markers in the telemetry data stream; here MaF is a major frame location, MiF is a minor frame location, and SF is a subframe location (see Section 6.4 for the definition of telemetry frames).
- Ways to encode the normal band, the caution band, and the alarm bands limits.
- A flag to specify if the processing computer should indicate crossing the caution and alarm band warnings when processing the data.
- The last value received from the payload.
- The time, or other sequencing indicator, when the base station received that value.

The operator's ability to enable or disable caution and alarm processing may seem undesirable. However, it is often operationally convenient to disable the warning temporarily when massive system reconfigurations are undertaken so that the operator is not flooded with unnecessary warning messages.

4.4.2.2 Data Timing

The data time value in Figure 4.8 is typically based on the telemetry minor frame (see Section 6.4.1.1 for a definition) or the telemetry packet (see Section 6.7 for a definition). The system designer may choose the data base time as either the time that the frame or packet arrives or a time code embedded in the minor frame or packet. The data processing software needs to know this so that it can properly process the data.

Often, the system places the data in the minor frame or the packet using a multiplexing method like that illustrated in Figure 6.3. When the system uses this method, the data base processing may index the sensor data time based on the value's location relative to the minor frame or packet start. In some instances, this may not be correct or sufficiently accurate. For example, a system may have correlated measurements that are taken at the same time and cannot appear at the same relative position in the minor frame or packet. In this case, the timing may be incorrectly rendered and it may cause a processing error. To remedy this situation, the system designer may use an embedded time code with the data, such as those found in Chapter 8.

4.4.2.3 Database Storage

The database holds at least one full cycle of all parameter samples in memory, or in a cache, for easy retrieval in building operator displays. Designers realize this architecture by using a circular memory buffer that the computer overwrites as new values become available from the payload. As the new values arrive, the software writes the older values to a permanent archive or storage for later analysis. The designer must organize the database storage architecture to ensure easy retrieval. The designer has options for commercial database software for this purpose and this software gives the designer a defined structure to the database, and a means to submit queries to the database for current or past values.

When the software stores the database values to the archive, the designer may consider applying data compression techniques that reduce the size of the archive. When compression is used, database designers consider techniques such as Lempel-Ziv encoding, or some form of delta modulation to transmit only changed values since the last saved data block. The usage depends upon how quickly the archive needs to be scanned for back values and how complicated the data retrieval becomes.

4.4.2.4 Telemetry Processing Levels and Unit Conversion

The telemetry processing software operational flow proceeds along the following lines:

1. Check incoming data for validity and errors to ensure that correct raw data are processed.

2. Separate the parameters in the current data segment from any included management or synchronization information; in Section 6.7.2, we will see the inclusion of fill data in the transmitted data set that the processing must remove at this stage.

3. PCM data are converted from a quantized output level to associated *engineering units*, where the term engineering units refers to the sensor voltage or current that was converted by the PCM process and now the values are in voltage or current.

4. The processing applies the sensor calibration to obtain actual measurements in real units, such as Kelvin, Watt, or Newton, from the engineering units; the software may need additional payload parameters or ancillary data accompanying the data set to complete processing at this stage.

5. The values are compared with the telemetry database entries for limit checks on caution or alarms.

Additional processing may include the construction of *derived parameters*, which are telemetry parameters computed by combining the basic telemetry parameters. The processing software inserts derived parameters as entries in the telemetry database as well.

We classify the extent of the telemetry processing done to the data at any given time by the *processing level*. Agencies or projects, such as the National Aeronautics and Space Administration (NASA) Earth Observing System (EOS) [EOSD10], may have specific definitions for their applications but they generally follow a format similar to the one given here. The processing levels indicate how close one is to the original source data and how easy it is to recover the original source data in case the processing algorithm makes any errors or if the analyst needs to revise the algorithm later. It is not necessary that every telemetry computer system execute these processing steps in real time. Depending upon system design and capabilities, the processing may occur either at another location or at another time as part of the post-reception data processing. The processing levels are as follows:

Level 00 — Data comprise raw sensor output prior to a standard word length or other formatting, i.e., raw PCM output from the ADC.

Level 0 — Data comprise uncalibrated sensor output that may be reformatted to a standard word length, e.g., a 12-bit PCM output is placed in a 16-bit word for data transmission. The data are correctly time ordered, and concatenated into nonredundant files with missing data recognizably flagged. Processing removes communications artifacts such as headers and synchronization data. The data may be in either "clear" or encrypted modes, as transmitted by the payload.

Level 1 — Level 0 data processed into calibrated engineering units. Additional system data necessary for full analysis may be associated with the Level 1 data products but not applied to the processing. Level 1 has two possible processing levels:

1. Level 1A: The analyst can recover exactly the raw sensor data after the Level 1 processing.
2. Level 1B: The analyst cannot recover exactly the raw sensor data after the Level 1 processing due to actions such as floating-point calibration conversion or data resampling to convert unequally spaced samples to equally spaced samples.

The Level 1 processor removes any source data encryption.

Level 2 — Level 1 data that are converted from raw engineering units to a more meaningful quantity; this may involve some limited combination with other parameters at the same data level.

Level 3 — The data set is fully converted and processed into useful production data. The processing may map the data onto relevant spatial or temporal grids to permit further processing or use by customers.

Level 4 — The analyst further processes the production data set through models or other high-level analysis.

We divide the Level 1 processing into two subtypes because the calibration process is sometimes irreversible. For integer data, the calibration is often reversible. For floating-point data, e.g., signal strength, the calibration curve does not usually allow the analyst to recover the original signal bit pattern unambiguously. Level 2 allows for the combination of several parameters into making a new parameter (a derived parameter) that is more meaningful. The Level 3 data are ready for analysis by those responsible for the payload.

Other data processing may be required in the telemetry computer to support the construction of the finished product. The system designer may consider the following processing options:

1. The removal of time reversal in the data due to the payload recording the data on a tape recorder and then playing the data back during transmission. When the payload records data magnetically, the operator usually does not rewind the payload recorder prior to playback; rather, the operator reverses the recorder's direction and plays the tape.

2. The removal of redundant data segments received from two or more locations (receiver diversity techniques) and the possible selection of one of the segments as more "correct" than the other due to transmission errors.

3. A retransmission request by the packet processor's Automatic Repeat Request (ARQ) protocol for payload data segments found to be in error.

Completion of these tasks may be part of the overall data set post-processing and not real-time processes.

4.4.2.5 Telemetry Packet Processing

As we will see in Section 6.3.2, we usually divide the telemetry transmission into two classes based upon the transmission methodology: frame-based transmission and packet-based transmission. System designers use the frame transmission mode with systems that need to maintain very close synchronization for proper data reception. Conversely, designers select packet modes when they are sure that the data channel delivers the payload's data properly despite small gaps in the channel activity. Packet telemetry is similar in concept to sending data over the Internet. When the designer formats the data transmission in this way, the designer often groups the data sensors in the payload into *virtual channels.* The virtual channel segments the data into logically related groups and

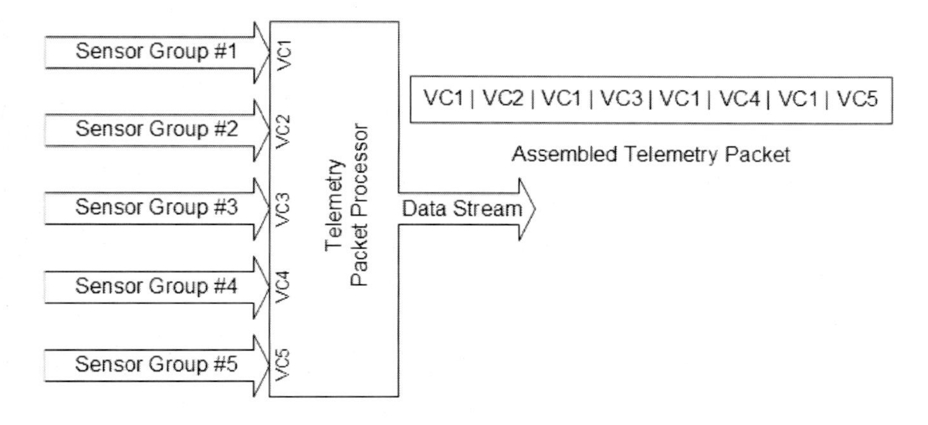

Figure 4.9: Virtual channel and packet telemetry organization in the payload.

transmits all the sensors in that group in the same manner. As Figure 4.9 illustrates, a dedicated packet telemetry processor in the payload may perform the processing for each virtual channel when this is done. The designer programs the operator's computer to recognize the sensor data according to each virtual channel and process it accordingly. The telemetry system processor may also need to know about the payload health and welfare data, also known as housekeeping data, to process the virtual channel sensor data. This technique works well for distributed data processing or in cases where the data sets need to be kept isolated.

4.4.3 Telemetry Displays

This section covers how to display individual numbers and additional information to help the operator detect and diagnose problems. The discussion proceeds along the lines of segregating data for easier understanding, additional display information for operator diagnostics, display interactions with the database, and some examples of these concepts. Figure 4.10 schematically illustrates how an individual display might look. This display has the following areas:

1. Main display area for viewing telemetry data values, messages, etc. The display designer partitions the telemetry database parameters by subsystem. The processing converts the displayed values into normal units (volts, amps, etc.). The telemetry parameter's color shows whether the system is in the normal, caution, or alarm region.

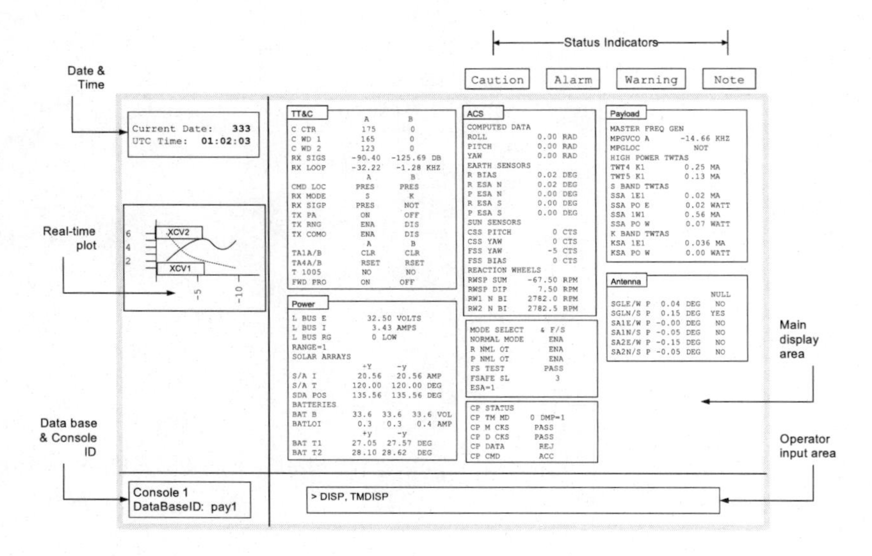

Figure 4.10: Example operator's telemetry display with real-time data, a real-time plot, and operator information and warning indicators.

2. An operator data-entry area, which provides a means for the user to enter commands that affect the display format, payload commands, or the configuration of the operator's equipment. The operator input area is immediately below the main display.

3. Status indicators to warn the operator of state changes within the telecommand and telemetry system. These are the small blocks along the top of the display (`CAUT`, `ALARM`, `INFO`, or `NOTE`). The figure shows useful items, such as the date and time, along the top of the display.

4. A real-time plot of a selected parameter in a strip chart display mode along the left-hand side of the display.

While the display example in Figure 4.10 contains a great deal of information, it does not take full advantage of the graphical programming currently available to the system designer. Figure 4.11 illustrates how the display designer does not always need to express the telemetry as a numeric value but may choose to use graphical indicators. Here, the display designer uses dipsticks, thermometers, and various gauges to express the data. The display processor paints the indicators to show the normal, caution, and alarm levels and include colors or other indicator

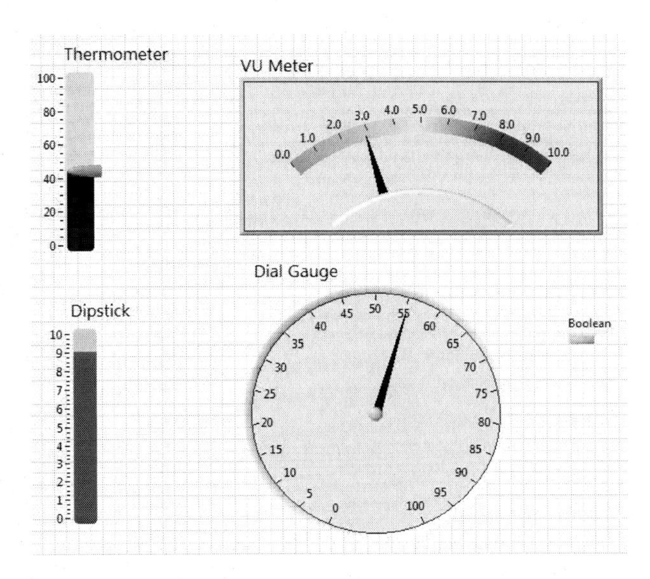

Figure 4.11: Examples of graphical display methods for data representation found in the LabVIEW® user interface palate.

markings on the graphic. Figure 4.12 illustrates how these displays are used as part of a graphical display methodology. In this type of display, designers keep the non-numeric displays always active. The designer groups text-based information and numeric information into smaller windows that the operator shrinks to an icon when not in use and then opens when needed to display contents or make an entry. The designer configures the contents of each grouping to meet the needs of an individual operator in a multiuser system. In this example, Figure 4.12 shows a traditional subsystem display with attitude-control parameters at the top and a real-time graphic in the middle, and an operator entry window at the bottom. To the right of the data entry window are several icons for windows that the operator shrank to keep them out of the way until needed. The advantage to this method is that the computer only needs to paint new versions of active windows in real-time upon a data change; this speeds up the computer's response. Also, when they are open, the display areas are larger in Figure 4.12 than in Figure 4.10 to provide better parameter visibility for the operator.

With smart payloads, one window in a display like that in Figure 4.12 may be for remote access over an Internet connection to a data server hosted on the actual payload computer.

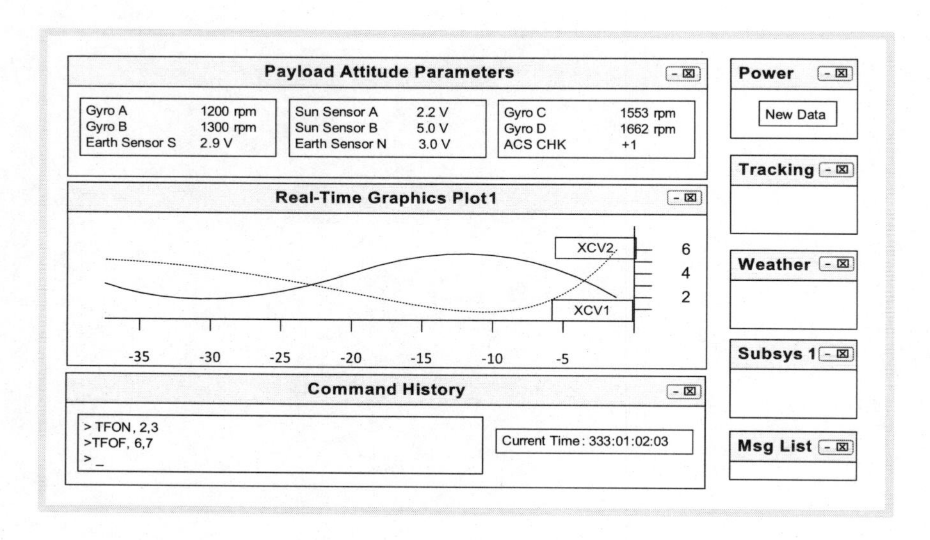

Figure 4.12: An operator display based on icons and pop-up windows to display information.

4.4.3.1 Telemetry Data Partitioning

Most telemetry displays for larger systems (more than 32 system parameters) produce more data than a single operator can absorb or concentrate on at any one moment. The display illustrated in Figure 4.10 had over 100 parameters out of a total database size exceeding 1200 parameters. As the reader may expect, the operator cannot inspect the entire database in a single view. To solve this information overload problem, display designers usually partition the telemetry data in some fashion. Typically, the designer partitions the data entries into logically connected subsets of the entire database. For example, all of the telemetry values related to payload battery voltages might be one partition.

Figure 4.13 illustrates one possible partitioning approach. It shows two data sources: the payload segment and the user segment. The latter source may seem to be trivial but in larger systems, such as satellite control stations, the control station itself generates a great deal of data about its current state. The designer further divides the database's payload portion into two subsets: the payload's measurement data and its health and welfare data, as we saw in Section 1.2. The payload's measurement data are those data the payload designers must collect to fulfill their mission. The payload health and welfare data describe

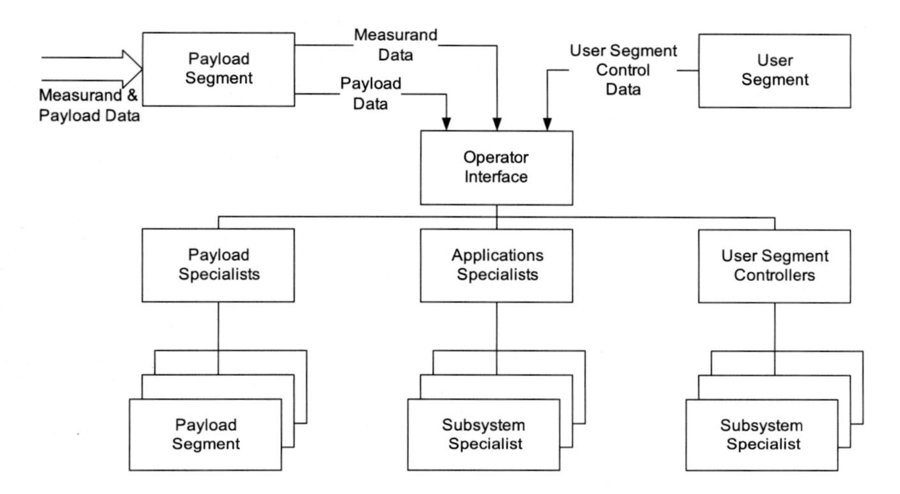

Figure 4.13: A schematic to illustrate possible database partitioning based on related subsystems.

the payload's operational state. Designers sometimes also call this the payload's *engineering data.*

For smaller systems, the designer realizes the data partitioning on a desktop system by using a single-screen windowing system. For large, complicated systems, the partitioning may involve multiple consoles managed by several subsystem specialists. In either case, the goal is to prioritize and sort the data to assist the operator in making an informed decision about the system. The data partitioning described here has the following characteristics:

1. Operators see only a portion of the entire available telemetry database at any time on the display screen, either as a full-screen display or within a multiwindow display. This is effective because operators either are subsystem specialists (on large systems) or are interested in only one small subset of the total database. Large systems usually have individualized displays for each subsystem specialist. The subsystem display may also show a block or circuit diagram illustrating where the data originates. In this case, the display shows the latest received values in the diagram so that the operator can see them in the overall subsystem context.

2. A warning system is required to inform operators of out-of-range conditions for parameters on their displays. Keeping the displays

manageable means the display designer partitions them into related groups of parameters that the operators access for trouble resolution. The display designer has the warning system alert the operator that other displays may need attention.

3. Larger systems usually include operations management procedures. Partitioning displays into manageable units allows other personnel to see relevant "snapshots" of data for operations management without having to search through the entire database.

The system designer will need to balance hardware capabilities with operator ease of use when making these kinds of choices.

4.4.3.2 Telemetry Status Indicators

For each parameter in the display, the display shows the operator the parameter along with a label and an indication of the parameter's unit. The display color-codes the value for normal or error conditions. The display designer codes the green/yellow/red status indication as follows. Green indicates that the parameter is within the telemetry database's normal band. Yellow indicates caution because the parameter is outside of the normal range but has not yet become in the range considered dangerous. Red indicates alarm because the parameter is in the operationally dangerous range.

The display software encodes the value according to these color cues so that the operator sees at a glance whether a parameter is within range or has a problem. Additionally, the user interface may log when a parameter changes status to give a permanent record for troubleshooting. This print record may show up both on a printer and as a log file that the user accesses on the display. In the Figure 4.10 sample display, the `CAUT`, `ALARM`, and `INFO` boxes along the top of the display all link to the log file. When the operator clicks on the box, the display shows the log file contents for that particular warning class.

4.4.3.3 Display Interaction with the Telemetry Database

The next important aspect of the design is how to link the telemetry database to the display. One technique is the database query method. In this method, the display sends a query to the database with a list of all of the display parameters. The database server responds with the values for each item in the list plus other related information such as status. The display shows the operator the current values in the appropriate

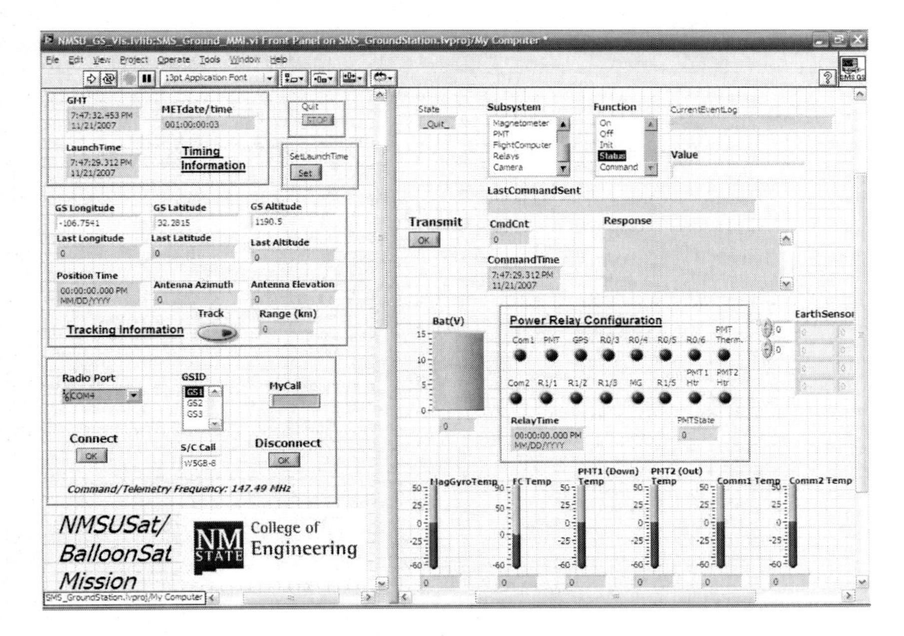

Figure 4.14: An example of a LabVIEW-based user interface display. Both payload and user base station data are displayed. The user is also supplied with as command entry interface.

template, and after the predetermined wait time, the display server sends an updated parameter list to refresh the display.

This client-server operational mode has definite advantages in large systems with many active displays. As the operator calls up new displays, the display sends the revised parameter list to the database server. The speed of new data arrival from the payload determines the display refresh rate. The refresh initiator is either the display client or the database server. Display designers also use this mode in Web-browser based display techniques so that the payload can send data over the Internet (see, for example Section 6.10.3).

4.4.3.4 Balloon Experiment Telemetry Display Example

Figure 4.14 illustrates a user interface for a telecommand and telemetry system used in a high-altitude balloon flight experiment. This interface was used for testing and simulations before flight and during the 24-h experiment at a 36-km altitude. We built this interface with LabVIEW® components. We added display elements as the hardware was integrated with the payload system to assist in the formal Integration and Test process. The left-hand section of the display contains information associated with the ground station equipment, while the right-hand section contains information related to the payload. The operator uses mouse-clickable push buttons to connect or disconnect the radio communications, enable antenna tracking, capture the launch time, and tell the interface to quit processing. The operator also uses text input to indicate a radio call sign and a pull-down indicator for selecting the serial port for communications and the ground station identifier. Other fields are to show current values such as real time, mission elapsed time, and the payload's position.

The operator inspects the real-time telemetry using a combination of dipstick and level indicators as well as actual values. The right-hand side has a pull-down menu for the operator to construct commands and a mouse-clickable push button to transmit the command. The display also presents the command echo from the telemetry data as well as command tracking and timing information. We will cover this in more detail in the next section.

4.4.4 Telecommand Interfaces

Since the operator needs to command the payload in an efficient (to correct anomalies) and correct (to avoid self-inflicted problems) manner, the system designer needs to take care in design of the telecommand system. The designer bases the command philosophies for the user interface portion of the telecommand and telemetry system on techniques utilizing command line input, graphical input, and function key input. We look at each of these in subsequent paragraphs. This section concentrates on the details of the command-process user interface. We will examine telecommand data transmission across the channel in more detail in Chapter 9.

Regardless of the technique used, the telecommand processing user interface must provide a translation between the raw input and the format the payload is expecting. The actual command to the payload may be a binary, ASCII, or mixed data string that may not be comprehensible to someone inspecting the command data. Generally, the operator prefers to have a command structure that is mnemonic in nature for ease of training and remembering the command functions at the input level. The top section of Figure 4.15 illustrates entering an example operator command. A command line interpreter program interprets, or parses, the operator's command entry. The command name may uniquely identify

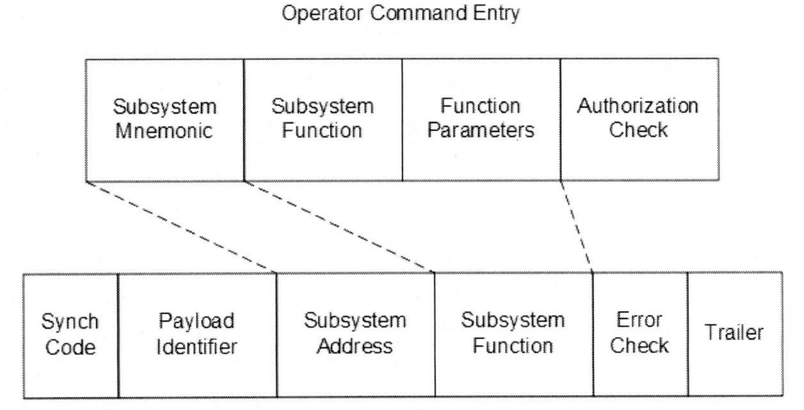

Figure 4.15: An example of the operator's command entry and the command sent to the payload.

the subsystem within the payload that acts on the command and the function the subsystem performs. The command may also need data to assist the command processor in executing the function and an authorization check so that the command processor is sure that a valid user has issued the command. The bottom section of Figure 4.15 shows the resulting command output fields.

The operator's workstation may automatically generate the authorization check before passing the requested command to the command processing software based on the user's login information.

4.4.4.1 Command Dictionary

System designers base the telecommand system design on a *command dictionary.* This dictionary specifies items such as

- The allowable commands for each subsystem, usually in a mnemonic form
- The allowable parameters for each command
- Any restrictions on specific operators or managers from executing the command

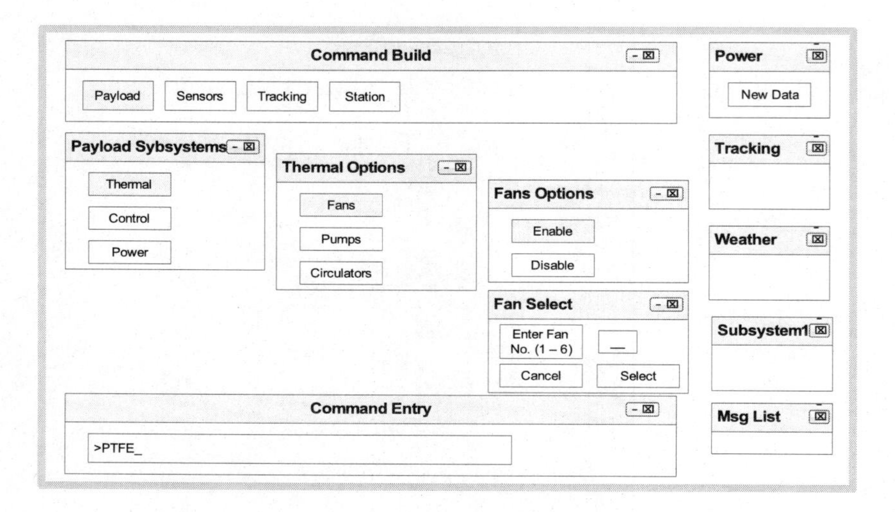

Figure 4.16: An operator command entry screen using both command line entry and menus.

- Any restrictions on times, payload state, or operational phase for sending the command

The command dictionary may also have a list of error codes that the payload command processor returns with the telemetry data to specify any detected transmission or syntax errors.

4.4.4.2 Command Data Input

The command input contains the command name and data to go with the command. This information needs to be in a form that the operator easily remembers. The simplest entry is a command line entry. However, interface designers use other tools such as pull-down menus to make this process less error prone. Here, we look at each of these techniques.

Command Line Input The command line method depends upon the operator knowing all of the individual commands with their subfields in the command dictionary for the system. A disadvantage is that the operator may have a large list of commands to remember. The advantage is that each command has a short list of parameters associated with it; thus, data entry is rapid and reliable. The lower part of Figure 4.16 gives an example where the operator input `PTFE` is the command line entry.

While commands may be mnemonic on the surface, interface designers can make them difficult to parse and to error check if the command fields are context sensitive. With this design, the entries in a given field take on different meanings depending upon the values of the preceding parameters. This design usually leads to longer training times and software maintenance difficulties.

Command Graphical Input The upper part of Figure 4.16 illustrates a design for using pull-down menus to build a command. The use of menus provides for easier operator training and makes for fewer commands that the operator must memorize. Additionally, the operator may be less prone to making typing errors by using the pull-down menu. For this example, the option `PTFE` was chosen which is parsed as *payload*, *thermal*, *fans* and *enable* was chosen. At this point, the operator needs to identify a fan number to complete the command. On each menu, the user selects a letter option either by clicking on it with the mouse or by typing the letter at the keyboard. It is necessary for the command design to have unique letter combinations, as the menus are traversed.

The main advantages of the menu-based system are apparent when training new operators to use the system or the operator needs to execute a rarely used command. The main disadvantage to a menu-based system is that data entry is slow if the operator must step through a large number of menu levels. Operators typically do not like this procedure when they try to enter a command rapidly after they become experienced with the system. Therefore, it is usually a good design feature to have a corresponding rapid entry method in addition to the pull-down menu to allow operators to enter the commands.

Command Function Key Input One way interface designers improve command line input is by using keyboard function keys (*F1* through *F12*) with `CONTROL`, `SHIFT`, and `ALT` variations. This allows the designer to map frequently used commands to key combinations to make entry quick and less prone to error. The design even allows for real-time insertion of command data into the command fields so that the entire command is not static. The insertion is done through pop-up dialog boxes or similar techniques typically included in commercial software packages. In addition, interface designers combine function keys with pull-down menus to allow for rapid data entry by experienced operators.

4.4.4.3 Command Processing

The command processing software has two major components: command parsing and validation. In this section, we look at issues affecting both of these software components.

Command Parsing The command parsing software includes the following processing functions:

1. Ensuring that the command name or means of calling it are correct for the system.
2. Ensuring that the operator has permission to enter the command if the system allots functions to multiple users.
3. Checking that the command fields have the correct syntax and command fields have any user-specified data limit checked.
4. Checking that the proper number of fields for the command is present.

The command parser is continuously active to ensure that any operator entry from the keyboard (or a mouse or verbally) is processed. The command parser program looks for these to either be fixed-length fields or use a delimiter, such as a comma, to separate the fields. After parsing the fields, the command processing software then maps the input to the correct payload format. The software uses a look-up table or a command database to form the appropriate command output. While this discussion has focused on the payload, system designers use the same strategy to configure the operator's support equipment.

Command Validation Command validation has two major parts: verifying that the command came from the correct operator and that the commands have proper syntax. In a large system, more than one operator may control aspects of the payload. If this is the case, then it is possible for an operator to issue an inappropriate command for their command responsibility. To prevent this, the command processor has a list of allowable commands for each operator position. If an operator attempted to execute an unauthorized command, the command processor blocks execution.

System designers include syntax validity checking in their design. The command processor must perform any possible validity checks on

Table 4.3: Start-up Command Macro for the Balloon Example Payload

Execution Time	Command	Action
3411824400	$$FCCDNOP*	; NO-OP
3411824401	$$RLOF*	; Turn OFF all power relays
3411824402	$$FCCDNOP*	; NO-OP
3411824403	$$CMON*	; Power up Communications
3411824404	$$FCCDNOP*	; NO-OP
3411824405	$$CMIN*	; Initialize Communications
3411824406	$$FCCDNOP*	; NO-OP
3411824407	$$GPON*	; Power up GPS
3411824408	$$FCCDNOP*	; NO-OP
3411824409	$$GPIN*	; Initialize GPS
3411824410	$$FCCDNOP*	; NO-OP
3411824434	$$FCCDSK0*	; Disable Schedule Processing

the user data entry to ensure that it is correct and appropriate (within established limits). The system designers need to specify the allowable data entry limits and conditions to trigger validity checking.

Command Files Operators find it tedious and error prone to enter the same command information repeatedly. For this reason, command files may be useful. A command file is a predesigned command sequence the command interface sends to the payload with little or no prior operator interaction required. Files are stored on disk or mapped to keyboard function keys. The payload processes the files as if the operator directly entered the file's contents. The technique has the advantage of requiring the operator to enter only a short sequence to transmit the file.

As Table 4.3 illustrates, the command file designer may use a command file format that includes an execution time (either absolute or relative to the time of the first command in the file), the command value with any associated data formatted using the normal command syntax, and optional comments. A standard delimiter, such as a semicolon, may separate any optional comments from the command.

Payload designers use command files to manage the payload by storing command files for specific instances within the payload computer system. Example applications of these command files include payload initialization or shutdown, critical event operations, and error recovery operations. Designers exercise these command files extensively as they develop the payload to ensure that the files operate safely.

4.4.4.4 Balloon Experiment Telecommand Interface Example

As we saw earlier, Figure 4.14 illustrates one method for command data input using the LabVIEW-programmed interface. The operator has two regions for entering commands: the left-hand partition controls the ground assets and the user interface, and the right-hand partition controls the payload.

The operator clicks on the push buttons in the left-hand partition to quit the interface, capture the payload launch time, start antenna tracking, and connect or disconnect the radio. When the operator clicks on these command buttons, the display software captures the data values and passes them to the specific functions for execution.

The operator selects options from the pull-down menus in the right-hand partition to build command values. Certain commands require operator-entered data values so there is a text entry field for these entries. When the operator desires to send the command to the payload, the operator clicks on the transmit button to capture the inputs from the menus and the free entry area and then the software packages it for transmission to the payload. The telemetry processing software displays command results returned from the payload in the region below the command build interface.

The command macro in Table 4.3 was used to start the payload. We designed this particular macro for the payload operator to either send it from the keyboard or have the payload automatically start it at power-up with the latter mode being the normal operational procedure.

Balloon Payload Command Dictionary Table 4.4 presents the flight computer portion of the command dictionary for the balloon experiment payload. Each command begins with a "$$" and ends with an "*". Each command in this example starts with "FC" to indicate that it belongs to the Flight Computer system. The next two characters specify the action to take. In this case, "ST" indicates status and "CD" indicates a command action. The remaining characters are parameters for the command. The error code is a value returned in telemetry. The payload transmits a code value of zero when it receives a correct command. The payload sends a nonzero value when it detects an improper command. The nonzero error codes in the command dictionary are unique for each subsystem. We use

Table 4.4: The Command Dictionary Covering the Flight Computer Subsystem

Command	Command Meaning	Error Code	Code Meaning
$$FCST*	Send Flight Computer Status (current relay state and ADC values)	0	command accepted
$$FCCDSK0*	Disable scheduled operations	0	Command accepted
$$FCCDSK1*	Enable scheduled operations	0	Command accepted
$$FCCDSCK*	Synch flight computer clock to GPS time	0	Command accepted
$$FCCDKLL*	Shut down flight computer	0	Command accepted
$$FCCDTXIfile*	Transmit image file where the full path needs to be given, e.g., C:\IM4082619TN.JPG	0	Command accepted
$$FCCDSKU*	Upload schedule file. This file is to be C:\SCHEDULE.CSV	0	Command accepted
$$FCCDAD1*	Read ADC bank 1	0	Command accepted
$$FCCDAD2*	Read ADC bank 2	0	Command accepted
$$FCCDNOP*	No-OP command	0	Command accepted
		100	Invalid FC function specified
		101	ON function not valid for FC
		102	OF function not valid for FC
		103	IN function not valid for FC
		104	Invalid CD specified for FC
		105	Invalid image file spec.for FC

Table 4.5: Example Function and Subsystem Mnemonics for the Balloon Payload

Function Mnemonics		**Subsystem Mnemonics**	
Code	**Meaning**	**Code**	**Subsystem**
ON	On	C1	Communications string 1
OF	Off	C2	Communications string 2
IN	Initialize	GP	GPS device
ST	Status	MG	Magnetometer/Gyro sensor
		PT	Photomultiplier tube sensor
		RL	Power relays
		FC	Flight computer

this format for the other subsystems in the experimental payload. Table 4.5 provides the mnemonics for the functions and subsystems found in the balloon payload example.

4.5 PAYLOAD COMPUTER SYSTEMS

The function of the payload software is the same as for the operator's computer system: to provide control and data processing functions. However, the payload's software has added maintenance functions not typically found in the operator's software system. Another difference is that the payload's processing level is not typically as detailed as that found in the operator's computer. Most designers only allow the payload to operate at the Level 00 or Level 0 processing levels. Figure 4.17 illustrates how a payload *software configuration* might appear in memory.

The initialization code establishes the initial payload configuration. This code section also has the ability to issue a reset based on a time-out condition in the payload timer. The payload timer, or *watchdog* timers, depending upon the reset strategy, is reset by the completion of one loop through the main processing software or the arrival of an external event such as receipt of a command from the operator. The timer's strategy ensures that the payload processor has not hung and that the link between the operator and the payload is still active. If the timer expires, then the event trigger for the timer did not happen and the payload needs a reset. A reset may involve a simple application task restart through a full system cold reboot, depending upon the designer's reset strategy.

Another function that system designers frequently include in the

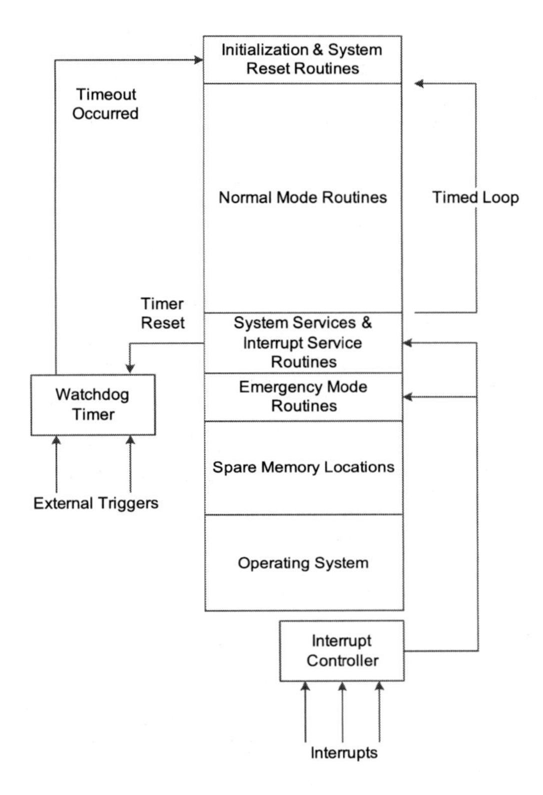

Figure 4.17: Routines for normal processing, system administration, and interrupt servicing in a payload memory configuration.

initialization code is the *Built-In Self-Test (BIST)* function. The BIST checks on the internal status of major payload systems by running self-diagnostic routines and reports the results to the operator. The BIST may also have the payload's sensors report measurements from internal sources to validate correct operation. The BIST functions can also be included in the list of available functions for operator commanding.

The normal-mode portion of the memory contains most of the software related to the payload control and data acquisition applications. This software runs as a continuous loop with the timing constraint. The designer predicts the time to complete the loop such that the software completes all tasks, such as gathering the data, packaging it for transmission, responding to operator commands, and performing any necessary control functions, in a worst-case time. At the end of the loop, the software resets the timer and then begins the loop again. If the timer expires before the execution loop completes, then the timer resets the payload software using the initialization routines.

A problem develops if the execution loop hangs at some point and execution is longer than expected so that the timer resets and the software resets. Even worse is an estimation error in the time to complete one loop. If this misjudged on the low side, then the payload resets on nearly every loop execution. This causes the payload to halt operations. Conversely, if the timing grossly overestimates the processing time, then the loop may spend a relatively long time in idle mode before resuming the processing list at the beginning. This is an inefficient use of processing resources.

The emergency procedures and system service routines are specialized software routines to process payload events that occur asynchronously. These events include payload fault detection interrupts, processing service interrupts, and operating system service calls. The designer configures the code based on the specifics of the payload architecture.

The payload may need to reserve a spare memory region for several reasons. As the payload matures, the normal mode program may grow and therefore needs room to expand. The payload memory may also develop bad locations that are no longer reliable. This is especially bothersome in space payloads where the operator cannot replace the payload because the payload cannot be physically retrieved. The operating program needs to have patches placed in it to jump around the bad locations into the spare locations and then back again to the normal-mode memory.

These bad locations typically are most prevalent in space payloads where cosmic ray events disturb the Random Access Memory (RAM) (Read Only Memory (ROM) is more robust to cosmic ray disruptions). When the particle hits the RAM chip, the energy may be sufficient to cause a *Single Event Upset (SEU)*. The SEU is manifested by a bit in memory changing (1 to 0 or 0 to 1). This is a major fault if the corrupted bit is the sign bit of a stored constant or is in an operating instruction. SEUs are usually detected by anomalous behavior reports. The operator then performs a memory dump to compare the payload memory with a map of the correct contents of the storage locations. If a sufficient number of SEUs occur in a memory chip, the upsets may damage the chip to the point where it no longer stores information reliably. Payload designers may use especially hardened memory chips to reduce the occurrence of SEUs.

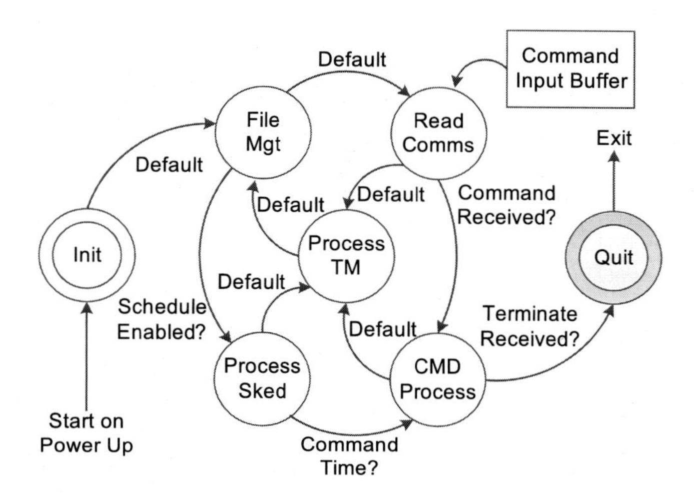

Figure 4.18: Example payload processing states for the telemetry and telecommand processing software.

4.5.1 Payload Command Processing State Diagram

Earlier we saw that we represent the user interface processing in terms of a state diagram. Figure 4.18 illustrates an example of a corresponding state diagram for the payload processing software. In this state diagram, the processing software may accept commands either from the operator or from a configuration file that stores timed sequences. The following states are in this processing state diagram:

Initialization (Init) to initialize the payload control computer and devices within the payload to start operations properly.

File Management (File Mgt) to configure processing schedule files, log files, etc. for processing and diagnostic purposes.

Process Schedule (Process Sked) to determine if it is time to execute a command from a pre-stored sequence schedule file.

Read Communications (Read Comms) to determine if a command has been sent from the operator and if it is present at the communications port.

Process Telemetry (Process TLM) to send the next group of telemetry data to the operator.

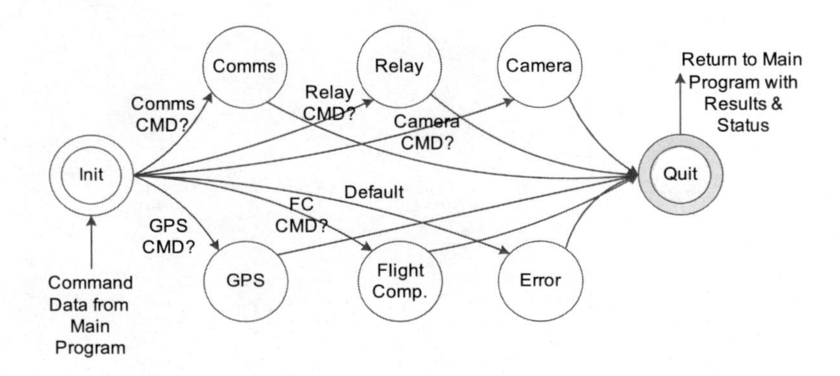

Figure 4.19: Command processing state diagram for the balloon payload example.

Command Process (CMD Process) to process the next command from either the operator or the schedule file.

Quit to terminate processing via an orderly shutdown.

It is legitimate to have a command to shut down the payload computer. However, the system designers normally include protection measure to ensure that a single command by the operator does not terminate the processing without an ability to restart it.

While Figure 4.18 is concerned with the payload processing, system designers develop a similar state diagram for the operator's command and telemetry processing as well.

4.5.2 Payload Command Processing

Figure 4.18 illustrated an example of an overall payload state machine processing. The command processing state is an example of how software designers develop a state machine to execute the processing within the individual states. Figure 4.19 illustrates a simplified version of the command processing state diagram. The overall processing flows as follows:

1. The *Init* state performs initial processing to determine if the command has the proper form and it is legitimate from the command dictionary.

2. If the software determines that the command appears to be le-

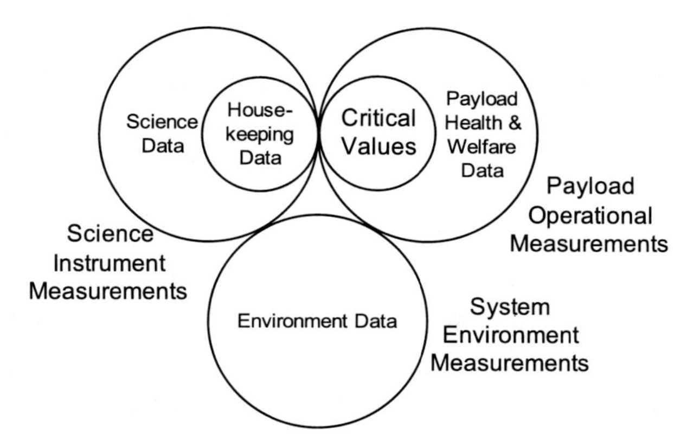

Figure 4.20: Data classes that may be present in an overall telemetry database.

gitimate, the processor chooses the next state from the available subsystem processing modules for the specific subsystems. In each module, the software examines the specific command fields to ensure that they correspond to legitimate options for the subsystem. If the software determines that the command is valid, then it executes the command and returns a success code to the main processing loop. If the software determines that the command is invalid, then the software returns an error code to the main processing loop that indicates the error type.

3. If the software detects an invalid command, then it passes control to a special error processing state for returning the error code to the main processing program.

4. After the command is processed, the state machine quits and the software returns control to the main processing loop.

4.5.3 Payload Telemetry Processing

Earlier in this chapter, we looked at the telemetry database in the user interface. To construct the database, the database designer uses the various data classes present in the payload. Figure 4.20 shows the relationship between the data classes found in the payload:

Payload Science Instrument Measurements These are the main informational measurements that are collected based on the payload's "mission." The instrument actually produces two data types:

1. Science data describing the specific knowledge measurements.
2. Instrument housekeeping data for describing the operating condition of the instrument itself.

(Note: depending upon the local usage, the terms "housekeeping" and "health and welfare" data may be used interchangeably.)

Payload Operational Measurements These data describe how well the overall payload is operating through measurements such as battery voltages, currents, accelerations, etc. There are two classes of operational data:

1. *Health and welfare* data which are the basic operational measurements the payload sends on a routine basis.
2. *Critical values* that are a subset of the health and welfare measurements and are especially important for debugging failure modes and they may be sent routinely, during special diagnostic events, or major configuration changes.

Payload System Environment Measurements These data describe the external environment through measurements of variables such as temperature, pressure, wind speed, etc. The environmental variables may affect operational considerations or measurement data processing so payload monitors them even if they are not part of the payload's "science" measurement list.

4.5.3.1 Payload Data Master Equipment List

Payload hardware designers describe the characteristics of the components in the payload with a *Master Equipment List (MEL).* The MEL includes items such as mass, volume, power requirements, etc. Along with the hardware MEL, the designer may also develop a data MEL that lists all of the input and output data for each device. Table 4.6 shows an example section of a data MEL where we arrange the sensors by subsystem and component within the subsystem. The port indicates "Input" for command data and "Output" for telemetry data. The "Digital" signal type is for numbers and "PCM" is for analog data that has been converted to a digital representation. We also give the number of samples per second for telemetry values and "Asynch" for values arriving asynchronously in

Table 4.6: Example Data MEL for Payload Subsystems

Sub-system	Sensor	Port	Signal Type	Samples/ Second	Bits	#
Science	PMT	Input	Digital	Asynch	16	2
	PMT	Output	Digital	1/10	24	2
	Temp	Output	PCM	1/10	16	2
Payload	FC Temp	Output	PCM	1/10	16	1
	Mag. Temp.	Output	PCM	1/10	16	1
	C. Temp.	Output	PCM	1/10	16	2
	Bat. Volt	Output	PCM	1/10	16	1
	Earth Sens.	Output	PCM	1/10	16	8
	Pressure	Output	PCM	1/10	16	1
	Camera	Output	Digital	Asynch	$\leq$500 kB	1

response to a command. The table lists the number of bits for each sample or command word along with the number of sensors of that particular type. The designer uses the complete table to size the database, and the data processing and distribution requirements in the payload. Additionally, the designer needs this data to size the telecommunications links and interface bus options.

4.5.4 Balloon Payload Computing System Example

Figure 4.18 provides the overall processing state diagram for the balloon payload example. The software designer for this experiment further divided several of the states in the overall state diagram into their own state diagram [Hora08]. Figure 4.19 illustrates the states for the command processing within the overall payload state machine.

Table 4.3 illustrates a command macro file used to initialize the example balloon payload. The payload executes this macro as soon as it starts the command processing software. The entry sequence is execution time, command value, and entry comment. The time is absolute with respect to the payload clock. Since this is an initialization file, the time is earlier than the payload power-up time so that the payload executes the commands immediately. The command sequence in this macro file ensures that all subsystems, except the Flight Computer, are unpowered. The sequence then turns on and initializes the communications subsystem, turns on and initializes the Global Positioning System (GPS) subsystem, and ends with turning off the scheduled command processing capability. A "no-op" command is used between each subsystem command to allow sufficient time for all processing to be completed.

The computer controls industry has developed a standard for computer-based acquisition, control, and processing known as the PC/104 standard. The concept is to use a standard PC-AT bus in a modular design, as Figure 4.21 shows. Each module performs a standard function such as analog-to-digital conversion or serial communications. Only the modules needed in the system are in the "stack." Figure 4.22 illustrates two methods of reading data in the flight computer module. The first panel shows using a LabVIEW® system call to read ten sensors sequentially from the ADC connected to the computer via the USB port. The second panel shows the computer reading a photomultiplier tube based on a command. The tube has internal electronics to convert the tube output into a 16-bit data value. In both reads, the software converts the digital values to text representations to permit easy transmission through the text-based radio communications network. Table 4.6 gives the data MEL for the balloon payload.

4.6 SECURE COMMUNICATIONS

System designers have many reasons to protect their telecommand and telemetry data from third parties. For example, in a medical telemetry system, patient confidentiality is a requirement so there must be some method for hiding that information from the public. One way is with a dedicated, protected transmission medium. This is generally not feasible for mobile entities in the system so the designer must use some shared access medium. In the modern networking age, system designers utilizing these shared media always have security as an issue in system design. Security will be a much larger challenge in the future as threats become more sophisticated. Technologies like Cloud-based data storage and software distribution open up new security boundaries.

In modern open communications, system designers use cryptographic algorithms implemented in hardware or software and some form of user authentication to achieve the desired information security. These processes have three stages: authentication that the users on each end are appropriate and valid for this transmission, encryption to hide the information content and make it look like random data, and the complementary decryption to recover the data. The encryption/decryption process requires a key to make the algorithms work. If someone compromises or guesses

(a) Flight computer board with processor and memory

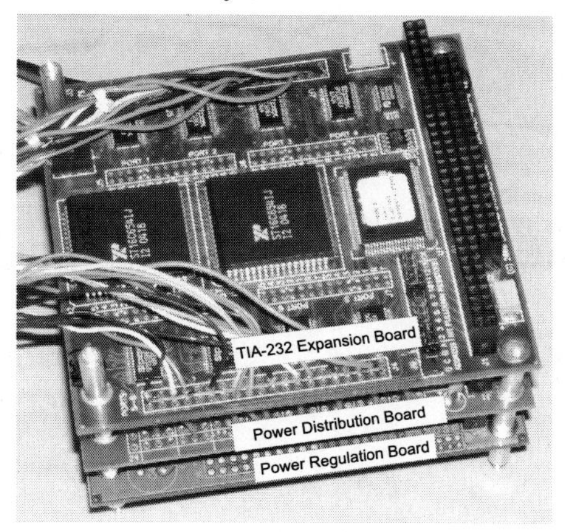

(b) Computer expansion boards

(c) Integrated balloon payload

Figure 4.21: The PC/104-based data acquisition system used in the balloon payload example. The assembled stack undergoing testing prior to full integration with the payload structure.

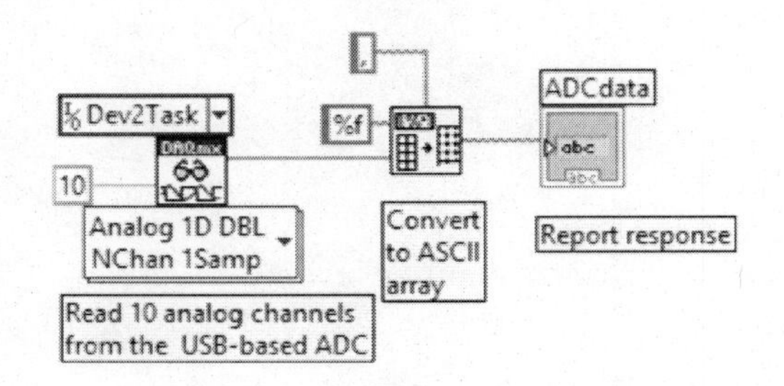

(a) Reading from the USB port

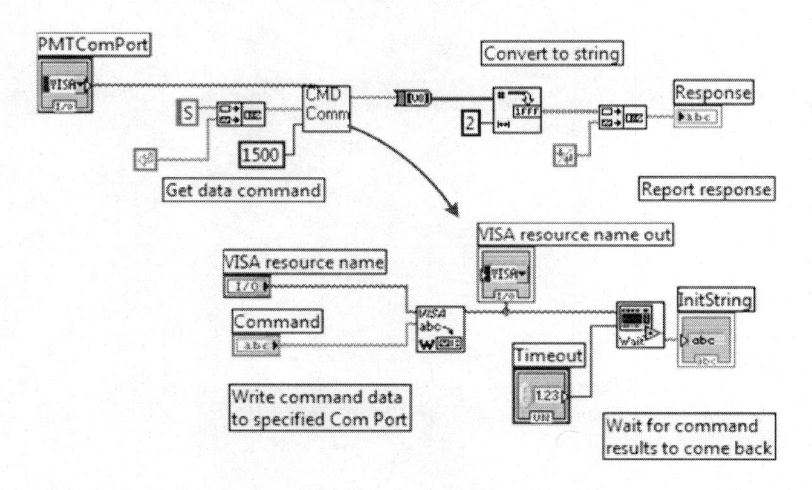

(b) Reading from the serial port. The "CMD COM" module is expanded to show reading the serial port.

Figure 4.22: Example data reading VIs for the balloon payload experiment.

the keys, then third parties can recover the data. If the keys are not secure, then third parties can even spoof the system and generate false commands or data values.

In this section we look at basic security needs, Cloud computing, key management, error effects, and hardware and software tools. The interested reader is encouraged to consult the extensive current literature on these topics. The reference [CCSDS3501] provides an overview of security threats to space missions that illustrates cases relevant to terrestrial applications as well.

4.6.1 Operating Modes

System designers can use secure communications techniques on both the telecommand and telemetry links. There is usually no requirement that both links be simultaneously secure at the same time. Like many other system states, the system designers usually allow the operators to switch between secure and open modes based upon operating circumstances. The security discipline may follow these states:

Mode — when the system is first initialized, both the telecommand and the telemetry links use the clear (nonencrypted) mode.

Secure Telecommand — the telecommand link is fully encrypted based on the system specification or operator commanding; this mode is to keep unauthorized users from accessing the command channel and interfering with operations or having access to command operational information.

Revert to Clear Telecommand Mode — if there is a problem with the telecommand link such as losing link connectivity for a defined time, the system reverts from encrypted mode back to clear mode until the link is properly reestablished.

Secure Telemetry Mode — the telemetry link is fully encrypted based on the system specification or operator commanding; this mode is to keep unauthorized users from discovering the telemetry data or interfering with its contents.

Revert to Clear Telemetry Mode — operators may command the payload into this mode if there are data reception problems or the payload may initiate this change itself if the telecommand link has reverted to clear mode.

Full Encryption — telecommand or telemetry links are encrypted, as dictated by the system operations rules.

The system designers have operational protocols to define the transition gates into and out of each mode. Certain payloads have "safe mode" operations where the payload automatically changes the telecommand and telemetry links to clear mode if the system enters certain defined fault states and the payload enters a self-recovery mode.

4.6.2 Cloud Computing

A logical extension of bulk data storage, the ubiquitous Internet, and the desire for access "anytime, anywhere" is the evolution of Cloud-based storage and user applications over the Internet. Cloud-based storage provides a means for automatic data backup to provide greater reliability in the enterprise. Clouds may be local/private, public, or hybrid configurations. Additionally, Cloud-based solutions allow a "many-to-many" data flow architecture where many data sources place their data products into the Cloud over communications networks for the many users of that data anywhere there is a connection. The movement of data storage from a specific disk storage location to a virtualized location also carries the risks of unauthorized access and modification, loss of data provenance, and premature distribution of data to non-project personnel and computing systems. Despite these risks, government and private sector entities appear to be moving rapidly to embrace Cloud-based solutions.

Figure 4.23 shows participants in a Cloud-computing environment. In this model, the data originator does not send individual data messages to the receiving users. Rather, the originator stores the data in the Cloud and the users subscribe to the Cloud storage to receive the data. The system security protocols must provide protection to the systems, the data produced, and the offered services. Security measures become especially difficult when telecommand and telemetry data flows over open networks like the Internet and users demand access to their data "anytime, anywhere" using laptops, smart phones, and tablet-based computing. In Figure 4.23, users may access the data asynchronously and may not utilize proper firewalls on their network communications. System designers frequently place security concerns over other concerns such as cost when dealing with sensitive data such as personal health data or controlling a major facility such as a power plant. As we move to the Internet of Things, users will have similar security concerns not only with the data

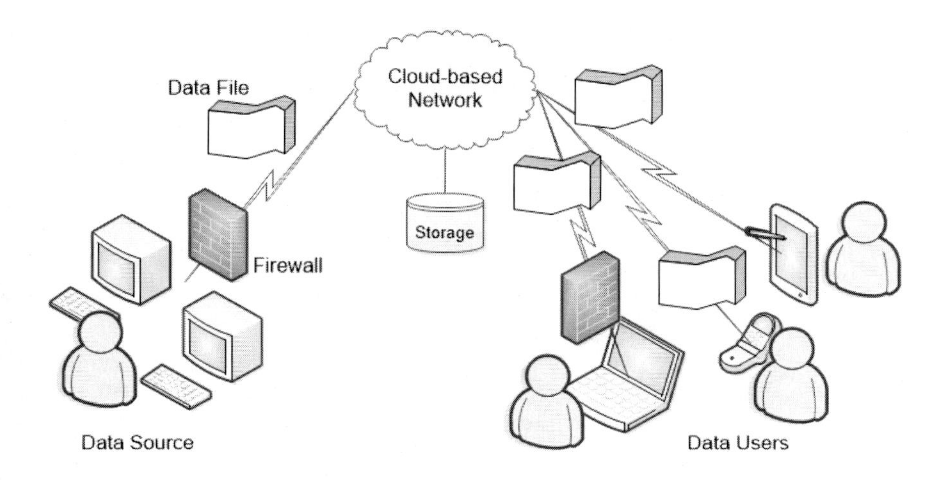

Figure 4.23: Devices in a Cloud-computing environment. The communications links are open, storage is assumed to be secure, and client users may or may not be trusted users.

from the source but also with the value-added services that the source may provide.

Data Provenance Associated with these basic security issues, system designers face the accompanying issue of data provenance for insuring that the telecommand and telemetry data comes from a valid source and has not been compromised either in transit or in storage. Data provenance is the principle of identifying and ensuring the ownership and sourcing of the data is as the customer expects or as advertised by the provider. Data users consider it essential that the integrity of the data is unquestioned as data is collected, shared, analyzed, refined, and updated (and the users repeat this cycle). This is becoming an increasingly important issue for designers using Cloud-based data delivery. As system designers utilize a variety of public and private data delivery methods or hosting, they require that the data's history be traceable as users share the data, process it, and convert it to production data products.

Data provenance provisions may require systems designers to append additional metadata with the raw data to capture the data's history. This process adds to the required overall transmission bandwidth. The system's telecommand and telemetry database designers must account explicitly for the metadata to assure provenance in the database structure.

4.6.3 Key Management

The encryption/decryption algorithm needs well-protected keys for the algorithm to be effective. System operators managing the trusted key distribution to the payload and control center utilize at least one of the following approaches:

- Using a single key for the mission's lifetime
- Pre-storing a key library and switching between keys on a basis determined by the mission protocol
- Computing new keys based on an algorithm using a well-defined variable such as time to make a new key
- Explicitly sending a key update via a secure communications channel.

The single key has the advantage of simplicity although it leads to problems if an unauthorized user discovers the key. However, this may be appropriate for short missions.

The pre-stored keys also represent a simple solution as long as the key dictionary does not become public knowledge. The key change method may be regular, e.g., once each day, or irregular, e.g., based on a specific transmission. In the first case, users should not repeat the keys. In the second case, the specific transmission may need to be properly encrypted to keep the change hidden.

Computing new keys assumes that the key value is well known and synchronized on both ends of the communications link. We assume the algorithm for computing the new key is public so we must have an additional secure algorithm to randomize appropriately the key for usage.

We predicate sending a new key on having a secure channel for key distribution.

System designers need to evaluate each approach and determine the right one for the mission architecture. Additional information on key management is available in [CCSDS3506] and references therein.

4.6.4 Communications Error Effects

Generally, the secure communications link user assumes that the link delivers the encrypted data without transmission errors. The Internet generally delivers the data without corruption so this is not a large concern for networked communications or any similar closed-link system. However,

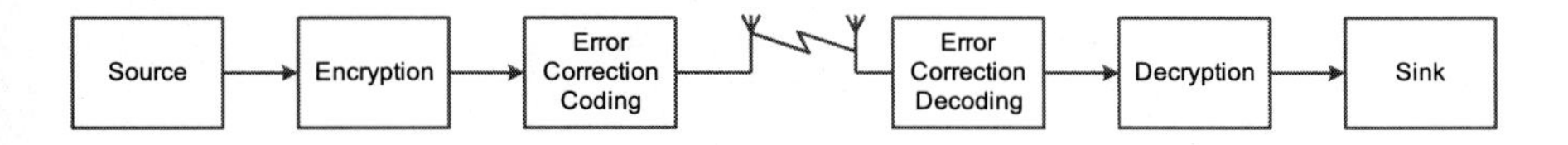

Figure 4.24: Data flow in an encrypted link showing error correction to protect the encrypted data from channel transmission errors.

this is not the case where encrypted data travels between two endpoints over a radio link. Wireless communications without error correcting coding have a much lower reliability than wired links. Therefore, wireless system designers use a data flow method like the one in Figure 4.24. Here, the error correcting code protects the encrypted data from transmission errors. Reversing the order of the encryption and error correction blocks does not protect the data.

In the system in Figure 4.24, the decryption algorithms need this error protection to deliver data to the end users. Modern encryption/decryption algorithms frequently use a multipass computation process based on a fixed-length data block. If the channel delivers even one bit error to the decryptor, the decryptor essentially scrambles the data block in the multipass decryption process giving a 50% error rate for the block.

4.6.5 Secure Hardware Systems

Modern hardware encryption devices are frequently developed around specific secure software algorithms such as the Advanced Encryption Standard (AES) (see below). Designers use these devices for message traffic encryption, disk encryption, and flash drive encryption. System designers may choose to use this network "appliance" to provide secure communications to devices on the network. As Figure 4.25 illustrates, the designer may attach a hardware encryption and/or decryption device to the host computer as an external device. A TIA-422 data interface transmits the clear and encrypted data as well as keys and control and status information. The encryption device has the AES, or comparable, algorithm in the device hardware. Some devices provide one-way encryption or decryption, while others provide both services in a single unit.

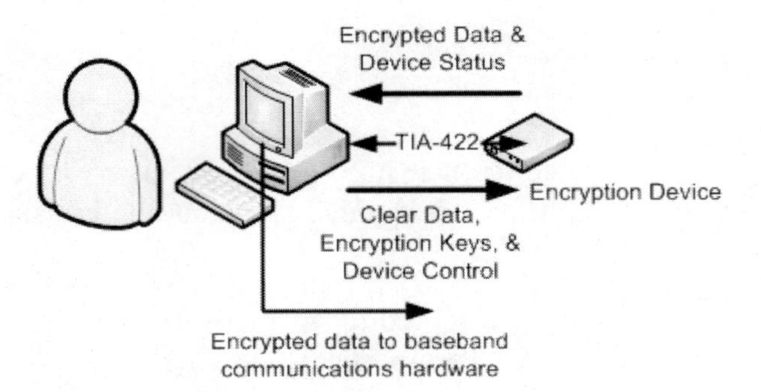

Figure 4.25: Example of an encryption device added to a computer to manage encrypted data transmissions.

4.6.6 Secure Software Systems

Systems designers may consider several software solutions for data encryption to protect data.

Advanced Encryption Standard The Advanced Encryption Standard (AES) [FIPS197] specified by the National Institute of Standards and Technology is a common algorithm utilized for data encryption. For example, the Consultative Committee for Space Data Systems recommends this algorithm with a 128-bit minimum key length and a reliable means of user authentication to provide space mission data security [CCSDS3520].

Virtual Private Networks The Virtual Private Network (VPN) is one common way to securely send sensitive information across public communications networks. The VPN creates a private end-to-end session that encrypts the data flowing between the ends. The VPN still requires a means of user authentication to establish the link between endpoints. Appropriate user login processes perform this function [CCSDS3504].

4.7 REFERENCES

[CCSDS3501] *Security Threats against Space Missions*. CCSDS 350.1-G-2. Consultative Committee for Space Data Systems. Washington, D.C., Dec. 2015. URL: `https://public.ccsds.org/Pubs/350x1g2.pdf`.

[CCSDS3504] *CCSDS Guide for Secure System Interconnection.* CCSDS 350.4-G-1. Consultative Committee for Space Data Systems. Washington, D.C., Nov. 2007. URL: https://public.ccsds.org/Pubs/350x4g1.pdf.

[CCSDS3506] *Space Missions Key Management.* CCSDS 350.6-G-1. Consultative Committee for Space Data Systems. Washington, D.C., Nov. 2011. URL: https://public.ccsds.org/Pubs/350x6g1.pdf.

[CCSDS3520] *CCSDS Cryptographic Algorithms.* CCSDS 352.0-B-1. Consultative Committee for Space Data Systems. Washington, D.C., Nov. 2012. URL: https://public.ccsds.org/Pubs/352x0b1.pdf.

[Cutl83] D. N. Cutler, Jr. R. H. Eckhouse, and M. R. Pellegrino. "The nucleus of a real-time operating system: a tutorial on the internals of RSX-11M." In: *RSX Support Papers.* Bedford, MA: Digital Equipment Corporation, 1983.

[EOSD10] National Aeronautics and Space Administration. *Data Processing Levels for EOSDIS Data Products.* 2010. URL: http://science.nasa.gov/earth-science/earth-science-data/data-processing-levels-for-eosdis-data-products/.

[FIPS197] *Advanced Encryption Standard (AES).* Federal Information Processing Standards Special Publication 197. National Institute of Standards and Technology. Nov. 2001. URL: http://www.nist.gov/manuscript-publication-search.cfm?pub_id=901427.

[Furh90] B. Furht et al. "Open systems for time-critical applications in telemetry." In: *Proceedings of the International Telemetry Conference.* Vol. XXVI. 1990, pp. 399 –419.

[Hora08] S. Horan. "Using LabVIEW to Design a Payload Control System." In: *Proc. International Telemetering Conf.* San Diego, CA, Oct. 2008, p. 84.02.02.

[Horo89] P. Horowitz and W. Hill. *The Art of Electronics.* 2nd Edition. New York, NY: Cambridge University Press, 1989.

[I2C14] *I2C-bus specification and user manual.* UM10204. Version Rev. 6. NXP Semiconductors N.V., Apr. 2014. URL: http://www.nxp.com/documents/user_manual/UM10204.pdf.

[IEEE100] Institute of Electrical and Electronics Engineers. *IEEE 100 The Authoritative Dictionary of IEEE Standards Terms Seventh Edition.* IEEE-STD-100. IEEE Press. Piscataway, NJ, 2000. DOI: 10.1109/IEEESTD.2000.322230.

[MS1553] *Aircraft Internal Time Division Command/Response Multiplex Data Bus. MIL-STD-1553B.* Washington, D.C.: Department of Defense, Sept. 1978.

[RS422] *TIA/EIA-422-B Electrical Characteristics of Balanced Voltage Digital Interface Circuits.* Telecommunications Industry Association. Arlington, VA, 1994.

[Shne87] B Shneiderman. *Designing the User Interface: Strategies for Effective Human-Computer Interaction.* Reading, MA: Addison-Wesley, 1987. ISBN: 0-201-16505-8.

[SPI04] *SPI Block Guide.* S12SPIV4/D. Version 04.01. Freescale Semiconductor, Inc., 2004. 40 pp. URL: `http://www.nxp.com/files/microcontrollers/doc/ref_manual/S12SPIV4.pdf`.

[USB00] *Universal Serial Bus Specification.* 2.0. Apr. 2000. URL: `http://www.usb.org/developers/usbtypec/`.

4.8 PROBLEMS

1. Consider a command packaged as *start_delimeter : command_name : command_function : command_data : stop_delimeter*. Here, the colon indicates a division between fields and is not part of the command. Design a state diagram to parse a received command in this format. The parsing needs to check for no transmission errors, correct field order, and allowable command names and functions. Indicate if the command appears correct or if you detect an error.

2. Write a computer program in a high-level language that places a data value on the screen at a fixed position with a four-character label of TEST preceding the value and a one-character label "V" following the value. If the data values as a function of time, $v(t)$, are generated by the equation

$$v(t) = 2\sin(t/2)$$

   Design the software so that the displayed color of the data values becomes green if the data are between -1 and 1; yellow if the data are between -2 and -1; and red if the data are between 1 and 2. Update the time value once per second and show how the display changes over many seconds.

3. Using the same function as in Problem 2, construct a running graph where the graph shows current time at the right-hand edge and the past 25 samples displayed consecutively to the left. As time

increments, the displayed value shifts one position to the left and the next "current value" appears at the right-hand edge. Add both x and y axes and label them. Indicate the normal, caution, and alarm levels on the graph.

4. Write a computer program in a high-level language that (a) parses an input string of a keyword followed by four parameters and (b) parses an input string of a keyword followed by either three or four parameters depending upon whether the first parameter is a y or an n.

5. Given the set of (x,y) pairs for position and associated measurement values: (0.00, 63), (1.25, 75), (2.00, 94), (2.9, 128), (4.10, 192), (5.00, 255). Convert the pairs to equally spaced pairs between 0.00 and 5.00 and round the values down to the nearest integer. Can you recover the original data from the uniformly sampled data values?

6. Develop a state diagram having between four and ten states for a familiar process. Include the necessary information to leave the start state. Identify the function for each state and the required information to transition to the next state. Include at least one fault detection input and one associated fault recovery state.

7. Develop a justification to support the common contention that the payload computer system should not perform data processing above Level 0; rather, have the user's computer perform all of the processing. Next, argue for why the payload computer system should form additional processing. Can you develop a strategy for deciding the trade-off on your arguments?

8. Suppose you are designing a medical telemetry system for use in an intensive care unit. What telemetry values would you include in your overall system? Generate a table that specifies (a) the telemetry measurement, (b) whether the measurement is a "science," "payload," or "environment" measurement, (c) whether the measurement is a regular measurement or a critical item measurement, and (d) a justification for the categorization.

9. Commands are normally echoed back to the sender as part of the telemetry data. Explain why one does this from a transmission quality point of view. Suppose there were virtually no transmission

errors on the link. Do you still insist on echoing the commands in telemetry?

10. The link between the host computer and the encryption device in Figure 4.25 is shown as a dedicated wire connection. Would a Wi-Fi or Bluetooth connection be appropriate? Justify your response.

CHAPTER 5

SIGNAL PROCESSING

5.1 INTRODUCTION

The field of signal processing as applied to telemetry systems is diverse. In this chapter, we will look at some of the limitations we need to place on signals and how we apply signal processing techniques to make a Pulse Code Modulation (PCM) telemetry system. The designer's ultimate goal is to produce the telemetry frame or data packet for the payload system. The payload transmits that data in such a way as to allow the receiving system to recover completely all signals in all of their detail. We will cover telemetry frame and packet formats in Chapter 6. However, to construct the data transmission technique, we need to understand the signal processing presented in this chapter.

As we saw in Section 1.1.3, a PCM system takes the bandlimited signal coming from the sensor package then samples and converts it into a digital signal prior to transmission. After reception, the system reconstructs and filters the sampled signal to replicate the input signal. In analog systems, the signal does not have this process performed; rather, the sensor output modulates the transmitter. The reason for going to a digital PCM system is because of the value-added processing that designers and users apply to the signal and the means for acquiring the data. The disadvantage is the required extra communications bandwidth relative to the analog signal.

In the PCM process, the data acquisition portion of the payload segment highlighted in Figure 5.1 generates the data and the highlighted user segment processes the data to prepare them for the operator's display. In this chapter, we discuss the necessary processing to support those functions. This chapter covers the following signal processing topics:

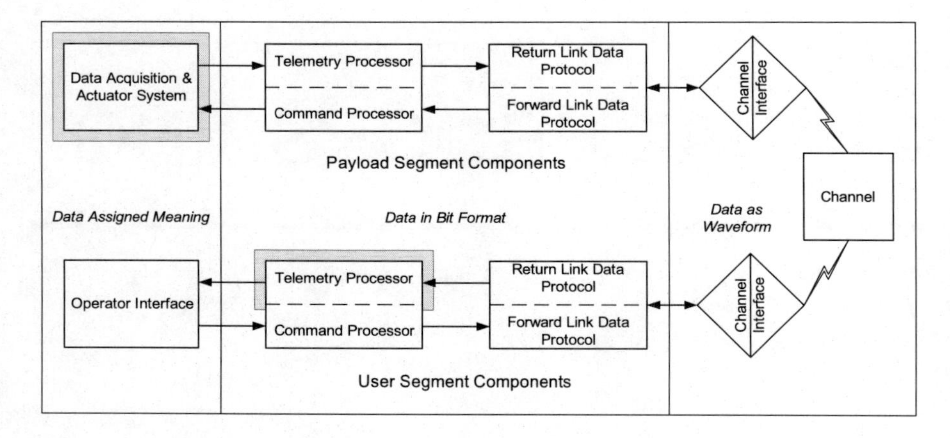

Figure 5.1: The highlighted blocks indicate the location of signal processing components in the end-to-end telemetry and telecommand system.

- Signal types
- Signal bandlimiting
- The sampling theorem
- Quantization effects
- Sampling hardware
- Signal reconstruction

These are the initial signal processing functions for PCM measurement systems.

5.2 OBJECTIVES

Our objectives for this chapter are to introduce the fundamental concepts of telemetry data acquisition. At the conclusion of this chapter, the reader will be able to

- Determine the effective bandwidth of sampled signals to allow for reasonable signal reconstruction.
- Determine the required sampling rate for signals based upon their effective bandwidths.

- Design a Low Pass Filter (LPF), High Pass Filter (HPF), and Band Pass Filter (BPF) for specified frequencies and attenuation levels.
- Determine the necessary number of quantizing levels to achieve a desired signal-to-noise ratio and the resulting output data rate.
- Describe the operating principals for the various types of sample-and-hold and analog-to-digital conversion circuits.
- Determine the total data volume to be transmitted for a measurement system.

With these skills, the designer can sample the signals and start taking the transmission process into consideration.

5.3 TRANSMITTING SAMPLED VERSUS CONTINUOUS DATA

Before discussing the details of sampled data, we first review the general signal classifications. The first group includes analog and PCM signals. We also look at time division signal multiplexing. For more information on signals, consult standard texts such as [Couc01; Skla01].

5.3.1 Continuous Analog Transmission

Most of the sensors considered in Chapter 2 produce a continuous analog signal. The signal's amplitude at any time is proportional to the measurand's strength as determined by the sensor. These signals are continuous both in time and in amplitude; however, the amplitude may be limited in magnitude due to technology limitations. Analog signals are also time-limited over the duration of some period of interest defined by the needs of the data system or user. The first three panels of Figure 5.2 illustrate examples of continuous analog signals generated via a LabVIEW® simulation. The figure labels these signals as *Signal 1* through *Signal 3*. We next use these three signals to consider frequency and time multiplexing and to motivate adopting PCM.

5.3.2 Multiplexed Analog Transmission

As we saw in Figure 1.4, one method the system designer may choose for transmitting the sensor information is to dedicate a specific channel, e.g., a specific transmission frequency, to each sensor. Therefore, the

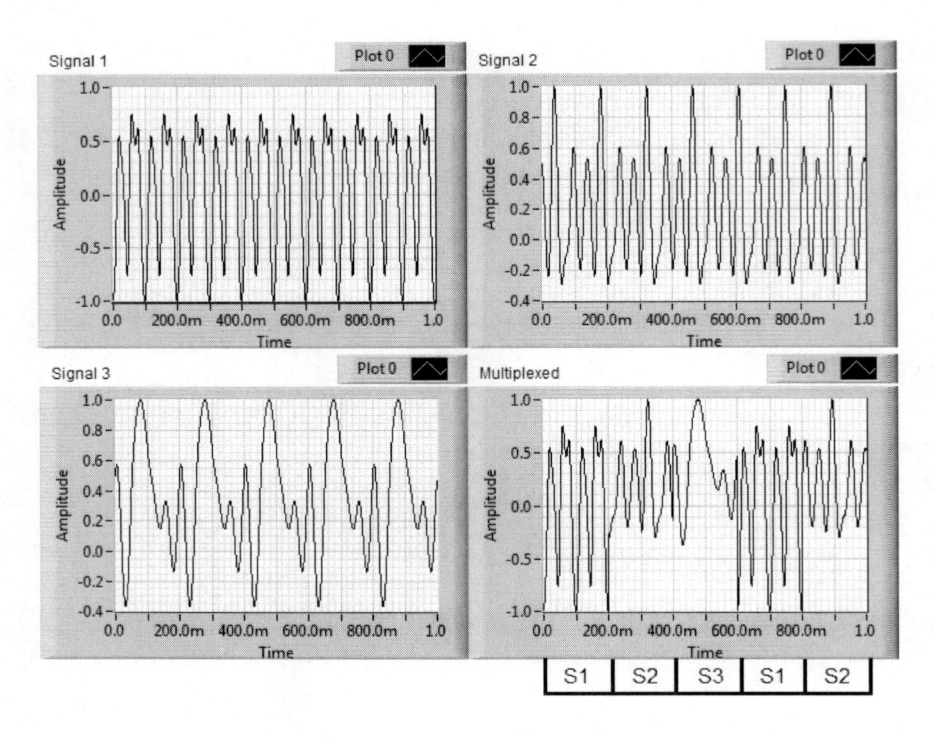

Figure 5.2: Time multiplexing of three signal channels for common transmission across a channel.

designer could take *Signal 1* through *Signal 3* in Figure 5.2 and use them to drive three analog modulators. This makes a Frequency Division Multiplexing (FDM) transmission approach. This approach captures all of the data from the sensor. However, the approach does not efficiently use the channel because a large number of sensors implies the need for an equally large number of discrete channel selections.

One way designers improve the transmission efficiency is to sample each analog channel in the system and then transmit the samples giving each sample a short time slice on a periodic basis as we also saw in Figure 1.4. This is a Time Division Multiplexing (TDM) approach to channel sharing. The bottom right panel of Figure 5.2 illustrates time multiplexing the three sensor signals on a common transmission channel. Within the time slice, the analog sensor output is still continuous. It is only in this case that the sensor does not have exclusive access to the channel. Another name for this process of periodically sampling the individual signals is *signal commutation*. The first problem for the designer is to determine the sampling discipline that ensures proper sampling of each sensor and efficient sharing of the transmission channel. Figure 5.2 also illustrates a second problem for the designer: how does the receiver know where each time slice for each sensor begins and ends?

5.3.3 Pulse Code Modulation Transmission

Since we are examining signal sampling, we make the next extension to sample the signal not only in time but also in amplitude. This is the basis behind Pulse Code Modulation (PCM) [Oliv48]. Figure 5.3 shows the PCM sampling process converting the original analog signal to a sampled signal. In this example, the PCM encoding of the signal converts the $-1\,\mathrm{V}$ to $+1\,\mathrm{V}$ signal amplitude to a number between 0 and 63. The mapping is linear.

The system designer has two variables to consider when configuring the PCM sampler: the rate at which the encoder samples the signal and the number of sampling levels the encoder uses to represent the signal. As Figure 5.3 shows, the sampling levels produce a small amount of jaggedness in the signal, which is one of the disadvantages to PCM. The mitigation for this effect is to use a larger number of conversion levels. We will look at both of these design variables.

Figure 5.4 illustrates the end-to-end PCM process for the data system. PCM signals start in the sensors as analog signals. The analog signals are processed through filters and an Analog-to-Digital Converter (ADC) to produce the digital result that Figure 5.3 illustrates. The number of bits used to encode the signal, N, depends upon the required quality of the conversion. The digital signal's amplitude ranges from 0 to $2^N - 1$. The source transmits the digital data stream over the channel where the receiver inverts the process to recover the best estimate for the original signal. First a Digital-to-Analog Converter (DAC) converts the digital signal to a full-scale analog signal that will still show the jaggedness of Figure 5.3. The final low-pass filter smooths that signal for presentation. As we proceed through this chapter, we examine how the designer selects the sampling rates, sampling resolution, and filtering processes to provide correctly sampled data.

This sampling does not remove the problem of knowing which samples belong to which sensor. That is resolved in Chapter 6 when we discuss how to package the samples for transmission.

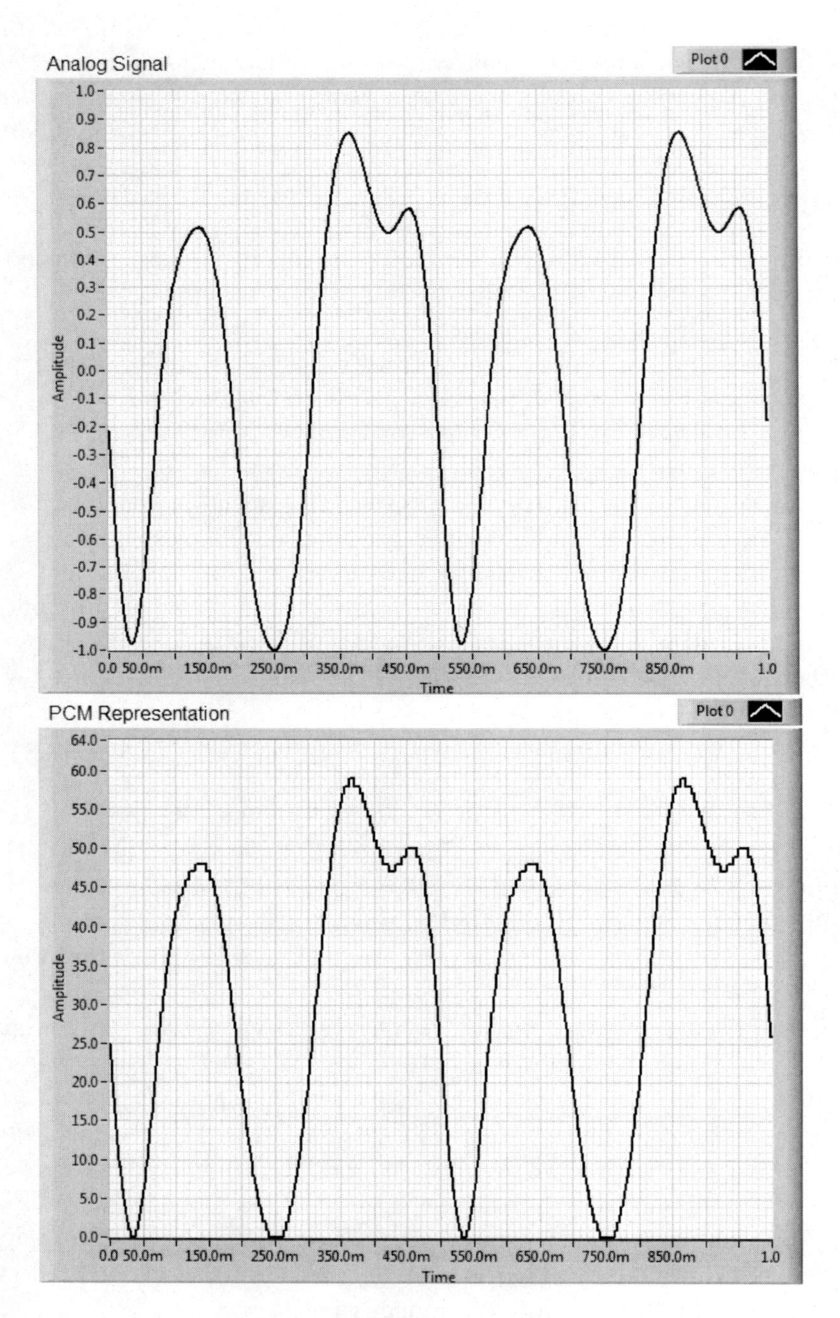

Figure 5.3: Analog signal and PCM representation of the signal. Note the jaggedness in the sampled version signal amplitude.

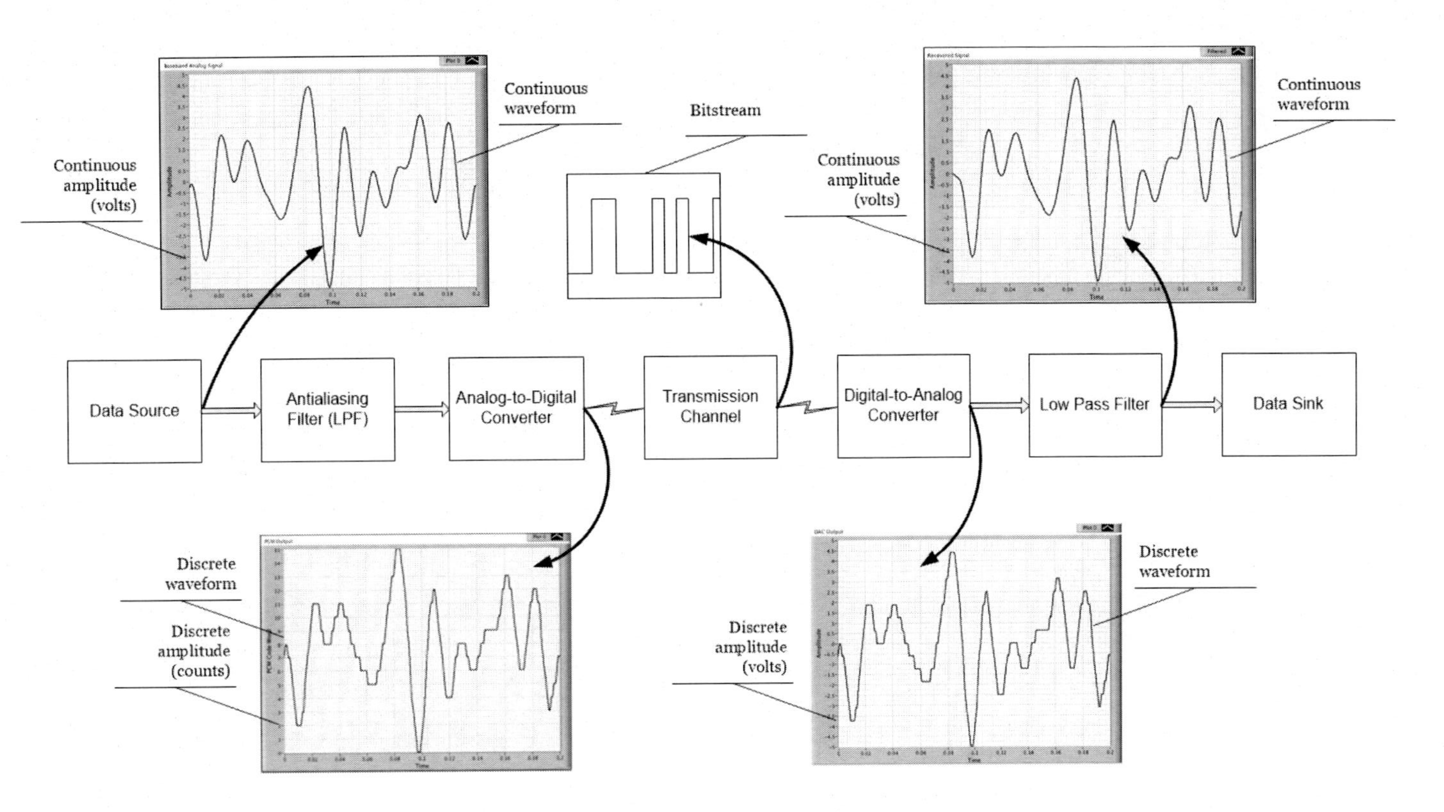

Figure 5.4: End-to-end PCM sampling and reconstruction process.

5.4 SIGNAL TYPES

The total PCM telemetry system uses several classes of digital signals in the transmitted data. The bulk of the transmitted data are the digitally converted analog signals. In this section, we classify the various data classes used in the telemetry data.

5.4.1 Pulse Code Modulation Signals

As we have seen, the PCM digital signals are the converted analog values. The data analyst transforms these signals back to their analog representation upon reception.

5.4.2 Digital Signals

System designers differentiate several types of digital signals that differ from the PCM signals. These signals start out as, and remain digital signals. We consider two types of digital signals: bi-level signals and discrete signals. Generally, these signals require no further processing other than, perhaps, averaging. They do not need filtering or special reconstruction.

5.4.2.1 Bi-level Signals

A bi-level signal takes on one of two values: 0 or 1. The top half of Figure 5.5 illustrates the point where a random bi-level data stream is generated using LabVIEW®. The bi-level value may also be designated as ON or OFF or logical high and low. Designers typically use the bi-level datum to indicate a switch setting or a payload state. Generally, only a single bit encodes the state of the variable.

5.4.2.2 Discrete Signals

The lower section of Figure 5.5 shows a discrete digital signal encoding a naturally occurring digital value in a LabVIEW® simulation of a discrete data stream. Designers use discrete variables to designate event counts, digital sensor output, e.g., elapsed seconds, and other digital data not associated with sampled analog sensors.

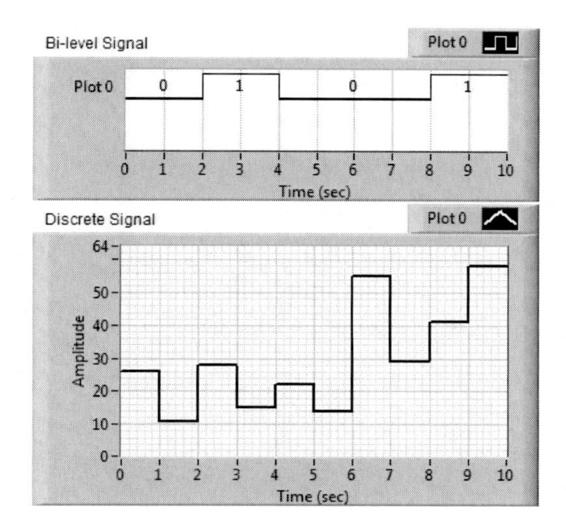

Figure 5.5: Examples of bi-level and discrete digital data.

5.5 BANDLIMITING

As we noted in Section 1.1.3, the PCM signal must be *bandlimited* for proper sampling and reconstruction. The question becomes how we determine the bandwidth for practical systems and how to sample properly to preserve the signal. In this section, we look at the necessary definitions in the frequency domain to provide the basis for proper sampling. We use the Fourier transform to define the frequency domain representation for the signals we sample.

5.5.1 Fourier Transforms

We use the *Fourier transform* to distinguish the frequency domain representation of a signal from its usual time domain representation. The two representations contain all of the signal information. The Fourier transform represents signals in their steady state mode and does not capture initial transients that may occur at start-up. In this section, we cover the definition of the transform and, since it is usually a complex function, the associated magnitude and phase representations.

5.5.1.1 Transform Definition

The integral in Equation (5.1) defines the *forward* (time domain to frequency domain) Fourier transform, $S(\omega)$, of a time domain signal, $s(t)$ [Blac59]:

$$S(\omega) = \int_{-\infty}^{\infty} s(t)\, \mathrm{e}^{-j\omega t}\, \mathrm{d}t \tag{5.1}$$

The frequency domain variable, ω, has the units of radians and relates to the typical frequency variable, f, measured in Hertz, via

$$\omega = 2\pi f \tag{5.2}$$

The integral in Equation (5.3) gives the *inverse* (frequency domain to time domain) Fourier transform:

$$s(t) = \frac{1}{2\pi} \int_{-\infty}^{\infty} S(\omega)\, \mathrm{e}^{j\omega t}\, \mathrm{d}\omega \tag{5.3}$$

Table 5.1 lists commonly used Fourier Transform pairs. In the table, $\mathrm{u}(t)$ is the unit step function where $\mathrm{u}(t) = 0$ for all $t < 0$ and $\mathrm{u}(t) = 1$ for all $t \geq 0$.

In telemetry systems, the measured signals, $s(t)$, are real-valued quantities, while the frequency domain representations, $S(\omega)$, are complex functions having real and imaginary parts. We represent the complex quantities in terms of magnitude and phase.

Mathematical conditions on $s(t)$ are required to ensure that the Fourier transform integrals exist. The signal types that telemetry systems sample generally meet these conditions so we do not worry about them here. Standard references, such as [Kamm00] derive and present various mathematical properties of the Fourier Transform.

5.5.1.2 Magnitude and Phase Spectra

We represent a complex quantity, σ, in two ways. The first way is as real and imaginary parts:

$$\sigma = a + jb \tag{5.4}$$

The second way is as a magnitude and phase:

$$\sigma = |K|\, \mathrm{e}^{j\theta} \tag{5.5}$$

Table 5.1: Representative Fourier Transform Pairs and Properties

Time Domain	Frequency Domain	Notes
$\mathrm{e}^{-at}\,\mathrm{u}(t)$	$\dfrac{1}{a+j\omega}$	$a > 0$
$\mathrm{e}^{-a\lvert t\rvert}$	$\dfrac{2a}{a^2+\omega^2}$	$a > 0$
$\delta(t)$	1	
1	$2\pi\delta(\omega)$	
$\cos(\omega_0 t)$	$\pi\left[\delta(\omega-\omega_0)+\delta(\omega+\omega_0)\right]$	
$\sin(\omega_0 t)$	$j\pi\left[\delta(\omega+\omega_0)-\delta(\omega-\omega_0)\right]$	
$\mathrm{e}^{-at}\sin(\omega_0 t)\,\mathrm{u}(t)$	$\dfrac{\omega_0}{(a+j\omega)^2+\omega_0^2}$	$a > 0$
$\mathrm{e}^{-at}\cos(\omega_0 t)\,\mathrm{u}(t)$	$\dfrac{a+j\omega}{(a+j\omega)^2+\omega_0^2}$	$a > 0$
$\mathrm{rect}\left(\dfrac{t}{\tau}\right)$	$\tau\,\mathrm{sinc}\left(\dfrac{\omega\tau}{2}\right)$	$\mathrm{rect}\left(\dfrac{t}{\tau}\right) = \mathrm{u}\left(t+\dfrac{\tau}{2}\right) - \mathrm{u}\left(t-\dfrac{\tau}{2}\right)$
$s(at)$	$\dfrac{1}{\lvert a\rvert}S\left(\dfrac{\omega}{a}\right)$	a is real
$s(t-t_0)$	$S(\omega)\mathrm{e}^{-j\omega t_0}$	
$s(t)\cos(\omega_0 t)$	$\dfrac{1}{2}\left[S(\omega-\omega_0)+S(\omega-\omega_0)\right]$	
$as(t)+bg(t)$	$aS(\omega)+bG(\omega)$	
$s(t)\otimes g(t)$	$S(\omega)G(\omega)$	$\otimes$ denotes convolution
$s(t)g(t)$	$S(\omega)\otimes G(\omega)$	$\otimes$ denotes convolution

The magnitude and phase of the complex quantity are

$$|K| = \sqrt{a^2 + b^2}$$
$$\theta = \tan^{-1}\frac{b}{a} \tag{5.6}$$

We apply these definitions for complex quantities to the Fourier transform at each frequency point, ω. A plot of $|S(\omega)|$ from Equation (5.1) versus ω is known as the *magnitude spectrum* or the *amplitude spectrum* of the signal. Similarly, a plot of $\theta(\omega)$ versus ω is known as the *phase spectrum* of the signal.

For example, Figure 5.6 illustrates the case where $s(t)$ is a damped sinusoidal signal $s(t) = \mathrm{e}^{-2t}\sin(8\pi t)\,\mathrm{u}(t)$. Table 5.1 indicates that the Fourier Transform of this signal is

$$S(\omega) = \frac{8\pi}{(2 + j\omega)^2 + (8\pi)^2}$$

Figure 5.6 plots the magnitude and phase of $S(\omega)$ as a function of frequency. We see from the figure that the most significant frequencies occur near 8π radians since this is around the frequency of the sinusoidal part of the original signal. We also see that the amplitude spectrum does not cut off at some maximum frequency but trails off to infinity with ever-decreasing amplitude. This behavior is typical for natural signals and the nature of this falloff is important when deciding the signal's bandwidth.

5.5.2 Signal Bandwidth

Engineers do not have a unique, mathematical definition for signal bandwidth. Signal bandwidth has several different definitions based upon the application's needs. As one may suspect, we cannot allocate an infinite-bandwidth channel to transmitting the signal as is required to transmit the entire spectrum of the damped sinusoidal signal in the previous example. To execute proper sampling, we need to determine a reasonable definition for a bandlimited signal.

5.5.2.1 Bandlimited Signals

Most signal processing is performed on signals that are (or may be considered to be) bandlimited. What does this criterion mean for actual signals? To be strictly bandlimited, a signal must meet the following

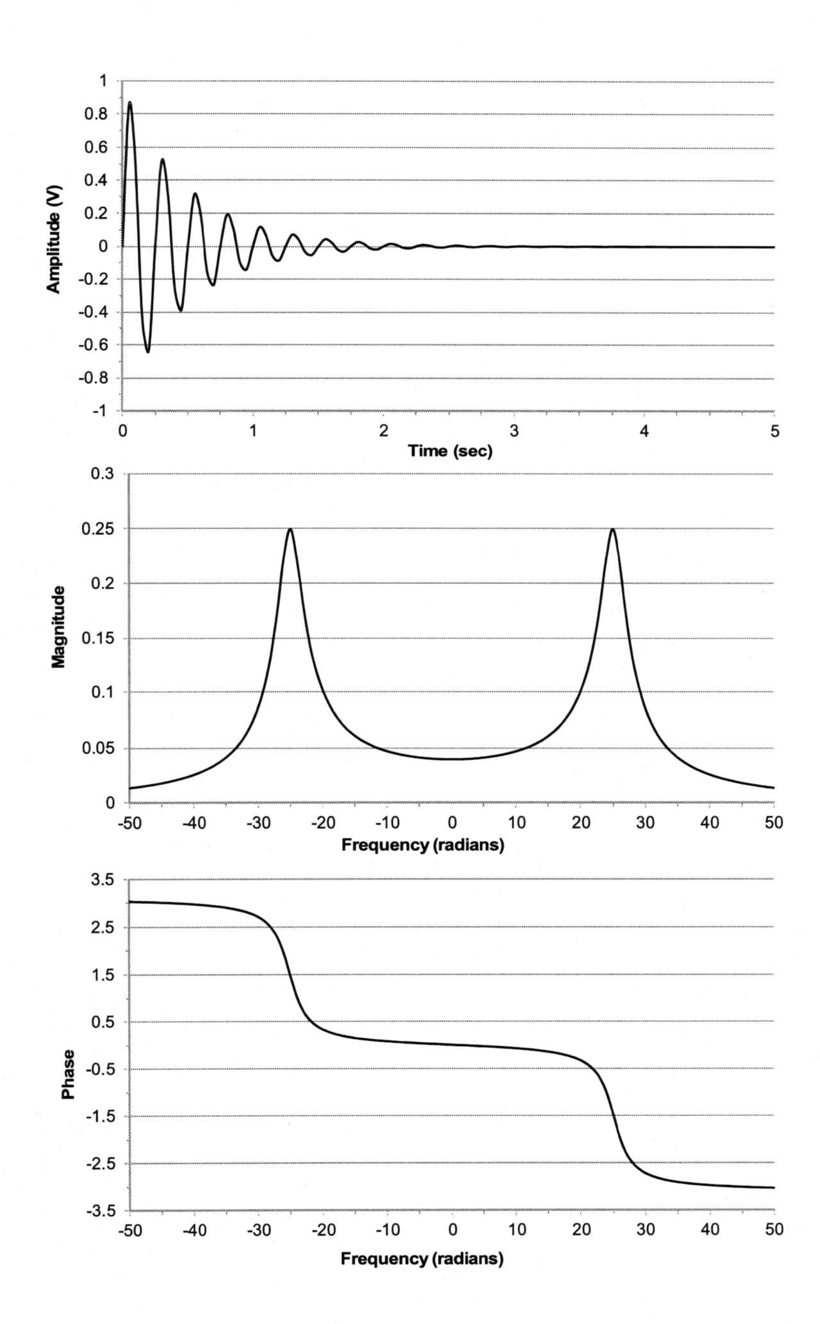

Figure 5.6: A damped sinusoidal signal and its associated magnitude (amplitude) and phase spectrum.

criterion on its amplitude spectrum: given that $s(t)$ and $S(\omega)$ are Fourier transform pairs, then the signal is *bandlimited* if $|S(\omega)| = 0$ for all frequencies ω in the range $\Omega < |\omega| < \infty$. The frequency Ω is termed the band limit.

In the time domain, there is a similar definition, namely a time-limited signal. That is, the signal $s(t)$ is time-limited if $s(t) = 0$ for all time t such that $|t| > T$ where T is the time limit. It does not really matter where $t = 0$ is chosen because all realistic signals can have an arbitrary time shift applied. These definitions have serious implications for telemetry systems because signals that are time-limited cannot be strictly bandlimited. Conversely, signals that are strictly bandlimited cannot be strictly time-limited because, to make that statement, we need to define the signal over all time (and not just the age of the universe - approximately 15 billion years). In telemetry systems (as in all natural experiments), we encounter strictly time-limited signals due to the finite-duration experiment. This means that to reconstruct an arbitrary signal exactly, we need an infinite-bandwidth system — not a good prospect for realistic hardware. Fortunately, we have ways to approximate the actual signals to whatever accuracy we deem necessary by using the concept of an essential bandwidth.

5.5.2.2 Essential Bandwidth Definition

Bandwidth is a frequency domain concept and we are able to use the frequency domain representation of the signal to determine the frequencies that are most important to reconstructing the signal after sampling. Even if the signal's amplitude spectrum does not mathematically go to zero, it generally decreases, and is close enough to zero after a certain point. If we neglect the signal's harmonic content from that point on, it does not greatly affect the signal's representation. We start by computing the signal's energy, E_S, in the frequency domain using

$$E_S = \frac{1}{2\pi} \int_{-\infty}^{\infty} |S(\omega)|^2 \mathrm{d}\omega \tag{5.7}$$

The *essential bandwidth* for a signal is that bandwidth, B, containing 95% of the total signal energy. We find this by solving the following equation for the bandwidth B:

$$0.95 E_S = \frac{1}{2\pi} \int_{-2\pi B}^{2\pi B} |S(\omega)|^2 \mathrm{d}\omega \tag{5.8}$$

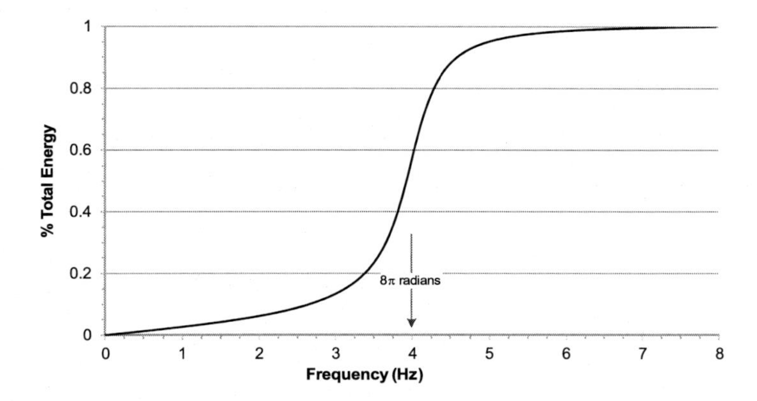

Figure 5.7: The signal's energy as a percentage of the total signal energy as a function of frequency for the damped sinusoid.

The essential bandwidth serves as a practical substitute for an absolute band limit for naturally occurring signals. The 95% metric may seem arbitrary, but it is consistent with regulatory bandwidth definitions where the 95% metric is often used.

For example, consider the damped sinusoidal signal from Figure 5.6. Figure 5.7 plots the signal energy versus bandwidth as a fraction of the total signal energy. The fraction of the total energy increases rapidly with bandwidth and then flattens out after 5 Hz. By making the computations in Equation (5.8), the 95% total energy bandwidth for this signal is 5.2 Hz. We see how well this band limit preserves the original signal by examining Figure 5.8. The figure displays the damped sinusoidal signal along with the signal filtered to include all frequencies up to 5.2 Hz.

Analysts also apply the essential bandwidth concept to signals that do not have an analytic equation describing the Fourier Transform. As an example, we use audio samples of heart sounds with LabVIEW® to make the necessary computations to compute the signal's amplitude spectrum and essential bandwidth. Figure 5.9 illustrates the unfiltered input, the amplitude spectrum, the energy distribution, and the signal filtered at the essential bandwidth. From the analysis, we find that the essential bandwidth for this normal heart signal is 247 Hz.

Next, we use the audio signal for a heart with a pulmonary condition and perform the same analysis with the results presented in Figure 5.10. From this analysis, we find that the essential bandwidth for this heart

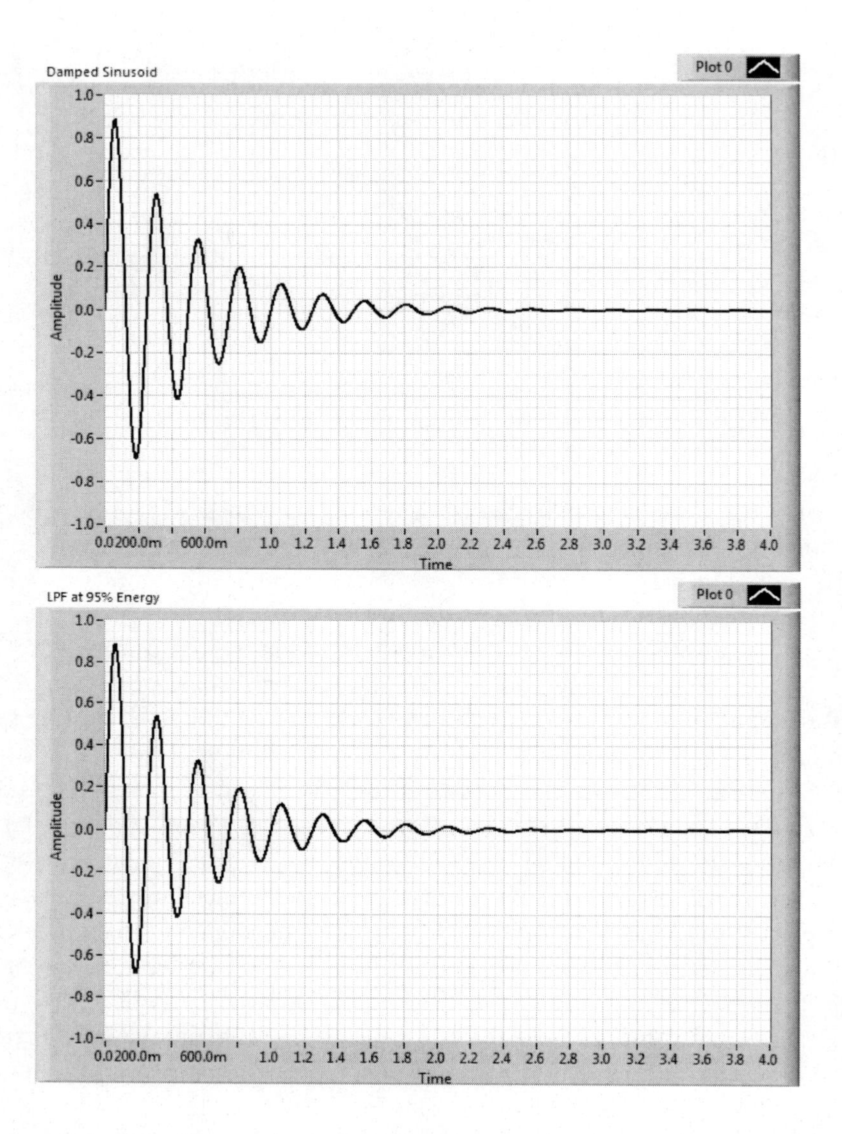

Figure 5.8: The damped sinusoidal signal and the damped sinusoidal signal filtered at the 95% essential bandwidth.

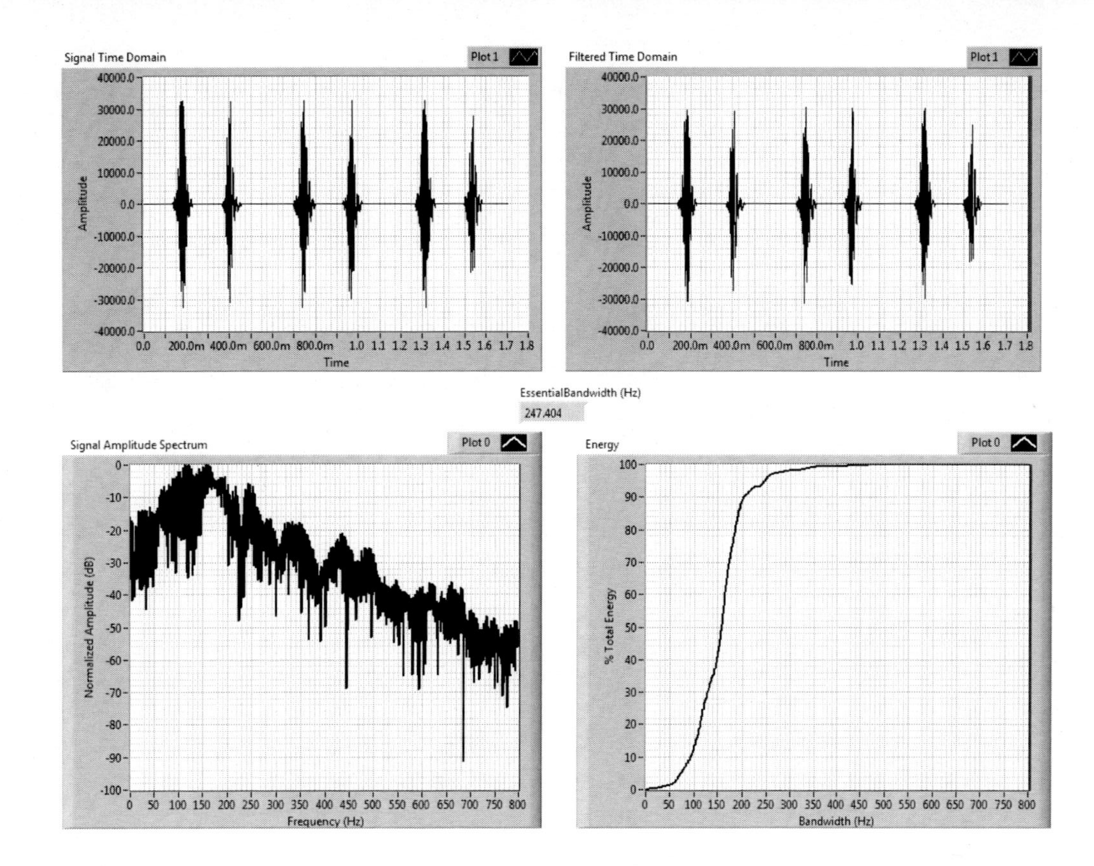

Figure 5.9: Estimating the essential bandwidth for the audio signal from a normal heartbeat. The essential bandwidth is 247 Hz for this example.

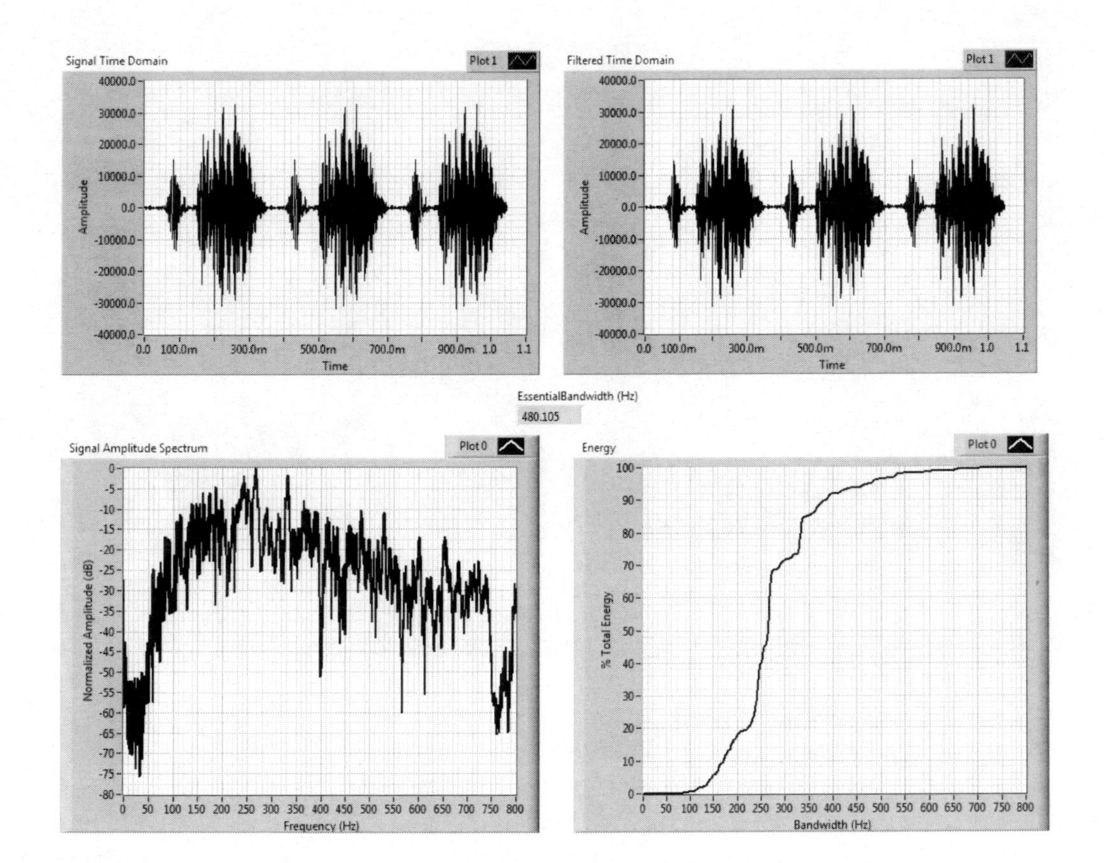

Figure 5.10: Estimating the essential bandwidth for the audio signal from a heart with a pulmonary condition. The essential bandwidth is 480 Hz for this example.

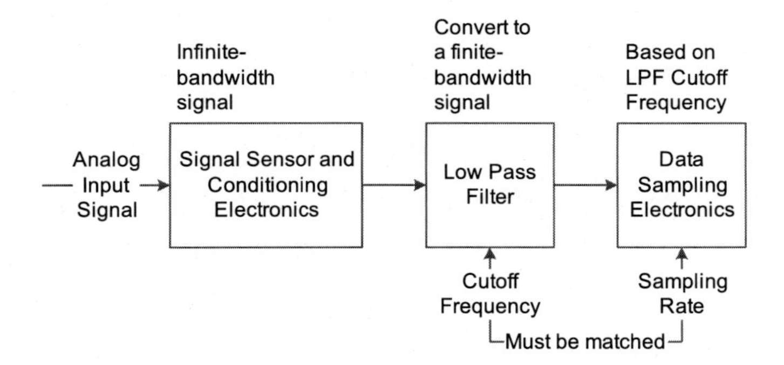

Figure 5.11: Signal sampling architecture with anti-aliasing filtering.

signal is 480 Hz. Notice that if we filtered this signal at the 247-Hz mark, we lose important spectral features. This emphasizes that the analyst must consider the full range of signals when deciding the natural signal's essential bandwidth.

5.5.3 Signal Bandlimiting Architecture

As we saw above, natural signals are never bandlimited in the mathematical sense, but we can define an essential bandwidth for these signals. To construct telemetry systems, we bandlimit the signals from the sensors using a LPF (also called an *anti-aliasing filter*) prior to actually sampling the signal and converting it to a PCM signal in a configuration like that shown in Figure 5.11. This LPF must include the signal's essential frequencies. This gives adequate signal reconstruction from a signal energy point of view [Kest04a]. In the next section, we will see how to sample and capture the essential frequencies properly.

5.6 SAMPLING

As we saw in Section 1.1.3, we must sample the desired analog signal to generate the PCM representation of the signal in preparation for transmitting and processing the signal. The job of the system designer is to determine the sampling rate and the sampling resolution to produce an accurate representation of the original analog signal. The usual model for the sampling process is to sample the analog signal at a uniform rate with a sample duration that is short compared with the sampling rate. Figure 5.12 illustrates the process where part (a) is the analog function

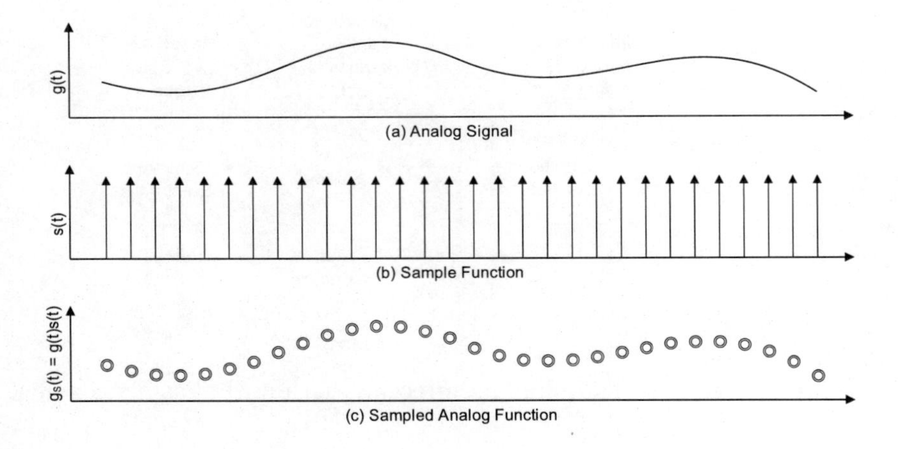

Figure 5.12: Sampling a continuous analog function.

$g(t)$ and part (b) is the sampling function, $s(t)$. The sampling function is based on a pulse of width of τ seconds and has a sample rate of f_S *samples per second* (sps) or a sample period of $T_S = 1/f_S$ seconds.

The sampling function $s(t)$ is a mathematical sequence of equally spaced delta functions convolved with a pulse of width τ seconds, or

$$s(t) = \left\{ \sum_{n=-\infty}^{\infty} \delta\left(t - nT_S\right) \right\} \otimes \text{rect}\left(\frac{t}{\tau}\right) \tag{5.9}$$

The sample times are every T_S s. The sample width, τ, is included because real sampling takes place over a finite time. Designers try to configure the sampling architecture so that the sample width is small compared with the sample period, T_S, so that any variations in the analog signal during the sampling interval are negligible. The sampling operation result produces replicas of the input signal at the sample times; no record is made of the input between the samples. Part (c) in Figure 5.12 illustrates the resulting output from sampling the analog function, $g_S(t)$. The result is the product of the analog sign with the sampling function, or

$$g_S(t) = g(t)s(t) \tag{5.10}$$

As noted, the sampled function really only exists at the sampling time and we should not use lines connecting the points. If the sample rate is high enough, then we retain all of the signal's significant features when we reconstruct the signal waveform. The sampling theorem answers the question "How fast is fast enough?"

5.6.1 Sampling Theorem

With bandlimited signals, we apply the *sampling theorem* to obtain the correct rate to sample the sensor's output. Based on the earlier work of Nyquist [Nyqu24] and Hartley [Hart28], Shannon showed that if we have bandlimited signals, then we may correctly recover those signals if we sample the signal at most every T_S seconds [Shan49]. The limited frequency band, B, of the original analog signal sets this rate by using the inequality

$$T_S < \frac{1}{2B} \tag{5.11}$$

Another way of stating this is that at the minimum sampling rate, f_S, must satisfy the inequality

$$f_S > 2B \tag{5.12}$$

The sampling rate in Equation (5.12) is usually called the *Nyquist sampling rate* and it has units of samples per second. We recover the original baseband signal from the sampled signal if we low-pass filter the samples.

These results are proven by applying Equations (5.9) and (5.10) with the convolution property of Fourier transforms (see Problem 3). Since real signals cannot be strictly bandlimited, we use the essential frequency bandwidth, based upon capturing 95% of the total signal energy in place of this absolute band limit. To achieve the 95% signal energy band limit, we must low-pass filter the sensor output as Figure 5.11 illustrated.

5.6.2 Oversampling the Nyquist Rate

The analyst generally has a difficult time determining whether any transients or interesting signal artifacts are present in a signal sampled at only the Nyquist rate. The sampling theorem is limited in real systems in one important sense: the theorem does not account for additive noise that may be present in the sampled signal. In systems with typical noise levels (Signal-to-Noise Ratio (SNR) on the order of 10), a more realistic criterion might be that the sampling rate should be a f_S in a range of $5B$ to $10B$, depending upon the noise level and experience with the signal conditions.

To illustrate how the density of sample points affects the perception of the signal, consider the damped sinusoidal signal sampled at the Nyquist rate and five times the Nyquist rate as Figure 5.13 illustrates. The essential bandwidth for this signal is 5 Hz, which gives a Nyquist rate of 10 sps. Figure 5.13(a) shows the signal sampled at this rate. Figure 5.13(b) shows the same signal sampled at ten times the essential bandwidth, which is five times the Nyquist rate, or 50 sps. If the analyst did not know that the sampled signal was a damped sinusoid, then the analyst might consider the minimally sampled signal to merely be noisy and not have the structure that we can see in the higher-sampled version.

Later, in Section 5.10.3.4, we will see another type of oversampling with the Sigma-Delta converter where the rate is very much higher than the Nyquist rate. The converter uses signal processing within the converter to keep the total amount of data generated restricted, while still effectively capturing the signal.

5.6.3 Aliasing

What happens if we acquire the samples of the analog signal less frequently than the Nyquist rate demands? The result is a phenomenon known as *aliasing* [Blac59; Kest04a]. Aliasing is the mapping of the high-frequency signal components into lower frequencies. Figure 5.14 shows this effect where we sample a high frequency signal (for example, the highest frequency component in the signal's essential frequency spectrum) at a rate below its Nyquist rate. As we see, there is a lower frequency signal that fits the sample points equally well, as does the intended component. Since we cannot tell which frequency is the correct one from the points alone, the aliasing has effectively converted a high frequency signal component to a low frequency signal component.

We see how this works for the entire signal by inspecting Figure 5.15. In Figure 5.15(a), the signal, $g(t)$, has been sampled properly with a sampling rate exceeding the Nyquist rate for the signal with an essential bandwidth B. Because the sampling process effectively makes multiple copies of the baseband spectrum, $G(f)$, in the frequency domain, these multiple copies are well separated from the baseband spectrum and they are removed by a LPF with a cutoff frequency larger than B Hz and less than $(f_S - B)$ Hz. Figure 5.15(b) shows the effect of sampling below the Nyquist rate on the signal's spectrum. The upper frequency components of the copy of the spectrum centered at f_S are aliased with the baseband

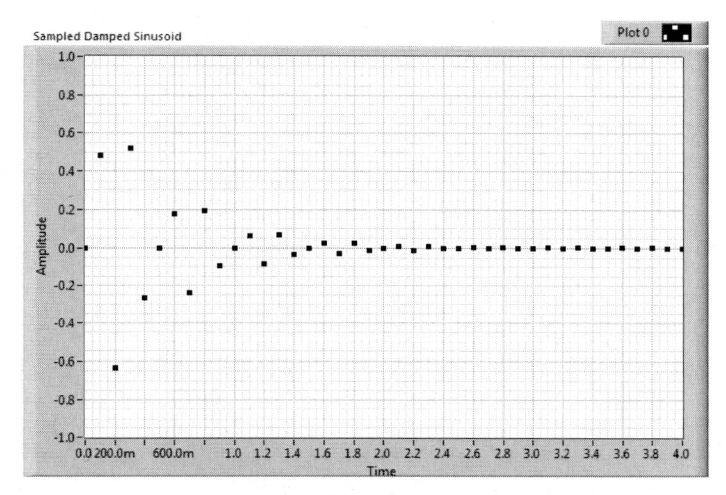

(a) Sampled at 2x the essential bandwidth

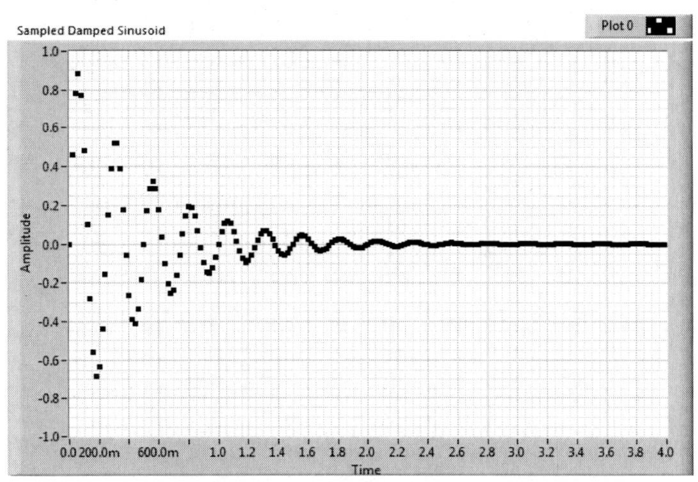

(b) Sampled at 10x the essential bandwidth

Figure 5.13: Damped sinusoidal signal sampled at 10 and 50 sps.

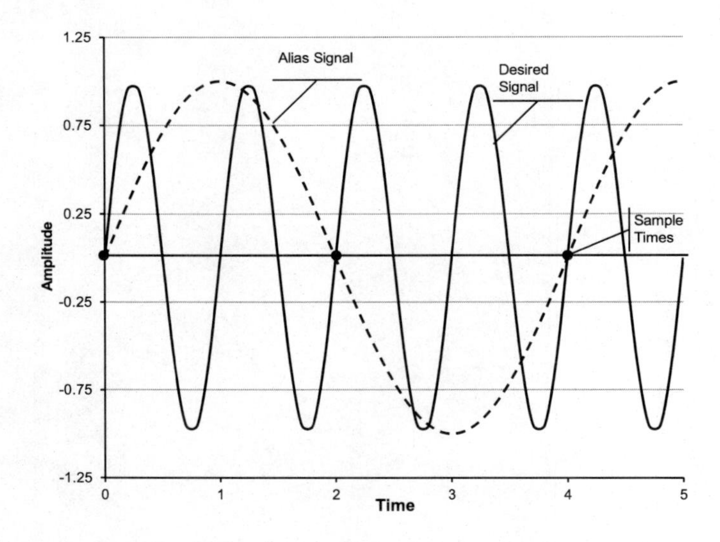

Figure 5.14: A signal sampled below its Nyquist rate and corresponding alias signal.

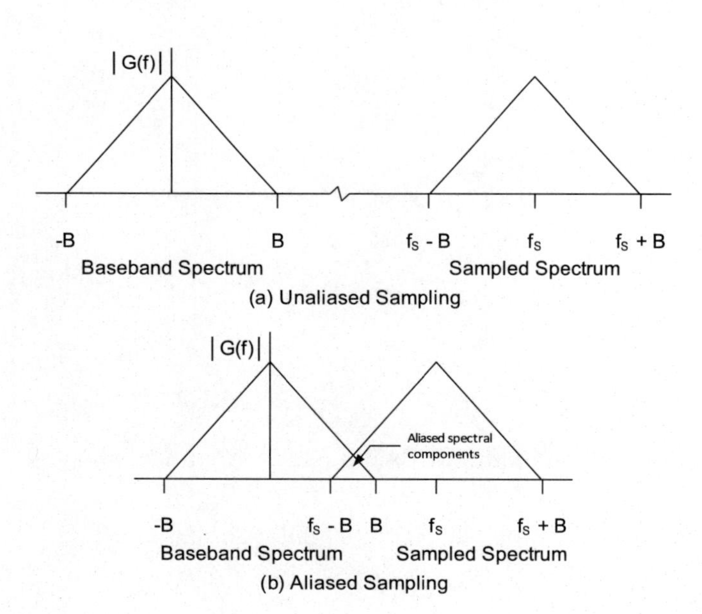

Figure 5.15: Effect of aliasing in the frequency domain.

components. In this case, the region of spectral overlap causes a distortion in the signal because the normal high frequency components now include additional copies of the spectral components, but at the wrong frequency.

Low-pass filtering at the essential bandwidth, B, does not remove the distortion since the frequencies from the copy of the spectrum centered at f_S are still mixed in with the baseband spectrum. To remove the aliased frequencies, the low-pass filter needs a cutoff frequency below $(f_S - B)$ Hz. The result of this filtering is that the baseband signal no longer has the entire essential frequency content so the signal representation is not faithful to the original, as we desire. The cure is restoring the sampling rate to one that at least matches the Nyquist sampling rate.

5.7 FILTER DESIGN

Many textbooks cover analog filter design theory, for example [Horo89; Hump70; John75; Meik90; Mosc81], so we do not attempt an extensive tutorial here. We present several rules of thumb for building filters for use with signals having frequency ranges up to 100 kHz. These filters use active components in a building block fashion to synthesize the desired characteristics. The goal is to provide a basic design that we can use with readily available components.

5.7.1 Reasons for Filtering

We have already seen two uses for LPFs. First, they serve as an anti-aliasing filter that establishes the essential frequency of an analog signal. Second, we use them to reconstruct the analog signal from the samples of the signal. Filters also remove noise in a data system or interfering signals from an undesired source. For each filter application, we must determine which portions of the frequency domain to preserve and which portions to attenuate.

5.7.2 Filter Types and Parameters

Figure 5.16 illustrates the standard filter types, namely

Low Pass Filter (LPF) — Preserves all of the frequencies below a certain point.

High Pass Filter (HPF) — Preserves all of the frequencies above a certain point.

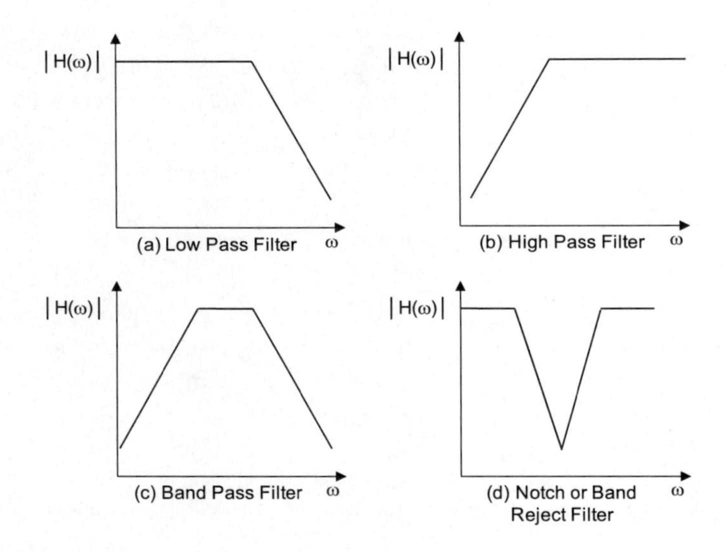

Figure 5.16: Standard filter type classifications as a function of the magnitude of their transfer function.

Band Pass Filter (BPF) — Passes all of the frequencies between two points and attenuates the others.

Notch or Band Reject Filter — Passes all frequencies except those between two points.

Figure 5.17 illustrates common features of interest using a LPF example. The *pass band* is that region in the frequency domain that the filter passes without attenuation. Usually, if the attenuation is less than 3 dB, designers consider the filter to be passing the signal without attenuation. The *cutoff frequency* defines the edge of the pass band. That is, the filter passes all frequencies lower than the cutoff frequency without attenuation. However, the filter attenuates those frequencies above the cutoff at some level. The *stop band* is that region in the frequency domain where the filter attenuates any signal to at least a maximum-specified amplitude, e.g., 40 dB, below the pass band. That provides a minimal attenuation level for all frequencies in the stop band. The *transition region* is that frequency region between the end of the pass band and the beginning of the stop band. The filter *roll off* specifies how quickly the filter transitions between the pass band and the stop band. The filter design specifications usually include values for these parameters.

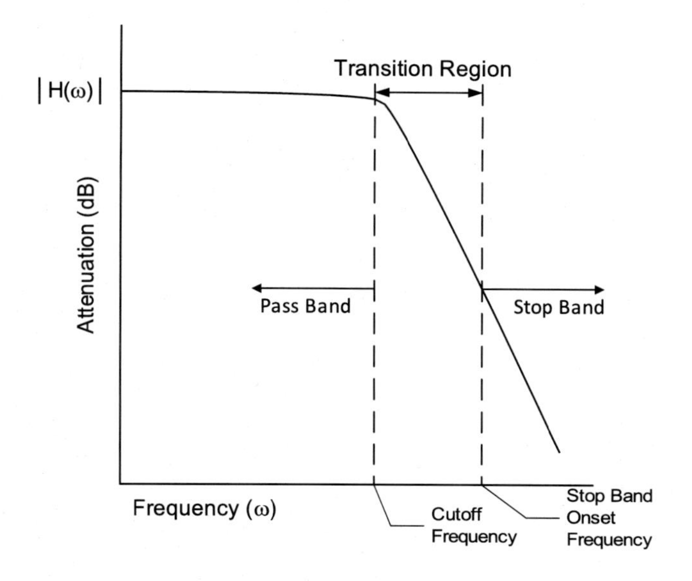

Figure 5.17: Standard parameters used to specify the filter transfer function.

5.7.3 Filter Transfer Functions

Engineers characterize a filter by its s-domain transfer function, $H(s)$, which describes the ratio of the signal output from the filter to the signal input to the filter. The nominal Laplace frequency domain variable $s = j\omega$ is used. Several standard filter families write their transfer function as a polynomial in frequency. In this section, we discuss the characteristics of Butterworth, Chebyshev and Bessel filter families and the polynomials that describe them. As we will see in the next section, designers use LPFs to synthesize other filter types so we only describe the low-pass polynomial here.

5.7.3.1 Ideal Filters

Before we examine specific filter realizations, we first look at what makes an *ideal filter* to serve as a basis of comparison. An ideal filter has the characteristics of

- No amplitude distortion in the pass band
- No phase distortion in the pass band

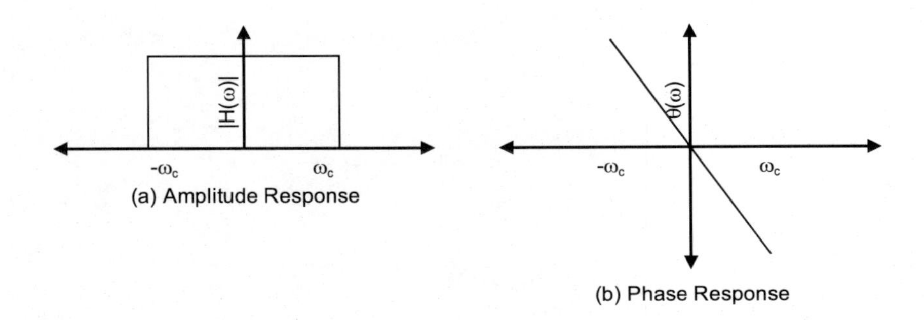

Figure 5.18: Amplitude and phase responses for an ideal LPF.

- Completely eliminating components outside of the pass band

Figure 5.18 illustrates the amplitude and phase response for an ideal LPF that provides distortionless transmission through the filter for the input signal's pass band frequencies. In the time domain, this is

$$y(t) = Kx(t - \tau_d) \tag{5.13}$$

where K is a frequency-independent signal gain and τ_d is a frequency-independent signal delay through the filter. The ideal filter has a pass-band, frequency-domain transfer function of

$$\begin{aligned} \mid H(\omega) \mid &= K \\ \theta(\omega) &= -j\omega\tau_d \end{aligned} \tag{5.14}$$

Circuit designers say that the filter has *amplitude distortion* if the filter gain is not sufficiently constant in the pass band. If we take the derivative of the phase in Equation (5.14) with respect to frequency, we see that the result is the constant delay τ_d. Circuit designers call this the *linear phase* condition where all of the signal components in the pass band arrive at the filter output in their proper relative phase. Nonlinear phase over the pass band gives rise to *phase distortion* in the output signal. The derivative of the filter transform phase with respect to frequency is the *group delay*. If the filter has linear phase, then the group delay is constant across all frequencies in the pass band.

Naturally, real filters made from real components do not show these ideal characteristics. The circuit designer needs to determine how much deviation from the ideal the system can tolerate in the application context.

5.7.3.2 Butterworth Filters

The Butterworth family of filters is maximally flat in the pass band. That is, the amplitude response in the pass band comes closest to the ideal of a perfectly flat response. For this reason, designers choose *Butterworth filters* for many applications. The transfer function with respect to frequency, ω, and filter order, n, is

$$|H_n(s)|^2 = \frac{1}{1 + (s/s_C)^{2n}} \tag{5.15}$$

The cutoff frequency of the filter in Equation (5.15) is s_C. For example, a second order Butterworth filter has a transfer function of the form

$$H_2(s) = \frac{1}{s^2 + \sqrt{2}s + 1} \tag{5.16}$$

Figure 5.19 shows the magnitude and phase for this transfer function. Note: in this figure, the plots are on a linear scale for both the x- and y-axes. On this plot, the amplitude remains relatively flat (falloff less than $1/\sqrt{2}$ in the pass band) and the phase remains relatively linear until frequencies near the cutoff frequency. After the cutoff, the filter attenuates the amplitudes, as desired, and the phase starts having nonlinear characteristics.

In Figure 5.19 and in subsequent filter transfer functions used in this section, we normalize the graph to have a cutoff frequency at 1 (unitless). By using frequency-scaling techniques, we are able to scale the transfer function to any desired cutoff frequency. In subsequent figures, we use a logarithmic scaling on both the frequency and the amplitude axes. We will use the customary approach and not plot the amplitude as a voltage from 0 V to its maximum voltage value. Rather, we will plot the square of the filter output voltage relative to its maximum voltage value, which is normalized to 1 V. The square of the voltage is proportional to signal power hence the *y-axis* represents a relative power *attenuation* or a relative *gain* in the voltage. In this case, engineers label the magnitude as an attenuation and make the relative amplitude plot in dB units.

Figure 5.20 illustrates the amplitude spectrum function for Butterworth LPFs of various orders. In the figure, the amplitude response for all filter orders stays relatively flat, especially with the logarithmic scale, until reaching the cutoff frequency. Butterworth filters of all orders pass through the 3-dB attenuation point at the cutoff frequency. As the filter order increases, the sharpness of the filter rolloff does indeed increase,

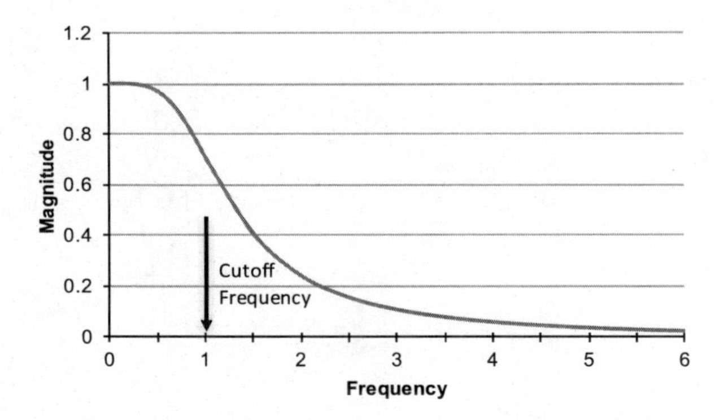

(a) Butterworth magnitude spectrum.

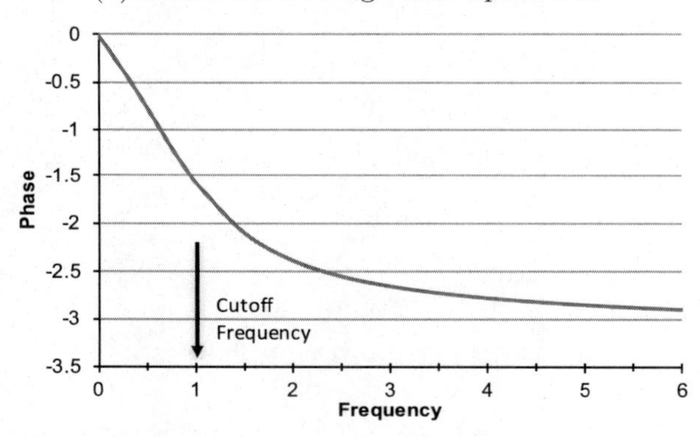

(b) Butterworth phase spectrum.

Figure 5.19: The magnitude and phase spectra for the second-order order Butterworth LPF.

as desired. Computing the phase response for all orders is an exercise at the end of the chapter.

5.7.3.3 Chebyshev Filters

If one exchanges absolute pass band flatness for a more rapid transition to the stop band, it is possible to accomplish a desired level of filtering with fewer components than with the Butterworth design. The Chebyshev filters meet such a design tradeoff in many cases. The Chebyshev filter's polynomial transfer function is

$$|H_n(s)|^2 = \frac{1}{1 + \epsilon^2 C_n \left(s/s_C\right)^{2n}} \tag{5.17}$$

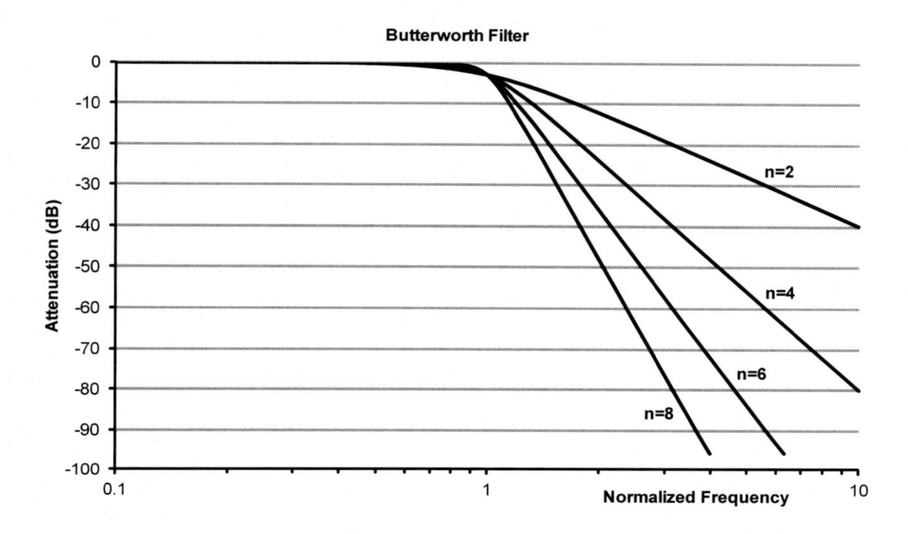

Figure 5.20: The relative magnitude of the transfer function describing the Butterworth LPF as a function of frequency and filter order.

The parameter ϵ in Equation (5.17) controls the amount of ripple in the pass band and $C_n(s/s_C)$ is the Chebyshev polynomial of order n. Figures 5.21 and 5.22 illustrate the amplitude responses for Chebyshev filters of varying amounts of ripple and differing orders. The filters shown have 0.5 dB, 1.0 dB, 2.0 dB, and 3.0 dB pass band ripple amplitudes. For the case of $n = 2$ and pass band ripple of 1 dB, the normalized, second-order filter transfer function becomes

$$H_2(s) = \frac{1}{s^2 + 1.097734s + 1.102510} \tag{5.18}$$

5.7.3.4 Bessel Filters

The Butterworth and Chebyshev filters discussed earlier did not use the phase response as a criterion in their development. If one plots the phase response for the filters, it is not exactly linear in the pass band. An attempt to design a filter with linear phase characteristics leads to the development of the Bessel filter family. The linear phase tends to prevent dispersion of the signal and can be good for digital pulses. The Bessel

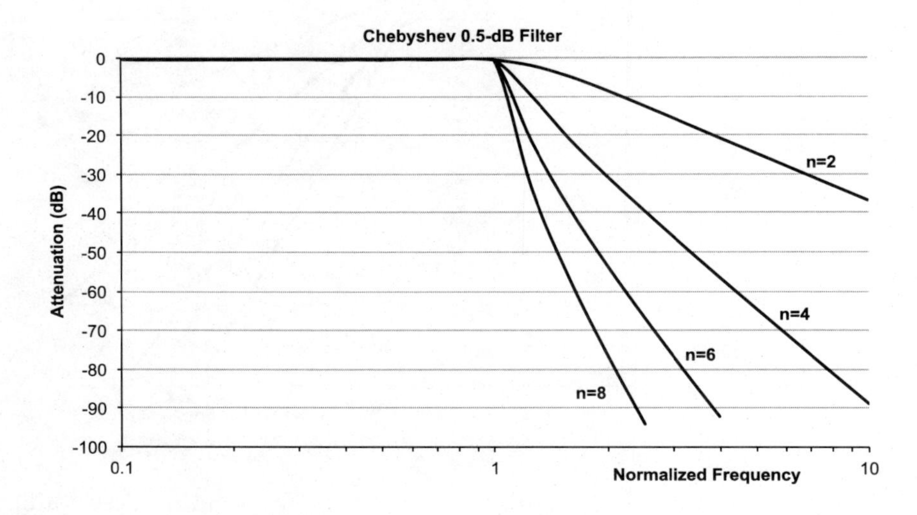

(a) 0.5 dB Ripple Chebyshev Low Pass Filter

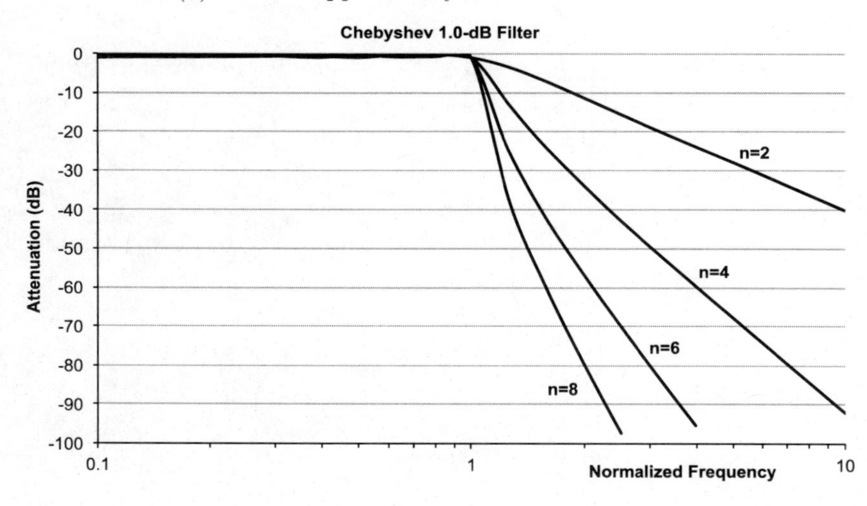

(b) 1.0 dB Ripple Chebyshev Low Pass Filter

Figure 5.21: The relative magnitude of the transfer function describing the Chebyshev 0.5 dB and 1.0 dB ripple LPF as a function of frequency and filter order.

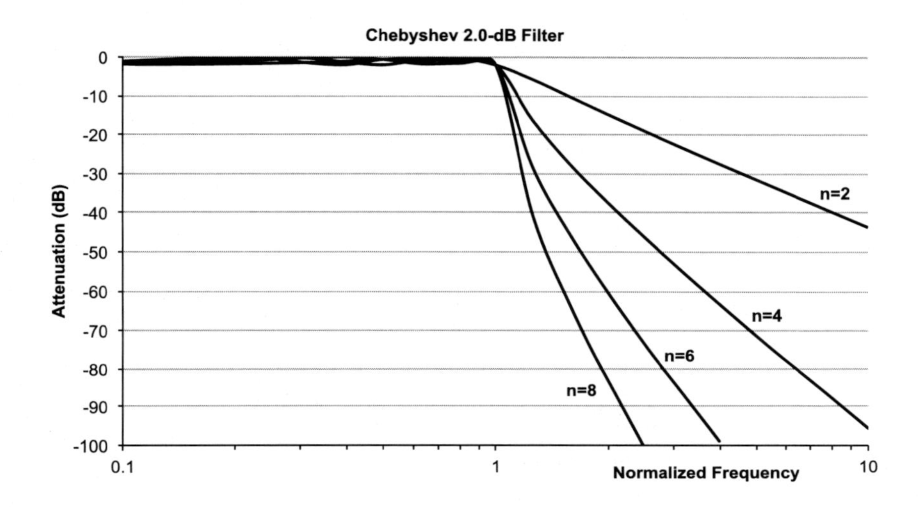

(a) 2.0 dB Ripple Chebyshev Low Pass Filter

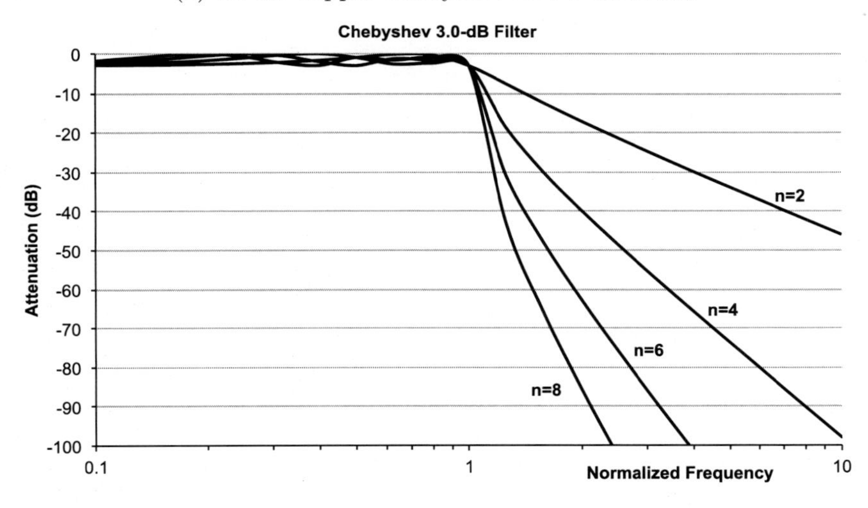

(b) 3.0 dB Ripple Chebyshev Low Pass Filter

Figure 5.22: The relative magnitude of the transfer function describing the Chebyshev 2.0 dB and 3.0 dB ripple LPF as a function of frequency and filter order.

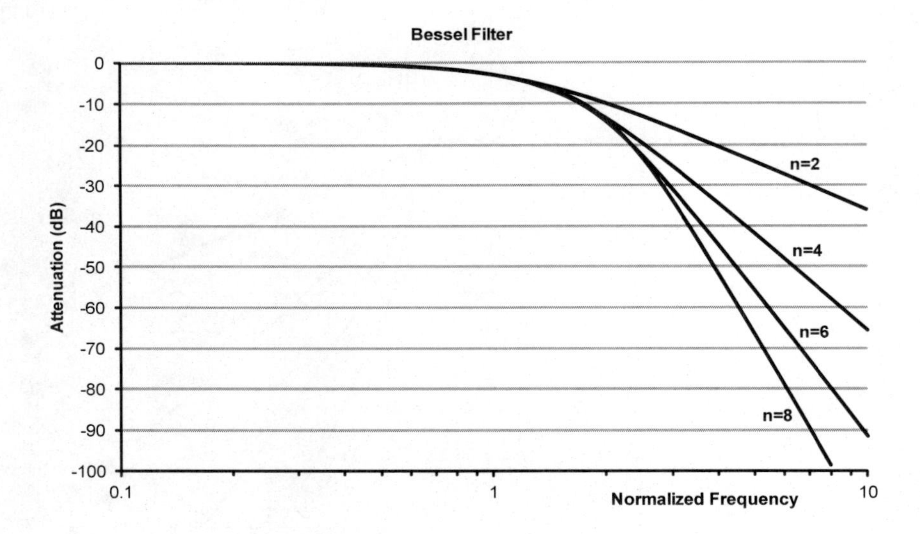

Figure 5.23: The relative magnitude of the transfer function describing the Bessel LPF as a function of frequency and filter order.

filter has a transfer function in the form

$$H(s) = \prod_{i=1}^{n} \frac{K}{s - p_i} \tag{5.19}$$

Figure 5.23 plots this amplitude response for even orders of n. The amplitude response is smooth like the Butterworth, but it suffers some attenuation in the pass band. The transition to the stop band is not as rapid as with the Butterworth. For the case of $n = 2$, the normalized, second-order filter transfer function becomes

$$H_2(s) = \frac{3}{s^2 + 3s + 3} \tag{5.20}$$

5.7.4 Analog Filter Design Method

There are a number of ways to realize the filter transfer functions in hardware with actual components. One common design method for accomplishing this task is to use the Sallen-Key LPF design [Sall55; Horo89]. This method uses two-pole operational amplifier circuits. To achieve higher filter orders, successive stages of two-pole building blocks

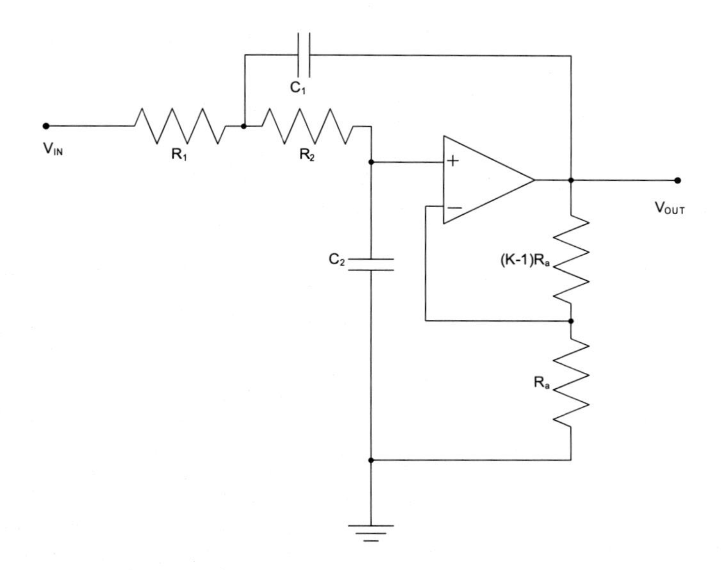

Figure 5.24: Sallen-Key filter design two-pole LPF building block.

are cascaded to synthesize the desired filter order. Naturally, the method generates only even-numbered filter orders.

5.7.4.1 Low Pass Building Block

Figure 5.24 shows the second-order, LPF circuit building block used in the Sallen-Key LPF designs. We also use the same low-pass building block to synthesize high-pass and band-pass configurations. This LPF circuit consists of two sections: an RC LPF followed by a noninverting output amplifier. By applying standard s-domain nodal circuit analysis, it can be shown that the transfer function, $H(s)$, relating the output to the input for this circuit, is given in terms of the resistors and capacitors in the filtering stage and the noninverting amplifier stage by the following equation:

$$
\begin{aligned}
H(s) &= \frac{K/R_1R_2C_1C_2}{s^2 + s\left(\frac{1-K}{R_2C_2} + \frac{1}{R_1C_1} + \frac{1}{R_2C_1}\right) + \frac{1}{R_1R_2C_1C_2}} \\
&= \frac{\Omega}{s^2 + a_1 s + a_0} \qquad (5.21)
\end{aligned}
$$

The a_0, a_1, and Ω constants in Equation (5.21) are functions of the R, C, and K values in the operational amplifier building block circuit.

From circuit analysis, we recognize the term $1/R_1R_2C_1C_2$ as the cutoff frequency for the filter and the constant in the first power of s term as the damping coefficient. If we normalize the filter to have a cutoff frequency of 1, let $R_1 = R_2 = 1\,\Omega$, and let $C_1 = C_2 = 1\,\text{F}$, then we can analyze the circuit and use frequency scaling techniques to place the cutoff at any desired frequency. By comparing the resulting transfer function with that for known filter families, we can make this structure into Butterworth, Bessel, or Chebyshev type filters.

By working through Equation (5.21) and then comparing the equation with the form for the Butterworth, Chebyshev, and Bessel filter transfer functions, one discovers that a frequency scaling factor, f_n, is required to make the Chebyshev and Bessel equations match exactly. The Butterworth filter does not need this factor.

This frequency scaling factor and the gain factor K are needed for each stage in the filter design. For example, if we need a fourth-order filter design, then we use two stages of the standard building block to realize the filter. Each stage has an associated gain and frequency scaling factors for Chebyshev and Bessel filters. Table 5.2 shows the necessary gain frequency scaling factors to synthesize filters to order eight from cascaded second-order sections. We elected to use only two-pole filter building blocks in the design procedure below because there are nearly as many components involved in a two-pole filter as are needed for a single-pole filter and design simplicity is added if only one standard building block is used in each step.

In the next paragraphs, we step through design of a LPF using the Sallen-Key design procedure and the operational amplifier basic building block circuit. In actual practice, we may need to iterate on this process several times to close on a design that meets all of the specifications and gives the minimum number of components. That is, the designer may find that both a Butterworth and a Chebyshev filter satisfy the requirements. However, the question is whether they both are of the same order. The filter with the smaller order is generally preferred. If the filter specifications are loose, several designs may meet the given specifications. If the specifications are too tight, the designer may not be able to satisfy all of the requirements with a design using this procedure. In that case, one or more of the specifications may need to be relaxed.

Table 5.2: Low Pass Filter Design Parameters

Number of Poles	Butter-worth	Bessel		0.5 dB Chebyshev		1.0 dB Chebyshev		2.0 dB Chebyshev		3.0 dB Chebyshev	
	K	K	f_n	K	f_n	K	f_n	K	f_n	K	f_n
2	1.586	1.268	1.272	1.842	1.231	1.955	1.050	2.114	0.907	2.234	0.841
4	1.152	1.084	1.430	1.582	0.597	1.725	0.529	1.924	0.471	2.071	0.443
	2.235	1.759	1.603	2.660	1.031	2.719	0.993	2.782	0.964	2.821	0.950
6	1.068	1.040	1.604	1.537	0.396	1.686	0.353	1.891	0.316	2.042	0.298
	1.586	1.364	1.689	2.448	0.768	2.545	0.747	2.648	0.730	2.711	0.722
	2.483	2.023	1.905	2.846	1.011	2.875	0.995	2.904	0.983	2.922	0.977
8	1.038	1.024	1.778	1.522	0.297	1.672	0.265	1.879	0.238	2.033	0.224
	1.337	1.213	1.832	2.379	0.599	2.489	0.584	2.605	0.572	2.675	0.566
	1.889	1.593	1.953	2.711	0.861	2.766	0.851	2.821	0.842	2.853	0.839
	2.610	2.184	2.189	2.913	1.006	2.930	0.997	2.946	0.990	2.956	0.987

5.7.4.2 Filter Type Determination

The first choice that the filter designer needs to make is the filter family. If the design specification calls for maximally flat pass band or permits very little amplitude ripple, then the designer chooses a Butterworth filter. If the specification permits some degree of pass band ripple or a rapid transition between the pass band and the stop band is required, then the designer may choose a Chebyshev filter. If the filter phase needs to be as close to linear as possible, then the designer chooses a Bessel filter.

5.7.4.3 Filter Order Determination

The designer determines the filter order after selecting the filter family. To make the filter order determination, the designer must normalize gain graphs. First, the designer normalizes the frequency range by dividing the cutoff frequency, f_c, and the frequency that specifies the start of the stop band, f_u, by the cutoff frequency value. That is, the cutoff frequency is set to 1 and the start of the stop band is made equal to $\widehat{f_u} = f_u/f_c$. Next, the designer marks the desired minimum attenuation in the stop band at the frequency $\widehat{f_u}$. The first curve that passes through or below the attenuation mark at the frequency $\widehat{f_u}$ determines the filter order. This process guarantees that the attenuation is at least the desired amount throughout the stop band. The designer reads the order of the filter n for this curve. The filter needs $n/2$ second-order active filter stages in its construction.

5.7.4.4 Resistor and Capacitor Selection

While there are a number of resistors and capacitors in the basic building block, we make the design procedure simple by standardizing resistor and capacitor values in each two-pole stage. Except for the Butterworth filter, the designer generally cannot use the capacitor values from one stage in subsequent stages.

The first simplification that is made is to set $R_1 = R_2 = R_a = R$ and $C_1 = C_2 = C$ in each stage; that is, the resistor and capacitor values are set to a common value in each stage. A convenient value for R is in the range of $10\,\text{k}\Omega$ to $100\,\text{k}\Omega$. A reasonable standard value is $47\,\text{k}\Omega$, which is in the middle of the range and gives the designer flexibility if a change is required later.

The capacitor selection depends upon the design's filter family. For

Butterworth filters, we choose a common capacitor value for all stages by

$$C = \frac{1}{2\pi f_c R} \tag{5.22}$$

The value for R in Equation (5.22) is the resistor value that was chosen above and f_c is the non-normalized cutoff frequency in Hz. For Bessel or Chebyshev filters, we choose the capacitor value in each stage by

$$C = \frac{1}{2\pi f_n f_c R} \tag{5.23}$$

Again, the value for R in Equation (5.23) is the resistor value that was chosen above and f_c is the non-normalized cutoff frequency in Hz. Additionally, the Bessel and Chebyshev filters have a frequency normalization factor, f_n, that is specific to each filter stage to make the realization match the filter polynomial. Table 5.2 lists the values for the Bessel and Chebyshev filters. Finally, we set the stage gain by reading K from Table 5.2 for each stage of the desired filter type. This value, along with the resistor above, sets the value for the resistor labeled $(K-1)R_a$.

For realistic filter construction, designers use standard component values for the resistors and capacitors. Since the exact values are not highly critical to the design success, choose the nearest standard value to the "correct" value in the design rule. One may need to iterate a bit on the resistor and capacitor values to bring them close to the standard values. If the selection of available capacitor values is small, the designer may reverse the order and first select a convenient capacitor value before solving for the necessary resistor values.

5.7.4.5 Sample LPF Design

As an example of this technique, we next design a LPF to have a cutoff frequency of 1 kHz with a nominal cutoff attenuation of 3 dB and the beginning of the stop band at 10 kHz with a minimum attenuation of 40 dB. The amplitude variation in the pass band is at most 1 dB. These specifications automatically prevent using a 2 dB or 3 dB ripple Chebyshev filter; however, we may consider other filter types at this stage. The normalized frequencies are then $f_c = 1$ and $\widehat{f_u} = 10$. From Figure 5.20, we see that the $n = 2$ Butterworth filter meets the specifications. We then choose the R component as 47 kΩ and solve for C using

$$C = \frac{1}{2\pi(10\,\text{kHz})(47\,\text{k}\Omega)}$$

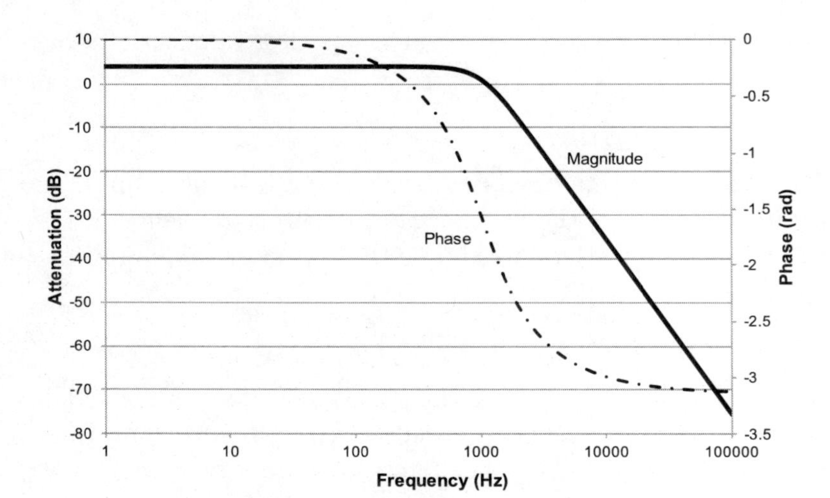

Figure 5.25: Amplitude and phase response of the simulated LPF design with a cutoff frequency of 1 kHz.

or 0.003 µF. The $(K-1)R$ resistor is determined using the value from Table 5.2

$$(K-1)R = (1.586-1)\,47\,\text{k}\Omega$$

or 27 kΩ. We use these component values in the LPF building block illustrated in Figure 5.24.

To verify that our component selections meet the design requirements, we simulated the amplitude and phase response of the filter circuit with commercial simulation software, as Figure 5.25 shows. The figure provides both the relative magnitude response (in decibels) and the phase response (in radians) as a function of frequency for the filter. For this simulation, the software uses ideal component models so this plot does not exactly match the results for a real circuit but it does give us the confidence that we are close to the desired performance.

5.7.4.6 Conversion to High-Pass Design

Designers use similar steps for HPFs and LPFs. Figure 5.26 shows the high-pass building block. The design is similar to the two-pole LPF but with the filtering resistors and capacitors interchanged. This effectively inverts the frequency axis and converts a low-pass transfer function into a high-pass transfer function. To use the LPF magnitude response graphs for high-pass design, we rotate the graphs around the cutoff frequency

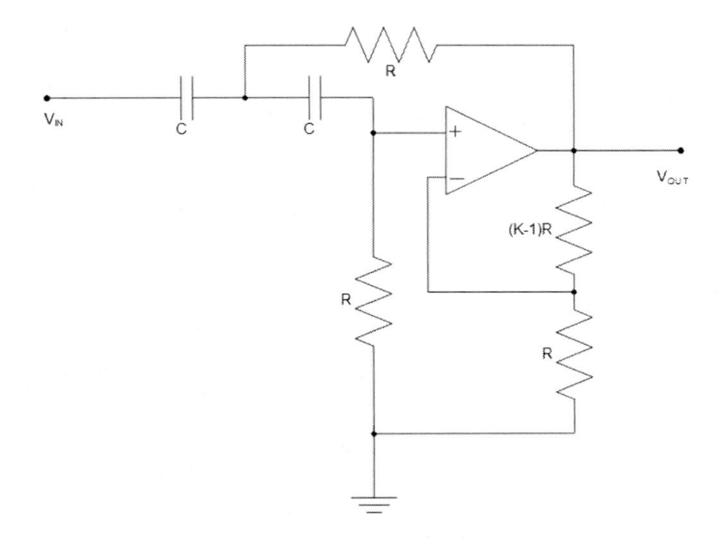

Figure 5.26: Building block for a HPF design based on the two-pole LPF.

point to produce a high-pass magnitude response. The gain factor, K, for each stage remains the same as in the low-pass case. We replace the frequency normalization factor, f_n, for each filter stage chosen from Table 5.2 by $1/f_n$ in Equation (5.23) since we are inverting the frequency axis to transform a low-pass design characteristic into a high-pass characteristic. Notice that we have dropped the subscripts on the resistors and capacitors because we use the same standardization method as we did with the LPF design.

Engineers do not normally use high-pass designs to filter the sensor output in telemetering systems. However, they are important for synthesizing the more commonly used BPF. We discuss this configuration in the next section. Designers may also use HPFs in the transmitter's Radio Frequency (RF) section. The high-frequency nature of this filter design requires specific RF design techniques.

5.7.4.7 Conversion to Band-Pass Design

Now that we have HPFs and LPFs, cascading the high-pass and low-pass sections synthesizes a BPF. Figure 5.27 illustrates an example of this concept. The filter is composed of a low-pass active filter building block followed by a high-pass active filter building block. The BPF designer chooses cutoff frequencies for the low-pass and the high-pass filter sections

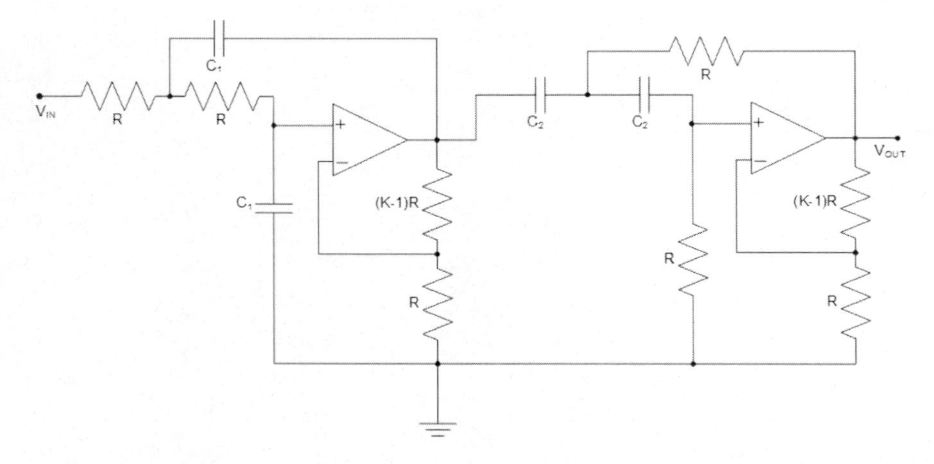

Figure 5.27: Synthesizing a BPF from LPF and HPF building blocks.

and treats them as two independent design problems. The low-pass stage has its cutoff frequency at the BPF upper cutoff frequency. The high-pass stage has its cutoff frequency at the BPF lower cutoff frequency. If we place the LPF section with its cutoff at the BPF lower cutoff and the HPF section with its cutoff at the BPF upper cutoff, then we have changed the filter into a band reject or notch filter.

As an example of BPF design, consider a specification that the BPF is to have a lower cutoff frequency of 1 kHz and an upper cutoff frequency of 25 kHz. We make no requirement on the rolloff characteristics or flatness at this time. We choose a two-pole Butterworth filter for both stages. As a starting point, choose the standard resistor to be 47 kΩ. The low-pass stage capacitor value is computed to be

$$C_1 = \frac{1}{2\pi(47\,\text{k}\Omega)(25\,\text{kHz})}$$

or 135 pF. The high-pass stage capacitor value is computed from

$$C_2 = \frac{1}{2\pi(47\,\text{k}\Omega)(1\,\text{kHz})}$$

or 3.39 nF. The gain control resistor $(K-1)R$ is computed from

$$(K-1)R = (1.586-1)\,47\,\text{k}\Omega$$

or 27 kΩ. We used these values in the filter illustrated in Figure 5.27 and

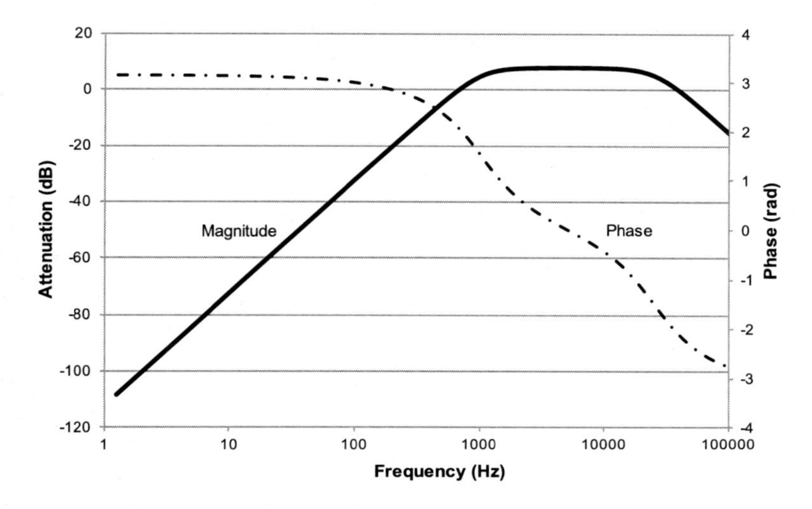

Figure 5.28: Amplitude and phase response for the simulated BPF.

then simulated using a standard circuit simulation package. Figure 5.28 gives the amplitude and phase responses from the simulation. Notice that the simulation shows actual gain in the BPF passband but the amplitude is attenuated by 3 dB from the maximum level at 1 kHz and 25 kHz.

5.8 SOFTWARE FILTER DESIGN

5.8.1 Digital Filter Equivalents

Designers do not always restrict their filter designs to analog hardware realizations. With the advent of digital signal processing devices as standard components in system design, designers frequently synthesize digital equivalents of analog filters in software. This has two major advantages in system design: reduced part count in the design and the changeable programming. The latter is difficult to do with fixed components as in the hardware filter. The software equivalent to the filters represented by the equations in Section 5.7.3 can be as complicated as the code space and processor speed allow. The standard texts on modern digital signal processing such as [Oppe10] cover digital filter design in detail so we will not cover them here.

5.8.2 Data Processing Filtering

Analysts use software for detailed data processing filtering and performing more sophisticated signal processing operations. The software filter can be a simple LPF to remove noise from the signal to complicated predictive analysis techniques such as Kalman filters.

The filtering process in hardware can only be a causal design. A suitable delay in the processing software produces a noncausal filter. In the filter realization, a causal filter only uses data from the current sample and previous samples. By using a selective delay, noncausal filters use not only the current sample and past samples, but they can also employ future samples beyond the current output sample point. When this is done, the output is not immediate and has a delay. The advantage is improved processing filtering improvement.

Two examples of software filtering for data processing filtering that we examine here are the moving average filter and the moving least squares filter. Both of these form LPFs, but they are not of the formal families used in the hardware design examples.

5.8.2.1 Moving Average Filter

The moving average filter works on a block of samples. Let $\{q_i\}$ be a block of N samples of a signal. The block is a subset of the overall data set collected from the sensor. Figure 5.29 illustrates how the software processes these points in a causal or a noncausal manner. If μ_n is the current block average, we find the causal average by using

$$\mu_n = \frac{q_n + q_{n-1} + \cdots + q_{n-N+1}}{N} \tag{5.24}$$

In Equation (5.24), we only use the current sample value, q_n, and the previously determined sample values $q_{n-1} \cdots q_{n-N+1}$. At the next average computation, the algorithm removes the oldest sample from the block and adds the newly arrived sample value to the block. We use a similar equation to find a noncausal average using

$$\mu_n = \frac{q_{n-\lfloor N/2 \rfloor} + \cdots + q_{n-1} + q_n + q_{n+1} + \cdots + q_{n-\lfloor N/2 \rfloor}}{N} \tag{5.25}$$

In Equation (5.25), we assume that N is an odd value and we take points symmetrically around the current sample value. If this is not true, the indices need adjustment. As in the causal case, the algorithm removes

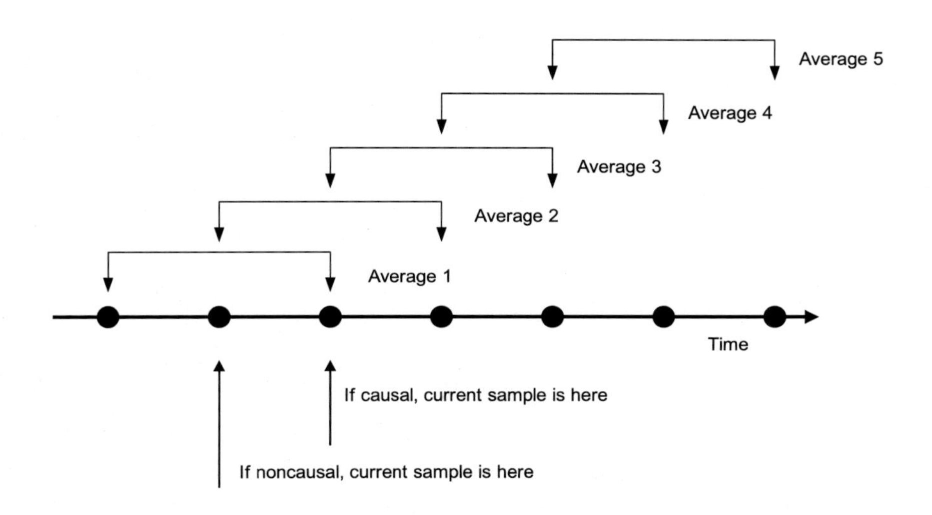

Figure 5.29: Forming either a causal or a noncausal three-point moving average of data points.

the oldest sample from the block when it adds the newly arrived sample value to make the next average computation.

Table 5.3 contains the data for both a three-point causal filter and a three-point noncausal filter to process. Notice that both filters do not start immediately but require a delay before producing the output. Figure 5.30 illustrates the results of the filtering process. Notice the typical behavior of both filters where the causal filtering tends to be below the measured line, while the noncausal filtering tends to track the measured values closer.

5.8.2.2 Moving Least Squares Filter

We can also apply the moving block technique to determining the least squares fit of a polynomial to a data set. For example, consider making a piecewise linear fit to a data set where we wish to determine the coefficients of a straight line fit over short data segments. We apply the following two equations to determine the slope and y intercept for the

Table 5.3: Software Filtering Using Causal and Noncausal Filters

Time	Unfiltered	3-Point Noncausal	3-Point Causal
1.00	0.929		
1.25	1.077	1.169	
1.50	1.501	1.423	1.169
1.75	1.692	1.768	1.423
2.00	2.112	2.008	1.768
2.25	2.220	2.267	2.008
2.50	2.469	2.462	2.267
2.75	2.696	2.714	2.462
3.00	2.975	3.032	2.714
3.25	3.425		3.032

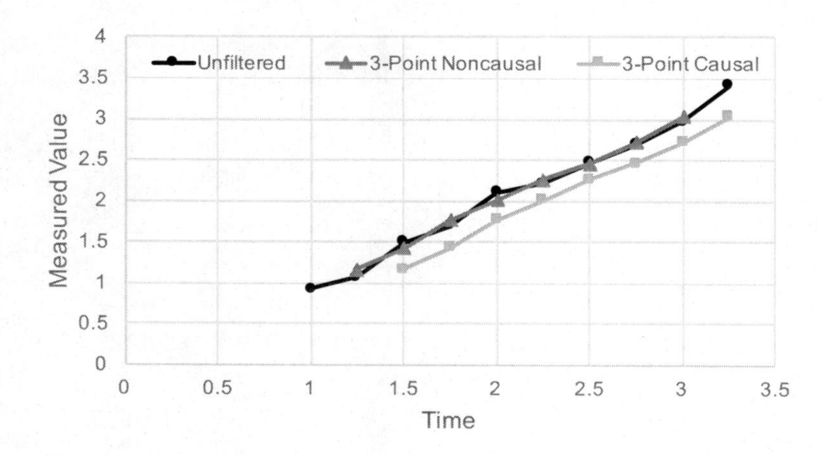

Figure 5.30: Example results of causal and noncausal software filtering.

straight line:

$$a_i = \frac{N \sum_{k=i-n}^{i+n} V_k t_k - \sum_{k=i-n}^{i+n} V_k \sum_{k=i-n}^{i+n} t_k}{N \sum_{k=i-n}^{i+n} (t_k)^2 - \left(\sum_{k=i-n}^{i+n} t_k \right)^2}$$

$$b_i = \frac{N \sum_{k=i-n}^{i+n} V_k \sum_{k=i-n}^{i+n} (t_k)^2 - \sum_{k=i-n}^{i+n} V - kt_k \sum_{k=i-n}^{i+n} t_k}{N \sum_{k=i-n}^{i+n} t_k^2 - \left(\sum_{k=i-n}^{i+n} t_k \right)^2} \tag{5.26}$$

The algorithm fits several points at a time and then moves to the next data block within the overall data set.

5.9 QUANTIZATION

As we saw in Section 5.3.3, the process of forming the digital PCM signal has two stages: sampling and conversion. The first stage of sampling is acquiring and holding the analog signal amplitude value, while the second stage converts the analog amplitude to a digital representation. Referring to Figure 5.3, when an analog signal is converted to a PCM signal, the resulting signal takes on discrete amplitude values as a function of time. Naturally, the sampling rate must be adequate to prevent aliasing and we assume that is the case for the remainder of this chapter. However, because we do not send the actual signal with infinite precision, we encounter some level of distortion — even if we sample at a rate higher than the Nyquist rate.

The lack of precision represents a loss of information about the analog signal. We represent this loss as a form of noise on the signal due to the quantization process. The system designer's job is to determine the acceptable level of this distortion so that the quantization process preserves the signal quality at a level that is useful to the measurement process. This section covers the quantization process in general and the ways to specify its performance. In the next section, we will look at hardware circuits to accomplish this task.

5.9.1 Quantization Process

As noted earlier in this chapter, quantizers typically operate with a uniform sampling rate. The system designer is not required to sample

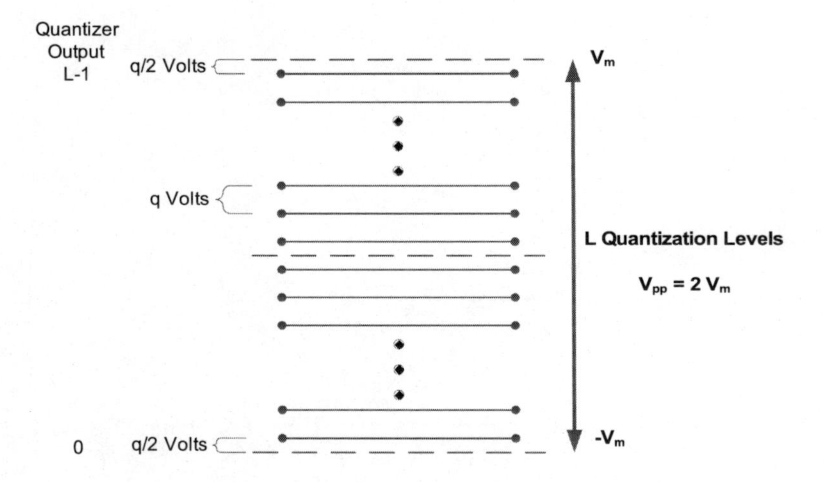

Figure 5.31: Uniform quantization voltage levels.

the analog signal at a uniform rate; rather, it is easier to do so with typical design techniques and components. Uniform quantization is also typical in telemetry systems and we assume this for the remainder of this chapter.

Uniform quantizing is the most straightforward method and is accomplished with the expected ADC structure. Uniform quantizing takes the full input peak-to-peak voltage range, V_{pp}, and divides it into some number of levels, L; each level is q volts wide. That is, the distance between each quantization level is

$$q = \frac{V_{pp}}{L} = \frac{V_{pp}}{2^N} \tag{5.27}$$

The voltage levels are numbered from 0 through $L - 1$. Normally, L is an integer power of 2 and N is the number of bits in the quantizer output so $L = 2^N$. Figure 5.31 shows the uniform quantization process. Typical values for the number of bits produced by the quantizers used in telemetry systems range from 8 to 16, depending upon the required voltage resolution, the speed at which the samples are taken, and the required transmission bandwidth.

If the quantizer accepts $-V_m$ as the minimum voltage level and the input signal's amplitude is V_{in}, then the code word, n, describing the

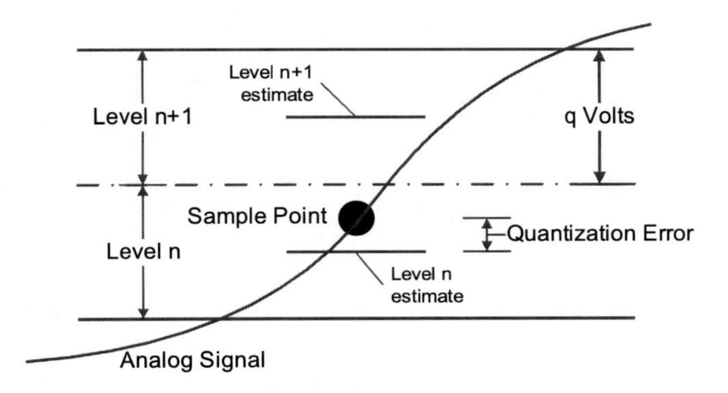

Figure 5.32: Quantization error when sampling an analog signal.

voltage level is

$$n = \left\lfloor \frac{V_{in} + V_m}{q} + \frac{q}{2} \right\rfloor \tag{5.28}$$

In Equation (5.28), the $\lfloor * \rfloor$ operation means round down the result to the nearest integer; n ranges between 0 and $L - 1$. If we wish to estimate the input voltage based on n, we can reverse the operation by

$$\langle V_{in} \rangle = nq - V_m + \frac{q}{2} \tag{5.29}$$

The $< V_{in} >$ in Equation (5.29) is to imply that this is an estimate for the input voltage and not the exact value because the rounding process in Equation (5.28) destroyed some of the information about the input signal. Therefore, as Figure 5.32 illustrates, the conversion produces a maximum uncertainty of $\pm q/2$ volts in the result.

5.9.2 Commutation

A telemetry system samples several analog channels to form the entire PCM data stream. To assist in this process, a modern sampling circuit often has a multiplexer as part of the chip to allow several analog channels to be sampled by a single device. This usually reduces the overall system complexity and power requirements. Engineers call the process of progressively sampling analog channels *commutation*. With control supplied by an external processor, the sampling does not need to be the same for each analog channel. Designers tune the commutation strategy to the Nyquist rates for each analog channel. If the system has several commutating devices, the designer must synchronize their results into a time-division format that allows the user to separate the channels. The telemetry frame formatter, covered in Section 6.4, performs this function.

5.9.3 Quantization Noise and Resolution

Quantization has a variety of degrees of *resolution*. The resolution the designer selects determines the quality of the reconstructed signal. As Figure 5.32 illustrates, the sampled analog signal is assigned a value, n, at each sampling instance. Notice that the quantizer assigns all of the input voltages in the range of Level n to the same estimated voltage level. This sampling process produces a small error that has a maximum value of $\pm q/2$ volts. The error between the actual analog signal and the quantized signal, which distorts the signal, is the *quantization noise*. This may seem to be a strange name, but engineers frequently refer to anything that degrades a signal as noise.

If we allow a maximum voltage error, $|e|$, for the quantization process, then we specify the maximum percentage error in the quantization process, $\%e$, based on the peak-to-peak voltage, V_{pp}, using

$$\frac{|e|}{V_{pp}} \leqslant \%e \tag{5.30}$$

This specifies the voltage level size in the quantizer, q. Perhaps a more useful design parameter is to specify the number of output bits that the quantizer needs. The number of bits, N, the quantizer should have is related to the percentage error, $\%e$, using

$$N(bits) \geqslant -\log_2(\%e) - 1 \tag{5.31}$$

So a smaller percentage error requires a greater number of bits in the quantizer.

5.9.4 Quantization Signal-to-Noise Ratio

The *quantization noise power*, N_q, for uniform quantization is a function of the level size, q, and we compute it using [Benn48]

$$N_q = \frac{q^2}{12} \tag{5.32}$$

The quantization noise is constant across all levels, and it is used to estimate the size of the level, q, to achieve a desired quality of result. We do this by defining a quantization signal-to-noise power ratio. The signal power is proportional to the square of the voltage value at each quantization level so it varies across the range of the quantizer. Therefore, we define the quantization SNR as either a peak or an average value. When the peak signal value is $V_{pp}/2$, then we use Equations (5.27) and (5.32) to find peak quantization SNR is

$$SNR_{peak} = \frac{L^2q^2/4}{q^2/12} = 3L^2 \tag{5.33}$$

Normally, the signal does not stay at its peak level all the time so the quantization SNR_{peak} may not be the best measure. A more representative signal level is the average or Root Mean Square (rms) level and the designer computes the average quantization SNR using

$$SNR_{average} = \frac{Average\ Signal\ Power}{q^2/12} \tag{5.34}$$

Designers have an equivalent way of writing this equation in dB units in terms of N (bits) when the signal takes the full conversion range as [Leon11b; Shei04]

$$SNR_{average} = 6.02N + 1.76\,\text{dB} \tag{5.35}$$

This leads to the definition of the Effective Number of Bits (ENOB)

$$ENOB = \frac{SNR_{average} - 1.76\,\text{dB}}{6.02} \tag{5.36}$$

If the sampling does not use the full conversion range, then the ENOB is decreased. In practice, this is the case. Systems designers normally select a conversion range that is larger than the expected input range. The design needs this "headroom" to account for possible overshoot caused by filters, unexpected signal amplitude swings on the input signal, and other unanticipated signal swings. (see Problem 12)

So far in this discussion, we assume that no channel errors occur during transmission. This is an important consideration if the probability of making a bit error during transmission without correction, P_e, is nonzero. If we allow for channel errors, then we can find an overall PCM signal-to-transmission-plus-quantization noise ratio. If $g(t)$ is the sampled and transmitted signal, then the SNR at the output of the receiver is a function of the number of quantization levels, L, the channel error probability, P_e, the mean-square signal amplitude, $\overline{g^2}$, and the square of the peak signal amplitude, g_p^2. The received SNR is [Lath98]

$$SNR_{output} = \frac{3L^2}{1 + 4\left(L^2 - 1\right) P_e} \left(\frac{\overline{g^2}}{g_p^2} \right) \tag{5.37}$$

The channel error, which we cover in Chapter 10, is a function of the modulation details.

5.9.5 Total Transmitted Data

While the number of bits used in the quantization process should be as large as possible to ensure the highest quality reconstruction of the signal, the transmission channel may not be able to support the desired data quantity. This becomes obvious when considering the aggregate data volume for the entire telemetry system. To gauge the magnitude of the situation, consider for the moment that each sensor in the telemetry system produces N bits per sample at a rate of f_S samples per second. The data rate for the ith sensor is

$$R_i = f_S N \tag{5.38}$$

If the telemetry system has M sensors, then the total number of bits per second, R, produced is

$$R = \sum_{i=1}^{M} R_i \tag{5.39}$$

The channel bandwidth required to support the transmission is on the order of R Hz.

Actually, the system must handle more data than this because transmission coordination requires management data to support the sensor data. The total number of transmitted bits per second must balance the following concerns:

1. The minimum Nyquist rate for each sensor to avoid aliasing
2. The required quantization resolution for adequate reconstruction
3. The channel bandwidth
4. The amount of management and framing overhead necessary to package the transmission and maintain coordination

The designer may need to iterate on specifications to achieve consistency between all of the competing requirements.

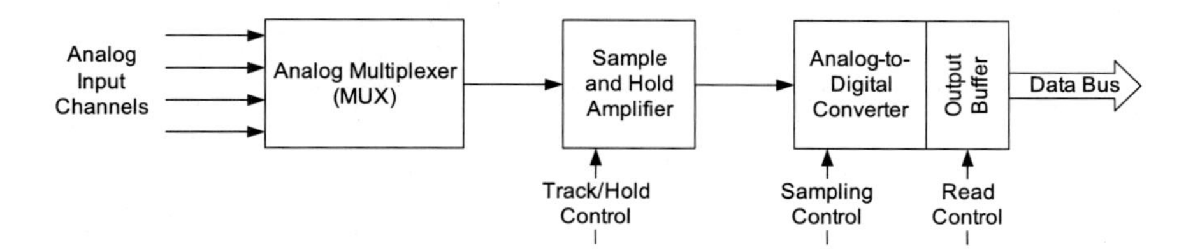

Figure 5.33: PCM sampling of analog data channels by using a sample-and-hold and analog-to-digital conversion circuitry.

5.10 SAMPLING HARDWARE

This section covers hardware associated with sampling signals. First we look at the sample-and-hold circuit, sometimes referred to as a Sample-and-Hold Amplifier (SHA), and then discuss the ADC needed to produce PCM data representations. Both devices are required to perform sampling. They may be preceded by an initial analog multiplexer (MUX), as Figure 5.33 shows. This allows for a decreased component count if the sample-and-hold and analog-to-digital converter circuits are fast enough to sample the required number of signals within their specified sample rates.

Manufacturers supply integrated MUX/SHA/ADC components in one package to reduce component count, power consumption, and, most importantly, provide very good component matching and guaranteed frequency response. In all cases below, the manufacturer assumes that the user pre-filters the signals to insure the correct cutoff frequency for the sampling rate used.

5.10.1 Process Timing

The fact that a manufacturer may supply a MUX that allows several channels to be sampled does not imply that the device is able to sample all of the channels in the time required to meet the Nyquist rate requirements of each channel in any given telemetry system. In the case of the cascaded MUX, SHA, and ADC configuration, the estimate for maximum time to performing a conversion, T_{conv}, is

$$T_{conv} \approx t_{mux} + t_{acq} + t_{setl} + t_{setup} + t_{conv} \tag{5.40}$$

The times in Equation (5.40) are the estimated process set up and operational times. Here, t_{mux} is the time to switch and propagate a signal through the MUX. t_{acq} and t_{setl} are the acquisition and settling times

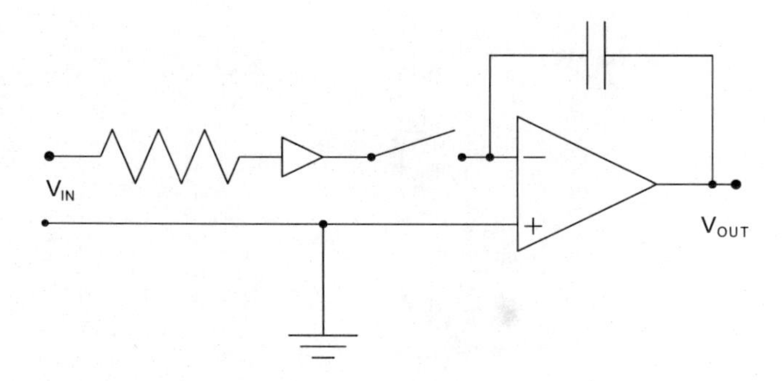

Figure 5.34: Sample-and-hold amplifier circuit components.

of the SHA. t_{setup} and t_{conv} are the setup and conversion times of the ADC. For the SHA and ADC, the two times listed are assumed to hold all of the significant delays in the conversion process. The final conversion time must match the signal sample rate to prevent aliasing, while also servicing all of the analog channels in the desired commutation time.

This is a worst-case situation because the MUX can be switched to the next signal in order to have the SHA start tracking it, while the ADC process is going on with the current data channel. This pipeline operation gives an improved performance. The amount of overlap possible between stages depends upon the exact characteristics of the devices used.

5.10.2 Sample-and-Hold Amplifiers

The SHA circuit's function is tracking the signal and, when commanded to do so, holding the last value at a stable level for the analog-to-digital conversion hardware to measure it [Gord80; Kest04a; Leon11a; Tewk80]. It then goes back into a tracking mode. In this process, we desire the output to change rapidly from track to hold modes, hold the voltage stable and exactly at the desired value, and then quickly slew to track when back in track mode. Most sample-and-hold circuits use an integrating/storage capacitor to smooth out any noise and to store the result.

Figure 5.34 shows a typical configuration for a sample-and-hold amplifier where a switched capacitor is used as a memory element to hold the voltage during the conversion process. Because the SHA is not perfect, we must rate the level of its imperfections. Figure 5.35 illustrates some of the typical quantities designers use. The primary characteristic is the *acquisition time* or the *transient response time*, defined as the

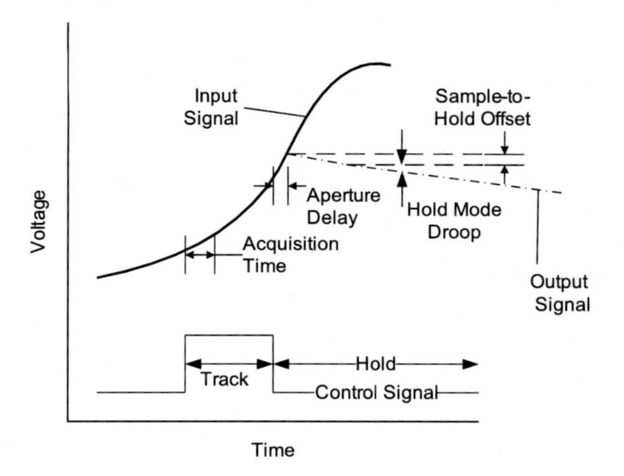

Figure 5.35: Sample-and-hold amplifier circuit operational characteristics.

time that the SHA requires for the hold capacitor to reach the full-scale voltage and remain within a specified error band around the final value. Typical specifications state that an input voltage can be acquired over a range of $\pm 10\,\mathrm{V}$ to within 0.01% within several hundred nanoseconds to 1 microsecond. This is important when the SHA is preceded by an analog multiplexer and must periodically switch from one input source to another. Other characteristics of SHA devices are as follows [Shei86; Shei04]:

Aperture Delay — The time elapsed from the command to hold until the switch is fully, electrically open; is caused by internal logic and switching delays. For a precise phase measurement, the control signal advances by this amount to ensure that the ADC samples the signal at the desired instant. This quantity is reasonably repeatable, but it does vary by the aperture jitter. Typical values for aperture delay are under 100 nanoseconds down to only a few nanoseconds.

Aperture Jitter — The variation in aperture delay from sample to sample; typically a few nanoseconds to less than 100 picoseconds. This results in errors in the sampling times and limits the maximum frequency that can be acquired by the SHA when sampling a single analog sinusoidal signal. In this case, for 1/2-LSB conversion accuracy, the maximum frequency that can be sampled due to aperture jitter is given in terms of the aperture jitter time, T_A, and the number of bits, N, to be used in the analog-to-digital conversion by the relationship [Leon97; Leon11b]

$$f_{max} = \frac{1}{T_A \pi 2^{N+2}} \tag{5.41}$$

Hold-Mode Droop — The rate of change in the output voltage during the "hold" period. Depending on the quality of the SHA, this parameter typically ranges from 0.01 µV/µs to 100 µV/µs.

Hold-Mode Feedthrough — The percentage of the input signal that appears in the output during hold mode. Manufacturers typically measure this parameter in dB and they list it as an attenuation or rejection over the valid input range. The value usually ranges from 66 dB to 85 dB.

Step Error or Sample-to-Hold Offset — The degradation in the output signal due to changing from sample mode to hold mode. The value is usually under several tens of millivolts. Many SHA circuits use an external resistor network to set this close to zero error.

5.10.3 Analog-to-Digital Converters

There are a number of different methods for performing the analog-to-digital conversion function. In this discussion, we examine three approaches to conversion: successive approximation converters, parallel or flash converters, and sigma-delta converters. Manufacturers classify the flash ADCs into two types: normal and dual conversion. In this section, we briefly look at the ways in which all three operate. We ignore refinements such as zero-point adjustments present in many commercial devices. These refinements give the devices greater accuracy but the exact method used and the amount of design required to realize the gains are very dependent upon the details of the chip structure. Usually the manufacturers' data sheets contain design examples for these support circuits.

5.10.3.1 Successive Approximation Converters

The successive approximation ADC family is one of the most widely used types of ADC and is available from a variety of manufacturers [Gord80; Kest04b; Shei86]. A successive approximation ADC attempts to synthesize the input signal within the converter. Figure 5.36 shows the

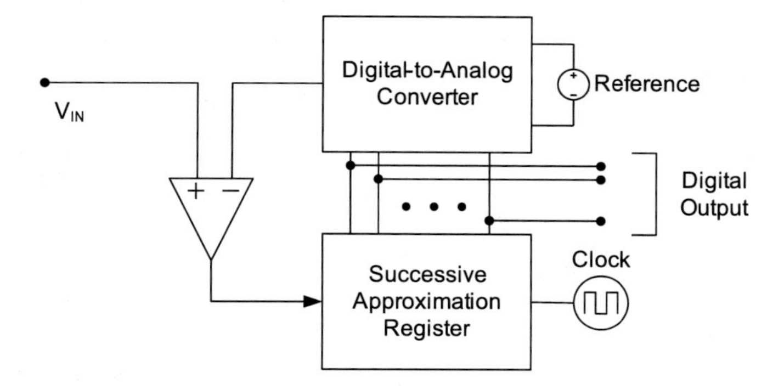

Figure 5.36: Successive approximation ADC elements.

synthesized signal cancels the input signal. The ADC is built to perform the quantization within N clock cycles where N is the number of output bits. At each clock cycle, the ADC adds $1/2^m$, $m = 1, 2, \cdots, N$ of the full-scale value to the previous value. If the resulting sum is less than the input voltage, then the converter keeps the value; if it exceeds the input voltage, the converter removes the last value.

Figure 5.37 illustrates the approximation process. Conversion speeds for off-the-shelf, 10-bit converters are approximately 10 to 100 µs. Many commercial converters allow output resolution to be user-selectable over a range, for example 8, 10, or 12 bits. Converters also have differing output modes. The data are available either as a parallel group of bits at the end of the conversion or as a serial bit stream. In the serial case, the most significant bit is available first and the least significant bit last, and each intermediate bit becomes available on successive clock cycles.

5.10.3.2 Flash Converters

Designers of high-speed applications like video processing typically use the parallel or flash ADC [Gord80; Kest04b; Shei86; Usha11]. Figure 5.38 illustrates that the circuitry consists of a series of op amp comparators, a precision-resistor voltage divider circuit, and a decoder circuit. The input signal is compared with the voltage stages in the voltage divider circuit. The quantization process takes only one clock cycle because all comparisons take place simultaneously. The ADC works by looking for the point in the voltage divider network where the input changes from being above the voltage drop to being below the voltage drop. Using this

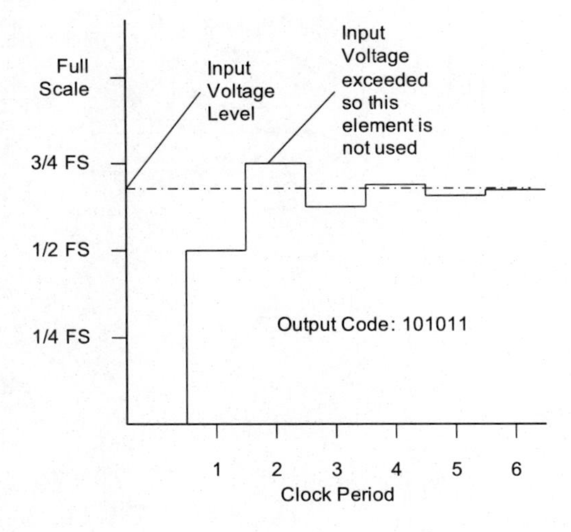

Figure 5.37: Successive approximation ADC operational process.

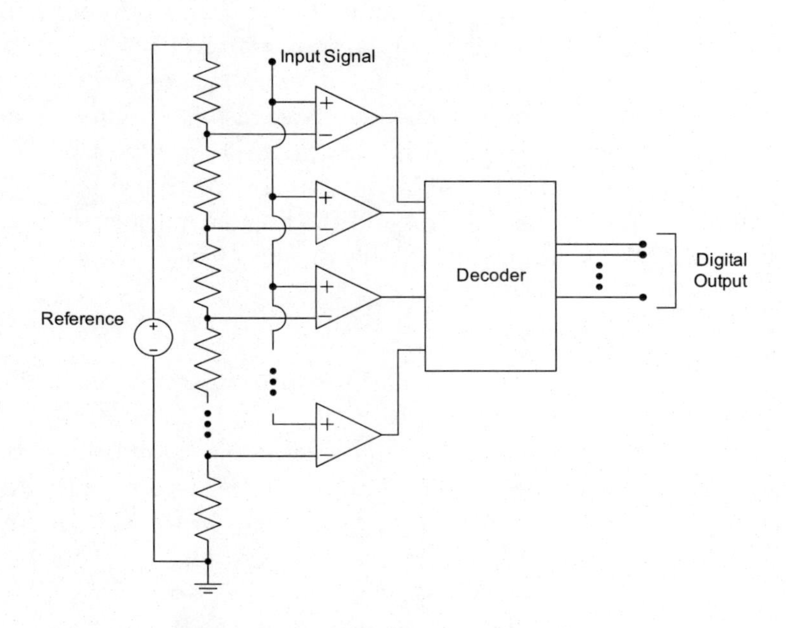

Figure 5.38: Parallel or flash ADC.

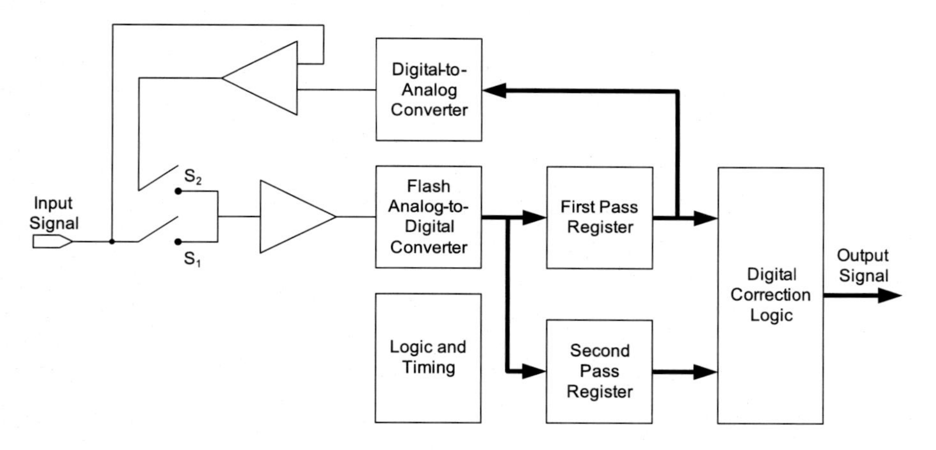

Figure 5.39: Dual-conversion flash ADC.

technique, 8-bit conversions achieve and can sustain 20 Msps rates. An 8-bit conversion requires 256 op amp comparators operating in parallel.

5.10.3.3 Dual Conversion Flash Converters

Single-pass flash ADC technology limits the number of useful levels in the converters. To achieve 16-bit conversions, for example, a dual-conversion flash ADC is needed. This converter requires a longer time to produce the output than a single-pass flash ADC; however, this is offset by the finer resolution this ADC produces. Figure 5.39 illustrates the typical elements of a dual-conversion flash ADC. On the first pass through the converter, the most significant bits are determined and stored. The converter also uses those bits as the input to a digital-to-analog converter that synthesizes the corresponding signal amplitude for the first conversion. The converter subtracts this amplitude from the input signal before the second pass through the converter. On the second pass, the converter determines the lower-order bits. The converter uses internal logic to control the timing through each pass and to mesh the results of the two conversion ranges to form a common output. The time to perform dual-conversion flash ADC is approximately 300 ns to 1 μs, depending on the technology and the number of bits in the conversion.

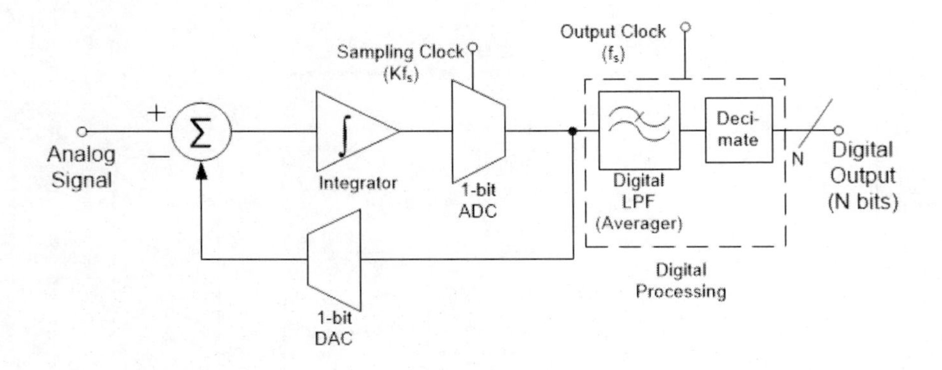

Figure 5.40: Block diagram for a single-bit Sigma-Delta ADC.

5.10.3.4 Sigma-Delta Analog-to-Digital Conversion

So far, we have been considering uniform sampling at something close to the Nyquist rate for the effective bandwidth of the signal. We also looked at the ENOB and how that ties to the full-scale SNR. The Sigma-Delta ($\Sigma\Delta$) ADC takes a different approach by combining a technique known as Delta Modulation with oversampling and digital post-processing to form the ADC [Kest04c; Gray87; Cand92].

Figure 5.40 illustrates a block diagram of a single-bit $\Sigma\Delta$ ADC. The key points to the ADC method are as follows:

- Using a differencing block to take the difference between the current input value and the accumulated input estimates to produce an error signal.
- Integrating the error signal.
- Sampling the error signal at a rate greatly exceeding the Nyquist rate for the analog input (Kf_s) with K about 100.
- Generating the difference signal via a DAC.
- Processing the oversampled signal using a digital filter (averaging) and decimating the output to bring the effective sample down to f_s.

This process provides the designer with several advantages. Notice that the ADC does not use an initial anti-aliasing filter or a SHA. The $\Sigma\Delta$ ADC runs in a continuous mode and the large oversampling rate moves any signal contribution above the effective bandwidth out of the band so it is not aliased. The analog signal's effective bandwidth still sets the desired sampling rate. A second important gain is that the oversampling process moves much of the quantization noise outside of the desired signal bandwidth. This gives a processing gain to the SNR and we have a better quantization signal-to-noise ratio.

Until recently, the $\Sigma\Delta$ ADC had a major downside: the analog input signal was generally limited to audio frequency ranges. Presently, designers find products with resolutions in the 16 to 24-bit range with up to 1 Msps sampling rate.

5.11 REFERENCES

[Benn48] W. R. Bennett. "Spectra of Quantized Signals." In: *Bell System Technical Journal* 27.3 (July 1948), pp. 446 –472.

[Blac59] R. B. Blackman and J. W. Tukey. *The Measurement of Power Spectra. From the Point of View of Communications Engineering.* New York, NY: Dover Publications, 1959.

[Cand92] J. C. Candy and G. C. Temes. "Oversampling Methods for A/D and D/A Conversion." In: *Oversampling Delta-Sigma Data Converters: Theory, Design, and Simulation.* Ed. by J. C. Candy and G. C. Temes. New York: IEEE Press, 1992, pp. 1 –25. NY.

[Couc01] L. W. Couch. *Digital and Analog Communications Systems.* 6th Edition. Upper Saddle River, NJ: Prentice Hall, 2001. ISBN: 0-13-081223-4.

[Gord80] B. M. Gordon. "Linear Electronic Analog/Digital Conversion Architectures, Their Origins, Parameters, Limitations, and Applications." In: *Data Conversion Integrated Circuits.* Ed. by D. J. Dooley. New York, NY: IEEE Press, 1980, pp. 32 –59. ISBN: 0-87942-132-0.

[Gray87] R. M. Gray. "Oversampled Sigma-Delta Modulation." In: *IEEE Trans. Commun.* COM-35.5 (May 1987), pp. 481 –489.

[Hart28] R.V.L. Hartley. "Transmission of information." In: *Bell System Technical Journal* 7 (3 July 1928), pp. 535 –563. DOI: 10.1002/j.1538-7305.1928.tb01236.x.

[Horo89] P. Horowitz and W. Hill. *The Art of Electronics.* 2nd Edition. New York, NY: Cambridge University Press, 1989.

[Hump70] D. S. Humpherys. *The Analysis, Design, and Synthesis of Electrical Filters.* Englewood Cliffs, NJ: Prentice Hall, 1970.

[John75] D. E. Johnson and J. L. Hilburn. *Rapid Practical Designs of Active Filters.* New York, NY: John Wiley & Sons, 1975. ISBN: 0471443042.

[Kamm00] D. W. Kammler. *A First Course in Fourier Analysis.* Upper Saddle River, NJ: Prentice Hall, 2000. ISBN: 0-13-578782-3.

[Kest04a] W. Kester. "Section 2.2: Sampling Theory." In: *Analog-Digital Conversion.* Ed. by W. Kester. Analog Devices, Inc., 2004. Chap. 2, pp. 2.23 –2.36. ISBN: 0-916550-27-3. URL: http://www.analog.com/library/analogDialogue/archives/39-06/data_conversion_handbook.html.

[Kest04b] W. Kester and J. Bryant. "Section 3.2: ADC Architectures." In: *Analog-Digital Conversion.* Ed. by W. Kester. Analog Devices, Inc., 2004. Chap. 3, pp. 3.39 –3.108. ISBN: 0-916550-27-3. URL: http://www.analog.com/library/analogDialogue/archives/39-06/data_conversion_handbook.html.

[Kest04c] W. Kester and J. Bryant. "Section 3.3: Sigma-Delta Converters." In: *Analog-Digital Conversion.* Ed. by W. Kester. Analog Devices, Inc., 2004. Chap. 3, pp. 3.109 –3.140. ISBN: 0-916550-27-3. URL: http://www.analog.com/library/analogDialogue/archives/39-06/data_conversion_handbook.html.

[Lath98] B. P. Lathi. *Modern Digital and Analog Communication Systems.* 3rd Edition. New York, NY: Oxford University Press, 1998.

[Leon11a] R. E. Leonard. *Picking the Right Sample-and-Hold Amp for Various Data-Acquisition Needs. DATEL Application Note AN-2.* Apr. 2011. URL: http://www.datel.com/data/ads/adc-an2.pdf.

[Leon11b] R. E. Leonard. *Data Converters: Getting to Know Dynamic Specs. DATEL Application Note AN-3.* 2011. URL: http://www.datel.com/data/ads/adc-an3.pdf.

[Leon97] R. E. Leonard. *Understanding Data Converters' Frequency Domain Specifications. DATEL Application Note AN-4.* Oct. 1997. URL: http://www.datel.com/data/ads/adc-an4.pdf.

[Meik90] Z. H. Meiksin. *Complete Guide to Active Filter Design, OP AMPS, and Passive Components.* Englewood Cliffs, NJ: Prentice Hall, 1990. ISBN: 0131599712.

[Mosc81] G. S. Moschytz and P. Horn. *Active Filter Design Handbook.* Chichester: John Wiley & Sons, 1981.

[Nyqu24] H. Nyquist. "Certain factors affecting telegraph speed." In: *Bell System Technical Journal* 3 (2 Apr. 1924), pp. 324 –346. DOI: 10.1002/j.1538-7305.1924.tb01361.x.

[Oliv48] B. M Oliver, J. R. Pierce, and C. E. Shannon. "The Philosophy of PCM." In: *Proc. IRE* 36 (11 Nov. 1948), pp. 1324 –1331. DOI: 10.1109/JRPROC.1948.231941.

[Oppe10] A.V. Oppenheim and R.W. Schafer. *Discrete-Time Signal Processing.* 3rd ed. Pearson Education, 2010. ISBN: 9780131988422.

[Sall55] R. P. Sallen and E. L. Key. "A Practical Method of Designing RC Active Filters." In: *IRE Trans. Circuit Theory* 2.1 (Mar. 1955), pp. 74 –85. DOI: 10.1109/TCT.1955.6500159.

[Shan49] C.E. Shannon. "Communication in the Presence of Noise." In: *Proceedings of the IRE* 37 (1 Jan. 1949), pp. 10 –21. DOI: 10.1109/JRPROC.1949.232969.

[Shei04] D. Sheingold and W. Kester. "Section 2.5: Defining the Specifications." In: *Analog-Digital Conversion.* Ed. by W. Kester. Analog Devices, Inc., 2004. Chap. 2, p. 2.103. ISBN: 0-916550-27-3. URL: http://www.analog.com/library/analogDialogue/archives/39-06/data_conversion_handbook.html.

[Shei86] D. H. Sheingold, ed. *Analog-Digital Conversion Handbook.* 3rd Edition. Englewood Cliffs, NJ: Prentice Hall, 1986. ISBN: 0-13-032848-0.

[Skla01] B. Sklar. *Digital Communications: Fundamentals and Applications.* 2nd Edition. Upper Saddle River, NJ: Prentice Hall PTR, 2001.

[Tewk80] S. K. Tewksbury et al. "Terminology Related to the Performance of S/H, A/D, and D/A Circuits." In: *Data Conversion Integrated Circuits.* Ed. by D. J. Dooley. New York, NY: IEEE Press, 1980, pp. 262 –269. ISBN: 0-87942-132-0.

[Usha11] R. K. Ushani. *Subranging ADCs. DATEL Application Note AN-5.* Apr. 2011. URL: http://www.datel.com/data/ads/adc-an5.pdf.

5.12 PROBLEMS

1. Find the essential bandwidth for a battery discharge curve where the output voltage is represented by $v(t) = 12\mathrm{e}^{-t/\tau}\,\mathrm{u}(t)$ volts. Here, τ is the battery time constant of 1 h. What is the minimum sampling rate to satisfy the Nyquist criteria?

2. Find the essential bandwidth for a square pulse. Plot the signal energy as a percentage of the total signal energy versus frequency.

3. Given that the output of a process that samples an analog function

$g(t)$ to produce the sampled function $g_S(t)$ that may be written as

$$g_s(t) = g(t) \left\{ \sum_{n=-\infty}^{\infty} \delta\left(t - nT_S\right) \right\} \otimes \text{rect}\left(\frac{t}{\tau}\right)$$

Find the Fourier transform of $g_S(t)$ in terms of τ and T_S assuming that $\tau = 0.01T_S$.
If f_m is the highest frequency of interest in the baseband Fourier transform of $g(t)$, then show from the Fourier transforms what must be the minimum relative ratio of f_S to f_m to avoid spectrum overlap in the frequency domain.

4. Sketch the magnitude and phase plots that correspond to Figure 5.18 for the ideal BPF and the ideal HPF.

5. Verify that the two-pole LPF has the transfer function given in Equation (5.21) in the text.

6. Rewrite Equation (5.21) in magnitude and phase form. Write down the total magnitude and phase equation if two stages of the basic building block are cascaded together to make a fourth-order filter.

7. Compute the phase response for Butterworth, Bessel, and 1.0 dB Chebyshev LPFs for 2nd, 4th, and 6th orders. How does the phase linearity change for the filters at the same order and as the filter order increases? Does the linearity region get smaller relative to the cutoff frequency as the order increases?

8. Design a filter that meets the following specifications and draw the circuit diagram for it:

   (a) Cutoff frequency = 10 kHz

   (b) Initial stop band frequency = 30 kHz

   (c) Minimum stop band attenuation = 50 dB

   Find the amplitude and phase response for the filter, and plot them versus frequency. Verify that the filter design meets the requirements. Use a computer aid such as Multisim® or MATLAB® to generate the amplitude and phase plots.

9. Design a filter that meets the following specifications and draw the circuit diagram for it:

(a) Lower cutoff frequency = 10 kHz

(b) Upper cutoff frequency = 80 kHz

(c) Lower transition region ends at 1 kHz

(d) Upper transition region ends at 500 kHz

(e) Minimum stop band attenuation = 40 dB

Find the amplitude and phase response for the filter, and plot them versus frequency. Verify that the filter design meets the requirements. Use a computer aid such as Multisim® or MATLAB® to generate the amplitude and phase plots.

10. For the signal in Problem 1, how many quantization bits do you estimate are required to achieve an average quantization signal-to-noise ratio of 30 dB for quantization over the essential bandwidth? How many bits per second does the Nyquist criterion imply? If you use a more realistic (five times the Nyquist rate) sampling frequency, how many bits per second are transmitted?

11. Derive the equation for ENOB from the full-scale SNR and the uniform quantization noise expressions.

12. Suppose the system designer wishes to have 4 dB of headroom in the conversion process over the design with a specified average quantization SNR. How should Equations (5.35) and (5.36) be modified to account for this extra range?

13. Consider an ADC configuration with a periodic waveform on the input, which uses the full-scale of the ADC. Derive an expression similar to that found for the peak signal-to-quantization noise ratio for the case of average signal-to-quantization noise for uniform quantization in terms of the number of quantization levels and the voltage width of the quantization level. Does the result vary if the ADC's input signal is a ramp signal or a sinusoidal signal?

14. Consider the signal $x(t)$ given by

$$x(t) = 5\sin(\pi t) + 5\sin(\mathrm{e}\,t) + 5\sin(\mathrm{e}\,\pi t)$$

where e and π are only to be considered shorthand for the frequency in radians/second. For this continuous-time signal, make the following computations:

(a) Sample $x(t)$ in increments of 0.05 s for a total of 100 samples.

(b) Quantize the samples using 4, 8, and 12 bits and plot the results.

(c) Compute the root mean square (rms) error between each of the quantization resolutions and the true signal value.

For each quantization level, what is the output bit rate? *Hint:* plot the unsampled function to determine the range of the function. It is at most 15 V but may be several volts less. Your quantization should cover this range fully (no clipping) and without many wasted bins.

15. For the data signal in the Problem 14, plot the output signal-to-noise ratio as a function of the number of quantization levels, L. Allow L to range from 64 to 512. Assume that the channel probability of error is 0.001. Repeat with a channel probability of error of 0.000 001.

16. Using the results of Problem 14, recover the original signal by low-pass filtering the sampled data. Plot the recovered signal with the original signal on the same axis.

17. Suppose a SHA is not used as part of the ADC sampling architecture. What is the maximum input frequency that a full-scale sinusoidal signal can have assuming a 10-bit conversion process with a sampling frequency of 100 ksps? In this sampling, the input signal cannot change by more than q during each sample period.

18. Consider a set of MUX/SHA/ADC devices with the following specifications to be cascaded together to form a data acquisition system. Show a timing diagram for converting four analog input channels to a digital data stream. Show the worst-case timing diagram (no overlapping of conversion cycles from one analog channel to the next). What is the maximum conversion rate in this mode? Can any parts of the process be overlapped? If so, draw a modified timing diagram and give the estimated maximum conversion rate in this overlap mode.

MUX

Analog input channels = 4
Switching time o $\leqslant$ 25 ns

Output settling time $\leqslant$ 225 ns

SHA

Acquisition time $\leqslant$ 1 µs
Aperture delay $\leqslant$ 30 ns
Aperture uncertainty = 1 ns
Aperture time = 25 ns
Hold mode settling time $\leqslant$ 185 ns

ADC

Parallel output bits = 12
Analog input setup time = 100 ns prior to start-conversion pulse
Start conversion pulse $\geqslant$ 100 ns
Output valid $\leqslant$ 250 ns after falling edge of start-convert pulse

II

DATA TRANSPORT, TIMING, AND SYNCHRONIZATION

Telemetry data transmission from the payload segment to the user segment and the commands from the user to the payload are typical communications problems encountered in many settings. Traditionally, the telemetering community has used a time-division method of packaging the data into link-level transmission data frames for transport over radio links. With the growth of the Internet, the telemetry data transmission process often looks like computer-to-computer communications as users find more ways to utilize computers and telecommunications for data acquisition and distribution. These data communications will occur over wireless and wired links as well as fiber optic links.

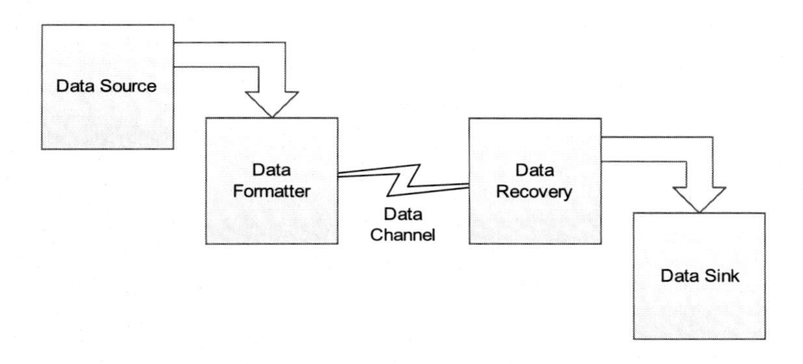

Data transmission for telemetry and telecommand data messaging.

In Part II, we investigate the following topics:

Telemetry Frames and Packets: An introduction to the data link layer data formatting used in reliable and unreliable links including methods for encoding the bit waveforms and the management information included in the transmission protocols

Data Synchronization: Methods for receivers to lock onto the transmitted data stream from the carrier waveform through final data values and the effects of errors on this process

Time and Position Determination: Methods for defining time, encoding the timing information in the data stream, and Global Positioning System (GPS) position and timing information

Telecommand Transmission Systems: Methods for users to generate commands and transmit them to the receiving system

Each format for data transmission has a matching method for data synchronization. The system requires this because the user segment must align itself with the initially unknown state of the payload's data clock. The techniques for time and position determination are useful for data logging, data analysis, and coordinating system operations. With the command system, we will also examine ways in which it interacts with the telemetry data to form a closed control system.

CHAPTER 6

TELEMETRY FRAMES AND PACKETS

6.1 INTRODUCTION

The payload transmits the data produced by the sensors to the operator's receiver. The payload's transmission system gives structure to the data transmission so that the receiver can synchronize itself to the transmitter and locate the individual sensor values. Traditionally, Pulse Code Modulation (PCM) telemetry systems have used a data structure known as a telemetry frame as the standard method for packaging the transmitted data when using radio techniques. Because of the Internet's growth, telemetry system designers also frequently use data packaging techniques based on computer-network communications standards. Designers also use these packet standards for command transmission.

The highlighted blocks in Figure 6.1 illustrate the locations of the subsystems to perform the payload's data packaging function. This chapter explains how the standard frames and packets are constructed. In particular, we examine the following issues common to all digital communications networks:

- Terminology
- Synchronization codes
- Signal commutation
- Channel layer data packaging
- Bit waveform formatting

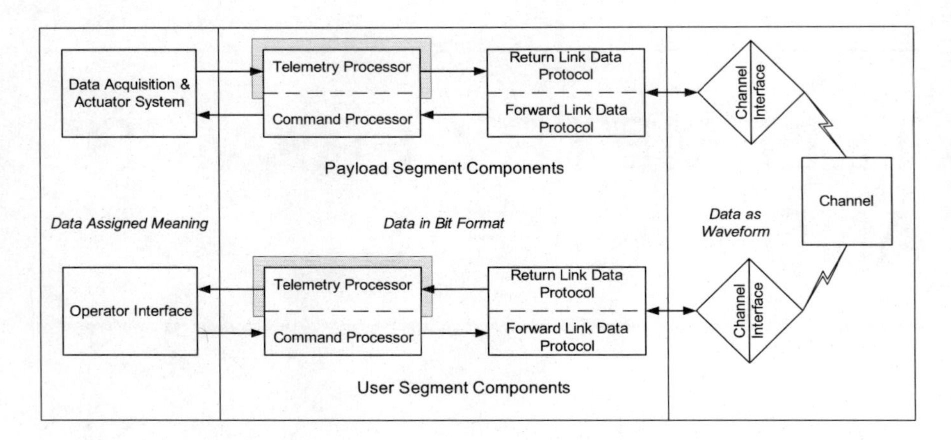

Figure 6.1: The highlighted blocks indicate the locations in the payload and operator segments involved in telemetry data transmission.

While these concepts are common to all telecommunications networks, the emphasis here is on telemetry and telecommand communications networks.

6.2 OBJECTIVES

The study of telemetering system frames and packets is similar to, and uses much of the same terminology as, non-telemetry data communications in digital communications networks. However, many specifics apply only to the telemetering field. Our objectives for this chapter are to enable the reader to

- Describe the standard Inter-Range Instrumentation Group (IRIG) frame structure and be able to define such terms as major and minor frame, and commutated, subcommutated, and supercommutated data.

- Identify synchronization methods within the frame structure.

- Efficiently present a data set in the IRIG frame format.

- Describe the strengths and weaknesses of packet telemetry versus frame telemetry systems and describe the basic packet structure requirements for packet telemetry systems.

- Describe the means, purposes, and results of data formatters used prior to digital data transmission.

The IRIG and packet formats have many commonalities with commercial networks even though they are formalized for telemetry and telecommand transmissions.

6.3 BACKGROUND

Before we examine the details of the data frame and packet structures used in telemetry transmissions, we first look at some common concepts to both techniques. These techniques are also useful for transmitting the telecommand packets that we will discuss in Section 9.6.

6.3.1 Context

The transmitting node has a definite advantage over the receiving node in the data communications network. Even if the payload and the operator are the only two entities in the network, the transmitter's job is easier than the receiver's job. This comes about because the transmitter is usually maintaining a transmission protocol it knows; it also knows where it stands in the protocol's time sequence. The receiver may know the protocol structure but it still needs to lock onto the protocol timing without much more information beyond the data. From the receiving node's point of view, the transmitter starts out as an asynchronous data source and the receiver must become synchronized with it. Hindering this process may be link artifacts such as

- Data dropouts or related link problems
- Errors in data transmission due to noise or other users
- Variable time delays in the transmission process

The payload designer's job is to help the receiving node in the synchronization process by the choice of "good" transmission protocols. All of these effects are potential problems for both frame and packet-based links.

Why do we make a distinction between frame-based and packet-based systems? The link types make different assumptions about link quality and reliability. The system designer's format choice depends on the designer's assessment of which method provides the desired quality of service in the data delivery over the actual link.

6.3.2 Data Link Layer Packaging

In the standard Open Systems Interconnection Reference Model, the telemetry data frame or data packet is a Level II or Data Link Layer construct. Level II is concerned with point-to-point data transmission and having the transmitter and receiver exchange a Protocol Data Unit (PDU). The PDU contains the necessary data structures, error checks, and management accounting to support the data transmission. In this view, the difference between frames and packets is simply the difference required to support the point-to-point link communications. In this chapter, we treat them not as separate entities but as variations on the same communications structure. In telecommunications terminology, the Level II PDU is often called a *frame* where a frame is defined as

> the sequence of contiguous bits delimited by, and including beginning and ending flag sequences. A frame usually consists of a specified number of bits between flag sequences and usually includes an address field, a control field, and a frame check sequence. Frames usually consist of the original data together with other bits that the receiver uses for error detection or control. Frame designers use additional bits for routing, synchronization, or overhead information not directly associated with the original data. In the multiplex structure of PCM systems, a set of consecutive time slots in which the position of each digit can be identified by reference to a frame-alignment signal. In a Time Division Multiplexing (TDM) *frame* system, a repetitive group of signals resulting from a single sample of all channels, including any required system information, such as additional synchronizing signals [ATIS].

This definition covers both telemetry frames and telemetry packets.

6.3.3 Commutation

A feature of both frame and packet telemetry is that the transmitted data channels are not continuous data transmissions. Rather, as we saw in Chapter 5, the payload produces the sampled data. Since the system transmits data from more than one sensor, there needs to be a method for time-sharing the data channel. Engineers call the process of sampling and supplying a time slot to each sensor *commutation*. We examine how designers approach commutation in both frame and packet modes.

6.4 TELEMETRY FRAMES

Telemetry frames are data link protocols designed to support TDM of sensor data over point-to-point link configurations. The goal of frame design is to provide data delivery over links that the designer assumes to be inherently unreliable due to poor link quality, have the potential for dropouts at inappropriate times, and/or need rapid resynchronization if the link is temporarily lost. Since the link is based on a point-to-point transmission mode, there is generally no need for address information within the frame unless a common collection point is processing several data sources simultaneously.

In many applications, such as measuring the status of a continuous process, the frame data is part of a continuous sequence and not a "one-time-only" measurement. In that case, the frame-based system is not designed for the ability to retransmit any data lost to link problems so the use of error-detecting techniques is not always found. Therefore, designers employ a strict TDM format that has a rich structure with well-defined locations for each sensor's information relative to the synchronization markers. This TDM structure has standard markers included in the frame to allow the receiver to resynchronize in a reasonable amount of time. This section discusses the general frame structure and the markers embedded in it to assist the receiver in synchronizing to the data structure. Naturally, if the data are unique or can only be gathered once, the designer may include some form of error detection and correction or retransmission if the system capacity allows it.

The general terms used with frames are standardized. However, the payload designer is still responsible for deciding the commutation strategy for the sensors because the standards do not cover that level of detail. The designer must consider sensor Nyquist rates and determine what management information is needed to ensure that accounting for the sensor values is possible.

6.4.1 Inter-Range Instrumentation Group Frame Definitions

As noted above, a TDM frame structure is composed of the data, synchronization markers, and management information necessary to find and account for all of the data. The designers of frame-based telemetry systems use this communications concept to organize the sampling of each sensor so that they transmit the entire data set in one cycle of the overall frame structure. Once the transmitter completes one cycle

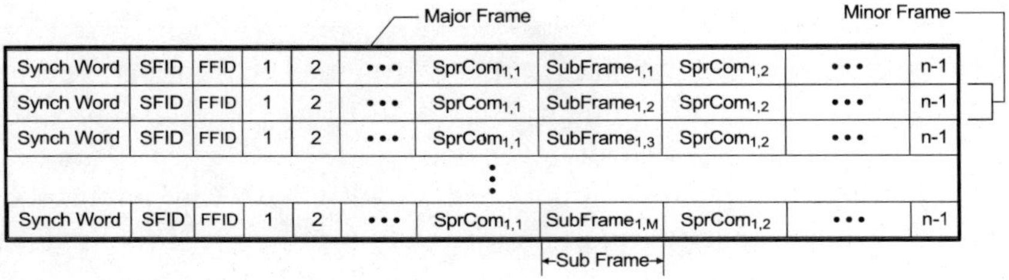

Major Frame

Minor Frame

Synch Word	SFID	FFID	1	2	•••	$SprCom_{1,1}$	$SubFrame_{1,1}$	$SprCom_{1,2}$	•••	n-1
Synch Word	SFID	FFID	1	2	•••	$SprCom_{1,1}$	$SubFrame_{1,2}$	$SprCom_{1,2}$	•••	n-1
Synch Word	SFID	FFID	1	2	•••	$SprCom_{1,1}$	$SubFrame_{1,3}$	$SprCom_{1,2}$	•••	n-1
							⋮			
Synch Word	SFID	FFID	1	2	•••	$SprCom_{1,1}$	$SubFrame_{1,M}$	$SprCom_{1,2}$	•••	n-1

Sub Frame

Figure 6.2: IRIG major frame structure containing minor frames, subframes, and synchronization information. The recommended locations for the Subframe Identifier (SFID) and Frame Format Identifier (FFID) are also shown.

of the data set, the next cycle starts over at the beginning of the frame structure. As we saw in Section 4.4.2, the system designers use details of this frame structure in configuring the operator's telemetry database.

Figure 6.2 shows the IRIG Standard 106 frame format [IRIG106]. While the developers formulated this standard for PCM telemetry systems at government test ranges, designers in industry use the terminology and definitions so we follow this standard in all discussions in this section. As we see in the figure, the data set held by the frame structure forms a large matrix. The payload samples all the sensors at least once within one cycle through the matrix. The payload may sample some sensors more than once depending upon criticality or Nyquist sampling requirements. The IRIG standard definitions relating to telemetry frames appear in the following paragraphs. For further information on these definitions, see [IRIG106].

6.4.1.1 Minor Frame

A *minor frame* is a fixed-length sensor data block plus any necessary synchronization markers and management information. When we view the structure in Figure 6.2 as a matrix, the minor frames are the rows in the matrix. The data block forming the minor frame is constrained to be a fixed number of bits in length, e.g., 8192 or 16 384. The frame designer divides these bits into an integer number of fixed-length (e.g., 8- or 16-bit) words. The structure that partitions the minor frame into synchronization markers, data, and management parameters remains fixed from one minor frame to the next. The structure fixes the information location relative to the minor frame's start.

The standard defines the minor frame length as the number of bits between the start of successive frame synchronization markers. Each marker, or *SynchWord*, is located as the first word in a minor frame. However, some older configurations place the synchronization marker at the minor frame's end. This may cause problems at the start of the synchronization process. The first data location after the synchronization marker is, by convention, numbered as data word 1. The last data location is numbered word $N - 1$ yielding a total length of N words in a minor frame when the synchronization word is included. The word numbering is independent of the number of bits per word at each location; i.e., both 8-bit words and 16-bit words may appear, for example, in a given minor frame format.

Within the minor frame, we describe the frequency with which the payload transmits the data relative to the minor frame rate. The minor frame rate is determined by the time between successive appearances of the initial synchronization marker. It is typical for the payload to sample parameters in the data set at a rate different from the minor frame rate. This occurs when the minor frame rate is either too fast or too slow for the signal's sampling requirements. In that case, we have various degrees of commutation to designate the frequency of sending a parameter relative to the minor frame rate.

Figure 6.3 illustrates data coming into a minor frame sampler from several sources. We divide the data into classes depending upon the transmission frequency relative to the minor frame rate. In later paragraphs, we will examine all of these sampling classes.

6.4.1.2 Major Frame

A *major frame* is an integer number of minor frames such that the payload samples each telemetry parameter at least once. The minor frame does not need to sample each sensor once per minor frame but the payload must sample all sensors at least once per major frame. In Figure 6.2, the major frame corresponds to the whole matrix. The IRIG standard sets the maximum major frame length to 256 minor frames. Some engineers use the nonstandard term "master frame" for the same concept as the major frame here.

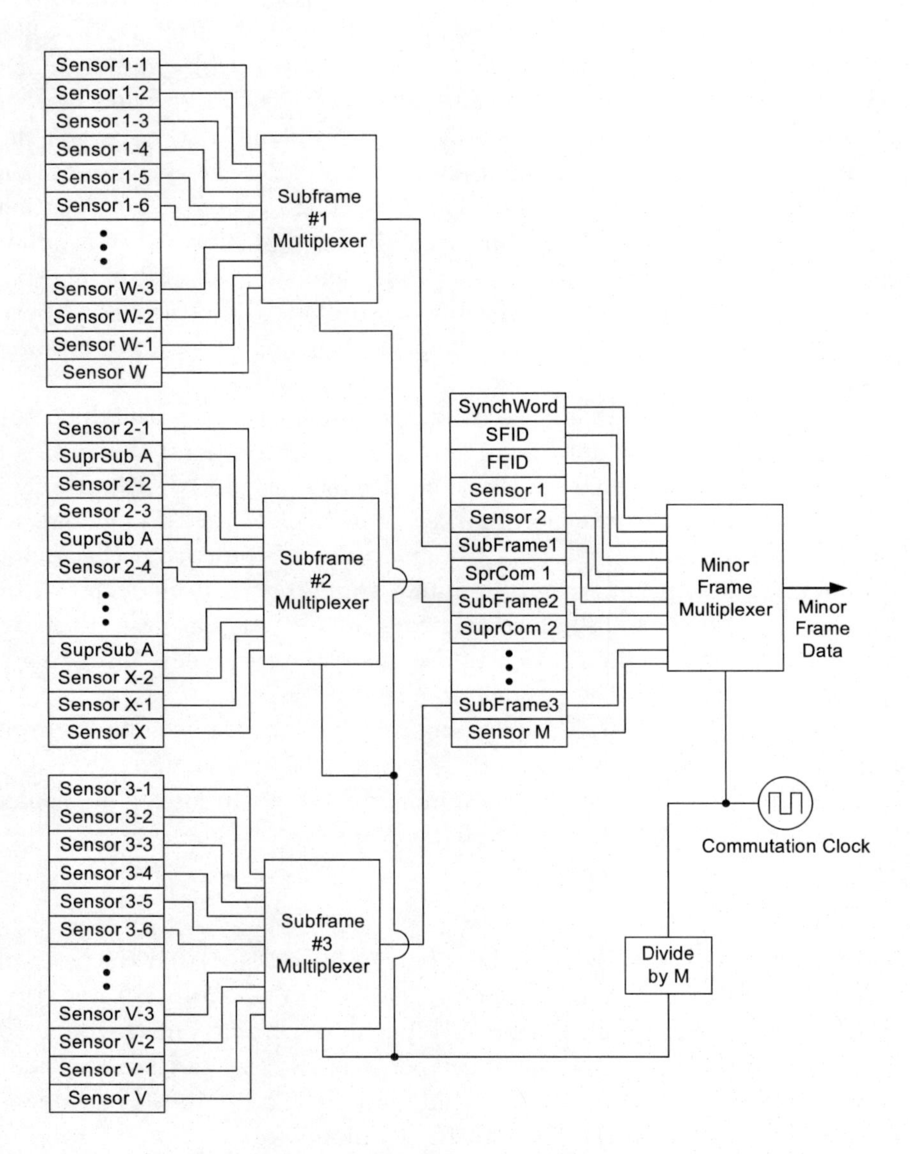

Figure 6.3: Sensor commutation to form a minor frame. Minor frame, subframe, and supercommutated sensor sampling with a Frame Format Identifier are included.

6.4.1.3 Commutated Data

Most payloads use a basic clock period and then sample all sensors at integer multiples either greater than or lower than this basic rate. This basic clock rate determines the minor frame rate. The payload sends *commutated* signals once per minor frame and they appear at the same location relative to the synchronization marker in each minor frame. In the IRIG frame illustration in Figure 6.2, these signals appear in the same column when the minor frames are stacked as a matrix. The columns labeled "1" and "2" in Figure 6.2 represent data from different commutated sensors. Figure 6.3 illustrates that, from a sampling point of view, the commutated signals are the signals from *Sensor 1*, *Sensor 2*, and *Sensor M* that the designer fed directly into the minor frame multiplexer on the right side of the figure.

6.4.1.4 Supercommutated Data

The payload may need to sample some signals at a higher rate than the basic minor frame clock rate to fulfill Nyquist rate requirements or to address some other critical need in the payload. *Supercommutated* signals are telemetry parameters that appear at a sampling rate that is an integer factor greater than the minor frame rate. They always appear at a fixed location relative to the synchronization marker within the minor frame.

Figures 6.2 illustrates a supercommutation example. The column holding $SprCom_{1,1}$ is the first appearance of the first supercommutated sensor, while the column holding $SprCom_{1,2}$ is the second appearance of the first supercommutated sensor. These readings represent two separate samples of the same sensor — and not two different sensors. The number of appearances of supercommutated data in each minor frame is not fixed in the IRIG standard. The frame designer must determine the supercommutation approach based on an analysis of the system's data transmission needs. Relative to the minor frame multiplexer on the right side of Figure 6.3, the supercommutated sensor is the signal labeled $SprCom_{1,1}$ and $SprCom_{1,2}$ with two inputs to the minor frame multiplexer.

6.4.1.5 Subframes and Subcommutated Data

Just as the payload needs to sample some sensors at a rate greater than the minor frame clock rate, the payload also has sensors that it does not sample as quickly as the minor frame clock rate. *Subcommutated*

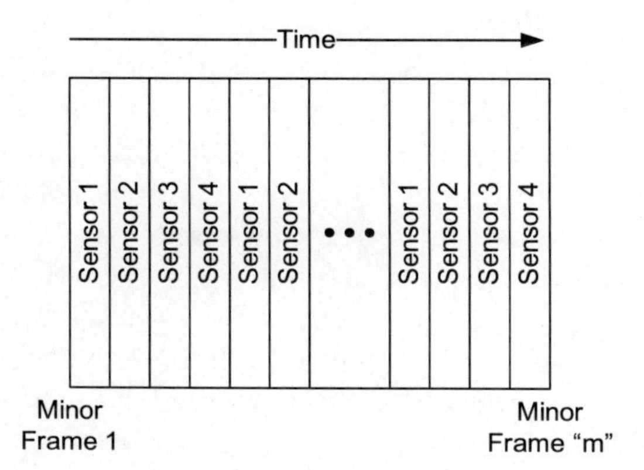

Figure 6.4: A subframe that is part of a major frame structure. This is one column in the matrix describing the major frame.

signals are telemetry parameters that appear at a sampling rate less than the minor frame rate; each parameter appears within a fixed subframe location.

Subframes correspond to fixed locations within the minor frame; they are columns within the major frame. Subframes are for signals coming from a common subsystem sampler such that subsystem samples all subframe values at least once per major frame. Three subframe multiplexers appear on the left side of Figure 6.3. Each of these multiplexers samples a group of sensors and sends the data values to the minor frame multiplexer. The payload gathers the output of the subframe multiplexer once per minor frame so the individual sensor reading appears less frequently than once per minor frame.

Figure 6.2 provides an example of a subframe where the subframe forms a column in the major frame matrix. In contrast with a commutated or supercommutated sensor that has the same sensor's data appearing in each minor frame location, the subframe has different sensors appearing at each row within the column of each minor frame. Figure 6.4 shows the subframe again where it is composed of the readings from four sensors. Each of those sensors has subcommutated data since their readings appear less frequently than once per minor frame.

The sensor readings cycle through the minor frame so, while they always appear in the same column of the major frame matrix, one needs to know the number of the exact minor frame row within the major frame to know which sensor's reading is currently being sent within the minor frame. The frame designer uses a SubFrame Identifier (SFID) to label each minor frame so that the sensor is identifiable within the subframe. The SFID is merely a counter that increments (or decrements) with each minor frame of the major frame. At the start of the major frame, the counter is set to its initial value, usually 1. It does not count continuously to give the total number of minor frames in the entire transmission. Rather, it counts to the end of the major frame and then resets. The IRIG standard recommends placing this subframe identifier immediately after the synchronization word and before any subframes in the minor frame structure. Locating the SFID in the first column after the synchronization word in Figure 6.2 is consistent with the IRIG recommendation.

6.4.1.6 Supersubcommutated Data

Supersubcommutated data are subframe data that appear more than once per subframe. This method is also referred to as "supercom on a subframe" because that is how the data are sampled. Figure 6.3 shows an example of supersubcommutated data. The *SuprSub A* sensor input appears multiple times on the multiplexer for the second subframe. Actually, the data in Figure 6.4 are also supersubcommutated because they appear more than once per subframe although they appear less frequently than once per minor frame.

6.4.2 Frame Examples

Figures 6.5, 6.6, and 6.7 show three examples of telemetry minor frames used in real data acquisition systems. In each frame, the frame designers gave parameter names as mnemonics that make sense to the data users, but the mnemonics are not meaningful for our use here. In Figures 6.6 and 6.7, the frames are relatively long so the parameters wrap from one line to the next until the whole minor frame structure is complete.

Figure 6.5 illustrates a single telemetry minor frame for a data acquisition system measuring the intensity of two satellite radio propagation beacons — one at 20 GHz and the other at 27 GHz [Rema93]. It also measures the intensity of the sky radiation at those frequencies using radiometers and measures local atmospheric parameters such as temperature, humidity, rain gauge output, etc. Finally, the system also monitors internal voltage levels in the hardware, settings of the programmable electronics, and temperatures of the electronics for calibration and fault

Time Stamp	20 GHz Beacon	27 GHz Beacon	20 GHz Radio	27 GHz Radio	Status Word #1	Status Word #2
4 bytes	2 bytes	2 bytes	2 bytes	2 bytes	2 bytes	2 bytes

Figure 6.5: The first telemetry frame example of a minor frame from a major frame with 60 minor frames. All minor frames use this same format.

VX	VY	O	TSP1	TSP2	VX	VY	SFID	TSP1	TSP2
VX	VY	LTEM1	TSP1	TSP2	VX	VY	LTEM2	TSP1	TSP2
VX	VY	LTEM3	TSP1	TSP2	VX	VY	LTEM4	TSP1	TSP2
VX	VY	SPNPR	TSP1	TSP2	VX	VY	PAIP	TSP1	TSP2
VX	VY	POWREL	TSP1	TSP2	VX	VY	VKR	TSP1	TSP2
							VCL		
FS1	FS2								

Figure 6.6: The second telemetry frame example of a minor frame from a major frame with 2 minor frames.

FS1	FS2	SCID	SFID	CMD CNT 1	CMD WORD 1	CMD WORD 2	CMD WORD 3
CMD CNT 2	CMD WORD 1	CMD WORD 2	CMD WORD 3	SF A	ACS 1	SF B	ACS 2
SF C	ACS 3	SF D	ACS 4	SF E	PWR 1	SF F	ANALOG
SF G	ANALOG	SF H	ANALOG	SF J	ANALOG	SF K	DIGITAL
SF L	SF P	SF Q	SF R	SF T	SF V	D1	SW
D2	D3	D4	D5	D6	D7	D8	D9
D10	D11	D12	D13	D14	D15	D16	D17
D18	D19	D20	D21	D22	D23	D24	D25

Figure 6.7: The third telemetry minor frame example for a major frame with 64 minor frames.

detection. The system has 60 minor frames per major frame. A time stamp, the measurement of the two beacons, and the two radiometers are commutated parameters since they appear in each minor frame. The two status words contain other parameters that the system measures on a schedule. The system measures the highly variable weather rainfall parameters once every six seconds, while it measures the slower-changing weather measurements and the status voltages once per minute. This frame structure does not have a SFID since the seconds portion of time tag is used for this purpose.

The second example (Figure 6.6) is for a system containing two minor frames per major frame. Each minor frame has fifty-two 10-bit words and contains four parameters that are supercommutated (*VX*, *VY*, *TSP1*, and *TSP2*) and one subframe. The subframe is one word wide and alternates between the parameters *VKR* and *VCL*. Notice the location of the SFID in the first line of the frame. The frame designer allocated ten bits to a 1-bit counter. It is typical for designers to slip other information into these words to use them more efficiently. Notice that the frame synchronization words, *FS1* and *FS2* are located at the frame's end.

The third minor frame example, in Figure 6.7, is from a major frame structure that contains 64 minor frames [NASA85]. Each minor frame is sixty-four 8-bit words long. This example has 16 subframes per minor frame and the subframes are designated *SFA* through *SFV*. The system commutes other parameters and they appear once per minor frame. The 16-bit synchronization marker (*FS1*, *FS2*) is located at the minor frame's start. Notice the command system feedback supported in the minor frame. *CMD CNT 1* and *CMD CNT 2* are counters to tell the operators how many commands the prime and backup command processors have received. The *CMD WORD* values that follow the counters hold the actual command data words received by each command processor so that the operators know which commands the payload is processing.

Telemetry frame processing works by inverting the construction process. For example, Figure 6.8 illustrates the telemetry frame for the balloon payload example given in Section 4.5.4. The frame is a mixture of PCM-encoded voltage values in the Analog-to-Digital Converter (ADC) blocks, integer values to encode the power relays' status, the command echo, and the command counter. The PCM data takes 5 V measurements and converts them to a corresponding 16-bit integer. This particular experiment communicates over a character-based radio modem so it first converts the telemetry data to American Standard Code for Information Interchange (ASCII) character equivalents of the hexadecimal numbers.

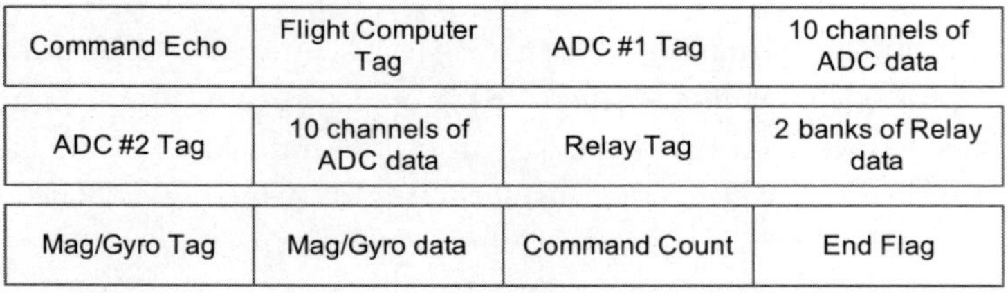

Command Echo	Flight Computer Tag	ADC #1 Tag	10 channels of ADC data
ADC #2 Tag	10 channels of ADC data	Relay Tag	2 banks of Relay data
Mag/Gyro Tag	Mag/Gyro data	Command Count	End Flag

Figure 6.8: Structure of the telemetry frames used in the balloon payload example.

At the ground station, a LabVIEW® Virtual Instrument (VI) reads the available data into a circular buffer from the computer's serial port. The VI scans the input data buffer between the *Synchronization Word* and the *End Flag* to find the next available minor frame for processing. The *End Flag* is used because the frame can have a variable length due to the variable-length *Command Echo* field. Level-zero processing is done to recover the PCM values, the command echo, and the integer values. The PCM and integer values are converted from their hexadecimal ADC output back into equivalent floating-point voltage values in the Operator Interface. Figure 4.14 displays the results.

6.4.3 Inter-Range Instrumentation Group Class I and Class II Telemetry

Designers use the format definitions described above in many data acquisition fields. The range telemetry community in the United States has developed further refinements on these concepts that we discussed in this section [IRIG106]. Other designers may use them as well since some of these refinements are especially applicable in troubleshooting and related management issues for payloads.

6.4.3.1 Standard Parameters

The Inter-Range Instrumentation Group (IRIG) standard specifies the parameters shown in Table 6.1, which are requirements for all telemetry data streams in frame format (Note: the various Non-Return to Zero (NRZ) and Bi-Phase (Biϕ) waveforms listed in the table are defined in Section 6.12 in this chapter). The IRIG standard allows some variation on this structure by defining *Class I* and *Class II* telemetry streams. Table 6.2 summarizes the class differences. The main difference between

Table 6.1: IRIG Standard Telemetry Parameters [IRIG106]

Parameter	**Specification**
Waveform encoding	BiΦ-L BiΦ-M (IRIG variation) BiΦ-S (IRIG variation) NRZ-L NRZ-M NRZ-S
Bit Rate Instability	Maximum of ±0.1% of specified rate
Bit Rate Jitter	Maximum of ± 0.1 of a bit interval averaged over 1000 bits
Frame Synchronization Word Length	16 bits to 32 bits
Major Frame Length	$\leqslant$ 256 minor frames
Minor Frame Counter	Recommended, especially for Class II

the two classes is in the level of complexity. The Class I standard is for basic, low-speed telemetry systems; that is, those producing fewer than 10 Mbps of data. The Class II system operates at higher speed and permits more complicated processing by having a variable frame format. The number of words or bits in a minor frame and the data bit rate are larger in Class II telemetry streams.

6.4.3.2 Format Changes

The standard permits Class II telemetry streams to utilize format changes for the entire stream if each minor frame contains an explicit marker that identifies the format and the marker code is unique under single-bit errors. The standard allows up to 16 different formats for each telemetry source, with each format having a unique Frame Format Identifier (FFID). The standard recommends placing the FFID after the SFID, as Figure 6.2 illustrates. The designer of the FFID encoding needs to produce a code that is immune to single-bit transmission errors and does not produce a valid FFID for a different format.

The frame designer must synchronize the format changes with the start of minor frames and not allow the receiver bit synchronizer to lose lock by having periods of no data transmission. This restriction may require the addition of fill bits to maintain synchronization. Two format

Table 6.2: IRIG Class I and Class II Telemetry [IRIG106]

Parameter	Class I Specification	Class II Specification
Minor Frame Length	Maximum of 8192 bits or 1024 words; fixed	$\leqslant$ 16 384 bits; may be variable length
Fragmented Words	Not Allowed	Allowed, up to 8 segments/word, in all but synchronization words
Tagged Words	Not Allowed	Allowed
Format Changes	Not Allowed	Allowed
Asynchronous Formats	Not Allowed	Allowed
Bit Rate	10 bps to 10 Mbps	>10 Mbps
Supercom Spacing	Evenly spaced	May be uneven
Data Format	Unsigned straight binary, discrete value, complement, BCD, gain-and-value	Other format
Word Length	4 to 32 bits	32 to 64 bits
Frame fill bits	Not Allowed	Allowed
Cyclic Redundancy Check	Not Allowed	Allowed; inserted at the end of the minor frame

changes are possible: changing the measurement list without changing the overall structure sizes and changing the format structure. In the first case, there is a change to the database mapping for decoding the meaning of each item in the major frame matrix. The second case is a more drastic change and may involve resynchronization of the overall structure from the bit synchronizer onward in the user base station system.

6.4.3.3 Asynchronous Embedded Format

Class II telemetry streams also have structural variations not found in Class I telemetry streams. The Class II standard allows for an *asynchronous embedded format* within the normal major frame format. This is

Minor Frame 1	S/W	1	2	3	4	
Minor Frame 2	5	6	7	8	9	
Minor Frame 3	10	11	12	13	14	
Minor Frame 4	15	S/W	1	2	3	
Minor Frame 5	4	5	6	7	8	
Minor Frame 6	9	10	11	12	13	
Minor Frame 7	14	15	S/W	1	2	
Minor Frame 8	3	4	5	6	7	
Minor Frame 9	8	9	10	11	12	
Minor Frame 10	13	14	15	S/W	1	
Minor Frame 11	2	3	4	5	6	

Embedded Data

Figure 6.9: An example of the asynchronous embedded format for telemetry frames. The designer allocated five words of embedded data in each minor frame.

a secondary data source riding along with the major frame stream. The IRIG standard recommends a maximum of two embedded streams in a major frame structure with the location of the embedded words fixed within the minor frame and transmitted with each minor frame.

With embedded streams, synchronization of the host telemetry does not allow for synchronization with the embedded data structure and must be determined separately. Figure 6.9 shows a 16-word embedded stream. The designer allocated five words within each minor frame to contain the embedded source data. The code synchronization word S/W starts each embedded minor frame. The embedded minor frame is not wholly contained within a single host minor frame. Hence, the receiver synchronizes it separately and decodes it from the host minor frame.

6.4.3.4 Tagged Data

The tagged data format is a method allowing the designer to transmit telemetry data within a fixed subframe size but without restricting the contents for each frame. Figure 6.10 illustrates how the tags, followed by the data value(s), represent a tagged datum within the frame. In this case, the designer allocates one byte to the tag and then the data field can be either a data block or individual values of varying lengths. The

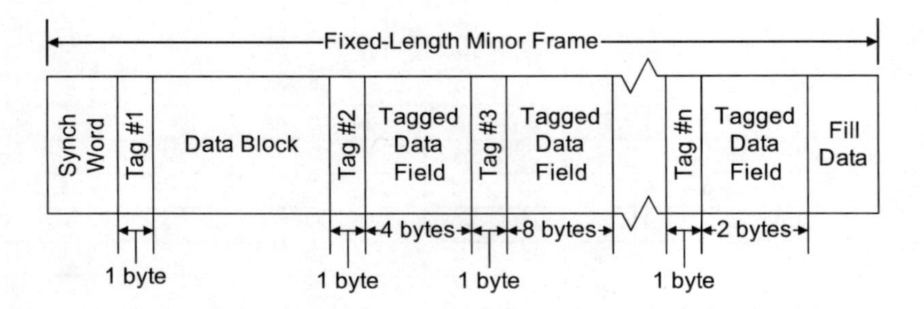

Figure 6.10: Examples of tagged data formats in a fixed-length telemetry frame.

designer adds fill data at the end of the frame to maintain the fixed frame size.

The IRIG standard allows for three modes of tagging data [IRIG106]. The first mode corresponds to that in Figure 6.10. The second mode is to allow transmission of MIL-STD-1553 packets within the telemetry frame. The third method is to allow the Aeronautical Radio-compatible bus data.

6.5 SYNCHRONIZATION CODES

Telemetry frames have two major synchronization markers: the minor frame *synchronization code* and the *subframe identifier*. This section covers the function of both markers. Section 7.7 describes the statistical characteristics of the minor frame synchronization code so that the frame designer chooses the proper length based on the expected link characteristics. The purpose of the two synchronization codes is to allow the data receiving station to lock onto the data pattern in the major frame. Because the designer fixes the minor frame's length, once the receiver locks onto the synchronization markers, it is a simple matter for the bit telemetry processing hardware to index into the received data stream to retrieve the desired data value.

As Figure 6.2 shows, each minor frame begins with the minor frame synchronization code or synchronization word. This word is a fixed bit pattern that repeats at a fixed interval in the data stream. When the designer arranges the major frame as a matrix, the synchronization code is the first column in the matrix. The standard notation is to consider the synchronization marker to be one word wide regardless of the number

Table 6.3: Optimal Synchronization Codes [IRIG106]

Length	Octal Pattern	Hex Pattern	Length	Octal Pattern	Hex Pattern
16	727100	EB90	25	762670400	1F2DC40
17	746500	1E6A0	26	764654600	3E9ACC0
18	746500	3CD40	27	765514600	7D69980
19	7631200	7CCA0	28	7536263000	F5E5980
20	7336100	EDE20	29	7536315000	1EBCCD00
21	7351300	1DD2C0	30	7657146400	3EBCCD00
22	74665000	3CDA80	31	77467650200	7F37D420
23	75346400	7AE680	32	77465450200	FE6B2840
24	76571440	FAF320	33	76723512230	1F74E9498

of bits used. The regularity in the frequency of the synchronization word allows the decommutation hardware to lock onto the minor frame structure and stay locked. These synchronization codes are usually at least 16 bits long with 32 bits being typical with modern computer processing.

Table 6.3 lists the standard optimal patterns as used by IRIG based on Hill, and Maury and Styles [Hill63; Maur64]. The synchronization codes appear both in octal (base 8) and hex (base 16) formats. These synchronization patterns are optimal in the following sense: if timing or framing errors in the receiver shift the code by one or two bits, either early or late, then the chance is small of mistakenly locking onto the code. A more mathematical way of saying this is that the codes have poor autocorrelation properties unless the bit timing is exact. The decommutator has the same chance of mistakenly finding all codes of the same length in a random data sequence, so the synchronization codes are not optimized in that sense. The codes in Table 6.3 may appear to be too long at first glance. For example, consider the optimal 17-bit code 746500_8, which translates to the binary pattern 11 110 011 010 100 000. The octal listing of the pattern appears to be 18 bits long, but we ignore the zero-pad bits in the final octal symbol, which is always a 0. The codes use a similar padding for the hex representations as well.

The purpose of the SFID is to indicate which minor frame of the major frame the transmitter is currently sending. That is, the SFID is really the row marker in the major frame matrix structure. Once the decommutator establishes the SFID, it uses the commutation structure within the subframe to determine which parameter the payload sent. The

SFID is either an up counter or down counter. The SFID should start at the smallest value when the design counts up and it should start at the largest value when the design counts down.

6.6 TELEMETRY FRAME DESIGN

While the IRIG standards cover many aspects of telemetry frame design, other practical aspects are not covered in the standard. In this section, we look at these aspects that the designer considers in a real system.

6.6.1 General Factors

There is no pre-defined, optimal algorithm to design the telemetry frame structure. The designer considers the following factors in the design process:

- The required sampling rate for each parameter (see Section 5.6).
- The required quantization resolution because this determines the minimum number of bits per sample per sensor (see Section 5.9).
- The required number of parameters of each type needed to be included: bi-level (single bit), sampled analog value, digital value.
- The error-checking and correcting code parameters to be specified, if used.
- The overhead information required for frame and subframe identification and synchronization, timing or counting parameters, command-link status and repetition, etc.
- The length of the individual subframes from the individual subsystems, which do not all have the same length so some scheme is needed to even them out.
- The required transmission rate and waveform encoding format for the data.
- The required modulation format and allowable channel bandwidth for the system.
- The required buffer sizes of the input computer or similar electronics for efficiently moving a frame or portion thereof through the synchronization and analysis hardware.

The designer considers each factor for each system configuration. The designer may need to reconsider the factors as the system usage evolves in time.

The designer must weigh the benefit as well as the cost for transmitting each telemetry parameter at a given time. For example, a status bit or word may only change upon failure of a component. The failure may never occur but if it does, the operators must react in a timely manner. How often that parameter is sampled and transmitted is a trade-off process. One extreme is for the designer to send only that parameter in the telemetry stream in which case the operator never misses the failure (and the status is always available). The other extreme is for the designer to send that parameter only once per major frame which is perhaps too infrequent for real-time error detection and correction. A designer's approach might be to transmit the parameter once per minor frame. More frequently might be better but that may make the minor frame too long for the channel transmission capacity.

The length of a minor frame and the number of minor frames per major frame involve costs. Each minor frame has associated data overhead in the form of synchronization markers, subframe identifiers, and error check codes. This management information does not carry useful sensor data and, in that sense, the management information carries a transmission cost or overhead. The more minor frames that the payload transmits, the higher the total overhead cost per major frame is. In the limit, one minor frame per major frame carries the minimum overhead penalty. However, the operator's receiving hardware includes several constraints for efficient processing.

The time to acquire and lock onto the telemetry stream is a function of the synchronization marker repetition rate. The telemetry processor usually examines several minor frames for the correct synchronization prior to processing the received data. This time may be very critical when trying to reacquire the link after an outage. A frame design with only one minor frame per major frame delays the synchronization time. In the limit, only sending synchronization markers achieve rapid synchronization at the cost of minimal data throughput. In addition, some computer hardware input/output systems work best with certain buffer sizes. Having an integer number of minor frames exactly filling this buffer aids in the acquisition process.

When designing the major frame structure, the designer may find that the initial mix of minor frame length, subframe length, and major frame size leaves gaps in the subframe where the payload transmits no real data. These gaps represent wasted transmission bandwidth and therefore represent an overhead penalty. In a perfect design, meaningful data fills each cell in the major frame matrix. In general, balancing each of these costs and benefits may require several iterations to obtain the best solution.

6.6.2 Management and Accounting Information

The payload system must transmit some management and accounting information. The synchronization code and SFID are included in every minor frame. The following list includes other types of management and accounting information often found in the data frame:

- Time codes to tag when the payload collects the information (see Section 8.4). The designer selects the exact timing resolution based upon the user's needs. A higher timing resolution requires more space in the minor frame.
- The telemetry processor may need identifiers for the payload if the processor analyzes the minor frames from several payloads in a common location. This identification assures operators that the telemetry processor is using the correct data stream.
- The SFID indicates the location within the major frame structure. Some payload designers also like to have some form of major frame counter if time codes are not used. The major frame counter is helpful in determining the length of data dropouts or for splicing together data collected at two or more operator receiving stations.
- The designer may add error check codes to the minor frame if the channel has a high probability of corrupting the data. We will deal with this explicitly in Section 7.7. If error-correcting codes are part of the design, the designer factors them into the transmission. One method for doing this is by adding extra information to the minor frame.

The frame designer must determine which additional items are required in each case.

6.6.3 Data Packaging

The way in which designers package the data in the telemetry frame also influences the transmission efficiency. The payload must transmit all of

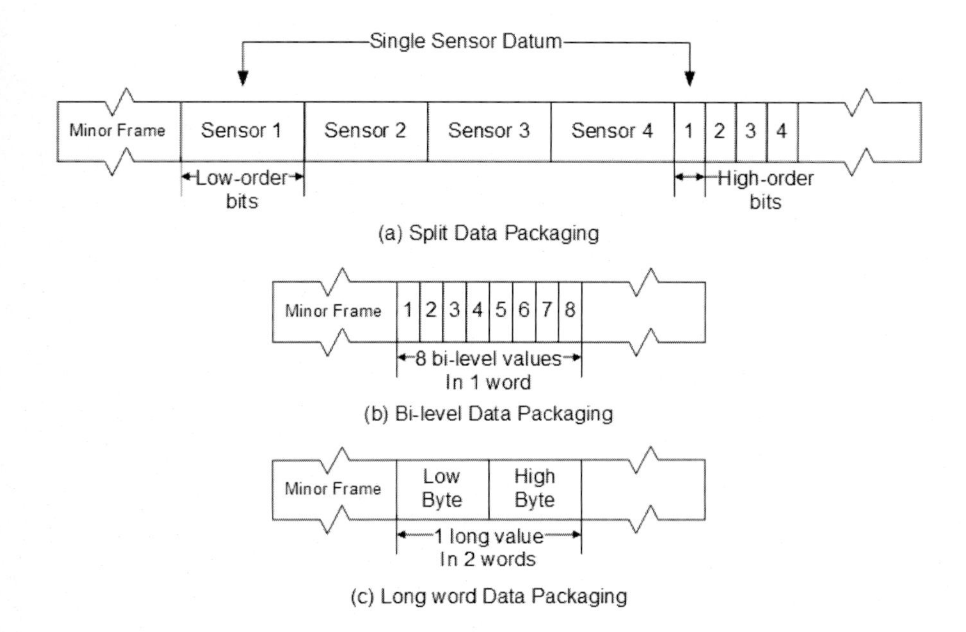

Figure 6.11: Options for data packaging within fixed-length words.

the data. However, the way the designer places the bits in the frame is sometimes open to the designer's interpretation. The easiest case to consider is a minor frame composed of an integer number of 8-bit words or octets. If all of the sensor analog-to-digital converters produce 8-bit results, then the obvious transmission protocol is to place one sensor reading per octet.

Figure 6.11 illustrates three examples of other possible packaging approaches for different word sizes. Figure 6.11(a) shows how 10-bit ADC outputs from four sensors are arranged for increased transmission efficiency by fragmenting the ten bits across two words rather than using two full 8-bit words for a 10-bit datum. The packaging places the lower eight bits from each sensor in a unique octet thereby using four octets. The packaging then concatenates the remaining two bits for each sensor into a sequence and places them in a shared octet. In this way, the minor frame uses five bytes rather than a worst-case packaging of eight bytes. The penalty comes in the form of more complex software processing at the transmission and reception nodes to pack and unpack the data.

Figure 6.11(b) illustrates how designers can package eight bi-level (0 or 1 digital signals, see Section 5.4) into a single octet. Designers use this packaging, for example, when the payload latches eight switch settings into a single data word. This also requires additional software processing to unpack the data, at least at the reception node. However, this is more efficient than placing each single bit into a unique word.

Figure 6.11(c) illustrates how long-word data, usually an integer number of octets in length, is packaged. It shows how a 16-bit analog-to-digital conversion places the result in two consecutive bytes. Notice that if the payload subcommutates the long-word data, then the payload sends single bytes in consecutive subframes rather than in consecutive bytes in a minor frame. The packager used this process in subframes in the first case of 10-bit data if the payload subcommutated each of the four sensor readings. It is important to notice the ordering of the bytes in the packaging.

6.7 PACKET TELEMETRY

Packet telemetry is a relatively new area in the telemetering field, as compared with the traditional frame formatting. The growth of the Internet has led many designers to look to that telecommunications infrastructure to provide the backbone for a variety of data acquisition needs. Additionally, there are non-Internet formats that practitioners developed over the years to supply a packet-based telemetry communications method. MIL-STD-1553 format is the commonly recognized data transfer format for avionic telemetry applications, but it does not comprise an entire telemetry system in that it really is concerned with obtaining data for a central master system from an instrument or subsystem. The standard does not specify how to format the data for the payload-to-ground link. This section covers packet structures and their future uses. We also look at standard communications packets that we transform into packet telemetry uses.

6.7.1 Packet Assumptions

The frame-based format assumes that the transmission channel is fundamentally unreliable (produces frequent errors and/or dropouts) so the PDU must be structured to allow rapid resynchronization to the data. Suppose that the transmission channel has a performance level approaching that of "guaranteed delivery" with an essentially error-free channel and infrequent dropouts. How might that change the concept of data transmission? Is there a means to utilize commercial telecommunications networks such as the Internet for data delivery? One answer to both questions is to use a packet-based system and allow the digital communications network to provide the required data transmission, especially in environments that have a very high probability of delivering each data packet to the operator correctly and with a low probability of packet loss. What are the advantages and disadvantages in packet telemetry? The advantages are as follows:

- The system is quite flexible and tailored to the needs of the individual subsystems. It allows for subsystem upgrades without reorganizing the rest of the system because the payload transmits the data as generated and therefore the payload packages the data when there is a significant quantity or when conditions warrant.
- The system uses standard communications protocols such as Transmission Control Protocol/Internet Protocol (TCP/IP) for which standard computer hardware and software exist.
- It is possible to interface the telemetry system with standard commercial carrier systems such as the Internet.
- The data stream has a structure that allows for error correcting and retransmission.
- The payload assigns the data packets a relative priority to allow more important data to be transmitted first.

Packet systems also have some disadvantages, for example,

- They may not be practical for small systems or for systems with little internal processing capability.
- They may increase the system complexity.
- They may contribute excessive overhead to the transmitted data than is acceptable given the available channel bandwidth.

In general, frame telemetry is best for use over relatively low quality, point-to-point wireless transmission links where the data format is relatively fixed. Packet formats are best over high quality links where the system distributes the data to multiple users or the system needs maximum flexibility. The system designer should examine in detail other types of links to weigh the costs and benefits of each transmission mode. In the

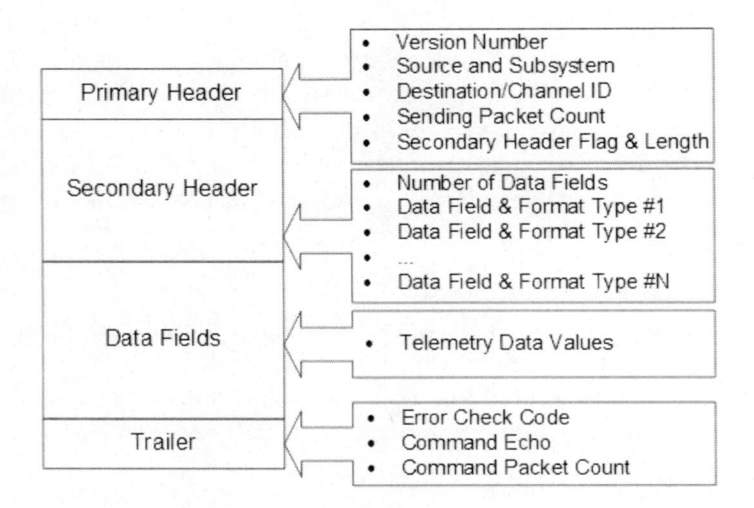

Figure 6.12: Overall structure of a telemetry packet.

following sections, we look at the packet structures for the MIL-STD-1553, Consultative Committee for Space Data Systems (CCSDS), data network protocols, and telecommunications standards as representatives of how designers apply packet standards to telemetry systems.

6.7.2 Protocol Data Unit Format

Figure 6.12 gives the overall transmission PDU format for packet communications, regardless of the protocol standard used. The packet has a header for source and destination addressing. The header may also have counters for packet accounting and flow control. The primary header is usually of fixed length as determined by the protocol's details. The protocol may also have a secondary header to give further information about the size and data formatting in the packet. The protocol details the secondary header's existence, format, and usage. Next are the data fields, and finally the trailer. The trailer usually contains an error check to aid the receiver in determining if the end-to-end system transmitted and correctly routed the packet. The routing in this sense is from the subsystem of the payload to the user base station.

The packet network may have a need to assist the operator's receiver in maintaining synchronization, especially over radio channels. One way to accomplish this is to have packets transmitted across the channel on a

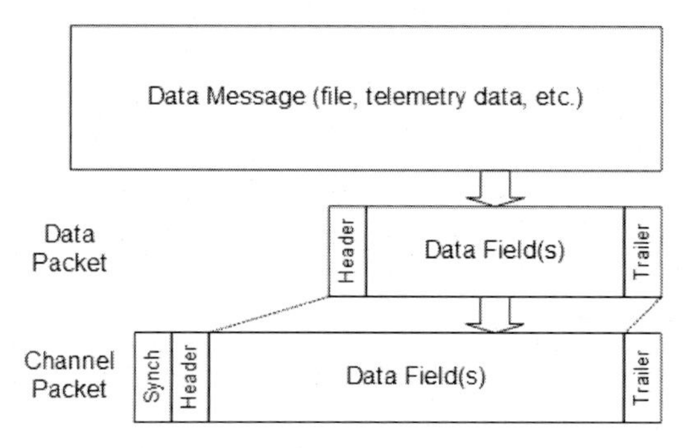

Figure 6.13: Channel PDU being filled by telemetry packets from a data file to be transmitted.

strict periodic basis. The packetizer inserts the actual payload data into the channel packets as the data arrive. This technique allows the payload to convert an asynchronous process into a synchronous process.

Figure 6.13 shows a packet insertion. Here, we envision the overall telemetry data as a file in the payload. The file is broken into manageable packets that identify the data set and contains other information necessary for processing. As these data packets become available, the packetizer inserts them into the channel packet. What happens in the channel packet if there is no data packet available when the packetizer is due to send the next channel packet? Normally, the payload inserts a fill packet to maintain synchronization. A fill packet looks like a regular data packet but it does not have valid data in the data field. The header also contains a marker flagging the packet as being a fill packet and its data field is not to be processed. The receiver discards the fill packets and only processes the actual data packets.

6.7.3 Packet Modes

Because the packet format does not require constant data transmission, as does the frame format, we invoke different disciplines in determining when to issue a data packet. Three easy-to-identify modes for packet telemetry transmission are the commutated, entropy, and virtual channel modes. We use these modes to describe how frequently to transmit the data.

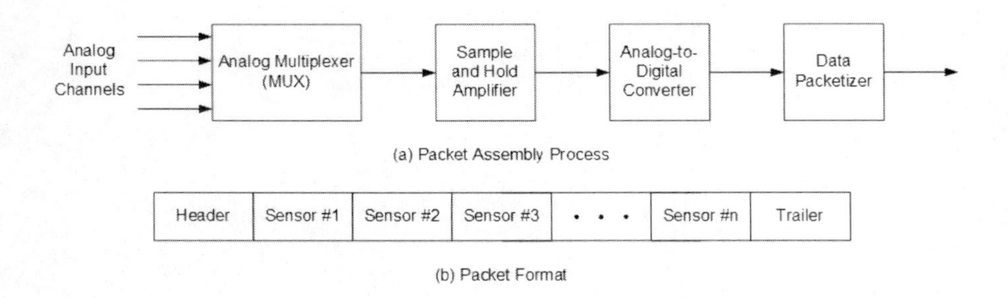

Figure 6.14: Commutated mode packet construction.

6.7.3.1 Commutated Mode

Figure 6.14 illustrates the *commutated mode*, which is very similar to the concept of a minor frame in a packet. We use this transmission to send routine (non-emergency) health and welfare information from the payload. A group of related sensors shares a common transport in the same packet.

For ease of design, one could group the sensors by Nyquist rate so that the payload samples all sensors at the same rate with one sample per sensor per packet and with no need for supercommutated sensors in a packet. The operator's receiver knows the sensor value by the position in the packet so the system has no need to transmit an individual sensor identifier before the reading. The header management information contains an identifier for the subsystem originating the packet and contains timing information.

6.7.3.2 Entropy Mode

We define the *entropy mode* (*entropy* implying surprise or new information) for sensors that do not change rapidly or change only in reaction to external events so that sending them once per data packet might be a waste of transmission bandwidth. In general, the data transmitted in this mode is asynchronous in nature and perhaps only described by their probability of occurrence. This is usually true of fault-detection logic or sensors that count slowly changing events. Coupled with this is the operator's desire to verify periodically that the state they are monitoring really is static. As Figure 6.15 illustrates, the payload sends packets in this mode only when one of the following conditions occur:

- A preset number of state changes occur

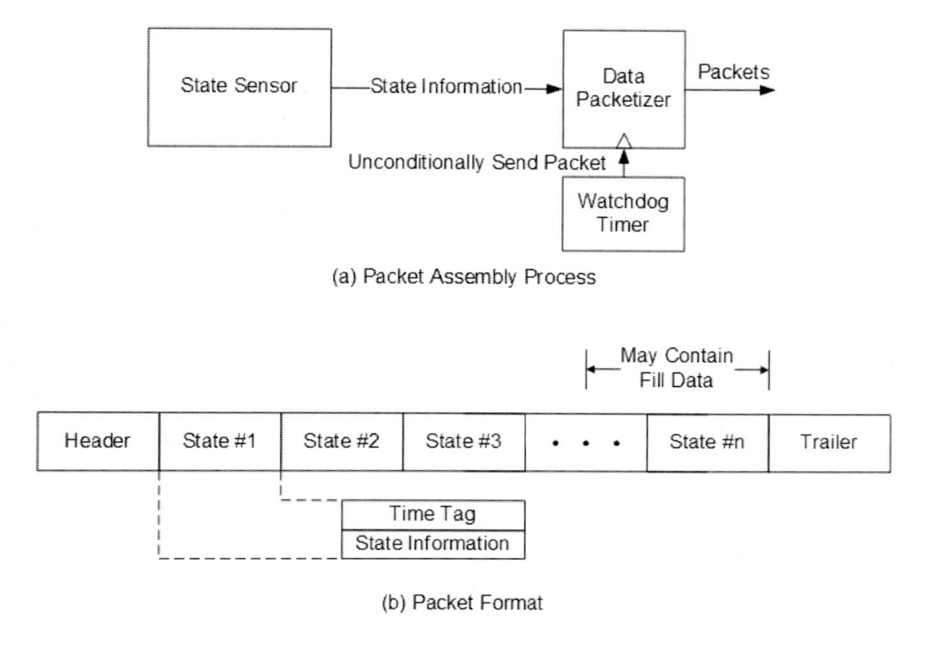

Figure 6.15: Entropy mode packet construction.

- The periodic watchdog timer expires
- The packetizer detects that a critical state change has occurred

Assuming that the packets are all the same size for transmission purposes, the last two conditions on packet transmission may cause the packetizer to pad out the packet length with fill data. If the first case occurs, then the payload resets the periodic timer. This mode has the advantage that it uses a lower transmission bandwidth under normal circumstances because this relatively static transmission mode only sends data to keep the ground apprised that all is well. However, with entropy mode packets, a transmission channel may need extra capacity because when errors occur, they quite often cause several sensors to detect problems all at once. This floods the channel with many emergency packets, thereby choking the channel unless the system designer makes a provision for spare capacity to meet this contingency. Designers use priority indicators on the emergency packets to suppress traffic and allow the emergency packets through.

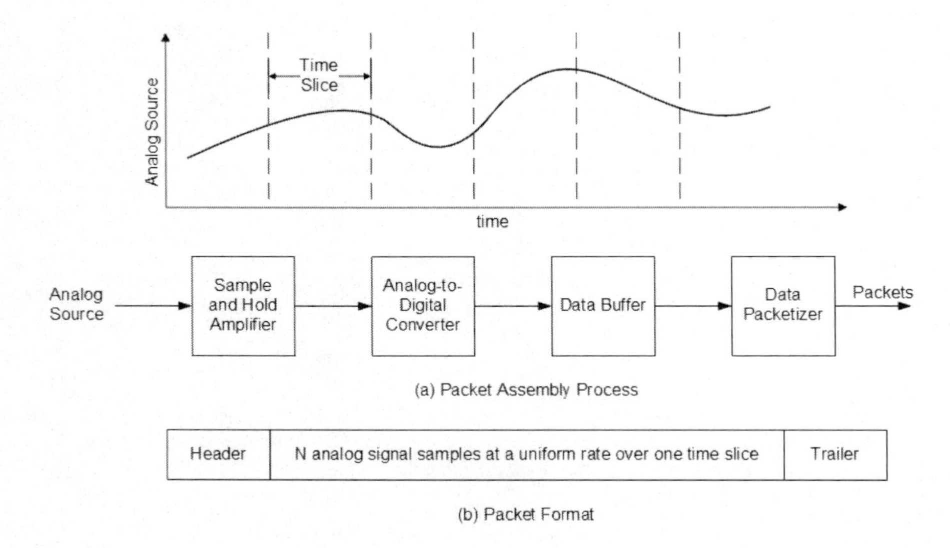

Figure 6.16: Virtual channel packet construction.

6.7.3.3 Virtual Channel Mode

Figure 6.16 illustrates the *virtual channel mode* where data from an individual sensor occupy the packet's entire data zone. The virtual channel mode is useful for payload science sensors that produce large streams of continuous data. We define this mode after the virtual channel concept used in normal packet communications where a packet path connects two nodes and acts like a dedicated circuit channel between the nodes. The end-to-end system treats each virtual channel, that is, each sensor, separately with respect to transmission priority, level of error detection and correction supported, and other transmission parameters that the designer builds into the transmission protocol. For example, an operating mode is to sample the sensor at a uniform rate over a fixed time slice. If the sensor is sampled at N samples per second and there are M bits per sample, then the data zone contains $N \times M$ bits representing the fixed duration sensor output.

The payload gathers the samples generated during the time slice and sequentially packages them in the telemetry packet. This maintains the correct time sequence within the packet. One special concern with the virtual channel mode occurs when the end-to-end system must maintain timing isochronicity over packet networks such as the Internet. This means the packets need to arrive at the receiving node in the correct time

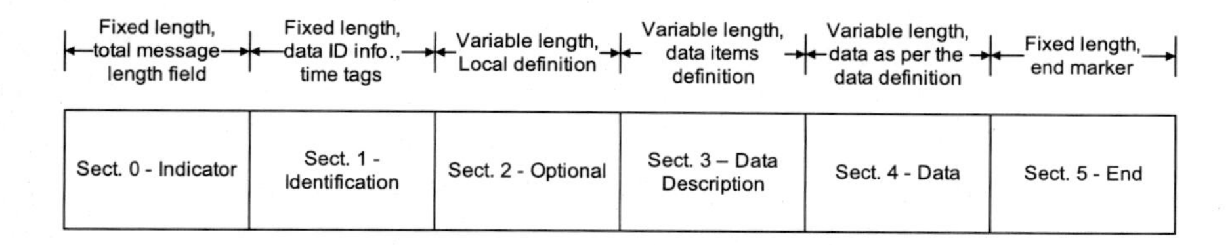

Figure 6.17: The BUFR message format for table-driven messages.

order and that end-to-end system maintains the inter-packet arrival time specification. This is especially important with streaming data such as video and voice data that require timing between packets. Video or audio packets arriving out of order or with large time gaps cause unwelcome effects at the receiving station. Designers configure the secondary header to provide sequence information.

6.7.3.4 Table Driven Format

Packet designers can extend the tagged data format we encountered in Section 6.4.3.4 to permit context-driven packet formats by using tables within the packet to define the packet contents. Designers refer to this method as a *table driven format.* This format has the advantage of permitting the system designer to tailor the data packet for the sensor needs of the moment. It has the disadvantage of requiring more sophisticated software that parses and reacts to the packet structure in real time as the data arrives. The system designer needs to provide specific table definitions to make the processing work.

An example of a table-driven message format is the Binary Universal Form for data Representation (BUFR) format for meteorological data messages [Drag07; Li11]. Figure 6.17 illustrates the overall BUFR message format. Each message contains six sections that are defined by the BUFR standard. As the figure illustrates, the fields in sections 0, 1, and 5 that generally describe the message are fixed length. The fields in sections 2, 3, and 4 that are data-specific are variable length. The optional field of section 2 is to support local data processing. Sections 3 and 4 describe the actual message data. The fields in section 3 describe the number of data segments in section 4, the data type for each segment, and data element descriptors. These fields are arranged as a table in section 3. The sender places the data items in section 4 according to the descriptions given in section 3. The processing software uses the contents of section 3

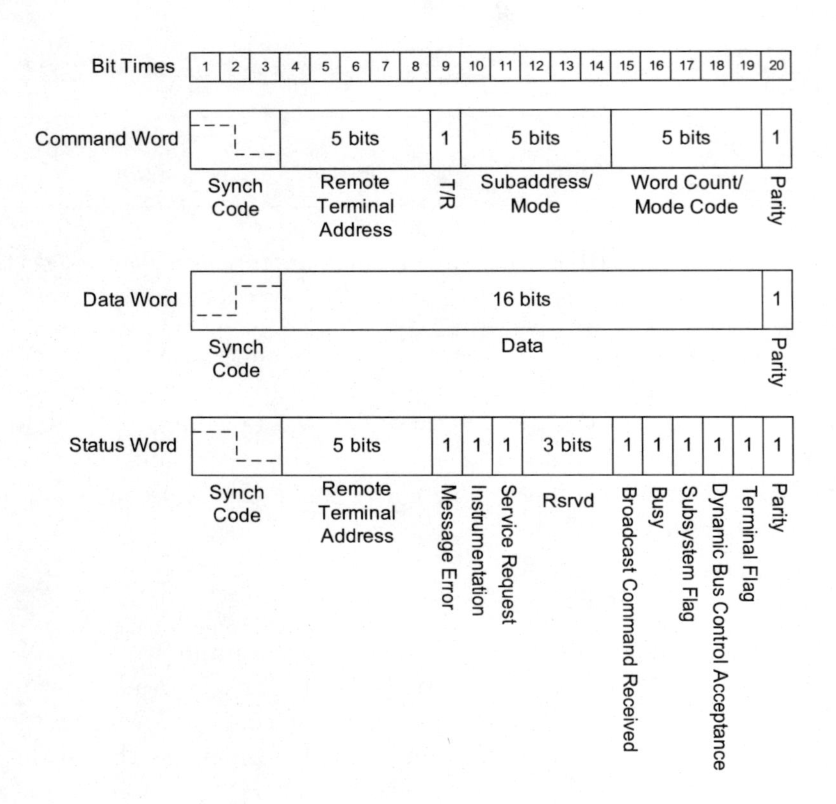

Figure 6.18: The MIL-STD-1553 packet structures for command, data, and status packets [MS1553].

to unpack the data values contained in section 4. The sender can use a different mix of data specifications in each message and the processing software can respond to each one as long as the message conforms to the message protocol.

6.8 MIL-STD-1553 PACKETS

In Section 4.3.2, we saw the bus configuration for the MIL-STD-1553 standard. In this section, we examine the standard's data protocol aspects. Figure 6.18 shows the three packet types delineated in the 1553 standard: command, data, and status packets [MS1553]. The standard packet length for each packet is twenty bit times in duration. The standard bases the bit timing on a 1 Mbps clock rate. The packet's first three bit times comprise a 2-bit synchronization marker whose timing and polarity are used to both mark the packet's start and type. Data are transmitted using a Manchester, or Bi-Phase-Level (Biϕ-L), waveform encoding. Users make messages by sequentially combining packets into a total message that has a maximum length of 32 packets.

The command packet is identified by a synchronization marker with a value of 1 that the sender has Biϕ-L encoded over the three bit times. Fields for a command packet include Remote Terminal (RT) and subaddresses, command direction (transmit or receive) and command data. The protocol uses a single parity bit for error checking.

The data packet is identified by a synchronization marker with a value of 0 that the sender has Biϕ-L encoded over the three bit times. The data packet contains up to 16 bits of data that the sender partitions into a single sensor value or into multiple sensor values, depending upon user needs. If the user needs fewer than 16 bits, the protocol permits having all zeros as the fill bits to keep the overall packet length fixed. Again, the protocol uses a single parity bit for error checking.

The status packet is identified by a synchronization marker with a value of 1 that the sender has Biϕ-L encoded over the three bit times. This is the same as the command packet's synchronization. This causes a problem if the Local Area Network (LAN) has a data dropout. However, if the LAN is operating properly, the command-and-response bus protocol keeps the command packet and status packet from becoming confused. Fields for a status packet include RT address and status bits. As with the other packet types, the protocol uses a single parity bit for error checking.

6.8.1 Inter-Range Instrumentation Group 106 Modifications

IRIG Standard 106 [IRIG106] defines the preferred methodology for using MIL-STD-1553 for range applications. The technique illustrates how a sender packages asynchronous data for a synchronous channel. The protocol still leaves most of the details for using 1553 to the instrumentation designers, but certain specifications and modifications allow standard usage. Figure 6.19(a) illustrates the 24-bit standard packet and the definition of its fields. The message packets hold command, sensor data, or timing information, which allows for resolution to 1 µs. Up to eight redundant bus pairs (1A and 1B through 8A and 8B) are permitted in the system; hence three bits identify the bus used for the packet. The identifier label, in bits 5 through 8, specifies the packet type and the

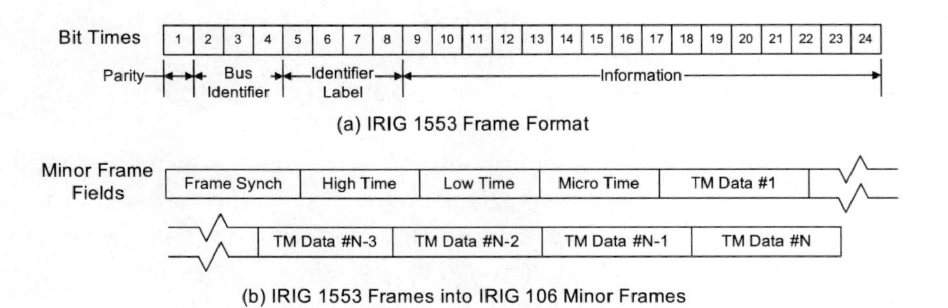

Figure 6.19: The IRIG 106 accommodation of MIL-STD-1553 packets in telemetry frames [IRIG106].

Table 6.4: Identifier Label Codes for the IRIG 1553 Frame Format [IRIG106]

Code	Meaning	Code	Meaning
1 1 1 1	Command A	0 1 1 1	High Order Time
1 1 1 0	Status A	0 1 1 0	Low Order Time
1 1 0 1	Data A	0 1 0 1	Microsecond Time
1 1 0 0	Error A	0 1 0 0	Time Response
1 0 1 1	Command B	0 0 1 1	User Defined
1 0 1 0	Status B	0 0 1 0	User Defined
1 0 0 1	Data B	0 0 0 1	Fill Word
1 0 0 0	Error B	0 0 0 0	Buffer Overflow

primary (A) bus or the backup (B) bus used in the hardware. Table 6.4 lists the sixteen possible identifier labels.

The designer time tags the command packets to the nearest microsecond with the time tag following the command in the output sequence. The standard suggests using the 0100 identifier code for time tagging a status response. However, if not used for this function, it is user-defined for use as in the codes 0011 and 0010. Section 8.4.2.3 discusses the details of the time tag structure.

The IRIG 106 standard allows for packaging the MIL-STD-1553 packets into frames for transmission between the payload and the receiving station using normal telemetry links. Each transmission frame has a length between 128 and 256 words, including a 24-bit synchronization word (76571440_8). Figure 6.19(b) illustrates this structure. The transmitted frames use Non-Return to Zero-Level (NRZ-L) data formatting and transmit the most significant bits first. The 106-type frames are usually synchronous, fixed-length frames, while the payload generates the 1553 packets asynchronously. The transmitter may have to send fill data to maintain the link synchronization.

6.9 CCSDS PACKETS

The Consultative Committee for Space Data Systems (CCSDS) organization has developed a series packet transmission protocols for file transfer, and Telemetry (TM) and Telecommand (TC) data transmission between users and their payloads. Figure 6.20 illustrates the overall packet data transfer process as a set of peer-to-peer relationships organized according to typical telecommunications networking layers. The structure is arranged around the following *networking layers* that are described in more detail in [CCSDS1300; CCSDS1302]:

Application — the user-level application such as a file transfer.

Network — the protocol for transmitting the data across the chosen network(s) and relieving the application process from knowing the details of the intermediate link protocols.

Data Link Packet — the Data Link Layer packet transmission protocol to frame the Network Layer frames in a manner appropriate to the specific network configuration.

Data Link Synchronization and Coding — the Data Link Layer method for achieving point-to-point data synchronization and channel error correction control coding across a specific network configuration.

Physical Layer — the channel-layer Radio Frequency (RF) modulation for point-to-point data transmission across the specific channel.

The lower-level of CCSDS protocols is broken into three families of protocols:

- TM and TC for basic packet data transmission
- Advanced Orbiting Systems for data transmission between major spacecraft and ground systems

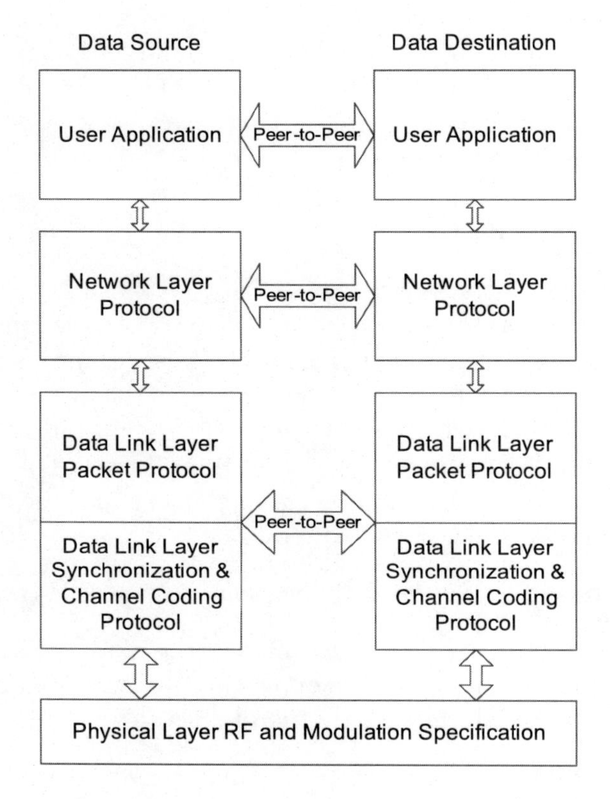

Figure 6.20: The CCSDS protocol stack architecture for peer-to-peer data transmissions.

- Proximity operations protocols for data transfer between spacecraft near each other

Here, we will concentrate on the TM and TC needs. The Networking Layer serves to encapsulate the user's data for transmission. The encapsulation service uses either Space Packets or Encapsulation Packets. We will concentrate on the former, while the latter is to wrap data from non-CCSDS networks for CCSDS-compatible networks. Refer to the CCSDS recommendations for more details on the protocols and the services they provide.

The Space Packet Protocol provides the interface between the possible Data Link Layer protocols and the various user applications identified through their application identifier. Figure 6.21 illustrates the framing structure including headers and user data fields. The Network Layer does

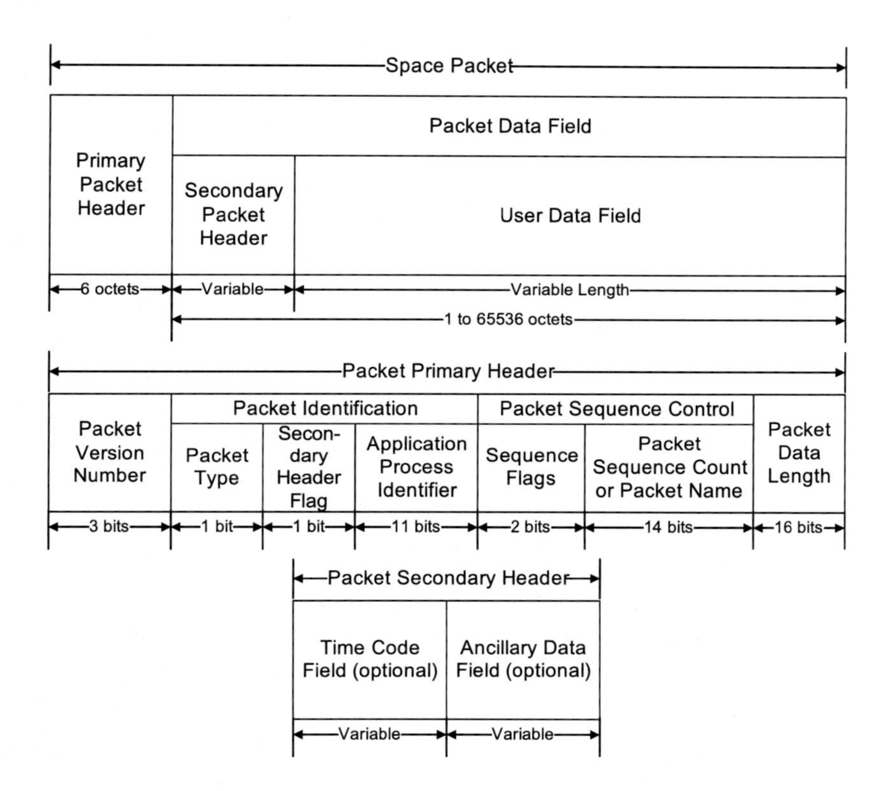

Figure 6.21: The CCSDS Space Packet Protocol Network Layer frame structure [CCSDS1330].

not guarantee end-to-end delivery of the data, removal of duplicates, or proper sequencing of the data. The calling application or a Transport Layer protocol must provide these functions.

The Data Link Layer provides point-to-point communications within a network. Figures 6.22 and 6.23 illustrate the CCSDS Data Link Layer transfer frames for our discussion, including the headers and user data areas [CCSDS1320; CCSDS2320]. The CCSDS synchronization and error control portion of the Data Link Layer define the allowable error correcting coding standards and usage parameters that the user may select for these data frames [CCSDS1310; CCSDS2310]. The TC frames are variable length because this feature facilitates rapid data transport. The TM transfer frames are fixed-length because this feature facilitates synchronization.

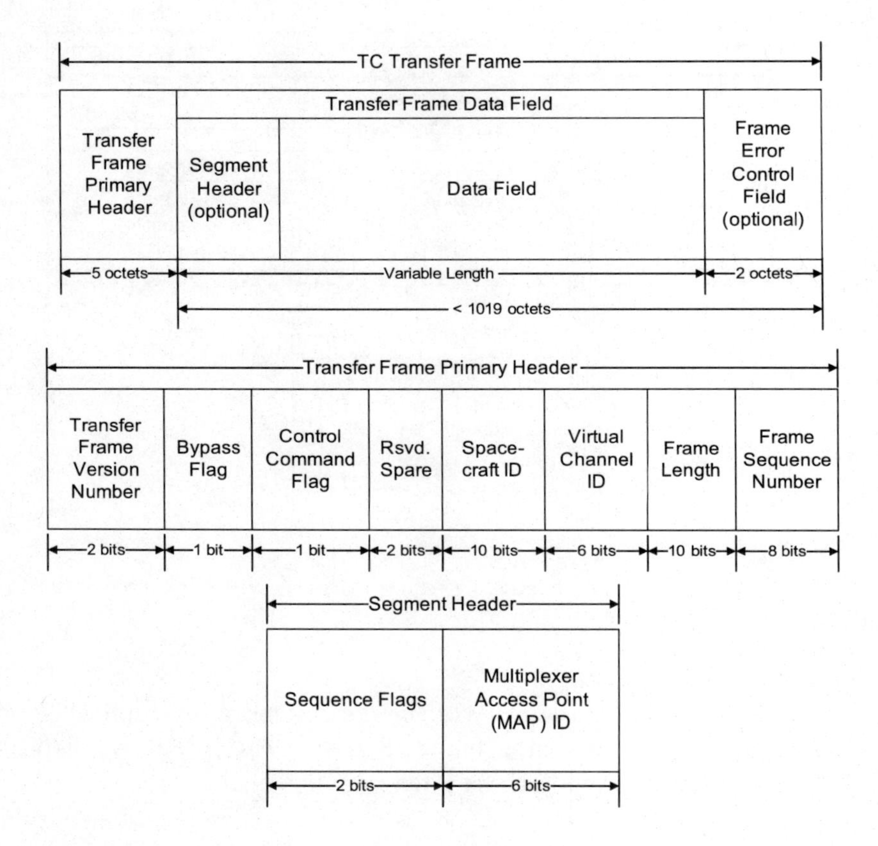

Figure 6.22: The CCSDS Telecommand Transfer Frame Data Link Layer packet frame structure [CCSDS2320].

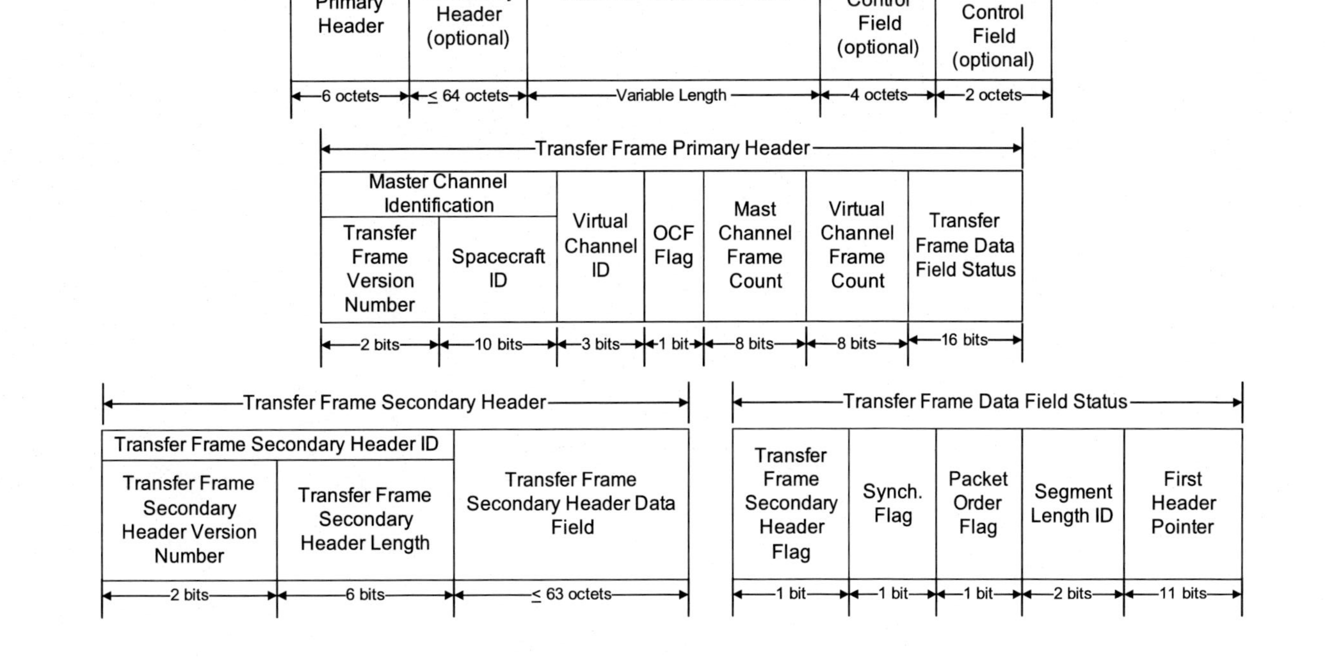

Figure 6.23: The CCSDS Telemetry Transfer Frame Data Link Layer packet frame structure [CCSDS1320].

The model for the packet transmission is similar to that given earlier in Figure 6.16 where a packetizer places the source data in a transmission packet destined for the end user. The overall process can assign each source its own virtual channel identifier. The protocols are designed to group multiple virtual channels into a common physical channel for transmission. In the protocol stack shown in Figure 6.20, the process defines the path between the data source (science experiment, physical subsystem of the spacecraft, or ground system user) and the data destination (a scientist, mission controller, spacecraft actuator, or instrument) as a virtual channel. The end-to-end system uses the virtual channel identifier in all routing and processing decisions. The packetizer places the virtual channel data into a Network Layer frame for transmission. The Data Link Layer passes these transfer frames between the peer communications nodes until the data arrive at the destination Network Layer peer. There, the destination application manages the data delivery to the user for each virtual channel.

6.10 DATA NETWORKING PACKETS

6.10.1 Background

The Internet has become the standard means for data transport in many environments. In this section, we examine how to accomplish this activity. System designers generally use either TCP/IP or User Datagram Protocol/Internet Protocol (UDP/IP) protocol suites as the primary means of data packaging for transport. The protocol designers did not necessarily specify these protocols for telemetry and telecommand data communications, but they are useful in supporting packet systems. The standard texts on computer networking, such as Stallings [Stal97], fully discuss the protocols. The advantage to using these protocols is that systems are designed using the standard hardware and software that is available to support the protocol and thereby ease their use in a new system.

The telemetry and telecommand packet links can run in two connection modes:

Connection Oriented where each endpoint in the transmission negotiates the transmission parameters and flow control before the data starts transmitting and all data is acknowledged as it is received at the destination.

Connectionless where the data sender transmits the packet without

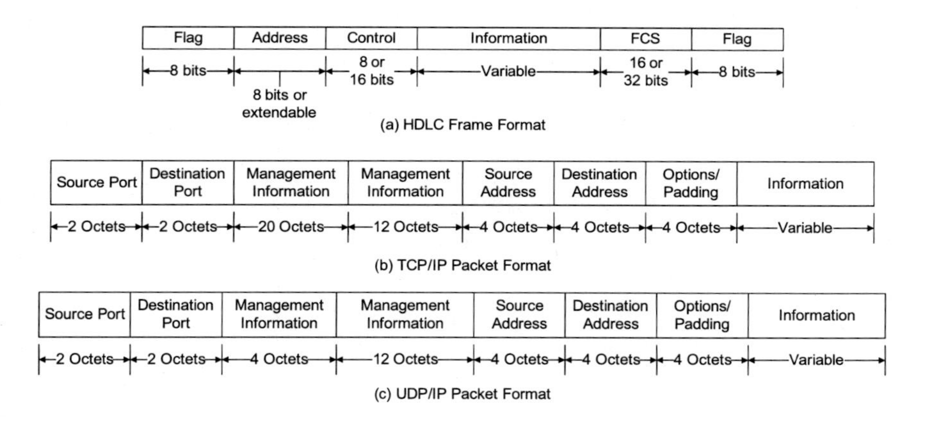

Figure 6.24: The HDLC, TCP/IP, and UDP/IP packet formats.

any pre-arrangements and expects the communications network to make a “best effort” to deliver the packet.

An example of the connection-oriented transmission is the File Transfer Protocol (ftp) file transfer program. An example of a connectionless transmission is a datagram such as an e-mail message. With connectionless transmissions, the system designer normally needs to include some form of data delivery management to ensure that the data are delivered as intended.

6.10.2 Packet Formats

The telecommunications packet formats are very similar to the packet telemetry formats that we described earlier. Figure 6.24 illustrates the High-level Data Link Control (HDLC), the TCP/IP, and UDP/IP packet protocols for data communications. These protocols have the advantage of the availability of standard hardware and software support that designers place directly in many computer systems. In these protocols, the packetizer places the telemetry or telecommand data in the packet's information field. In each case, as far as the packet protocol is concerned, the data are unstructured. Therefore, the system designer must specify a well-designed interface specification that spells out the packet structure. These protocols just provide for transport through the communications network. The individual user must still unpack the data from the packet. This flexibility allows each destination station to have its own method for unpacking the data, which may lead to configuration problems.

The HDLC protocol is really a link layer data packet for transmitting data between two points. System designers can use this for point-to-point telemetry and telecommand transmission. When used by itself, the system designer is assuming that the data will arrive successfully and not be diverted to another node. Therefore, HDLC is good for transmitting data that has either a management protocol above it or does not require much management such as an IRIG frame format.

The TCP/IP protocol family provides the system designer with an end-to-end data transport across a multinode network or over a network where packet flow and packet accounting are required. For example, the system designer may choose a connection-oriented protocol like ftp or Delay Tolerant Networking (DTN) [RFC4838] to guarantee the delivery of the main payload science data packets or telecommand command files.

The telemetry frames in Section 6.4.1 are datagrams that correspond to connectionless transmission so a UDP/IP data packet protocol may be an appropriate choice for the system designer. For example, health and welfare data packets or single command words may be sent connectionlessly using datagrams.

6.10.3 Data Servers

Using a particular packet format is not as important as controlling and accessing the software running on the computers using the protocols. Figure 6.25 illustrates one example of how the control software may be configured [Fan00; Jian00]. The data acquisition computer system integrates the sensor array, the associated data acquisition electronics, and the supporting software. The collected data includes Global Positioning System (GPS) time and position, accelerometer, atmospheric pressure, and radio propagation measurements. The computer in this example collects the data and passes it over the Internet to a data server computer using the LabVIEW® data socket support software. The data server formats the data into a format where they are accessed using a standard Web browser. The data user on the client computer then accesses the data server over the Internet and the data acquisition computer sends the data to the user's Web browser. While this example uses LabVIEW®, designers use other software as well. The common feature is that the designer utilizes standard TCP/IP services to support the data transfer.

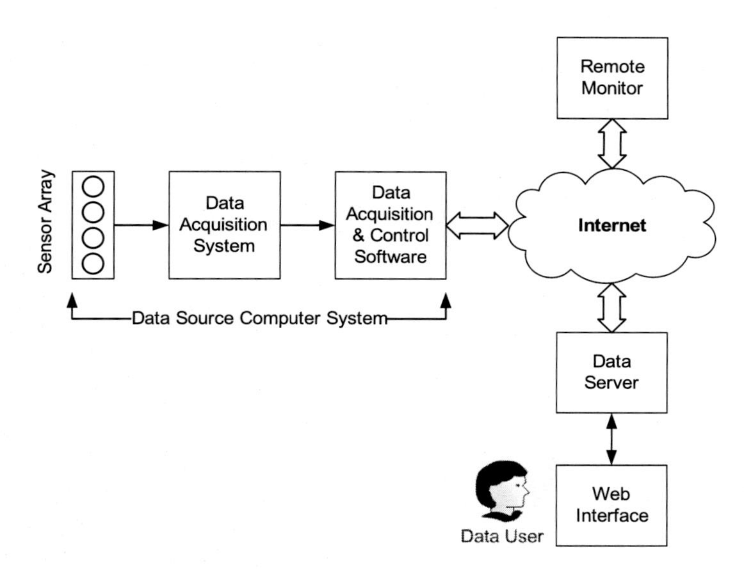

Figure 6.25: Block diagram of a data server configuration using LabVIEW to host the data service software.

6.10.4 Data Throughput Issues

While the use of TCP/IP for data transport may seem like a good way to leverage existing resources, the designer cannot apply the protocols blindly. The packet format supports data transport with no problem. However, the protocol is more than the packet format. It also includes the protocol management software and the interactions with the operating system. We also need to include interactions with the transmission channel. This becomes important because the protocols assume relatively error-free channels such as LANs and closed telecommunications links. When the system transmits data over a radio link or other relatively unreliable links, the protocol does not necessarily work as desired.

Link designers need to consider the following protocol issues:

Channel Bit Error Rate — channel bit errors can cause lost packets and will eventually require a method to resend the lost data.

Flow Control — protocols such as TCP/IP utilize flow control mechanisms that are influenced by networking delays and networking errors.

Unbalanced Link Data Rate — the telecommand and telemetry data tend to have different date rates and this rate difference can influence effectiveness of the flow control algorithm.

Packet Size — smaller packets can have more rapid transport through the network but a greater percentage of the total data flow is tied up in management data so the designer may need to balance these effects.

Contact Management — are the network endpoints always mutually visible or are there dropout times? How will the dropouts be managed and how will any data lost in the process be recovered?

Protocols such as DTN (also known as Disruption Tolerant Networking) are under development to address these types of issues that the standard TCP/IP protocols do not address.

6.10.5 Inter-Range Instrumentation Group 106 Packet Encapsulation

Earlier, in Section 6.8.1, we saw how the IRIG-106 standard permits encapsulating MIL-STD-1553 packets into the minor frame format for transmission. The IRIG-106 standard also permits packaging standard networking packets in the minor frame for transmission using a similar technique [IRIG106]. The standard imposes several restrictions on the minor frame's design:

- The minor frame synchronization word is 32 bits long and takes the value of $0xFE6B2840$ when there is no frame-length error correction applied and $0x1ACFFC1D$ when the option of Reed-Solomon error correction is applied.
- The minor frame length is restricted to be $223 * N$ bytes where N is between 1 and 8.
- If Reed-Solomon error correction is used, it must be included with the total frame length and not added as an extra length.

The link designer encapsulates building on the Asynchronous Embedded Data concept in Section 6.4.3.3 and the Tagged Data format of Section 6.4.3.4. The designer needs to keep three concepts in mind:

1. The standard does not require the data system to place an embedded packet in each minor frame.

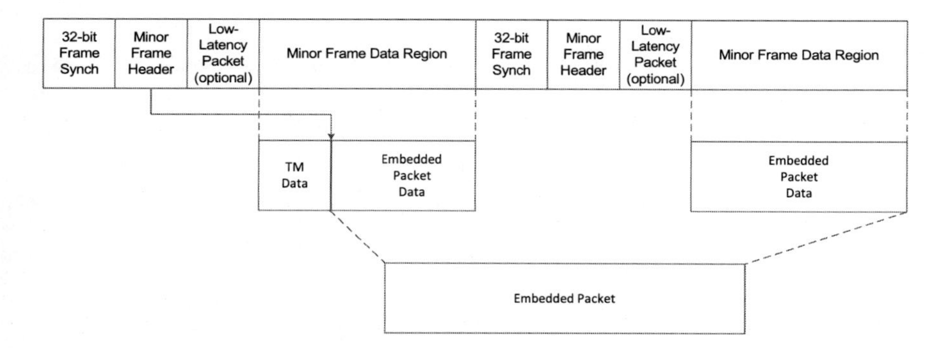

Figure 6.26: The IRIG encapsulated packet format for placing packet data into minor frames. Both low-latency packets and standard packets are placed in the minor frame.

2. The standard provides two versions of encapsulation: standard packet encapsulation and low-latency packet encapsulation.

3. The standard requires that low-latency packets must fit within a single minor frame, while the designer may stretch the packet placement over several frames with standard encapsulation.

Figure 6.26 illustrates how the standard performs the encapsulation process. The frame designers use the minor frame header to identify the data stream the minor frame is sending, the presence of low-latency packets, and the index into the data region where the encapsulated packet data starts. The 12-bit datum for the last two header elements are encoded with a 12-bit error correcting code. The encapsulated packet also has a 12-bit header protected with a 12-bit error correcting code. This header specifies the following fields:

- The data type appearing in the packet such as application data, fill data, Internet Protocol (IP) packet, etc.

- The packet fragmentation state of complete packet: the first segment of a packet, the middle segment of a packet, or the final segment of a packet

- The packet length in the minor frame data region (up to $0xFFFF$ bytes)

Low-Latency packets are simpler. The standard allows them to be optional and vary in length up to the size of the minor frame data region. The standard does not allow them to span multiple minor frames. The packets end with a one-byte end word: $0xFF$ if another low-latency packet follows the current packet and $0x00$ if no further packets follow the current packet. Notice that the standard does not define a header or structure because that is left to the link designer.

6.10.6 Telemetry Data Streaming

Applications in the commercial world commonly provide data streaming services to support audio and video data distribution. System designers can also use these types of services for telemetry data distribution. Major concerns that designers need to address with streaming service approaches are as follows:

- Maintaining propagation delays within the bounds of the system's needs (the data must be delivered within the time bounds required for the mission).
- Controlling the timing jitter in the data transport because the variability in the data delivery is often more important than the actual delay itself.
- Ensuring that the effective data rate for the data transport network can sustain the input data rate without dropping data due to buffer overruns on the input.

These concerns are in addition to the expected concerns over packet data errors.

Telemetry system designers often need to configure the design to accommodate widely distributed system elements. In this case, the data receiver may be located many kilometers from the end users. The question now arises as to how to maintain connectivity in this architecture. One approach is suggested by the government test range community in the form of a data streaming service described by the Telemetry over Internet Protocol (TMoIP) standard [RCC218]. Figure 6.27 illustrates this concept.

The goal of the TMoIP standard is to provide seamless data transport with the only artifact of the process being the propagation delay. The overall TMoIP architecture has two major segments: the payload and the Ground Network. The payload is a PCM data source starting at the sensor and includes the necessary signal processing and formatting electronics.

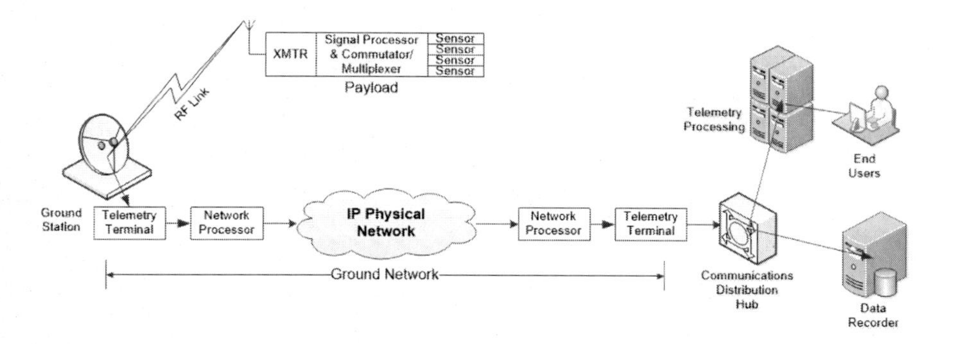

Figure 6.27: Data flow process for the TMoIP standard.

The payload ends with the digital transmitter and antenna. The standard does not require any specific data formatting on the payload-to-ground-station link.

The Ground Network is often composed of the following elements:

Ground Station — the receiving antenna, and associated RF and tracking electronics

Telemetry Terminal — the electronics to interface between the devices on the input and output of the Ground Network and the Network Processor

Network Processor — the electronics to format and manage the data streaming over the IP network

The Telemetry Terminal does not perform any processing of the actual telemetry data. Frame synchronization, decommutation, and processing algorithms all reside with the Telemetry Processing node. The Communications Distribution Hub is a switch to route the data to the relevant nodes.

The Network Processor utilizes standard IP datagrams such as UDP/IP packets in the network data transfer. The networking designer should allow multicast operations over the IP network when providing data distribution to multiple sites. The Network Processor maintains the real-time nature of the telemetry stream service. This includes managing the IP packet flow, packet delivery ordering, recovering from lost or missing packets, and maintaining a nearly constant propagation delay.

6.11 COMMAND PROCESSOR INTERFACE

As noted earlier, the frame formatter or packet formatter has an interface to the command processor. The payload does not need the formatters to process command data. Rather, the payload uses the interface to feed back the received command information to the operators. With this, the ground station can verify that the payload received the commands properly. When we discuss payload commanding in Section 9.6, we will see how this feedback loop works in more detail. At this point, one should be aware of the following traits of the interface:

- The command processor should not process commands faster than the minor frame rate to prevent losing commands in the feedback loop (unless the command word return is supercommutated).
- Due to its importance, the system designer typically designs the command processor with both a prime and a backup processor. The packetizer must query both command processors if both are active, or, if only one is active, then the packetizer must provide an indication of which command processor it is using.
- The command processor may also have a count associated with it for use with the feedback loop in order to help identify sequences of the same command. The packetizer must also transmit this counter.
- The command processor may need to interface with the telemetry processor to supply copies of the command for the command echo data field and report any error codes due to the command content.

The command interface acts like a digital data source for the frame formatter. A major difference from normal payload data is that the system design may place constraints on how quickly and often the packetizer samples this source to insure correct payload operation.

6.12 DATA WAVEFORM FORMATTING FOR TRANSMISSION

Once the data are gathered, the payload must make them ready for transmission. This process involves the specifications of the transmission hardware used at both ends of the transmission link. It may also include regulatory concerns dealing with the occupied bandwidth. In this section, we look at these issues.

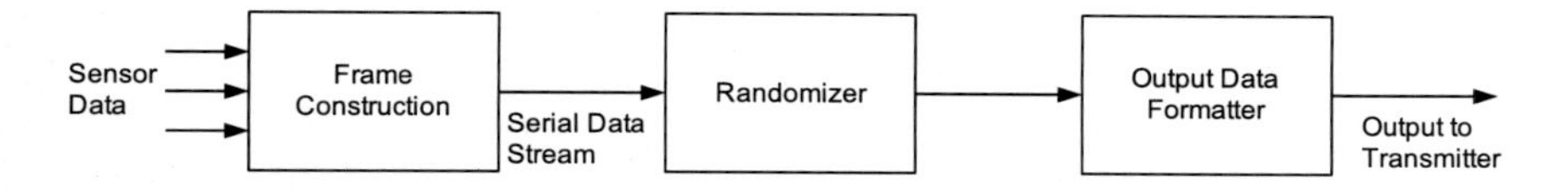

Figure 6.28: The process of formatting the data for transmission and the interface to the communications system.

6.12.1 General Structure

In general, an interface specification is required for the data transmission that defines the correspondence between the logic level and the actual voltage level used in the hardware and the waveform encoding used to transmit a logic zero and a logic one. The first specification is to ensure proper operation of the transmitter and receiver. The second has a more subtle use: to assist in clock extraction at the receiver and to provide desired transmission bandwidth characteristics in the transmitted signal.

Figure 6.28 illustrates this process as part of the telemetry transmission system in the payload. The same considerations exist on the command link to the payload, so the designers need a similar subsystem there. The major component is the randomizer that breaks up the long strings of 0s and 1s. The second provides the data format for transmission. The voltage levels at the output of this module must match the transmitter's specifications. We look at these components below.

6.12.2 Data Randomizers

The function of a *data randomizer*, sometimes called a *data scrambler,* is to provide an approximately equal number of 0s and 1s in the data stream by systematically breaking up long consecutive strings of 0s and 1s. For example, the CCSDS recommends that any data sequence have no more than 64 bits of consecutive 0s or 1s. This is further refined by recommending a minimum of 125 or 275 bit transitions in any sequence of 1000 consecutive bits, depending upon link distance [CCSDS4010].

A randomizer does not make random changes to the data nor does it encrypt the data. Because it is a logic state machine, the randomizer is a deterministic process that the receiver reverses. Figure 6.29 illustrates a typical data randomizer, as described in [IRIG106]. We configure the randomizer with a bit code indicating cells in the shift register enter the exclusive OR gates. The procedure is to label each cell with the cell the most delayed in time corresponding to the most significant or

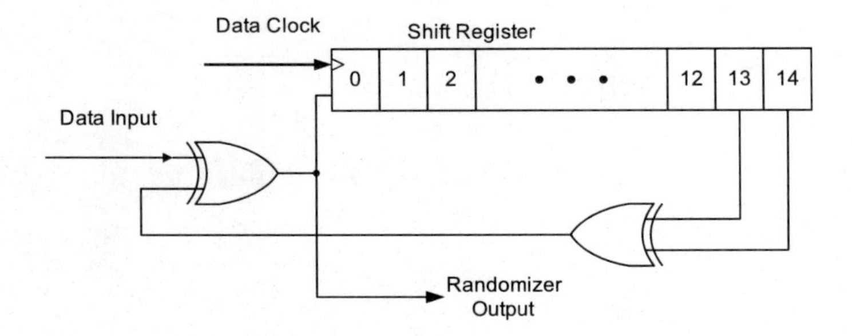

Figure 6.29: The shift-register based data randomizer.

leftmost bit in the pattern. A 1 indicates that the cell is part of the exclusive OR and a 0 indicates no connection. The randomizer always uses the current input data bit as the last cell in the shift register. For example, a connection of the 14th and 13th cells is represented in base-2 and octal form as 110000000000000_2 and 60000_8, respectively. This is the case illustrated in the figure.

6.12.3 Data Format Specification

Designers format digital data for transmission by several methods [Lind73; Skla01; IRIG106]. Every format has advantages and disadvantages. Figure 6.30 illustrates the formats typically found in telemetry systems. The description of how they work appears below. As to notation, a -*L* identifier indicates that the format is a function of the current logic level; -*S* and -*M* identifiers denote that a level change occurs if the current logic symbol is a *space* (a logic 0) or a *mark* (a logic 1), respectively.

Note: In specifying the voltage levels below, we must distinguish between the logic levels (0 and 1 that correspond to false and true, respectively) and the signal voltage level. Unless specified otherwise, we use the convention that a $-1\,\text{V}$ signal level represents logic 0 and a $+1\,\text{V}$ signal level represents logic 1. This is only a convenient convention. We could just as validly have made the opposite assignment or another convenient designation as long as the designers document it in the system documentation.

We give the details for each of the waveforms below. We provide two definitions of Biϕ formats. The first is the typical definition given by most references. The second definition is used in the current edition of

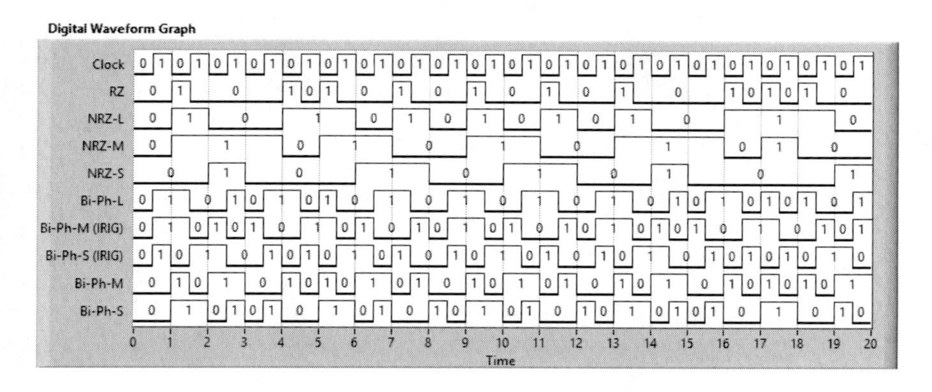

Figure 6.30: The standard logic waveform encoding formats used in telemetry systems. Use the NRZ-L line as the uncoded input-data reference line.

the IRIG 106 [IRIG106]. The IRIG definitions are easier to synthesize in actual hardware than the traditional definitions.

Return to Zero (RZ) in which a logic 1 has its voltage level asserted for the first half of the bit period and then the waveform "returns to zero" (the logic 0 level) for the second half of the bit period. For a logic 0, the waveform makes no voltage change throughout the bit's duration.

Non-Return to Zero-Level (NRZ-L) has logic 0 or 1 represented by the appropriate voltage level for the entire bit duration.

Non-Return to Zero-Mark (NRZ-M) produces a voltage transition from the previous level if the current data bit is logic 1 and no transition if the current bit is logic 0.

Non-Return to Zero-Space (NRZ-S) produces a voltage transition from the previous level if the current data bit is logic 0 and no transition if the current bit is logic 1.

Bi-Phase-Level (Biϕ-L) Biϕ-L, or Manchester encoding, represents a logic 1 by its voltage level for the first half of the bit period and by logic 0 voltage level for the second half of the bit period. The encoding represents a logic 0 by its voltage level for the first half of the bit period and by logic 1 voltage level for the second half of the bit period.

Bi-Phase-Mark (Biϕ-M) represents logic 1 by a voltage transition at the beginning of the bit period as well as a voltage transition in the middle of the bit period. A logic 0 generates a voltage transition only at the beginning of the bit period.

Bi-Phase-Space (Biϕ-S) represents the logic 0 by a voltage transition at the beginning of the bit period as well as a voltage transition in the middle of the bit period. The logic 1 generates a voltage transition only at the beginning of the bit period.

IRIG Bi-Phase-Mark (Biϕ-M) the IRIG 106 variant has a voltage transition at the midpoint of each bit. Logic 1 has no voltage change at the start of the bit period, while logic 0 has a voltage change at the start of the bit period.

IRIG Bi-Phase-Space (Biϕ-S) the IRIG 106 variant has a voltage transition at the midpoint of each bit. Logic 0 has no voltage change at the start of the bit period, while logic 1 has a voltage change at the start of the bit period.

Of these formats, the NRZ and Biϕ formats are standard for range telemetry. However, many telemetry processors allow for the other formats. There are additional formats common in the telecommunications industry but not widely used in the telemetry field.

6.12.4 Data Format Generation

Standard digital devices like latches and ADCs typically do not produce all of the various data patterns. These devices produce waveforms that have a NRZ-L format, although a few single-chip devices do convert NRZ-L to Manchester encoding. We can convert from NRZ-L to the other patterns by using standard logic gates and delay lines. Figure 6.31 shows several example logic circuits to perform this task. We assume that the initial phase of the clock is the logic 0 state for the first half of the clock period. If the first half of the clock period is logic 1, then some of the inverters in the logic equations change. As the figure shows, delaying and exclusive ORing the data stream with itself produces the differential NRZ formats. Biϕ-L is produced by exclusive ORing the NRZ-L data stream with the data clock.

The differential Biϕ waveforms are logic functions of the delayed data stream, current clock state, and current data state. The logic to produce the correct waveform at twice the clock rate is more complicated.

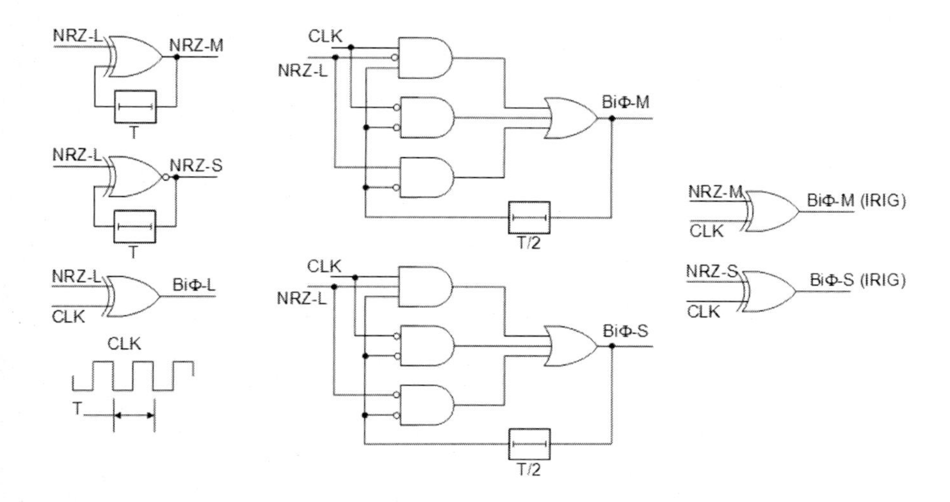

Figure 6.31: Logic circuits to realize waveform encoding based on the NRZ basic format.

Interestingly, the IRIG 106 formats for Biϕ-M and Biϕ-S are generated in a manner similar to Biϕ-L. The logic circuit generates the Biϕ-M format by exclusive ORing the clock with the NRZ-M waveform. In a similar way, the logic circuit generates the Biϕ-S format by exclusive ORing the clock with the NRZ-S waveform.

6.12.5 Inter-Range Instrumentation Group Differential Encoding

If the system designer chooses to transmit the data using an offset-type of quadrature modulation like the Shaped Offset Quadrature Phase Shift Keying (SOQPSK) that we will examine in Section 10.4, the IRIG standard requires that the designer use a *differential bit encoding* in the modulation process [IRIG106]. The differential encoding assigns the even-numbered bits to the "*I Channel*" and the odd-numbered bits to the "*Q Channel*." The IRIG differential encoding is given by the following logic equations using the *I-Channel* and *Q-Channel* symbols and the current bit value, b_n:

$$I_n = \begin{cases} b_n \oplus \overline{Q}_{n-1} & \text{for } n = 2, 4, 6, \ldots \\ I_{n-1} & \text{for } n = 1, 3, 5, \ldots \end{cases} \tag{6.1}$$

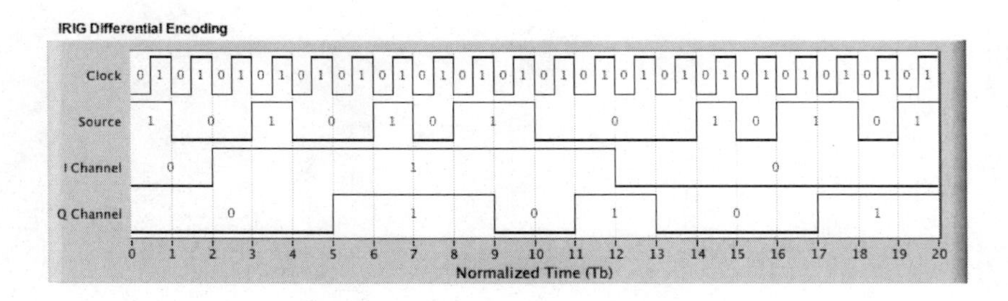

Figure 6.32: The IRIG differential encoding of NRZ-L data for quadrature transmission.

$$Q_n = \begin{cases} Q_{n-1} & \text{for } n = 0, 2, 4, \ldots \\ b_n \oplus I_{n-1} & \text{for } n = 1, 3, 5, \ldots \end{cases} \tag{6.2}$$

Figure 6.32 illustrates an example of the differential encoding with a random bit stream. Notice that each I or Q symbol covers two bit times.

6.12.6 Usage Characteristics

For the designer to choose the correct data format for the data transmission, the designer must know three important characteristics of the data: the occupied bandwidth, the ability to assist in data clock recovery, and the DC power level. The designer should also know whether the sender transmits the data at baseband or in a band pass (modulated) mode. When the sender transmits data at baseband, the sender transmits the bits as they come from the source, that is, from the frame formatter or packet construction hardware. If the sender transmits the data in a band pass mode, the bits modulate a carrier.

An example of the former includes transmission of data packets over a LAN, while an example of the latter is the transmission of data over a radio link. The designer must know the waveform's bandwidth and the DC characteristics because most data links restrict the occupied bandwidth. These characteristics also restrict the way the designer physically transfers the data into the modulator. The data clocking will be discussed further when we cover synchronization. Examples of how the differing data formats appear in their power spectra are described below.

The waveform encoding Power Spectral Density (PSD) determines the necessary transmission bandwidth. The PSD frequency-domain bandwidth is a function of the number of bits per second transmitted and the number of voltage transitions per bit. Increasing both parameters increases the bandwidth of the PSD. The PSD functional forms for the NRZ and Biϕ formats are given in most digital communications references such as [Lind73; Skla01]. These waveforms are applicable to the -L, -M, and -S equally. The normalized baseband NRZ PSD as a function of frequency, f, and the bit period, T_b, is

$$S_{NRZ}(f) = \frac{\sin^2(\pi f T_b)}{(\pi f T_b)^2} \tag{6.3}$$

The normalized baseband Biϕ PSD as a function of frequency, f, and the bit period, T_b, is

$$S_{Bi\Phi} = \frac{\sin^4(\pi f T_b/2)}{(\pi f T_b/2)^2} \tag{6.4}$$

Figure 6.33 illustrates these PSD functions for the NRZ and Biϕ formats. We normalized the plots with a 1 bps data rate. Other data rates scale the plots proportionally. The functional formats for the power spectral densities take on differing shapes. Each has advantages and disadvantages. The NRZ format has significant power levels close to DC frequencies that may cause problems if the system transmits at baseband. The Biϕ format has a null at DC that may allow for good baseband transmission in addition to its self-clocking property. Because these spectra have differing shapes, the band-pass transmission bandwidth differs. Usually, the transmitter filters at the first spectral null. In this case, we see that the NRZ data formats have their first null around the data rate, while the Biϕ format has its first null around twice the data rate.

One way to rate the transmission bandwidth efficiency for the waveform encoding is with the Power Out-of-Band (POB), which is [Math74; Simo76]

$$POB(f_o) = \frac{\int_{-\infty}^{-f_o} |S(u)|^2 du + \int_{f_o}^{\infty} |S(u)|^2 du}{\int_{-\infty}^{\infty} |S(u)|^2 du} = \frac{\int_{f_o}^{\infty} |S(u)|^2 du}{\int_{0}^{\infty} |S(u)|^2 du} \tag{6.5}$$

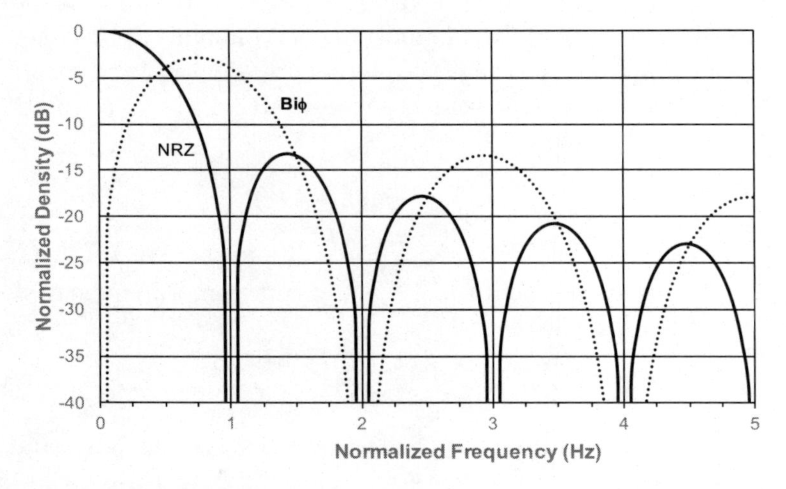

Figure 6.33: Normalized theoretical PSD for NRZ and Biϕ waveform encoding.

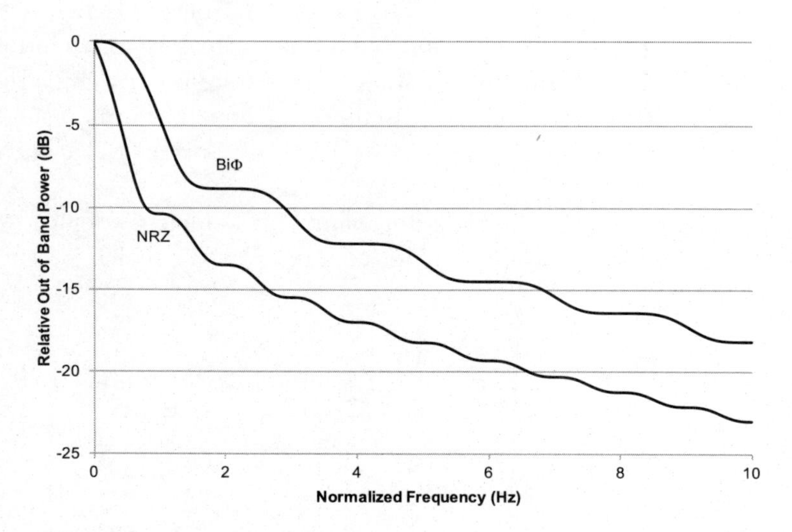

Figure 6.34: The power out of band as a function of frequency for NRZ and Biϕ waveform encoding.

This measures the contribution of the PSD, $S(f)$, above a certain frequency, f_o, as compared with the total PSD for the signal. If we want a 99% power point for the signal, which is a common requirement for regulatory measures, we find the frequency that gives a POB of 1%. Figure 6.34 is a plot of the POB for NRZ and Biϕ waveforms as a function of frequency with a 1 bps data rate.

Doubling the bandwidth for Biϕ waveforms, relative to the NRZ waveforms, is the trade for better clock extraction characteristics. All of the Biϕ waveforms have at least one level transition per bit, while with the NRZ waveforms there is no such guarantee. Hence, a data clock is derived more easily from the Biϕ waveforms, which aids in data synchronization. We say that the bandwidth is doubled for Biϕ over that for NRZ because the first null is at 2 Hz rather than at 1 Hz in the normalized plot given.

The usage of a differential waveform (-M or -S) rather than a level waveform (-L) does not provide an advantage or disadvantage from a transmission bandwidth perspective. Figure 6.35 compares the -L and -M formats for simulated random data streams. The NRZ-L and NRZ-M formats have the same shapes and their spectral nulls are at the same frequencies. The Biϕ waveforms behave in a similar manner, albeit at twice the bandwidth of the NRZ waveforms.

We note that the differential waveforms (-M and -S) have the advantage of being immune to polarity reversals in transmission. This frequently occurs when transmitting digital data with a modulated carrier. The disadvantage to differential waveforms is the slightly greater complexity in generating and recovering the data and the higher transmission error rate. The transmission complexity results from the added delay logic that the designer customizes to the data period. The system designers must standardize the state of the first bit sent prior to the actual data transmission. This initial bit resolves the differential coding ambiguity and its value makes no difference as long as both sides agree on it. The higher transmitted error rate occurs because, with a differential waveform, the current level is a function of both the present and the previous data points. When a bit error occurs, the next bit is also in error due to the memory in the encoding process (see Problem 12).

6.13 REFERENCES

[ATIS] Alliance for Telecommunications Industry Solutions. *ATIS Telecom Glossary 2011*. 2011. URL: http://www.atis.org/glossary/.

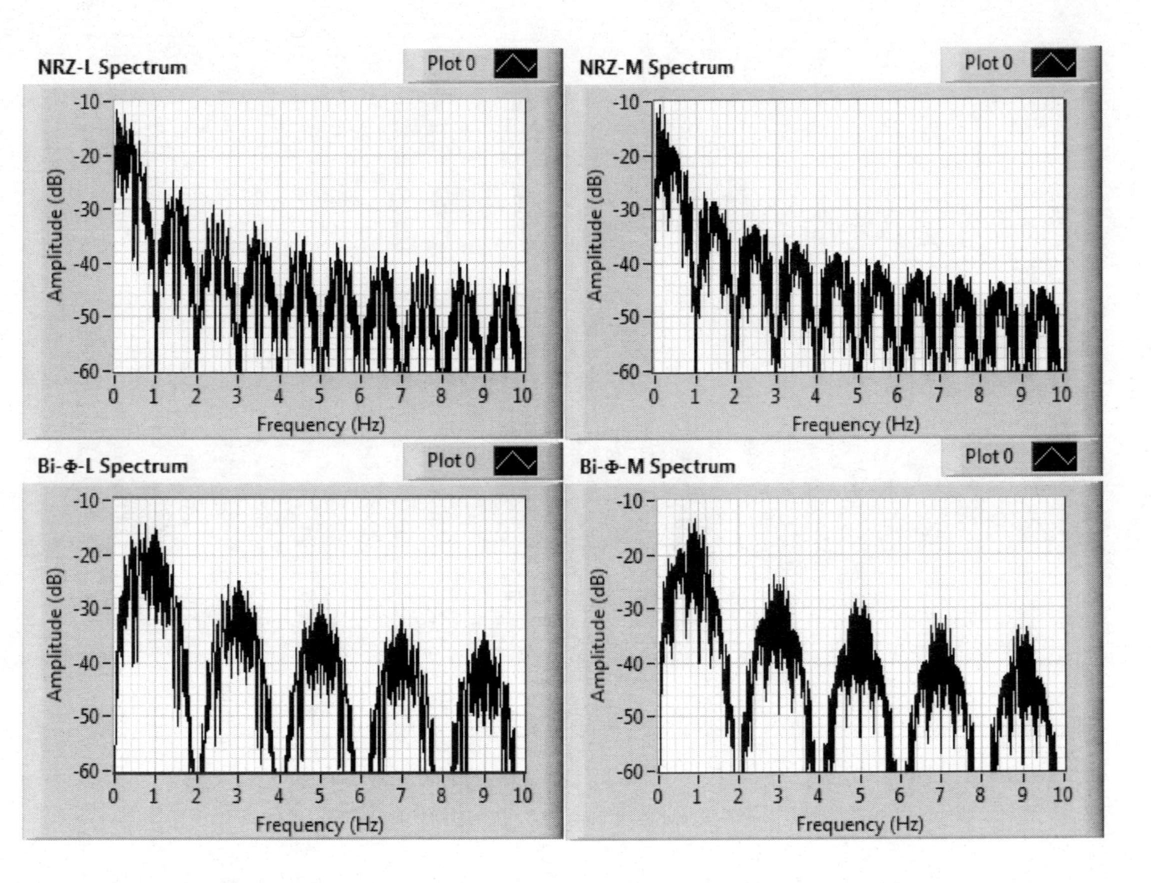

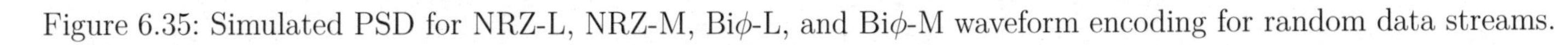
Figure 6.35: Simulated PSD for NRZ-L, NRZ-M, Biϕ-L, and Biϕ-M waveform encoding for random data streams.

[CCSDS1300] *Overview of Space Communications Protocols.* CCSDS 130.0-G-3. Consultative Committee for Space Data Systems. Washington, D.C., July 2014. URL: https://public.ccsds.org/Pubs/130x0g3.pdf.

[CCSDS1302] *Space Data Link Protocols – Summary of Concept and Rationale.* CCSDS 130.2-G-3. Consultative Committee for Space Data Systems. Washington, D.C., Sept. 2015. URL: https://public.ccsds.org/Pubs/130x2g3.pdf.

[CCSDS1310] *TM Synchronization and Channel Coding.* CCSDS 131.0-B-2. Consultative Committee for Space Data Systems. Washington, D.C., Aug. 2011. URL: https://public.ccsds.org/Pubs/131x0b2ec1.pdf.

[CCSDS1320] *TM Space Data Link Protocol.* CCSDS 132.0-B-2. Consultative Committee for Space Data Systems. Washington, D.C., Sept. 2015. URL: https://public.ccsds.org/Pubs/132x0b2.pdf.

[CCSDS1330] *Space Packet Protocol.* CCSDS 133.0-B-1 Cor. 2. Consultative Committee for Space Data Systems. Washington, D.C., Sept. 2012. URL: https://public.ccsds.org/Pubs/133x0b1c2.pdf.

[CCSDS2310] *TC Synchronization and Channel Coding.* CCSDS 231.0-B-2 Cor. 1. Consultative Committee for Space Data Systems. Washington, D.C., Apr. 2013. URL: https://public.ccsds.org/Pubs/231x0b2c1.pdf.

[CCSDS2320] *TC Space Data Link Protocol.* CCSDS 232.0-B-3. Consultative Committee for Space Data Systems. Washington, D.C., Sept. 2015. URL: https://public.ccsds.org/Pubs/232x0b3.pdf.

[CCSDS4010] *Radio Frequency and Modulation Systems – Part 1 Earth Stations and Spacecraft.* CCSDS 401.0-B-26. Consultative Committee for Space Data Systems. Washington, D.C., Oct. 2016. URL: https://public.ccsds.org/Pubs/401x0b26.pdf.

[Drag07] M. Dragosavac. *BUFR User's Guide.* European Centre for Medium-Range Weather Forecasts. Reading, UK, 2007. URL: https://www.wmo.int/pages/prog/gcos/documents/gruanmanuals/ECMWF/bufr_user_guide.pdf.

[Fan00] J. Fan and S. Horan. *Implementing Real-Time Radio Propagation Measurements Over the Internet.* NMSU-ECE- 00-004. New Mexico State University. Las Cruces, NM, Apr. 2000.

[Hill63] E.R. Hill. "Techniques for Synchronizing Pulse-Code-Modulated Telemetry." In: *Proc. 1963 National Telemetering Conf.* Albuquerque, NM, May 1963, pp. 3.3.1 –3.3.13.

[IRIG106] Telemetry Group. *Telemetry Standards.* 106-15. Secretariat, Range Commanders Council. White Sands, NM, July 2015. URL: `http://www.wsmr.army.mil/RCCsite/Documents/Forms/AllItems.aspx`.

[Jian00] H. Jiang and S. Horan. *Wireless Telemetry and Command (T&C) Program.* NMSU-ECE- 00-003. New Mexico State University. Las Cruces, NM, Apr. 2000.

[Li11] R. Li, M. Hu, and J. Olson. *BUFRPrepBUFR User's Guide.* Version 1.0. 2011. URL: `ftp://ftp.cpc.ncep.noaa.gov/hwang/HW/BUFR/BUFR_PrepBUFR_User_Guide_v1.pdf`.

[Lind73] W. C. Lindsey and M. K. Simon. *Telecommunication Systems Engineering.* Englewood Cliffs, NJ: Prentice Hall, 1973.

[Math74] H. Mathwich, J. Balcewicz, and M. Hecht. "The Effect of Tandem Band and Amplitude Limiting on the Eb/No Performance of Minimum (Frequency) Shift Keying (MSK)." In: *IEEE Transactions on Communications* COM-22.10 (Oct. 1974), pp. 1525 –1540. DOI: `10.1109/TCOM.1974.1092108`.

[Maur64] Jr. J. L. Maury and F. J. Styles. "Development of Optimum Frame Synchronization Codes for Goddard Space Flight Center PCM Telemetry Standards." In: *Proc. National Telemetering Conference.* Houston, TX, June 1964, pp. 3.1 –3.10.

[MS1553] *Aircraft Internal Time Division Command/Response Multiplex Data Bus. MIL-STD-1553B.* Washington, D.C.: Department of Defense, Sept. 1978.

[NASA85] *Space Network TDRSS Data Book.* TDRSS Program Division, Code TX. Washington, D.C.: National Aeronautics and Space Administration, 1985.

[RCC218] Telecommunications and Timing Group. *Telemetry Transmission Over Internet Protocol (TMoIP) Standard.* 218-10. Secretariat, Range Commanders Council. White Sands, NM, Oct. 2010. URL: `http://www.wsmr.army.mil/RCCsite/Pages/Publications.aspx`.

[Rema93] P. W. Remaklus. "Data Collection with the ACTS Propagation Terminal." In: *Proc. 17th NASA Propagation Experimenters Meeting (NAPEX XVII) and the Advanced Communications Technology Satellite (ATCTS) Propagation Studies Miniworkshop.* JPL Publication 93-21. Pasadena, CA, June 1993, p. 267.

[RFC4838] V. Cerf et al. *Delay-Tolerant Networking Architecture.* RFC 4838. Apr. 2007. 35 pp. URL: `https://tools.ietf.org/html/rfc4838`.

[Simo76] M. K. Simon. "A Generalization of Minimum-Shift-Keying (MSK)-Type Signaling Based Upon Input Data Symbol Pulse Shaping." In: *IEEE Transactions on Communications* COM-24.8 (Aug. 1976). DOI: 10.1109/TCOM.1976.1093380.

[Skla01] B. Sklar. *Digital Communications: Fundamentals and Applications.* 2nd Edition. Upper Saddle River, NJ: Prentice Hall PTR, 2001.

[Stal97] W. Stallings. *Data and Computer Communications.* 5th Edition. Upper Saddle River, NJ: Prentice Hall, 1997. ISBN: 0-02-415425-3.

6.14 PROBLEMS

1. Design an IRIG-compatible major and minor frame structure that meets the transmission requirements for the sensors given in the table. That is, determine

   (a) The minor frame length

   (b) The number of minor frames per major frame

   (c) The design of any subframes that you think are reasonable

   (d) The packaging of the data within the word boundaries

   Use minor frame per second as the normalized base sampling clock frequency for the sensor array. Use 8-bit word lengths for each word in the minor frame. Assume that you applied a 16-bit synchronization code to each minor frame. Determine the number of bytes per second that need to be transmitted.

   Telemetry Sensor Database

Sensor Type	Number of Sensors/Type	Individual Sensor Rate Relative to Minor Frame Rate	Number of Bits/ Individual Sensor Sample
I	8	1	8
II	32	1	1
III	32	0.25	8
IV	4	1	10
V	2	2	12
VI	4	0.5	4

2. The optimal synchronization codes listed in Table 6.3 are not unique

for each length. Determine at least four other 16-bit synchronization codes that have the same correlation properties for 1-bit and 2-bit shifts (both early and late) as the listed optimal code. Assume no transmission errors are present.

3. The IRIG standard requires that the synchronization marker, the FFID word location relative to the synchronization marker, and the frame bit rate shall not be changed during the transmission. Justify this restriction.

4. The standard requires the FFID be constructed so that a single bit error does not produce another valid FFID. Consider 16 possible frame formats in a system. What is the minimum length of the FFID to encode the 16 possibilities and provide for the error protection? Develop a table to show that your technique works.

5. Design a packet structure using the HDLC format to transmit the data given in Problem 1. Assume that each sensor class generates its own HDLC package. How many bytes per second does the packetizer generate? Assume 8-bit addressing and control, and 16-bit frame check sequence on each HDLC packet.

6. Consider a telemetry system where each major frame contains 64 minor frames. Each minor frame is 1536 bytes long. Each minor frame can be composed of either PCM sensor values or segments of file data. The file data frames are composed of data from three independent sources that equally share the frame contents. The transmitter sends either minor frame type depending upon the availability of queued data. For this system, answer each of these questions:

   (a) What is your recommendation for how to differentiate between the two minor frame types?
   (b) What is your recommendation for the format of the sensor frame including the identifier and any relevant accounting/-management data?
   (c) What is your recommendation for the format of the file frame including the frame identifier, file content identifier, and any relevant accounting/management data?

   Illustrate your recommended formats.

7. Design the logic to generate entropy-mode packets. Include the watchdog timer and timer reset, and a means to fill data packets with fill data under the correct circumstances.

8. Consider that you are part of a conceptual design team for a Saturn probe. The probe needs to transmit its science, environmental, and payload state data to the main spacecraft over a 3-hour period. As the probe enters Saturn, it will switch between two different high-rate sensors for 1 minute at a time. Each sensor will generate housekeeping data that amounts to 5% of the science data. The probe itself has 256 housekeeping sensor values for monitoring the probe state and environment at a relatively low rate. There is a subset of 16 critical sensor values within this housekeeping data. Design a packet system that specifies

   (a) The packet protocol method, e.g., TCP/IP or CCSDS
   (b) The transmission mode type from Section 6.7.3 for each data type
   (c) How you will identify the data type and mode type within the protocol structure

9. Draw the configuration for a randomizer with tap settings at 024_8. For the randomizer, (a) show the structure of the randomizer including the placement of the tap settings, and (b) show how the process works for the binary input data stream 1 010 101 011 111 100 000 010.

10. Is a randomizer needed for the Biϕ family of waveform encodings? Explain your reasoning.

11. Format the binary input data stream 1 010 101 011 111 100 000 010 in the NRZ-L, NRZ-S, Biϕ-L, and Biϕ-M formats. Use both the standard and the IRIG 106 definition for the Biϕ-M format.

12. Using the binRY data stream 1 010 101 011 111 100 000 010, show that a NRZ-M data encoding is correctly decoded if the polarity of the data stream is inverted. Show that single bit errors in transmission cause two errors in reception when the data are NRZ-M encoded.

13. Develop the logic circuits necessary to recover the data stream from the IRIG differential encoding found in Equations (6.1) and (6.2).

Use a simulator to demonstrate that your logic correctly assigns the even- and odd-numbered bits to the *I*- and *Q-Channel*, encodes and decodes the differential encoding of the bit stream, and then reassembles the two channels into a single serial channel.

14. Determine the minimum frequency to give an out-of-band power of 1% for NRZ and Biϕ waveform encoding.

15. Develop (or use a commercial simulation package) a simulation scenario to emulate entropy mode and virtual channel mode packets on the same link. Give the entropy mode packets the highest priority and have two priority levels of virtual mode packets, both of which are below the priority of entropy mode packets. Investigate the amount of extra capacity to transmit all of the entropy mode packets fully and the amount of delay it gives to the virtual channel traffic. Make the simulation conditions varied enough to investigate several modes in this parameter space.

16. Suppose you have an asynchronous embedded telemetry stream to transmit with the nominal telemetry stream. The host has sixty-four 8-bit words per minor frame and sixty-four minor frames per major frame. The embedded stream is thirty-two minor frames long with each minor frame being twenty-eight 8-bit words long. The host stream allocates 16 words per minor frame for the embedded telemetry stream. Sketch the flow of the embedded stream through the host major frame. Assume that both streams are initialized to start with the first minor frame of each synchronized and thereafter no attempt is made to force synchronization. How often will the two streams have their first minor frames resynchronized?

CHAPTER 7

DATA SYNCHRONIZATION

7.1 INTRODUCTION

In the transmitter/receiver pairing, the transmitter has the advantage because, at every moment, it controls what data it is transmitting, what the carrier phase and frequency are, and where the communications are in the Protocol Data Unit (PDU) structure. However, the receiver needs to determine this information on its own based only on what information is transmitted. Synchronizing the data flow between the transmitter and the receiver must be performed when either device is initialized or reset, and when the communications link fails. The receiver performs the following operations on the received data stream to make the synchronization process work:

1. Recovers the data clock rate and synchronizes to the bit pattern.
2. Achieves synchronization with PDU structure.
3. Achieves synchronization with the PDU internal data.

Only after the receiver synchronizes to the data can it reliably process the data. The highlighted blocks in Figure 7.1 illustrate the locations of the components in the user segment and the payload segment to perform these operations. The radio receivers in both the user and the payload segments need to perform all of these functions. The synchronization functions are not an artifact of the telemetry link only but occur for telecommand links as well. This chapter covers these three tasks and

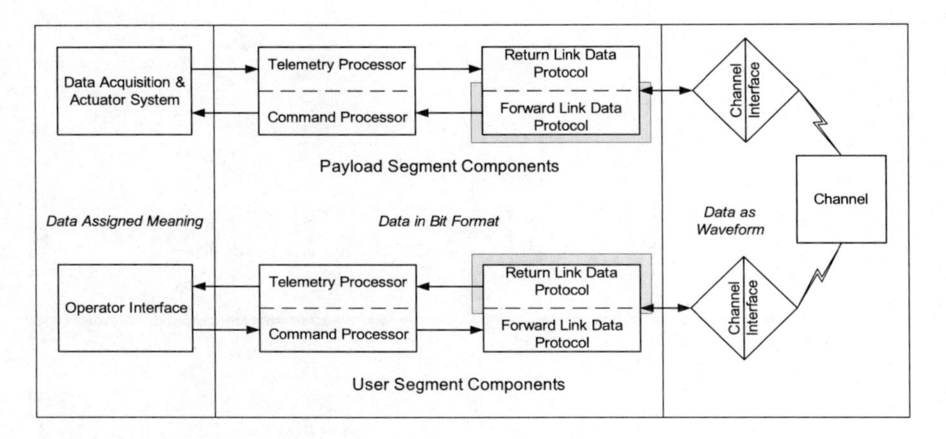

Figure 7.1: The highlighted blocks correspond to the components in the user segment and the payload segment involved in data synchronization.

some of the techniques involved. We also examine data-stream error detection and reassembly.

7.2 OBJECTIVES

The data-stream synchronization concepts are fundamental to understanding how the receiving system orients itself to the input data's unknown phase. Our objectives for this chapter are to have the reader be able to

- Describe the operating principles for a bit synchronizer in general and the variations on the method.
- Draw state diagrams for PDU synchronization.
- Compute the required length of the minor frame synchronization code for a given false lock rate and compute the probability of missing a valid frame synchronization marker given a channel probability of a bit error.
- Explain the issues involved with data sequence reassembly.
- Discuss error detecting and correcting codes for telemetry frames, packets, and telecommands.

While we discuss many of these processes in terms of frame telemetry,

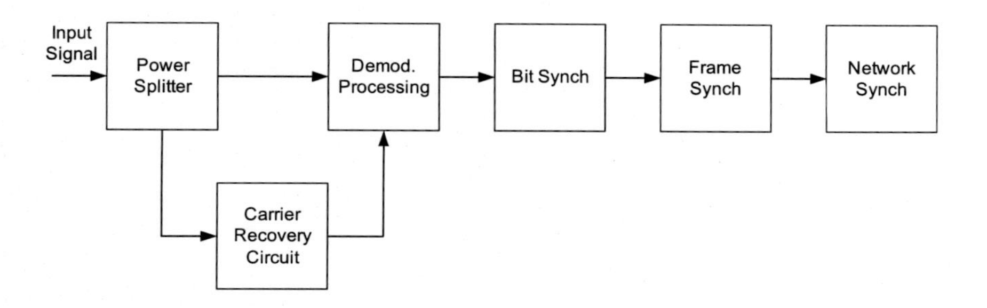

Figure 7.2: Overview of the steps in the synchronization process.

they have their analogs in packet telemetry and telecommand links, even if not explicitly mentioned.

7.3 SYNCHRONIZATION PROCESS

Figure 7.2 shows a synchronization process applicable to any communications system. The process requires five synchronization levels before the receiver produces reliable data. The states are

1. Carrier synchronization
2. Demodulate the data from the carrier
3. Bit synchronization
4. Frame synchronization
5. Network synchronization

The receiver must step through each of these states using only the information in the data signal. Here, we assume that the receiver knows the transmitter's modulation format and data rate so the receiver will not need to determine them from the received signal characteristics. The time at each level depends upon the data's quality and the time until the receiver finds the synchronization markers. Figure 7.3 shows these steps happening over time. The major difference in some systems is that they operate only at baseband and therefore do not need the carrier synchronization state. In all cases, there is no back channel information from the transmitter to the receiver to assist in the process.

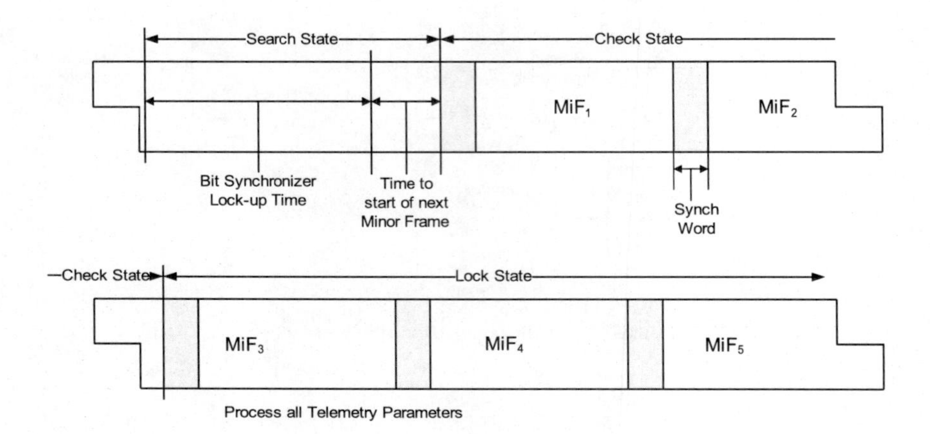

Figure 7.3: The telemetry frame synchronization timing process. Packet and telecommand synchronization require fewer steps.

7.4 CARRIER SYNCHRONIZATION

If the data system is a baseband system, it does not require *carrier synchronization*. If the data system uses radio techniques with a modulated carrier signal, the receiver must synchronize to the signal's carrier frequency. The receiver considers the carrier signal coming into the radio to have a random phase angle. Additionally, the carrier signal may have a slightly different frequency than the one to which the radio is tuned. This "off tuning" may be caused by a relative Doppler shift between the transmitter and receiver or frequency drift in the transmitter due to issues with the electronics. The carrier synchronization problem is beyond the scope of this chapter but we will see this again as part of the modem structure in Section 10.4. The receiver's radio electronics require the equivalent of 100 to 500 bit times, depending upon the signal's quality, to complete the carrier synchronization process [Ande99].

7.5 BIT SYNCHRONIZATION

Bit synchronization's goal is to produce stable data and a stable data clock. This is true whether the system sends data over a carrier-based transmission or a baseband transmission. In this section, we explore the issues related to bit extraction, bit formatting, and data timing. The

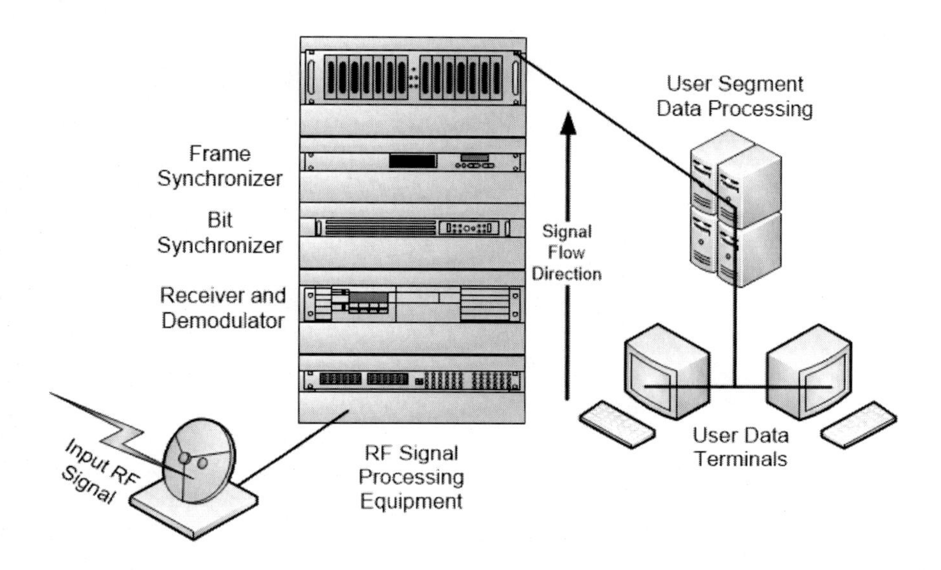

Figure 7.4: Electronic components involved in the synchronization process. The signal flow from the RF input to the formatted data output. The electronics also contain extensive software components in their operations.

receiver needs to complete these functions to interpret the data in the subsequent data synchronization steps.

7.5.1 General Functions

Locking the receiver to the transmitted bit stream structure is the first step the receiver must perform to begin processing the actual data in any context [ElMo80; Simo70a; Simo70b]. In radio systems, the Radio Frequency (RF) electronics converts the input modulated signal to its baseband waveform and passes that waveform to the bit synchronizer to synchronize with the individual bits. Figure 7.4 illustrates the bit synchronizer's placement in the receiver electronics chain. It functions between the demodulation electronics and the PDU synchronization electronics.

Figure 7.5 illustrates the functional components found in many commercial bit synchronizers. Any one specific bit synchronizer may not have all of these components but this illustration gives the desired functions. The components in the figure perform the following functions:

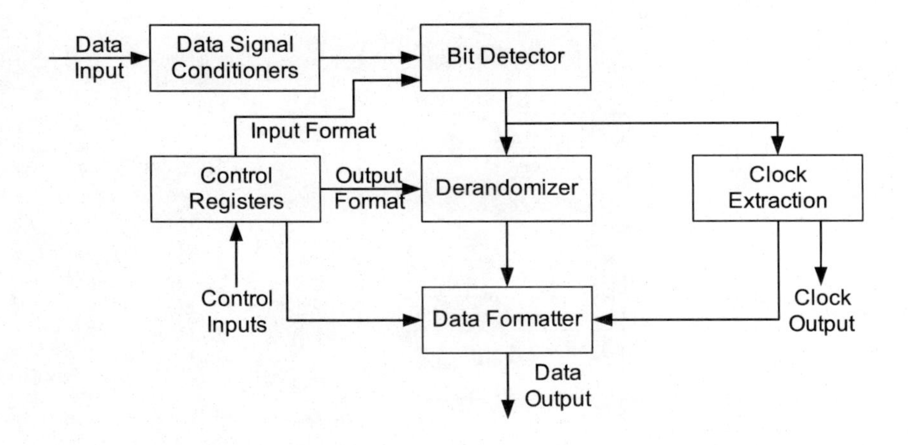

Figure 7.5: Functional components found in a bit synchronizer.

- The data signal conditioners provide signal gain control and conditioning to supply a standard-level signal to the bit detector.
- The bit detector recognizes the bit pattern for the input data stream's format.
- The control registers supply configuration data to operate the bit synchronizer correctly with this data coming from either a remote controlling computer or a front-panel switch bank.
- The clock extraction circuitry provides a synchronizing clock signal slaved to the data transitions for use by the internal data formatter and the external circuitry.
- The data formatter provides translation, if desired, between the standard format of the input data stream and the output data stream; for example, Manchester or Bi-Phase-Level (Biϕ-L) coding on the input data stream is translated to Non-Return to Zero-Level (NRZ-L) format on the output and all the bits complemented because of the way in which the data was transmitted.
- The derandomizer provides the occasionally needed function of removing the randomization added in the transmitter to give a good mixture of 1s and 0s. This occurs when the data source produces long strings of either 1s or 0s.

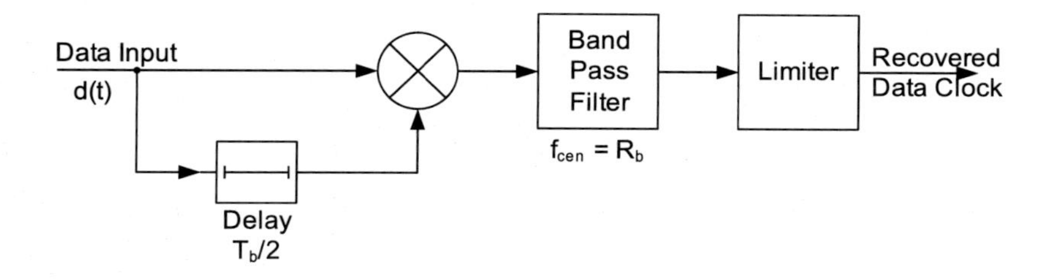

Figure 7.6: Components used in an open-loop clock extractor.

The most critical function is the production of the uniform clock slaved to the data stream. In the balance of this section, we discuss ways to do this operation and discuss the formats for the various digital data types and the function of the randomizer/derandomizer circuits.

7.5.2 Data Clock Extraction

The bit synchronizer's most difficult job is extracting the synchronized clock from the random data stream. In general, the user needs a clock signal that transitions exactly in phase with the data, is present even when the data form a string of all 1s or 0s, and is immediately valid as soon as the unit has power applied. Clock extractors cannot satisfy all of these conditions simultaneously for all data formats. As we will see, the data transmission format we choose makes a difference in how well the clock extractor meets these objectives. Additional details are in [Ande99; Skla01].

7.5.2.1 Open-Loop Clock Extractors

Open-loop clock extractors have characteristic circuitry that estimates the clock transition time based upon transitions in the current data stream level [Gill63]. These are called "open-loop" because no feedback information from previous symbols is used to derive the current clock rate and transition time. Figure 7.6 shows an example open-loop clock extractor. The circuit mixes a copy of the input bit stream with a delayed version of the same data. The delay is one-half of the bit period, T_b. A narrow band pass filter with a center frequency, f_{cen}, at the bit rate, R_b, filters the mixer output.

This extractor produces clock transitions only when the data changes state at a reasonably high rate. Long strings of constant data levels often

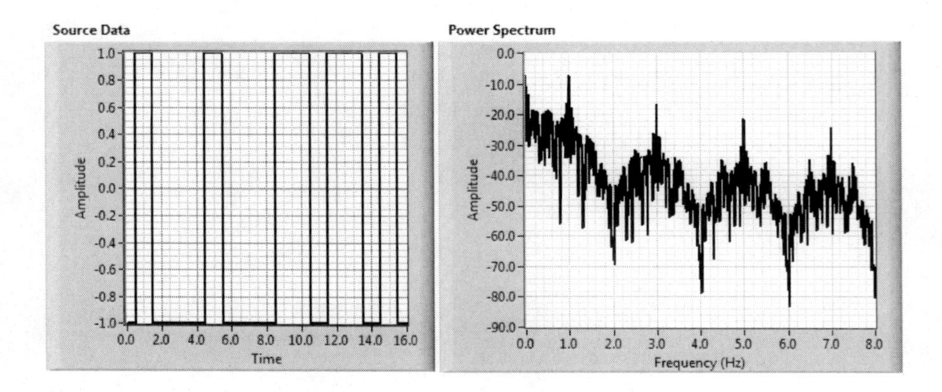

Figure 7.7: LabVIEW simulation of a sample random data waveform and its associated power spectral density with its input data rate normalized to 1 bps. Note the clock harmonic at the bit rate and subsequent odd integer multiples of that rate.

cause the clock to "disappear" because the output of the mixer becomes a DC voltage level in this case. Figure 7.7 shows how the extractor works for NRZ-type data. Mixing the input data stream with a delayed version of itself produces a Return to Zero (RZ) data format that has a discrete harmonic at integer multiples of the data clock frequency. The first panel shows the input data, while the second panel shows the spectrum of the signal at the mixer output.

The band pass filter in Figure 7.6 isolates this discrete harmonic. The harmonic at the band pass filter's output is a sinusoidal waveform that a hard limiter converts to a square wave. Harmonics exist in the power spectrum at odd intervals of the clock rate. If the data stream contains a long run of 1s or 0s, this harmonic disappears from the band pass filter's output and the clock along with it. Thus, a randomizer is necessary to ensure a sufficient clock transition density.

7.5.2.2 Closed-Loop Clock Extractors

Closed-loop clock extractors use feedback information to provide a more accurate clock rate and clock transition time [Simo70a; Simo70b; Lind73; Skla01]. Usually some form a Phase Locked Loop (PLL) produces the clock signal and the circuitry slaves the PLL output to the incoming data stream. While this clock extractor provides a more stable clock, especially during runs of constant data, it requires a longer startup time

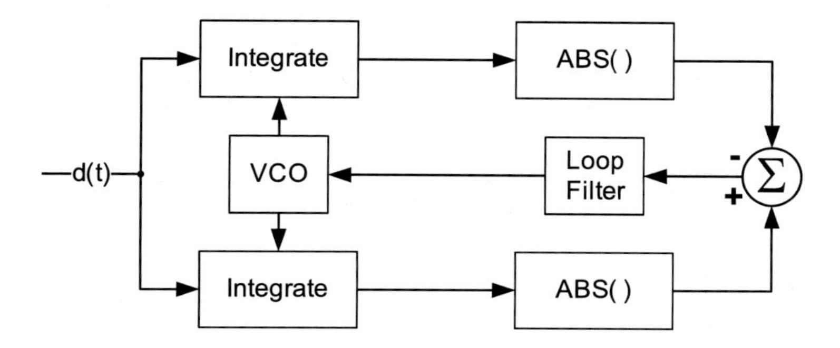

Figure 7.8: A closed-loop clock extractor, also known as an early-late gate, circuit [Simo70a].

to slave the PLL to the data stream.

Figure 7.8 shows a closed-loop synchronizer. This synchronizer performs two integrations within a bit period. The integrations are staggered within the bit period and the total time from the start of the first integration until the end of the second integration is one bit period. The absolute values of the two integrations are then subtracted and the difference forms an error signal that is used to drive the Voltage Controlled Oscillator (VCO) of the PLL. The error signal captures the phase difference between the incoming data stream and the VCO.

If the integrations are the same, then the VCO is in synchronization with the data stream. As expected, circuit designers realize open-loop clock extractors in the electronics more easily than closed-loop synchronizers, but the closed-loop synchronizer's added complexity may be justified when improved performance is required.

7.5.3 Data Formats

A bit detector must have the logic to recognize the waveform encoding pattern representing the bit waveform formats discussed in Section 6.12.3. Manufacturers allow a user to specify the incoming waveform encoding as part of the bit synchronizer's initialization options. This is an error source if the receiver does not match the transmitter's waveform encoding pattern. For a differential encoding pattern, such as Non-Return to Zero-Mark (NRZ-M) or Non-Return to Zero-Space (NRZ-S), there is some ambiguity in the initial logic state. The designer may wish to use a

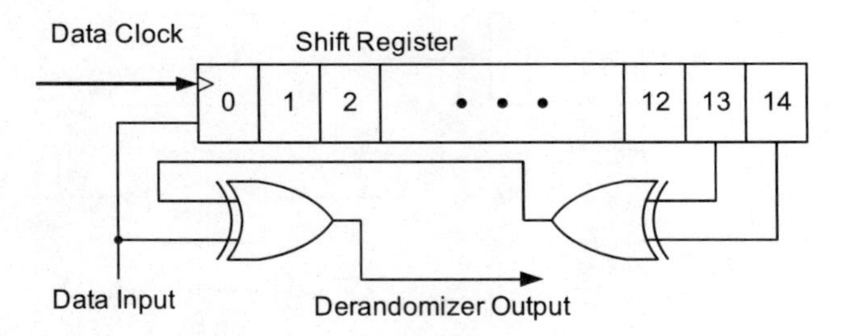

Figure 7.9: A derandomizer circuit.

short acquisition sequence to force the receiver into a known state before decoding the actual data.

7.5.4 Derandomizer

As we saw in Section 6.12.2, we use a randomizer circuit to break up long sequences of 1s and 0s in Non-Return to Zero (NRZ)-encoded data. The randomizer is merely a finite state machine consisting of a shift register and a multiple-input exclusive OR gate. The derandomizer is the complementary machine to remove the randomization effect and restore the original data stream.

The randomizer/derandomizer circuit has memory. As a result, if a transmission error causes a bit to change, the derandomizer's output has more than one error and these errors continue until the affected bit has completely cleared the derandomizer. This penalty must be borne in mind when weighing the good effects of having uniform mixtures of 1s and 0s on the data stream. Figure 7.9 illustrates a derandomizer to match the randomizer given earlier in Figure 6.29 [IRIG106]. We configure the derandomizer with the same bit code used in the randomizer to indicate which cells in the shift register connect to the exclusive OR gates.

The derandomizer configuration needs modification if the transmitter plays the data in time-reversed mode. This situation may occur when flight tape recorders capture the data in real-time, while the payload is out of range of the base station. When it is time for playback, the tape is not rewound prior to playback; rather, the recorder reverses the drive direction and the tape is "dumped" at the record speed. This reverses the time ordering of the data and the ground station hardware must remove this reversal in real-time as the data are received.

7.6 PROTOCOL DATA UNIT SYNCHRONIZATION

After the bit clocking has been determined, the synchronization begins to place meaning on the data. Frame synchronization's purpose is to lock onto the data structure embedded in the received Protocol Data Unit (PDU). In this section, we look at synchronization for telemetry frames, and packets for telemetry and telecommand. We also look at probabilities of incorrectly synchronizing.

7.6.1 Telemetry Frame Synchronization

Telemetry frame synchronization starts once bit synchronization has occurred. Bit synchronization slaves the receiving system to the payload's clock. The frame synchronization process slaves the receiver to the payload's data structure. Without the bit synchronization process, this process cannot be performed.

The frame synchronization process assumes that the data are flowing continuously from the transmitter. The system users generally are not concerned with losing a small amount of data to start the process so transmitter does not send a special data acquisition sequence to initialize the reception.

Frame synchronization is broken into three logical states that designate the levels of knowledge about the payload's data structure [Ande99; Skla01]. Figure 7.10 is a state transition diagram for the three states:

Search State where the synchronizer examines the incoming data for the initial structural benchmarks.

Check State where the synchronizer makes its preliminary indexing into the data state and refines the estimate.

Lock State where the synchronizer has locked with the data structure and full processing data proceeds.

We next look at each of these states in detail.

7.6.1.1 Search State

In the search state, the frame synchronizer looks for the synchronization marker that delineates the start of a minor frame. The synchronizer may stay in the check state for several occurrences of the synchronization word to verify that it has found the word correctly and that the word repeats

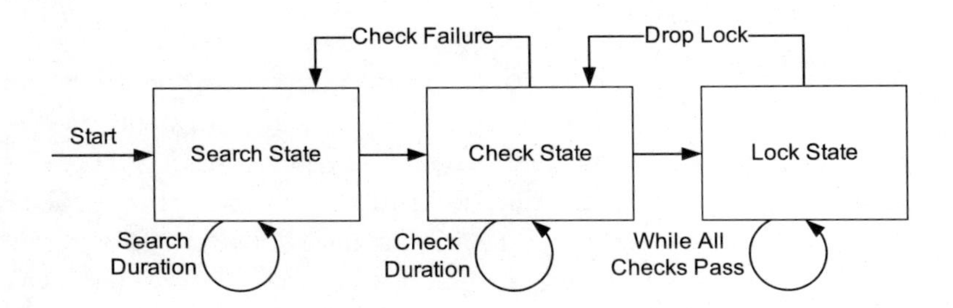

Figure 7.10: The states in the telemetry frame synchronization process. The durations are usually programmable during search and check modes.

at the correct frequency. For the frame synchronizer to perform this job, it must know both the synchronization word value and the repetition length of the minor-frame structure. The designer accomplishes this by programming a set of control registers at initialization. If the synchronizer finds the synchronizing word occurring at the correct frequency and for the number of times specified in the control register, it proceeds to the check state.

Figure 7.11 illustrates the search strategy, which passes the incoming data stream into a correlator [Ha90]. The individual data bits are clocked into a shift register at the start of each bit time. During each bit period the correlator checks the contents of the shift register against the prestored synchronization word pattern by looking for a one-to-one match between the shift register and the stored pattern. The designer sets a predetermined number of bits in the synchronization word to match the prestored pattern exactly. When the correlator detects this number of bits, it emits a pulse to indicate that it has found the synchronization word. Designers use the optimum codes presented in Section 6.5 as the synchronization word patterns because they have low correlations, unless the received code is exactly aligned in the correlator with the desired pattern. Other codes, for example all 1s (or all 0s) or alternating 101 010, etc., show correlations when offset from exact alignment. The correlation function, $R(m)$, is expressed as

$$R(m) = \sum_{n=1}^{N} w(n)d(n-m) \tag{7.1}$$

Here, $w(n)$ is the desired N-bit synchronization word pattern and $d(n-m)$ is the input data stream shifted by m bits from exact align-

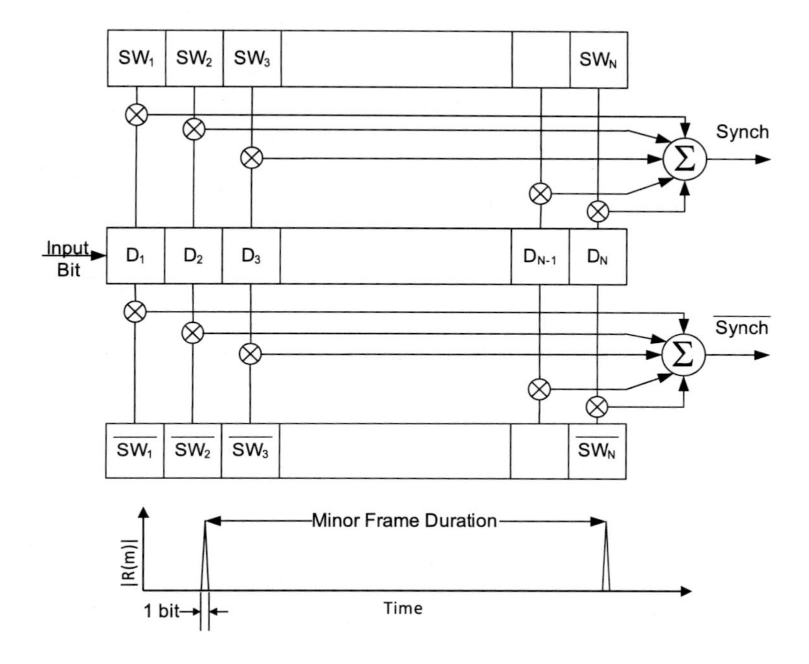

Figure 7.11: Correlator seeking the bits in the stored synchronization word and the complement to the synchronization word. The correlator has one correlation per minor frame period.

ment with the synchronization word. Figure 7.11 illustrates the resulting correlation function. The correlation has a maximum once per minor frame time when the synchronization word aligns with the data stream. There should be little correlation between the desired synchronization word and the random data at other times.

The correlator illustrated in Figure 7.11 normally has the ability to search for the complement of the desired synchronization word in parallel with searching for the specified synchronization word. The designer configures the correlator with this additional mode because the digital communications process may invert the data during the transmission/reception process. The complementary synchronization pulse indicates the occurrence of such a data inversion so the processing hardware can correct the received data for this condition.

7.6.1.2 Check State

In the check state, the frame synchronizer assumes that the synchronization word has been correctly located in the serial data stream. If the transmitter is using the minor frame/major frame structure, the check state determines the current location within that structure. Check state processing determines the current value of the SubFrame Identifier (SFID) and that the SFID is incrementing properly. If the SFID appears to take improper action, the state of the frame synchronizer may revert back to the search state.

The frame synchronizer must know beforehand the location SFID and its range in the data structure. Once the check state has determined that the SFID is behaving properly, the frame synchronizer moves to the lock state. While the counter check is progressing, the synchronizer still checks the synchronization word for the proper value and repetition frequency. However, in the check and lock states, the designer may not require the synchronization word to have all bits exactly match the specified pattern. This allows the data stream to undergo a bit error without requiring the frame synchronizer to revert to the search state. Once the frame synchronizer enters the check state, it is permissible to begin processing commutated and supercommutated telemetry data, but not subcommutated data.

7.6.1.3 Lock State

Once the frame synchronizer has aligned itself with the frame structure in the check state, it moves to the lock state. The telemetry processor can now process all data, including the subcommutated data, because the synchronizer knows for sure the subframe identifier. The telemetry processor knows the supercommutated data and commutated data locations by their positions relative to the subframe identifier within the minor frame.

After the lock state is achieved, the user segment processor fills the matrix-like structure for the major frame shown in Figure 6.2 because the SFID is now established and the subcommutated data are found. While in the lock state, the synchronizer still checks for valid synchronization word contents and repetition frequency as well as the correct incrementing by subframe counter. It will also check the Frame Format Identifier (FFID) (see Section 6.4.3.2) to determine if any frame structure changes occur. If the synchronizer detects anomalies, it may revert to the check state.

If the link is lost entirely, the frame synchronizer reverts to the search state.

7.6.2 Packet Synchronization

The system designer must consider several differences from frame processing with packet telemetry and telecommand transmissions. The major difference is that packet systems assume higher link reliability so the excessive checks are not required in the receiver. Another major difference is that the system designer may not be using a continuous data flow on the link or may permit large gaps between transmissions. In this instance, the designer may include an acquisition sequence in the packet synchronization process. This acquisition sequence may be a string of alternating 0s and 1s ($0x55$ or $0xAA$) to cover the bit synchronization interval.

Once the receiver finds the packet synchronization marker, the receiver assumes the start of a valid packet and extracts the data. A recheck for the synchronization mark, as done in frame telemetry, is not possible because the receiver treats each packet independently with no assumptions made about the spacing between packets. As we saw in Figure 6.12, the packet has the following parts:

- A packet header
- A secondary header, depending upon the protocol
- The user's data in either a fixed-length or a variable-length field
- The packet trailer

The packet header may be a standard within the system for all payload sources. It identifies the sending sensor system and perhaps a destination processing system. It also contains accounting information and keys on how the packet is to be processed. All of these are in fixed fields at defined offsets from the synchronization code. One of these fields is the packet count which serves the same function as the SFID so the receiver can use the received incrementation sequence to determine link quality or missing packets.

The packet header may have a flag field indicating the existence of a secondary header. However, the secondary header may be context-sensitive and the existence, size, and fields may be a function of the payload's state, the individual sensor, and the type of sensor data. This is similar to a Class II telemetry frame where the FFID indicates frame format changes.

With packets, each major subsystem packages its data differently and uses header codes to describe which protocol format is in use with that packet. The packet processor needs a database to parse these headers and point to the header-specified data location within the packet's data field. The data fields' offset locations usually are relative addresses referenced to the packet's start. A sophisticated software system is required to support this processing.

Frame telemetry fills the transmission channel and provides a continuous data stream. Packet systems may not guarantee a full transmission channel. In fact, their optimization of data flow process may make for gaps in the transmission due to sampling optimized for the sensor Nyquist rate.

If the transmitter sends packet data over a closed communications network such as the Internet, maintaining bit synchronization is not a large problem. However, if the transmitter sends packet data via radio, the carrier synchronization process mentioned above comes into play. Because the carrier synchronization may take over 100 bits per instance, the link designer tries to keep the link active and synchronized even when the transmitter is not actively sending data. This keeps the receiver from losing the first part of each packet for carrier resynchronization.

The system maintains link synchronization by inserting fill packets into the transmission process. The fill packets have a special marker in their headers to indicate that the packet contains fill data instead of real data. The transmitter inserts telemetry data packets into a channel data-link packet for transmission over the link. To maintain synchronization, the transmitter must produce the channel packets at a uniform rate. Figure 7.12 shows that the transmission interface needs to have a buffer to hold the telemetry packet data until a full packet is available for transmission.

If the telemetry or telecommand data are not ready for transmitting a data packet, the transmitter uses a fill packet in the next channel packet in place of an actual data packet. Figure 7.13 shows that the resulting data transmission needs a fill packet inserted between actual data packets. The receiver needs to recognize the appearance of the fill packets and discard them upon reception. The receiver should not try to process the fill data. The receiver also needs to concatenate the real packets, after it removes the fill packets, to generate the complete data set.

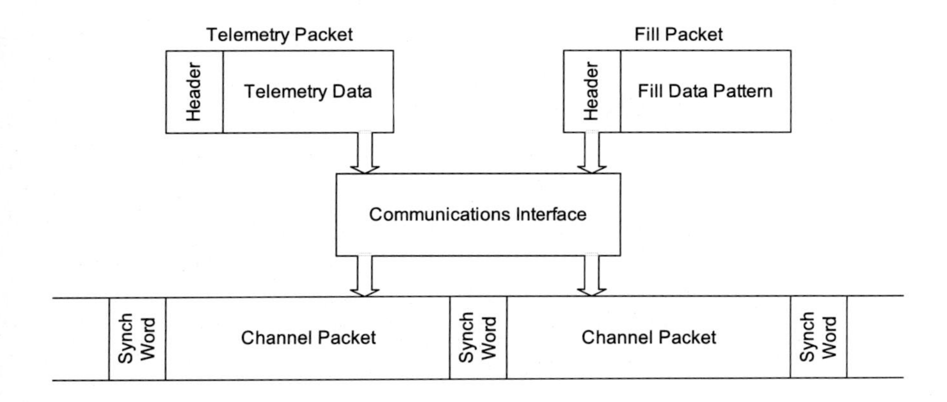

Figure 7.12: Packet-telemetry channel interface to allow the insertion of fill data.

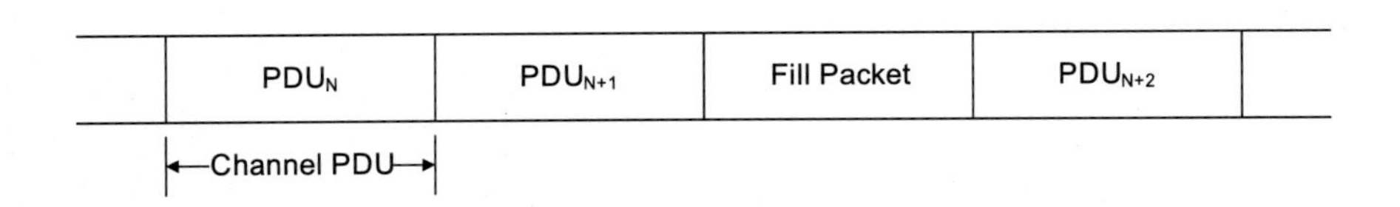

Figure 7.13: Appearance of fill data in the channel packet data stream.

7.6.3 Network Synchronization

As computer-based systems use common telecommunications networks such as the Internet, they need a higher level of management integration to ensure that the components properly interact and coordinate their actions.

Once the receiver synchronizes to the incoming data, an additional level of synchronization may occur depending upon the protocols used in data transmission. For example, when Transmission Control Protocol/Internet Protocol (TCP/IP) is used, the system performs end-to-end data synchronization. This ensures that the collection of individual data segments knit together to form the complete data set.

Another component of network synchronization is the coordination between the entities based on common timing using protocols such as the Network Time Protocol. This allows the components to initiate actions based upon a common timing standard.

7.6.4 Statistical Measures

The primary statistical measures used in association with the frame synchronizer are, first, the probabilities of falsely locking onto a random data pattern and believing it to be the synchronization word and, second, the probability of missing a synchronization word in the data stream due to an unacceptable number of bit errors in the data. The first is a function of the synchronization word length and not the channel bit-error rate. The second is a function of both parameters.

7.6.4.1 False Lock Probability

Figure 7.14 illustrates the false lock problem for a PDU of arbitrary length. A false lock occurs if the correlator indicates a correlation at any time other than when the synchronization word aligns perfectly in the correlator. If the correlator examines N bits in the shift register and allows up to k differences from the ideal pattern, the probability of a false lock, P_{FL}, caused by data patterns accidentally matching the true synchronization pattern is

$$P_{FL} = \frac{1}{2^N} \sum_{j=0}^{k} \binom{N}{j} \tag{7.2}$$

This probability is the average number of bits between false locks

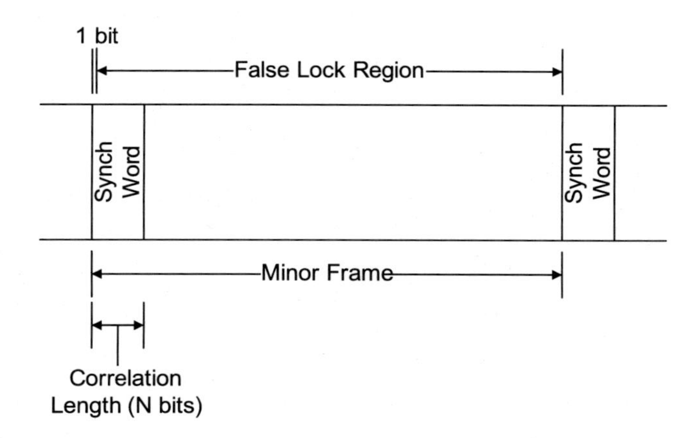

Figure 7.14: The false lock region in a sample minor frame. A false lock occurs outside the one-bit region at the start of the synchronization word.

in a statistical sense. For example, consider a 0.0001 probability of false lock and a data rate of 2048 bps. This gives a false lock, on the average, approximately every 10 000 bits or 10000/2048 = 4.88 s. If a transmission uses a 32-bit synchronization code and the performance specification allows two errors in the synchronization pattern, then the probability of false lock is

$$P_{FL} = \frac{1}{2^{32}}\left[\binom{32}{0} + \binom{32}{1} + \binom{32}{2}\right] \tag{7.3}$$

This gives a probability of false lock of 1.23×10^{-7}. A data rate of 2048 bps yields, on the average, 0.000 252 false locks per second, or 3964.4 s between false locks. Figure 7.15 shows that if errors are allowed in the received synchronization pattern, a longer pattern is required to achieve the same probability of false lock as that for allowing zero errors.

For an initial search for telemetry frames, the number of allowed errors, k, might be set to zero so the probability of a false lock is only 2^{-N}. This gives the maximum probability of finding the synchronization pattern. The receiver may need to find the correct pattern for several cycles before it allows synchronization pattern errors and the synchronizer remains locked. A typical specification for a false lock probability is the desire for a chance of a false lock of once per day or once per year. From this specification and the data rate, the designer determines the required synchronization word length.

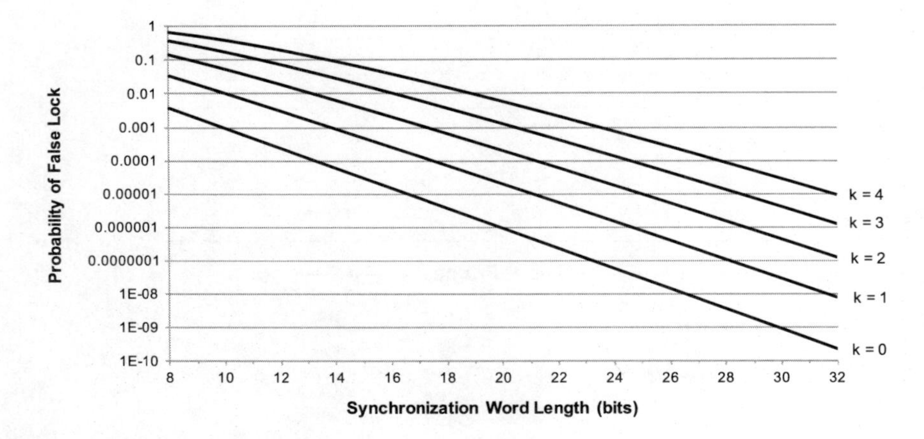

Figure 7.15: Probability of synchronization false lock as a function of the synchronization word size and the number of misses allowed in detecting the synchronization word.

7.6.4.2 Missed Synchronization Probability

The probability of missing a synchronization word in the data stream relates directly to the number of bit errors the synchronization word incurs due to channel errors. If the correlator allows for k or fewer errors in the received synchronization code of length N bits, the probability of a missed synchronization code, P_M, when the channel has a bit error rate, p, is

$$P_M = \sum_{j=k+1}^{N} \binom{N}{j} p^j (1-p)^{N-j} \tag{7.4}$$

For example, if a correlator requires 2 out of 32 bits to maintain the locked state and the channel Bit Error Rate (BER) is 10^{-5}, the probability of a missed synchronization is

$$P_M = \sum_{j=3}^{32} \binom{32}{j} (0.00001)^j (1-0.00001)^{32-j} \tag{7.5}$$

The P_M in this example evaluates to 4.96×10^{-12}. Figure 7.16 illustrates the probability of a synchronization miss as a function of channel BER and the number of missed bits in the synchronization word.

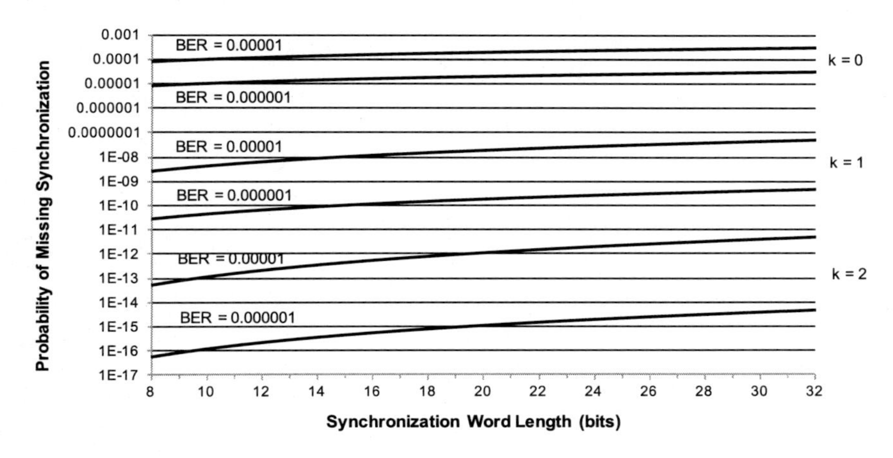

Figure 7.16: Probability of missing the synchronization as a function of the synchronization word size and the number of misses allowed in detecting the synchronization word and the channel error rate.

7.7 CHANNEL ERROR DETECTION

Knowing that the channel may induce errors is an important consideration for data synchronization and processing. Mathematical techniques exist to help detect and correct channel errors.

So far, in the frame synchronization process, we have checked just a few specific locations for being correct: the synchronization code and the SFID in the case of frame telemetry and the synchronization code in the header for packet systems. In most cases, the PDU is several hundred to several thousand bits long, so this level of checking corresponds to just a few percent of the total. Since both of these checks occur at the beginning of the PDU, this leaves the majority of the data without validity checks and in which errors may occur. There are techniques available to detect errors and even correct some errors in the received data in real time. The techniques that we consider here are

- Data post-processing
- Application of a block error detection and correction code
- Improving channel performance

The first two techniques require more sophisticated processing. The first is not strictly real time; the second is real time. Error detection and correction codes require greater transmission bandwidth. Improving channel performance may imply costly upgrades to equipment. The criteria system designers use to select an appropriate technique include the following points:

- What are the implications for real-time data acquisition? Does a transmission error compromise the real time nature of the payload or does it cause an operational impact such that operators spend a great deal of time trying to correct conditions that are not relatively important or can be corrected by post-processing?
- What is the cost of the added bandwidth to transmit the data and the required error codes? This cost may approach 100% to 200% in extra overhead in some strategies and may violate system constraints in other areas.
- What is the cost for added electronic and software complexity? Is the error detection and correction system too expensive to justify usage, especially at high data rates?

In many cases, it is just not worth the added costs for real time error detection and correction, for example, in digital voice transmission and slowly changing, noncritical data. Errors in digital voice transmission appear as "static" in the transmission. The receiver processing easily detects the latter case by averaging several samples and flagging the obvious bad cases. Other situations where it is desirable not to perform real time error detection and correction include high quality science data that comes from a source with a great deal of redundancy such that the experimenter processes the data after reception to efficiently remove any transmission errors. Examples of data that may require error detection and correction in real time are those that affect human health and safety or are critical to payload integrity. Such cases are special situations within the data transmission.

7.7.1 Probability of Error

Another part of the error detection problem is an estimation of the probability of an actual data error. We compute the expected number of errors in a data block if we know the channel BER and the length of the data block. We compute the probability of k or fewer errors, P_k, in a block of N bits (minor frame length or packet length, for example) by

$$P_k = \sum_{j=0}^{k} \binom{N}{j} p^j (1-p)^{N-j} \tag{7.6}$$

For example, for a channel BER of 0.0001 and a data block length of 512 bits, the probabilities of exactly zero, one, and two errors are computed individually from

$$
\begin{aligned}
P_0 &= (1 - 0.0001)^{512} = 0.9501 \\
P_1 &= (512)(0.0001)(1 - 0.0001)^{511} = 0.04865 \\
P_2 &= \frac{(512)(511)}{2}(0.001)^2(1 - 0.0001)^{510} = 0.001244
\end{aligned}
\tag{7.7}
$$

The probability of a total of zero and one transmission error is 0.987 35. The probability of a total of zero, one, and two transmission errors is 0.999 98.

7.7.2 Post-Processing Error Correction

Processing the received data to detect and even correct transmission errors is oftentimes the most efficient error recovery technique. Some efficiency occurs because of the simpler electronics for transmission and reception. If the channel is essentially error free, so that the actual errors are rare and become obvious, the analyst detects transmission errors by "eye balling" the data. There is no need to perform real time correction because the transmission errors are "massaged out" during subsequent processing which, in itself, is quite extensive.

Most of the post-processing work amounts to some form of filtering. The analyst uses a moving time averaging of the data, for example, averaging the last five received values as we discussed in Section 5.8.2. To illustrate this technique for a set of voltage samples, v_i, we compute the noncausal estimate for the correct value, $\langle v_i \rangle$, from the current value, the two previous values, and the next two values using

$$
\langle v_i \rangle = \frac{1}{5} \sum_{j=i-2}^{i+2} v_j \tag{7.8}
$$

This processing minimizes any minor transmission errors. The filtering process is also a form of persistence checking on a state level. For example, consider the case of a payload with attitude control thrusters. Should the controller take immediate, corrective action if a thruster suddenly becomes enabled "on its own"? That is, the user interface indicates that the system has changed its state without operator intervention. In most cases, it is better to check for two consecutive occurrences of the fault before taking action thereby discerning whether a payload anomaly has really occurred or the problem was only a transmission error.

7.7.3 Error Detection and Correcting Codes

Because data transmission is never perfect, link designers often transform the data into a form that has a lower overall probability of error than uncoded data. Adding extra bits to the data to provide transmission redundancy accomplishes this transformation. The goal is to have a lower probability of error for both the information bits and the redundancy bits than with the information bits alone.

This section presents an overview of these coding techniques. For more details, see standard references such as [Lin04; Skla01]. One of the advantages of modern signal processing electronics is that standard error-correcting encoders and decoders are readily available as components from the vendors.

7.7.3.1 Error Detection Codes

Block codes are error-checking codes that designers place on fixed-length blocks of transmitted data. In their simplest form, they detect but not correct errors. Codes that are more powerful both detect and correct errors. An example of a simple block code is the single *parity bit* added to single-character American Standard Code for Information Interchange (ASCII) transmissions in data communication. This single-bit code is only capable of detecting odd numbers of bit errors in the data block. For ASCII code, the receiver computes the parity over only six or seven bits. Senders use the code on relatively noise-free channels. Hence, it is powerful enough to detect most transmission errors over relatively reliable channels. The parity bit cannot correct transmission errors.

For a longer data stream, like a minor frame or a packet, a single parity bit does not suffice since it is highly probable that an even number of errors occurs. If all we desire is a quick check of the transmission quality, the checksum is a stronger measure than a single parity bit. The *checksum* is the sequential addition of the data words in the block without using carry bits in the process. The sender appends the resulting sum to the data block. Figure 7.17 shows an example of this process for a minor frame. The receiver recomputes the checksum over the data and compares it with the value sent. If they agree, then the data are assumed to be valid. If not, the operator knows that the current minor frame contains suspect data. Even if the checksum cannot correct a transmission error, the fact that the receiver flags the data quality is important in operations. The checksum has the advantage of easy computation as components

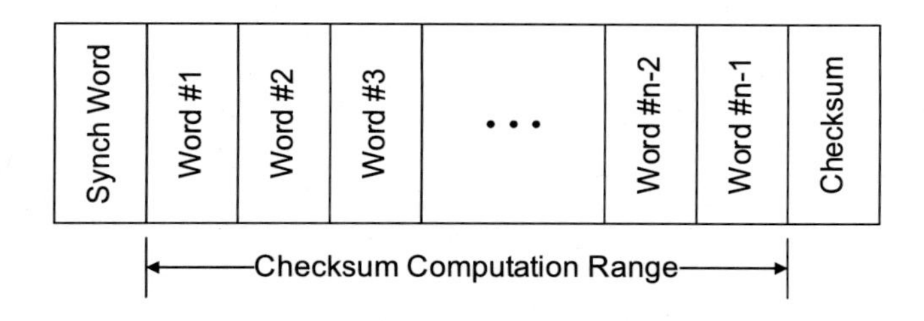

Figure 7.17: Location of the checksum in the data block.

both generate and receive the data so systems make the check in real time.

While the checksum is superior to the parity bit, it is not adequate for long data strings because there are multiple ways to generate the same checksum and errors can be undetected in the receiver. The Inter-Range Instrumentation Group (IRIG) telemetry standard recommends using a Cyclic Redundancy Check (CRC) error detection code for minor frames [IRIG106]. The telemetry standard recommends using one of the telecommunications standard codes

CRC-16-ANSI: $x^{16} + x^{15} + x^2 + 1$

CRC-16-CCITT: $x^{16} + x^{12} + x^5 + 1$

CRC-32: $x^{32} + x^{26} + x^{23} + x^{22} + x^{16} + x^{12} + x^{11} + x^{10} + x^8 + x^7 + x^5 + x^4 + x^2 + x + 1$

The standard recommends placing the CRC code at the end of the minor frame and not including the frame synchronization code in the computation.

7.7.3.2 Block Error Correction Codes

We make the error-checking concept more useful by adding more parity bits so that the receiver both detects and corrects any errors. The additional parity bits result in a larger transmitted data block. If the initial block size is k bits in length and the transmitted block is n bits in length, then $n - k$ parity bits have been added and the code is referred to as a (n, k) code [Hamm50]. The encoder forms the transmitted code word, c, by multiplying the uncoded data word, u, by a generator matrix,

G, using modulo-2 arithmetic

$$c = uG \tag{7.9}$$

Normally, block codes are designed to be systematic, that is, the transmitted code word is partitioned into the original data block of k bits and the $(n - k)$ parity bits added to either the front or the back of the uncoded data block. For a systematic block code, the generator matrix is

$$G = \begin{bmatrix} I & P \end{bmatrix} \tag{7.10}$$

where I is a $k \times k$ identity matrix and P is the parity check matrix. One property of block codes is that a parity check matrix, H, checks the transmitted code word for errors. If transmission errors are not present in the code word, c,

$$cH = 0 \tag{7.11}$$

The parity check matrix, H, is related to the generator matrix, G, by

$$H = \begin{bmatrix} P \\ I \end{bmatrix} \tag{7.12}$$

If there is an error in transmission, the received code word is not equal to the transmitted code word. We view the transmission process as generating an error vector, e, which has a zero in every location except for bit positions where an error has occurred. The positions where an error occurs are equal to a one in the vector. The received code word, r, is the modulo-2 addition of the error vector, e, with the transmitted code word, c

$$r = c \oplus e \tag{7.13}$$

When there is a transmission error, the parity check computation is non-zero. Rather, it produces a syndrome vector

$$rH = (c \oplus e)\, H = 0 \oplus eH = eH \tag{7.14}$$

The syndrome is processed to determine where the error occurred.

We use the (7, 4) Hamming code as an example of a block code. Most texts on digital communications and coding theory cover the derivation of Hamming codes so we take those results as a given [Hamm50; Hamm86; Lin04; Swee91]. The Hamming code encodes every block of four bits as a block of seven bits, including three parity bits. The advantage of this code is that every single-bit error in the block is detected and corrected. The generator matrix for the code is

$$G = \begin{bmatrix} 1 & 0 & 0 & 0 & 1 & 1 & 0 \\ 0 & 1 & 0 & 0 & 1 & 0 & 1 \\ 0 & 0 & 1 & 0 & 0 & 1 & 1 \\ 0 & 0 & 0 & 1 & 1 & 1 & 1 \end{bmatrix} \tag{7.15}$$

There are several equivalent generator matrices so the exact encoding is not unique. The parity check matrix is

$$H = \begin{bmatrix} 1 & 1 & 0 \\ 1 & 0 & 1 \\ 0 & 1 & 1 \\ 1 & 1 & 1 \\ 1 & 0 & 0 \\ 0 & 1 & 0 \\ 0 & 0 & 1 \end{bmatrix} \tag{7.16}$$

Figure 7.18 illustrates a LabVIEW® virtual instrument to make the Hamming code computations. Each incoming data byte is broken into two 4-bit nibbles and encoded. The two 8-bit code words joined to form a 16-bit transmission word. Because the generation of the code words is a deterministic process, the necessary matrix multiplications can be precomputed and the code words are stored in a lookup table. The uncoded data become addresses to index into the table to produce the desired code word. This process is efficient for simple codes that do not require a great deal of memory.

The LabVIEW® decoder illustrated in Figure 7.18 works by computing the syndrome for each pair of transmitted bytes. The syndrome indexes into an array containing error patterns for each syndrome. The error pattern is bit-wise modulo-2 added to each received byte. The decoder strips the lower four bits from the corrected byte and concatenates the pairs of 4-bit segments to produce the recovered data byte.

The Hamming code is easy to compute but its performance is not exceptionally strong. The cyclic codes are a more powerful group of block error correcting codes in which the valid code words are cyclic shifts of other valid code words. The encoder uses a generator polynomial, $g(D)$, with the uncoded data $u(D)$. D is a delay generator, which a shift register can realize in the electronics. The code word, $c(D)$, is generated by

$$c(D) = u(D)g(D)mod\left(D^N - 1\right) \tag{7.17}$$

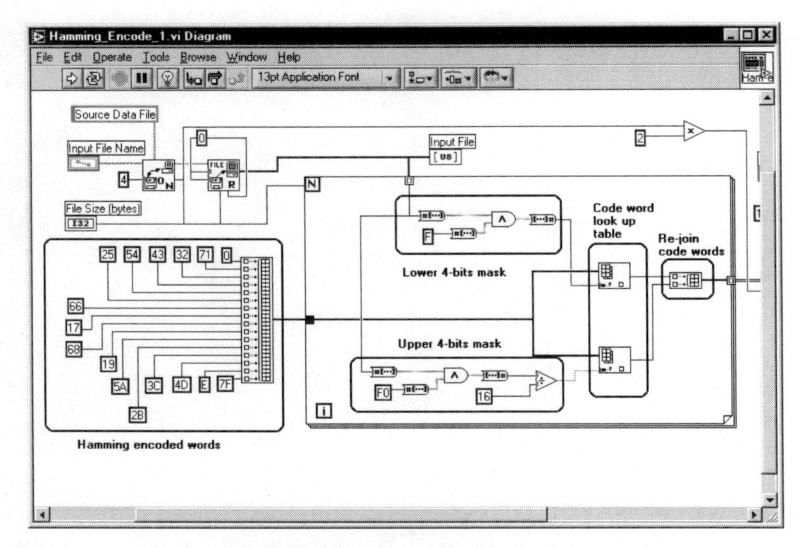

(a) Encoder for Hamming Code

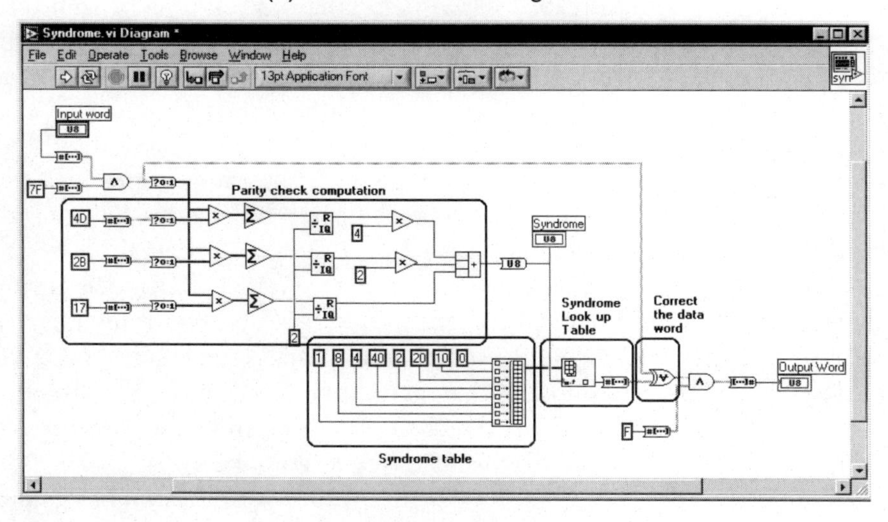

(b) Decoder for Hamming Code

Figure 7.18: A LabVIEW VI to encode and decode data with a Hamming error correcting code.

The (7, 4) Hamming code is represented by one of the valid generator polynomials. The Bose Ray-Chaudhuri Hocquenghem (BCH) codes use algebraic polynomials for binary alphabets. The Reed-Solomon (R-S) codes use algebraic polynomials for nonbinary code word alphabets. The Low Density Parity Check (LDPC) codes are block codes found in current telemetry and telecommunications systems such as satellite television because they are efficient with very good error-correcting properties. For performance comparisons, see, for example [CCSDS1301]. These codes are now common building blocks in communications design.

7.7.3.3 Convolutional Codes

Convolutional codes represent another major classification of error correcting codes. They are covered in detail in the standard references such as [Lin04; Swee91]. We examine the basic properties here. A convolutional code is a state machine that takes the current input symbol and functionally combines it with the previous input data symbols to produce the output symbol. The current encoder output symbol remembers the influence of previous input symbols. The output symbol has a greater length than the input symbol. The code rate is the ratio of the input symbol length to the output symbol length.

The state machine uses simple shift register and adder building blocks. Figure 7.19 illustrates a standard convolutional encoder [Lin04]. This encoder takes a single input bit and produces two bits on the output; hence the code rate, R, is $1/2$. In addition to the current incoming data bit, the encoder remembers the previous six bits as well so the output is a function of seven data bits. This is known as a constraint length seven ($K = 7$) encoder. The code designer needs to select the cells of the shift register to combine into the output bits. Researchers have developed standard configurations with good performance qualities and manufacturers commonly use these configurations. Convolutional encoders can also use more than one bit on the input and produce more than two bits on the output.

The decoder for the convolutional encoder is more complicated than the encoder is. Viterbi [Vite67; Forn73] developed an algorithm that performs an optimal search through the potential states to determine the one solution that has the maximum probability of being correct when the channel has transmission errors. Both the convolutional encoders and decoders are now standard designs in many design packages.

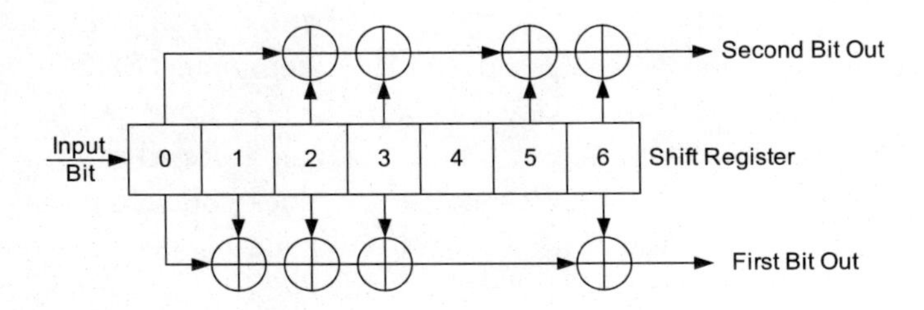

Figure 7.19: A shift-register based, rate-1/2 convolutional encoder with constraint length 7.

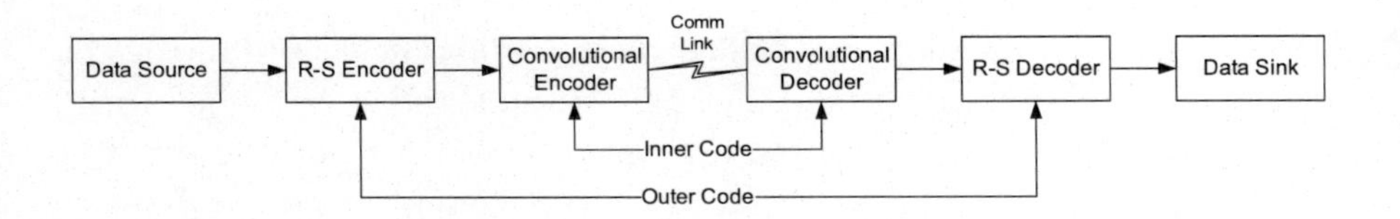

Figure 7.20: A concatenated coding design with a R-S outer code and a convolutional inner code.

7.7.3.4 Concatenated Codes

It is typical for link designers to need a stronger coding method than either a block code such as R-S or a convolutional code produces individually. Additionally, convolutional coding decoders are known to be prone to occasionally producing bursts errors in the decoding process. One solution to achieve improved performance is to use a *concatenated code* structure with a convolutional code as the inner code and a R-S code as the outer code [CCSDS1301; CCSDS1310]. The block code, with a data interleaver, corrects the burst errors from the convolutional code. Figure 7.20 illustrates this process. Here, the R-S code is applied first to the data and then the convolutional code is applied to the R-S encoded data. The system transmits the twice-encoded data by appropriate techniques such as radio. The receiver reverses the process. First the decoder for the convolutional encoder is applied to recover the R-S encoded symbols. Next, the R-S decoder recovers the original data.

7.7.4 Channel Improvement

Improvement of the channel noise characteristics can come by one of two methods: improving the channel or by improving the strategy for sending the data through the channel. In general, we cannot physically improve the channel unless we enclose it in an environment like a laboratory. Free space channels have to live with the restrictions placed on them by nature. To a certain extent, transmitting with a greater power level improves the channel. However, this is expensive and increases the payload's cost or size (or the transmitting station) beyond that which the customer can afford. Improving the channel transmission strategy comes with adding error correcting or detecting codes.

These codes increase the number of transmitted bits without adding new data to the system. The redundancy necessary to detect or correct an error amounts to a significant overhead in the transmission system. To maintain the same throughput with this additional burden, we increase the data transmission rate proportionally. As we will see in Chapters 10 and 11, this burden may violate, or make worse, the channel error constraints. The bottom line is that the system designer may need to iterate through several variations of the system design to accomplish all of the design goals for equipment, electronics and software, and data quality.

7.7.5 Coding Gain

One way to rate the coded transmission system is via the amount of *coding gain* the error correcting code supplies. In other words, how much additional power beyond the coded data at a constant BER, or quality of service, is needed to transmit the uncoded data? We think of coding gain as an improvement in the quality of service at a constant input Signal-to-Noise Ratio (SNR).

The concatenated encoding technique mentioned above gives a coding gain up to 7 dB when the channel BER is 10^{-5}, depending upon the decoder's configuration [CCSDS1301]. Using only the convolutional encoder gives a coding gain of approximately 5 dB itself. This implies that, with the concatenated code, the receiver tolerates a lower SNR than the uncoded channel to achieve the same quality of service on the recovered data. The convolutional code only allows a reduction of the channel SNR by 5 dB. Conversely, keeping the input SNR the same, we achieve a much higher quality of service on the channel. With this encoding technique, the residual probability of bit error is much smaller than the 10^{-5} uncoded channel BER available to typical radio channels. Table 7.1 summarizes the coding gain available with several coding techniques found in telemetering systems.

Table 7.1: Example Error Control Coding Gain at BER = 10^{-5}

Code	Rate	E_b/N_0 (dB)	Coding Gain (dB)
Uncoded	1	9.5	0
Convolutional	1/2	4.25	5.25
R-S	(255, 223)	6	3.5
LDPC	0.875	3.75	5.75
Concatenated		2.75	6.75

Generally, the coding gain computation is rather difficult for complicated coding structures. The (7, 4) Hamming code illustrates the principle with an easier set of computations. The Hamming code is able to correct all single-bit errors in the code word. This means that the transmission channel needs to produce two or more errors in the transmitted word before the code breaks down. We compare the performance with the uncoded data as follows. The uncoded data have a problem when each four-bit block has one or more errors so the uncoded word error probability, P_W^{uc}, is

$$P_W^{uc} = 1 - (1 - p)^4 \tag{7.18}$$

The coded word is received correctly whenever there are fewer than two transmission errors in the block so the coded word error probability, P_W^c, is

$$P_W^c = 1 - \sum_{n=0}^{1} \binom{7}{n} p^n (1 - p)^{7-n} \tag{7.19}$$

If the probability of a coded word error is less than the probability of an uncoded word error, then the encoder has a coding gain for this channel.

We extend this concept for other coding techniques with typical channel error rates. For white noise channels using typical digital modulation methods, the upper bound for the coding gain, G_C, is [Ande99; Skla01]

$$G_C \leqslant 10 \log \left[R(t + 1)\right] \tag{7.20}$$

R is the code rate and t is the number of errors that the decoder can correct in each block of transmitted data.

It is possible to have too much coding gain. Consider a link design using an error correcting code that reduces the necessary SNR to achieve

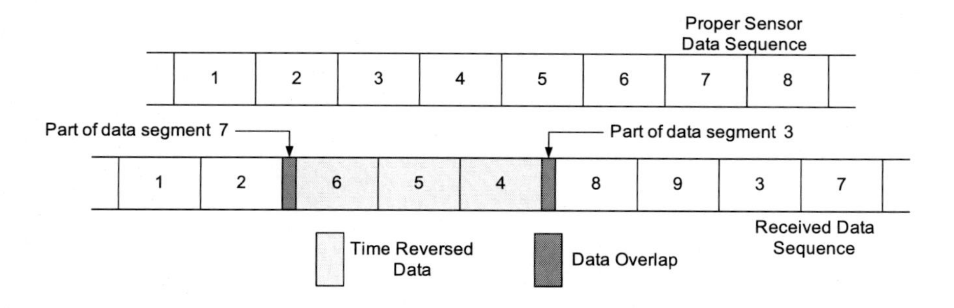

Figure 7.21: The arrival of data without proper time ordering prior to processing.

a channel BER of 10^{-5} from 10 dB to a signal to noise ratio of 1 dB. At this low value of SNR, the receiver may have trouble locking onto the incoming data carrier and therefore the receiver passes no data to the data synchronization and decoding process. The system designer believes that the coding overcomes the poor channel conditions but the receiver tracking loops still need to maintain their integrity for a reliable data reception.

7.8 DATA SEQUENCING

As part of the user segment's synchronization process, data processing often includes the data's time sequence reassembly [Carp90]. That is, the data may arrive out of normal time order and the user segment may need to return it to its proper time order. This may occur for several reasons, including the following ones:

- The payload recorded the data on a tape recorder and the tape playback of the data introduces a time reversal of the data segments.
- A retransmission of certain data segments occurred due to requests from the user segment for data that were received in error.
- The transmitted data may have repeated segments due to both of the previous conditions.
- The system used a transmission channel, such as the Internet, and this channel does not guarantee time-ordered delivery of the data.

Figure 7.21 illustrates the result of this scrambling process where the user may not receive the correct, time-ordered output of the sensors when the data are initially examined. The user segment then must sort out all of the above effects. The data processing in the user segment also must flag data gaps due to link dropouts or missing segments. The data transmitted from the payload may include time tagging or sequence numbering to assist the user segment in the process of correctly time ordering the data. The payload encodes the time tags using a format like the one we describe in the next chapter. The payload places this time tag in the minor frame as a normal minor frame word.

In packet telemetry, the payload typically places the time tag in the packet header. The designer uses a sequence counter in the same way and the counter represents an enlarged version of the minor frame SFID. In this case, the count most likely increments with each major frame but not with each minor frame. The SFID takes care of the minor frame relative sequencing. Again, the payload transmits the major frame sequence number as a minor frame word. The sequence counter is part of the packet header. As we saw with the TCP/IP packets, the header already contains a counter for flow rate control.

Figure 7.22 shows an example of this process for transmitting Joint Photographic Experts Group (JPEG) encoded images from a nanosatellite communications system [Hora01]. The intent was to have the data received at many locations and returned to the main control center using an electronic mail process. This means that the data arrive at irregular intervals out of correct time order. Each mail message needs a means to identify the originating satellite JPEG file the message comes from, and where this data fits into the overall picture. Part (a) of Figure 7.22 shows how the minor frames have identification data embedded to allow reconstruction of the data set. Part (b) shows the initial segments of several image minor frames containing the identifiers. In user-segment computer processing, the processor sequentially orders the minor frames, removes the identification information, weaves the image pieces back together, and reconstructs the original image.

After the user segment receives the data, the data-processing system usually permanently records the data in some manner as part of a permanent archive. As the recording process occurs, it is common to place a received time code with the data. For example, the data processor inserts the time stamp at the end of each minor frame, major frame, or packet. The processor also magnetically records the time tag as a separate channel.

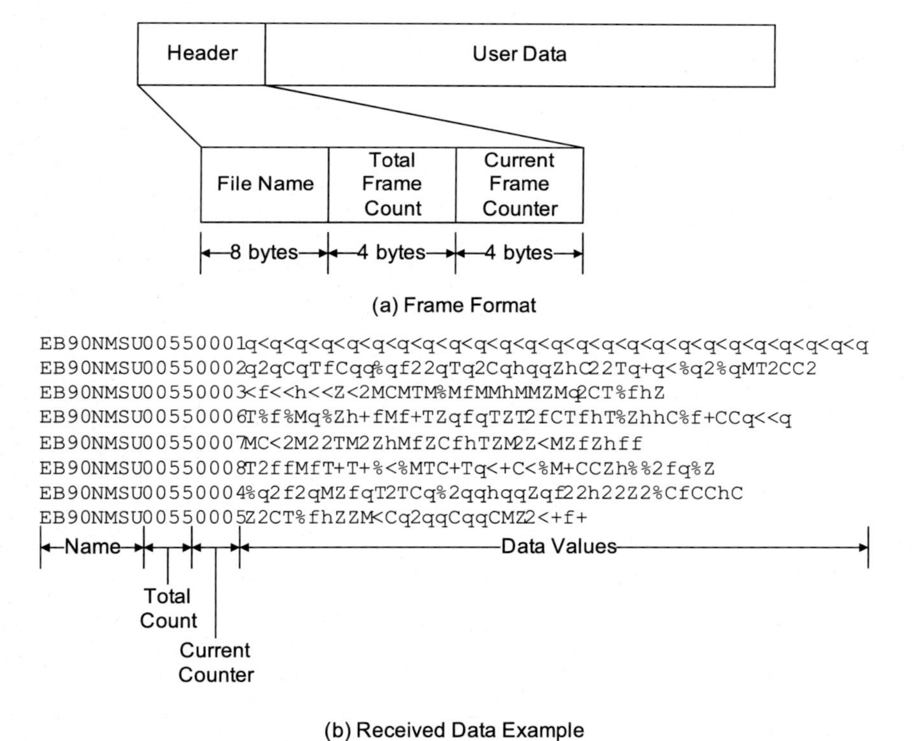

Figure 7.22: An example of minor frames with sequence counters and file identifiers to allow data reconstruction.

7.9 REFERENCES

[Ande99] J. B. Anderson. *Digital Transmission Engineering.* Piscataway, NJ: IEEE Press, 1999.

[Carp90] R. D. Carper and W. H. Stallings. "CCSDS Telemetry Systems Experience at the Goddard Space Flight Center." In: *IEEE Network Magazine* 4.5 (Sept. 1990), pp. 17–21. DOI: 10.1109/65.59305.

[CCSDS1301] *TM Synchronization and Channel Coding – Summary of Concept and Rationale.* CCSDS 130.1-G-2. Consultative Committee for Space Data Systems. Washington, D.C., Nov. 2012. URL: https://public.ccsds.org/Pubs/130x1g2.pdf.

[CCSDS1310] *TM Synchronization and Channel Coding.* CCSDS 131.0-B-2. Consultative Committee for Space Data Systems. Washington, D.C., Aug. 2011. URL: https://public.ccsds.org/Pubs/131x0b2ec1.pdf.

[ElMo80] A. El Moghazy, G. Maral, and A. Blanchard. "Digital PCM Bit Synchronizer and Detector." In: *IEEE Transactions on Communications* COM-28 (8 Aug. 80), pp. 1197 –1204. DOI: 10.1109/TCOM.1980.1094812.

[Forn73] G. D. Forney. "The Viterbi Algorithm." In: *Proceedings IEEE* 61.3 (Mar. 1973), pp. 268 –278. DOI: 10.1109/PROC.1973.9030.

[Gill63] W. Gill and J. Spilker. "An Interesting Decomposition Property for the Self-Products of Random or Pseudorandom Binary Sequences." In: *IEEE Transactions on Communications Systems* 11 (2 June 1963), pp. 246 –247. DOI: 10.1109/TCOM.1963.1088753.

[Ha90] T. T. Ha. *Digital Satellite Communications.* 2nd Edition. New York, NY: McGraw-Hill, 1990. ISBN: 0070253897.

[Hamm50] R. W. Hamming. "Error Detecting and Correcting Codes." In: *Bell System Technical Journal* 29.2 (Apr. 1950), pp. 147 –160.

[Hamm86] R. W. Hamming. *Coding and Information Theory.* 2nd Edition. Englewood Cliffs, NJ: Prentice Hall, 1986. ISBN: 0-13-139072-4.

[Hora01] S. Horan et al. *Dual Transceiver User's Manual.* UN2-MAN-31114-404. New Mexico State University. Las Cruces, NM, Sept. 2001.

[IRIG106] Telemetry Group. *Telemetry Standards.* 106-15. Secretariat, Range Commanders Council. White Sands, NM, July 2015. URL: http://www.wsmr.army.mil/RCCsite/Documents/Forms/AllItems.aspx.

[Lin04] S. Lin and D. J. Costello. *Error Control Coding: Fundamentals and Applications*. 2nd Edition. Pearson Prentice Hall, 2004. ISBN: 978-0130426727.

[Lind73] W. C. Lindsey and M. K. Simon. *Telecommunication Systems Engineering*. Englewood Cliffs, NJ: Prentice Hall, 1973.

[Simo70a] M.K. Simon. "Nonlinear Analysis of an Absolute Value Type of an Early-Late Gate Bit Synchronizer." In: *IEEE Transactions on Communication Technology* COM-18 (5 Oct. 1970), pp. 589 –596. DOI: 10.1109/TCOM.1970.1090410.

[Simo70b] M.K. Simon. "Optimization of the Performance of a Digital-Data-Transition Tracking Loop." In: *IEEE Transactions on Communication Technology* COM-18 (5 Oct. 1970), pp. 686 –689. DOI: 10.1109/TCOM.1970.1090401.

[Skla01] B. Sklar. *Digital Communications: Fundamentals and Applications*. 2nd Edition. Upper Saddle River, NJ: Prentice Hall PTR, 2001.

[Swee91] P. Sweeney. *Error Control Coding*. Hempstead: Prentice Hall International, 1991. ISBN: 0-13-284118-5.

[Vite67] A. J. Viterbi. "Error Bounds for Convolutional Codes and an Asymptotically Optimum Decoding Algorithm." In: *IEEE Transactions Information Theory* IT-13.2 (Apr. 1967), pp. 260 –269. DOI: 10.1109/TIT.1967.1054010.

7.10 PROBLEMS

1. Using a simulator such as LabVIEW® or MATLAB®, develop an open-loop clock extractor. Verify the power spectrum for the mixer's output when you use a random data set as the input. Plot the output of the band pass filter and the limiter. Suppose that the input data becomes a long string of 1s or 0s. Show that the clock signal is lost.

2. Given a random data set: 0, 1, 1, 1, 0, 0, 0, 1, 0, 1, 0, 1, encode the data using a NRZ-S format. In decoding the data, make a random determination for the initial state of the decoder. Does the decoder recover the data set properly? Try using an acquisition sequence such as 0, 1, 0, 1 and repeat the decoding process with a random initial state. Does the decoder properly recover the data portion of the transmission this time?

3. Random data in a telemetry minor frame may look like a synchronization code. What modifications to the correlator, which searches

for the synchronization word, do you need to add to the circuit for it to gate the correlator to only look for the synchronization word at a time near the expected time? What do you need to do for correlator initialization?

4. For a system with a 1 kbps data transmission rate, how long must the minor frame synchronization code be to have a probability of false lock with perfect correlation of the synchronization code of less than once per day? Once per month? Once per year?

5. For the synchronization word lengths found in Problem 4, what is the probability of missing the synchronization word if the channel bit error rate is 10^{-3}? If the BER is 10^{-4}? If the BER is 10^{-6}?

6. If the system in Problem 4 is allowed to accept up to two errors in the synchronization pattern, what must the synchronization word length be for a false lock rate of less than once per day? Once per year?

7. For the system in Problem 4, if two errors are allowed in the synchronization pattern, what is the probability of missing the synchronization word if the channel BERs are 10^{-3}, 10^{-4}, and 10^{-6}?

8. In the case of small bit error rates, compute the expected time to achieve full synchronization with a telemetry stream given

   (a) A bit rate of 2 kbps
   (b) A requirement of 200 bits for reliable bit synchronization
   (c) A minor frame synchronization length of 16 bits
   (d) A minor frame length of sixty-four 8-bit words
   (e) A major frame length of sixty-four minor frames
   (f) A requirement for two correct frames to leave the search state
   (g) A requirement for one correct frame to leave the check state

9. Which of the following proposed synchronization patterns, when received as part of a data stream, has the best correlation characteristics? Best is in the sense of being the least likely to generate a false lock when compared against a stored copy of the pattern for the cases of a 1-bit shift and a 2-bit shift (consider both right and left shifts). **Hint:** consider a 1 to be a signal level of +1 V and a 0 to be a signal level of −1 V.

(a) 10111000
(b) 11100010
(c) 01000111
(d) 11100011
(e) 00011101

10. Given the following block of data points, perform a five-point average (current value plus the two previous and the two subsequent data points):
2, 2, 2, 3, 3, 4, 5, 5, 5, 5, 6, 6, 3, 7, 8, 8, 8, 7, 7, 6, 6, 5, 5, 4, 4, 3, 3, 2, 2, 2.
Which point probably suffered a transmission error? What is your best estimate for the correct value of that data sample? Which bit was probably corrupted?

11. Given the data and answers from Problem 10, do your answers change if you use a five-point average formed from the current data point and the previous four points? What are the differences in the two with respect to initialization time and time at which the data appears?

12. Given the following data sequence of hexadecimal values (assume each value is 8 bits and represented as two 4-bit nibbles), compute the 8-bit checksum for transmission:
01, AA, 23, 58, F4, 23, 24, 67, 69, 62, 64, 62, 73, 71, 66, 61, 61, 77, 16, 16, 01.

13. Explain why the receiver processes supercommutated and commutated data when the frame synchronizer is in the check state; that is, before the synchronizer enters the lock state. Why must subcommutated data wait until the synchronizer has entered the lock state before they are processed?

14. When you use a checksum to detect errors in the telemetry frame, why do you exclude the synchronization word from the checksum computation?

15. For a derandomizer with tap settings at 024_8, show the structure of the derandomizer. Suppose the received data stream is 1 010 101 010 111 100 000 010 and is the input to the derandomizer. What is the output of the derandomizer? Indicate the bits in error on the output of the derandomizer.

16. Show how to modify the derandomizer to correct for time-reversed data. Verify the operation with a sample data stream.

17. Devise a logic circuit to open-loop extract data clock from BiΦ-L and BiΦ-S data formats. Show by example that the circuit works.

18. Devise a logic circuit to open-loop extract data clock from the IRIG BiΦ-M data and convert the output to NRZ-L format.

19. Devise a logic circuit to open-loop extract data clock from the IRIG BiΦ-S data and convert the output to NRZ-M format.

20. What must the worst-case channel BER be to use a single-bit parity check in a minor frame of length 512 bits? For this to work, at an acceptable missed error detection rate, the chance of two or more errors in the block of data is at most 10^{-4}.

21. If you used a fill packet in a transmission system, how did you encode the data field for the packet? Did you use all 1s or all 0s? Explain your reasoning.

22. If a channel has a bit error probability of 10^{-4}, estimate the probability of a word error for a 4-bit data block. If a (7, 4) Hamming code is used, what is the new probability of word error?

23. A packet data system uses the following parameters: 10 240 bits/packet, a transmission rate of 1 Mbps, and a propagation delay of 250 ms. If the packet's ground-processing time is 10 ms, how long must the packet sequence counter be so that the ground station and payload accurately track the number of packets in transmission and receive a reception acknowledgment?

CHAPTER 8

TIME AND POSITION DETERMINATION

8.1 INTRODUCTION

In addition to the time records that are part of the "science" data that the payload sends, we find time information in many places in telemetry and telecommand systems. It is often necessary to determine the timing of specific events such as when a processor received a particular data set, when an operator entered a command to change a payload state, when a payload schedule changed configurations, etc. The time information used in answering these questions comes from standard time signals broadcast by time services or from a network time service. We may also be interested in the elapsed time between two events such as when an event triggered an alarm and how long until the system corrected the condition. In this case, we are interested in the linear measurement of a time interval, not in the exact "when." This chapter explores time in both senses.

Timing information may also appear in the system as an explicit datum that the system collects or displays. The payload data system may use timing information as a parameter in the telemetry data stream to time tag the data transmitted to the user. The user's data processing computers may archive the received data with a reception time stamp. Operator displays typically show the current time as local time or universal time, or perhaps, the elapsed time since the experiment began. Timing data collection is like the collection of any other data in the system. It must be standardized, calibrated, and distributed.

We have come a long way from one of the earliest successful time

Figure 8.1: Time synchronization ball at the Royal Observatory in Greenwich, England.

synchronizations services, which Figure 8.1 illustrates. The photo shows a portion of the old Royal Observatory in Greenwich, England, just south of London. The large ball on the weather vane dropped each day at a certain time to allow the ships on the Thames River to synchronize their clocks as an aid to navigation. Precise clock information allowed for precise determination of longitude, also reckoned from the Royal Observatory [Sobe95].

Positional information is also very important. If we wish to point an antenna at a moving spacecraft, we must know where that spacecraft is in relationship to the antenna. Alternately, consider a remote meteorological network of movable sensors where each station sends the current conditions to a central receiving station. If the stations broadcast their current position as part of the data stream, the database automatically reflects the new positions.

This chapter first examines the definitions of time and how the different definitions are related. This includes discussions of absolute and relative time measurement. Next, we investigate the ways in which timing information is encoded. Several standards have been set by the International Organization for Standardization (ISO), Inter-Range Instrumentation Group (IRIG), National Institute of Standards and Technology (NIST), and the Consultative Committee for Space Data Systems (CCSDS). After a discussion of timing information, we move to a discussion of position determination. Historically, the development of accurate navigation aids has followed this progression. After precise time measures were developed, it became possible to prepare accurate maps showing the placement of land areas at their proper longitudes.

8.2 OBJECTIVES

Our objectives for this chapter are to understand the ways of encoding both absolute and relative time information and standard position location services. At the end of this section, the reader will be able to

- Explain the differences between the various time systems.
- Encode a time based on ISO, IRIG, NIST, and CCSDS standard formats.
- Explain the fields used in Global Positioning System (GPS) navigation messages.

These time codes and formats are frequently encountered in telemetry and telecommand systems.

8.3 DEFINITION OF TIME

It may seem odd that there might possibly be more than one definition of time encountered in telemetering systems work. People have measured time and made clocks for hundreds of years. Today, the nontechnical users joke about the addition of "leap seconds" to standardize time. For most people, the only "problem" with time is which direction to change the clocks during the transitions from daylight savings time to standard time. The local time, as derived from the radio or our cell phones, is sufficient to organize our lives for most uses. This time system uses a 24-h day that proceeds in a uniform manner each day. However, for exact metrology, the variations in the Earth's rotation rate, the exact operation of atomic clocks, and similar corrections must be applied.

From a modern perspective, it is more correct to think of time in the coordinate sense as is done with geographic positions. Two quantities are involved in measuring time: the precise determination of when "now" is in an absolute sense and the elapsed time between two events. This is analogous to measuring the precise location on Earth and a distance between two specified locations. The former sense of time allows users to synchronize operations between remote stations. Experimenters use the latter sense of time as an experimental quantity to mark events or derive a new quantity such as distance. The determination of "now" is referenced to standard time services in the United States. Typically, a counter of some sort measures the elapsed time. In the paragraphs below, we look at both issues.

8.3.1 Absolute Time

To determine what the time is, we must first determine in which time system we wish to express the answer [Lomb02; Nels05; Mcca91; Quin91; MICA05]. Do we need a coordinate in time or a named time in units of seconds? To illustrate, several types of time now in use are

- International Atomic Time (TAI)
- Sidereal Time (ST)
- Terrestrial Time (TT)
- Coordinated Universal Time (UTC)

Fortunately, algorithms allow us to convert between the different types of time. The exact form the user employs is a function of the measured quantity and the measurement precision. For example, some time systems make explicit allowances for the irregularities in the Earth's rotation.

We also encounter different nomenclatures in different measurement communities. For example, Zulu Time is typically another way of saying Greenwich Mean Time (GMT) — a time based on the exact local time of Greenwich, England. Users sometimes refer to GMT as a synonym for Universal Time, although that practice is now obsolete. Another example is the use of the term Julian Date (JD) that many use as the day number of the year (a number from 1 to 365 or 366). To the astronomical community, the JD is the count of astronomical days beginning at 12.0 h in Greenwich, England on 1 January 4713 B.C. as reckoned by the Julian proleptic calendar. It therefore becomes necessary to obtain the exact definition when moving to a new user community.

8.3.1.1 International Atomic Time

The international standard definition of a *second* is 9 192 631 770 periods of the radiation for two atomic levels of cesium 133. This definition is the basis for TAI in the International System of Units (SI). The cesium atom's "ticks" are the coordinate bases for time, while the definition of the *second* is scaled to the system humans use. The metrology community officially adopted TAI on 1 January 1972 although atomic time scales had been in use since 1955. The primary standard for this measurement is based at the *Bureau International des Poids et Mesures* in Paris. Atomic time accumulates at the rate of 86 400 s/d.

8.3.1.2 Sidereal Time

Historically, people have related time to the positions of the sun and stars on the sky — not to cesium atom oscillations. Sidereal Time (ST) is a direct measurement of the Right Ascension coordinate position of the stars currently on the observer's meridian. This coordinate is measured from 0 h to 23 h. The period of a day, as measured with respect to the "fixed" stars, is 23 h, 56 min, and 4.09 s; the length of the year is 365.24 d. For most of history, the 24-h approximation to the time between successive alignments between the Earth, sun, and stars has been sufficient for most uses. However, the system of approximations, including the addition of leap years, is not sufficiently accurate to track the Earth's true rotation and keep the time system coordinated with the Earth's observed motion relative to the stars.

8.3.1.3 Terrestrial Time

Terrestrial Time (TT) is a theoretical time scale that is independent of the Earth's rotational irregularities. TT is measured in SI seconds and is defined at the Earth's mean sea level. TT is defined relative to TAI by $\mathrm{TT} = \mathrm{TAI} + 32.184\,\mathrm{s}$.

8.3.1.4 Coordinated Universal Time

The Universal Time (UT) system uses a 24-h mean solar time format tied to the Earth's motion with respect to the stars as measured from the Earth. Most civil time systems use the UT system as a basis for their time distribution services. Two variations on UT make a difference at the seconds' level. Time measurements obtained from stellar observations and uncorrected for the periodic components in the Earth's motion are designated as UT0, while those that have been corrected for the Earth's periodic motion are designated UT1. These corrections occur because the Earth does not rotate in a uniform manner. The uniform motion approximation has both periodic and nonperiodic irregularities added to it to give an overall variable rate over a period of time.

Most time services broadcast Coordinated Universal Time (UTC). This practice began on 1 January 1972. The difference between UT1 and UTC (in the sense of UT1 - UTC) is then broadcast as the signal DUT1. The time service keeps this difference to within 0.90 s by the introduction of *leap seconds*; typically, the time service performs this correction at the end of June or December.

The time code service provides fine corrections to the difference on the order of 0.1 ms. The difference between TAI and UTC is kept to an integer number of seconds; the initial difference was 10 s. As of 1 June 2017, this difference is 37 s with the inclusion of 27 leap seconds since 30 June 1972. The leap second correction is available on NIST's Web site: `http://www.nist.gov/pml/div688/grp50/leapsecond.cfm`. The time service cannot predict the exact dates when the service will add the differences far in advance due to the irregular variation in the Earth's rotation rate. One must consult the tabulations in listed current publications on time standards or at the NIST Web site.

8.3.1.5 Julian Date

We compute the Julian Date (JD) from the current calendar date and UTC. The JD begins at noon UTC. If we wished to include all possible dates from 4713 BC to the present, the algorithm would become very complicated because of the calendar reforms. However, the following algorithm, which was developed for use between 1801 and 2099 and simplified for use after 28 February 1900, is used for most current computations. To compute the JD at a given year, Y, month, M, date, D, and time UT, we use the equation [Meeu88]

$$JD = 367Y - \left\lfloor \frac{7\left(Y + \left\lfloor \frac{M+9}{12} \right\rfloor\right)}{4} \right\rfloor + \left\lfloor \frac{275M}{9} \right\rfloor + D + 1721013.5 + \frac{UT}{24} \tag{8.1}$$

The $\lfloor * \rfloor$ is the *floor* operator; that is truncate and not round the answer. For example, the JD of 4 July 1976 at 14:30:22 UTC is

$$\begin{aligned} JD &= 725192 - 3459 + 213 + 4 + 1721013.5 + 0.604421296 \\ &= 2442964.104 \end{aligned} \tag{8.2}$$

This computation is also on the Web for any date at `http://aa.usno.navy.mil/data/docs/JulianDate.php`. The software included with the *Multiyear Interactive Computer Almanac* makes this computation [MICA05].

Because the JD is very large at the current time, it is convenient to use the Modified Julian Date (MJD), which is the full JD minus 2 440 000.5 and equivalent to resetting the JD epoch to be 24.0 May 1968 (the fractional day indicates beginning at 0 h UTC). Other common modified JDs correspond to a reset epoch of 1.0 January 1958, which is the start of TAI, or 1.0 January 2000.

8.3.2 Elapsed Time

The time difference between two events is easier to measure at high resolution than the absolute time, particularly when comparing measurements made in differential mode versus those made in absolute mode. The measurement of relative time occurs primarily in distance measurements using a time-of-flight timer. It is also useful for arbitrary event timing and for deriving secondary quantities. For example, metrologists use the time to charge a capacitor or heat a substance to a certain temperature to derive physical properties and measures.

The basic means designers use to find the elapsed time is to measure the number of pulses coming from a precision oscillator in a digital clock with an arrangement such as Figure 8.2 illustrates [Hora84]. The gating $START$ and $STOP$ signals delimit the count duration and directly control the pulses from the oscillator. As we saw in Section 2.4.2.2 and 2.4.5, a refinement on this technique is to use either a time-to-amplitude converter or a Vernier timer to attain higher resolution over a short duration.

With Figure 8.2, we are concentrating on timing long durations. In this case, the experimenter is concerned with the clock's frequency stability and frequency drift. The *frequency stability* is the ratio of the measured frequency to the nominal frequency and the *frequency drift* is the rate at which that ratio changes. Table 8.1 provides representative values for various commercial crystal-based oscillators. Figure 8.3 illustrates a commercial atomic clock, also included in Table 8.1, that provides a

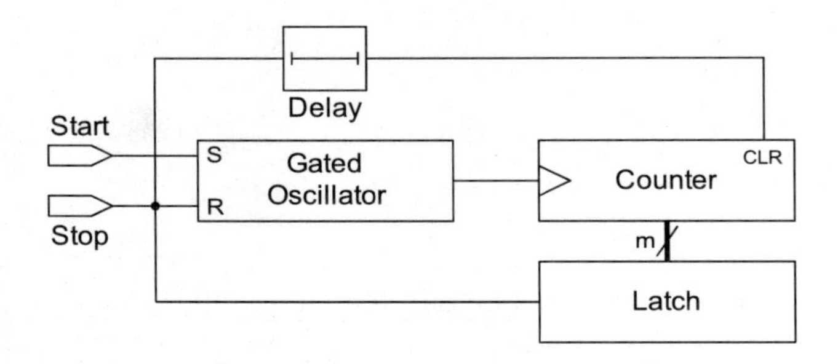

Figure 8.2: Relative time measurement with counter-based timers.

Figure 8.3: A modern, commercial chip-scale atomic clock providing a 10-MHz oscillator with a stability exceeding crystal oscillators.

10-MHz oscillator and a 1-PPS output having a stability of 2.5×10^{-10} over 1 second and an aging less than 10^{-9} per month.

8.4 TIME CODE FORMATS

Designers choose from several available standard time code formats in data systems to mark when events occur. This section discusses the ISO, IRIG, NIST, and CCSDS standard formats that allow the expression of time in varying degrees of precision and, in some cases, the specification of control signals for use with the time code.

8.4.1 International Organization for Standardization

The International Organization for Standardization (ISO) has defined standard formats for expressing dates and times for use in applications [ISO8601]. These formats do not dictate the means for distributing the

Table 8.1: Representative Oscillator Characteristics at the Unit Level.

Oscillator Type	Stability	Drift	Frequency Range	Power
Quartz Crystal (XO)	10–100 ppm	10 ppm (10 year)	0.010–700 MHz	125 mA; 3.3 V
Voltage Controlled Crystal (VCXO)	20–100 ppm	5 ppm (10 year)	15.5–800 MHz	150 mA; 3.3 V
Phase Locked Crystal (PLXO)	2.5 ppm	1 ppm (1 year)	5–500 MHz	200 mA; 3.3 V
Temperature Compensated Crystal (TCXO)	1 ppm	1 ppm (1 year)	10–50 MHz	2 mA; 3.3 V
Programmable PLL Crystal	20–50 ppm	5 ppm (10 year)	5–1000 MHz	80 mA; 3 V
Oven Controlled Crystal (OCXO)	0.5 ppb	0.5 ppm (1 year)	10–50 MHz	150 mA; 3 V
Chip Scale Atomic Clock (CSAC)	0.3 ppb	1 ppb (1 month)	10 MHz, 1 Hz	36 mA; 3.3 V

timing information itself. The date encoding standard is based on the current Gregorian calendar dates. The year (YYYY), month (MM), and day of the month (DD) are encoded in either "complete" or "extended" format as Table 8.2 illustrates. The extended format uses a dash delimiter between the fields. The day within the year is encoded in either month and day form or as an Ordinal Date (the day number of the year). The standard also defines an abbreviated format, e.g., month and year only, based on this standard representation. The calendar date may also be expressed in terms of the week number within the year and the day of the week with Monday being day 1. The week encoding uses the "W" after the year to mark the start of the week encoding as Table 8.2 shows.

Time encoding is also expressed in either complete or extended format. The time can be expressed in either local time or UTC. The time may also be encoded with a fractional second with either a comma (preferred) or a period (also called a stop) to delineate the fractional part. If UTC timing is encoded, the encoder uses a "Z" in the time field. The user may also indicate if the local time is either ahead (+) or behind (−) UTC by appending the hours and minutes difference to the local time encoding. Table 8.2 shows these time encoding options along with encoding both the date and the time in a single grouping.

8.4.2 Inter-Range Instrumentation Group

Inter-Range Instrumentation Group (IRIG) time standards are primarily used on military test ranges and the time code follows several different formats for differing resolutions. The IRIG standard has two formats. One is a full-frame format and the other is a timing word format for inclusion with other telemetry data. As we will see later, NIST also uses one of the frame formats in its standard time service broadcasts. The IRIG timing word format has two related variations: the Pulse Code Modulation (PCM) word format and the IRIG MIL-STD-1553 word format.

8.4.2.1 Inter-Range Instrumentation Group Time Frame Formats

Table 8.3 shows the lengths and rates of the IRIG time-code frames [IRIG200]. The user must specify the following parameters for the IRIG format:

Table 8.2: ISO 8601 Standard Encoding of Date and Time [ISO8601]

Encoding	**Complete**	**Extended**
Calendar Date	YYYYMMDD	YYYY-MM-DD
Ordinal Date	YYYYDDD	YYYY-DDD
Week Date	YYYYWwwD	YYYY-Www-D
Local Time of Day	hhmmss,ss	hh:mm:ss,ss; or
	hhmmss.ss	hh:mm:ss.ss
UTC Time of Day	hhmmss,ssZ	hh:mm:ss,ssZ; or
	hhmmss.ssZ	hh:mm:ss.ssZ
Local Time Relative to UTC	hhmmss±hhmm	hh:mm:ss±hh:mm
Local Date and Time of Day	YYYYMMDDThhmmss	YYYY-MM-DDThh:mm:ss
UTC Date and Time of Day	YYYYMMDDThhmmssZ	YYYY-MM-DDThh:mm:ssZ

Table 8.3: Lengths and Rates for the IRIG Time Code Frames [IRIG200]

Format	Bit Rate	Frame Rate	Bits/Frame	Control/Frame
A	1000 bps	10 fps	78 bit	18 bit
B	100 bps	1 fps	74 bit	18 bit
D	1 bpm	1 fph	25 bit	9 bit
E	10 bps	6 fpm	71 bit	36 bit
G	10 000 bps	100 fps	74 bit	27 bit
H	1 bps	1 fpm	32 bit	9 bit

- The overall frame format designator from Table 8.3, i.e., A, B, D, E, G, or H
- The transmission format, i.e., pulse width modulation digital format, Bi-Phase (Biϕ) (also called Manchester) encoded digital format, or amplitude modulated carrier format
- The carrier frequency or resolution
- The available data fields used in the overall frame, i.e., Binary Coded Decimal (BCD) encoded time information, control information, and Straight Binary Seconds (SBS) information

Figure 8.4 illustrates how the IRIG time code options are identified by alphanumeric codes. The BCD_{TOY} is a BCD-encoded time in days, hours, minutes, seconds, and sub-seconds, depending upon resolution. The BCD_{YEAR} is a BCD-encoded year. For the example given in Figure 8.4, `B230` specifies a *B* frame format, Biϕ data encoding, 10-kHz clock rate, and use of BCD_{TOY}, Control Functions (CF), and SBS data fields. Not all options are available for all frame formats. The baseband-data format portion of the IRIG format has three types of information: position identifiers, timing data logic 0 or 1 levels, and control data bits. Additionally, the standard specifies using index markers for those bit locations where there is no instance of the three information types. Table 8.4 specifies how each pulse type is encoded using Pulse Width Modulation (PWM) where T_b is the bit duration in the time code frame. The PWM encoding is a return-to-zero format with the datum 0, 1, or markers being differentiated by different pulse widths.

The standard identifies the time code's frame start by two position identifier pulses in succession and labeled P_0 and P_r. The position identifier pulses then repeat every 10 bits after P_r and are numbered P_1, P_2,

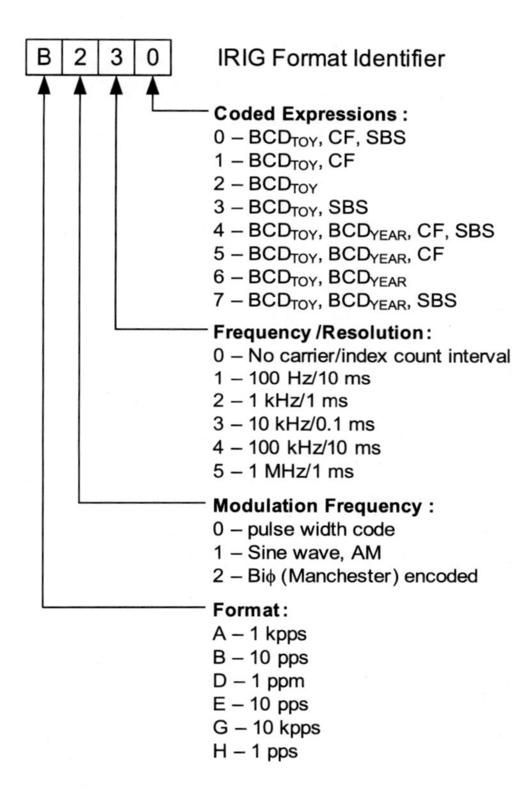

Figure 8.4: IRIG time-code format identifier options [IRIG200].

Table 8.4: Pulse Width Modulation Formats [IRIG200]

Signal Type	**One** (T_b)	**Zero** (T_b)
Position Identifier	0.8	
Timing Data	0.5	0.2
Control Data	0.5	0.2
Index Marker	0.2	

Table 8.5: IRIG Frame BCD Time Encoding Specification [IRIG200]

	BCD Format (bits per field)				Sub-	SBS
Code	Days	Hours	Minutes	Seconds	seconds	Seconds
A	10	6	7	7	4 (0.1)	17
B	10	6	7	7		17
D	10	6				
E	10	6	7	3		
G	10	6	7	7	4 (0.1) 4 (0.01)	
H	10	6	7			

etc. with P_0 as the last position identifier in a frame. Next comes the time code data between the position identifiers. Finally, the user may place control information between position identifiers at the frame end.

Table 8.3 lists the number of control bits the standard allows in a timing frame. The user has discretion over how the control bits are used and their meanings are not parts of the standard. The designer presents the timing information within the frame in either, or both, of two formats: (1) days, hours, minutes, seconds encoded in BCD format and (2) seconds since midnight encoded as a binary count using SBS format.

Table 8.5 gives the timing data information for each of the IRIG formats. The entries in each column show the number of bits used to encode the timing information. The A and B formats cited in the table allow the user to express the time of day as an integer number of seconds in addition to the BCD encoding of days, hours, minutes, and seconds. The timing is referenced to 0^h UTC. For example, consider how to format 11:33:25 UTC in both BCD and SBS modes. The BCD encodings using the number of bits in the IRIG fields are

11 hours: `01 0001`

33 minutes: `011 0011`

25 seconds: `010 0101`

Notice that normal BCD encoding uses four bits for every digit and the IRIG BCD encoding uses a number based upon the largest digit to encode. As the number of seconds since 0^h UTC, the time is 41 605 which is encoded in binary as 1 010 001 010 000 101.

Table 8.6: Placement of Timing Information within a Frame [IRIG200]

Decimal Digits	Preceding Position Identifier	Position Indices
Units of Seconds	P_0	1-4
Tens of Seconds		6-8
Units of Minutes	P_1	10-13
Tens of Minutes		15-17
Units of Hours	P_2	20-23
Tens of Hours		25-26
Units of Days	P_3	30-33
Tens of Days		35-38
Hundreds of Days	P_4	40-41
Tenths of Seconds		45-48
Hundredths of Seconds	P_5	50-53
For Codes A, B, and E	P_5	
Tens of Years		51-54
Hundredths of Year		56-59
For Code G	P_6	
Tens of Year		66-69
Hundredths of Year		56-59
SBS	P_8	80-97

Table 8.6 shows the locations of timing information within the frame. The pulses, P_i, are intraframe markers to reference the start of timing fields and are located at integer subrates of the frame rate. For example, Figure 8.5 shows the IRIG B timing frame with a pulse-width modulation baseband format, for example.

Both the carrier-modulated and Biϕ-encoded forms of the IRIG time code frame use the baseband pulse-width modulated signal to generate the transmitted signal. The carrier-modulated form uses the pulse-width encoded waveform to modulate a sinusoidal carrier. As expected with an amplitude modulation technique, the carrier amplitude depends upon the state of the baseband signal; however, the standard does not use on-off keying. Rather, the standard indicates the pulse-on and pulse-off states in the baseband signal by different nonzero amplitude levels. Typically, the relative carrier amplitude during the time the pulse is on state is 10. When the pulse is off, the relative carrier amplitude is 3. The allowable

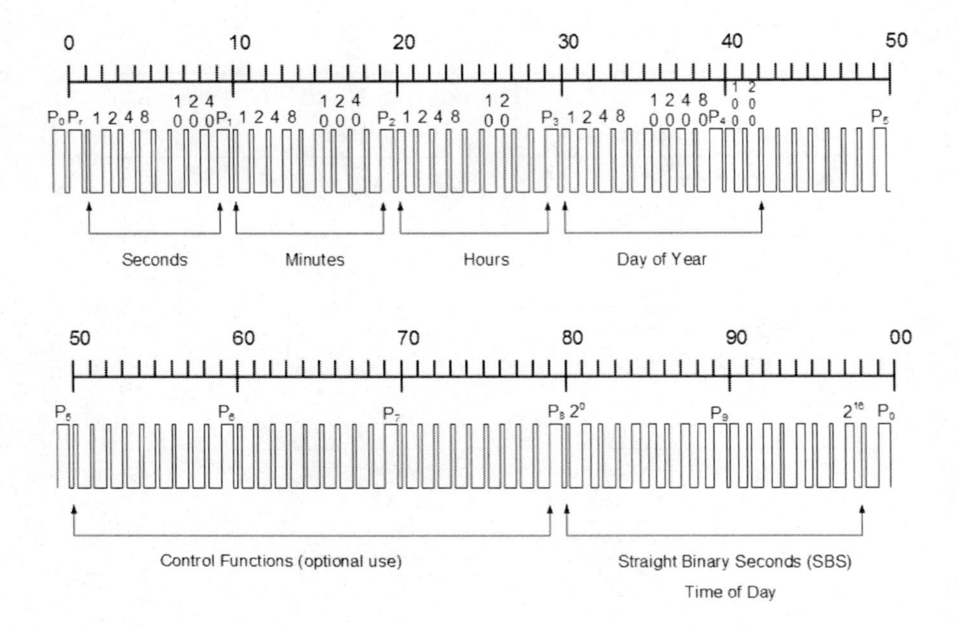

Figure 8.5: Time encoded using the IRIG B frame format.

standard range for the carrier amplitudes is between 3:1 and 6:1.

The carrier frequency is usually 10 times the bit rate. Table 8.7 lists the typical standard carrier frequencies for the IRIG time code formats. Figure 8.6 is an example of this encoding. Five random bits appear at the top of the figure. The normal NRZ-L encoding of the bits is below the random bits. The figure shows the amplitude-modulated carrier in the bottom section. The carrier frequency is 10 times the bit rate. A 10:3-carrier amplitude ratio is used.

The Biϕ encoding format is a digital analog to the carrier amplitude modulation format. Here, the pulse-width modulated signal is exclusive ORed with the clock signal. The clock signal is not slaved to the bit rate, but to a higher rate clock that is an integer multiple of the bit rate. Table 8.7 gives the allowable rates. Figure 8.6 also shows Biϕ encoding of the random bit stream.

8.4.2.2 Inter-Range Instrumentation Group Pulse Code Modulation Timing Word Format

IRIG has defined a standard timing word format for use in PCM telemetry frames [IRIG106]. The standard does not specify the entire frame with

Table 8.7: Typical IRIG Time Code Format Options [IRIG200]

Format	Carrier Frequency	Frequency/ Bit Rate Ratio	Coded Expressions
A	10 kHz	10:1	A130, 132, 133, 134
B	1 kHz	10:1	B120, 122, 123, 127
D	100 Hz	6000:1	D111, 112
	1 kHz	6000:1	121, 122
E	100 Hz	10:1	E111, 112
	1 kHz	100:1	121, 122, 125
G	100 kHz	10:1	G141, 142, 126
H	100 Hz	100:1	H111, 112
	1 kHz	1000:1	121, 122

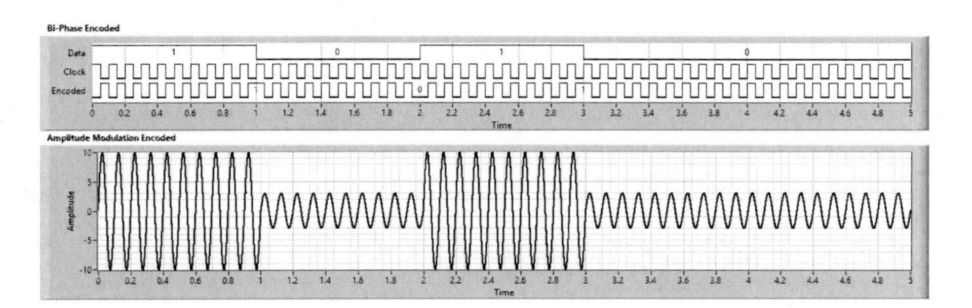

Figure 8.6: The IRIG data encoding for Biϕ (top panel) and amplitude-modulated carrier (bottom panel) based on a random bit pattern.

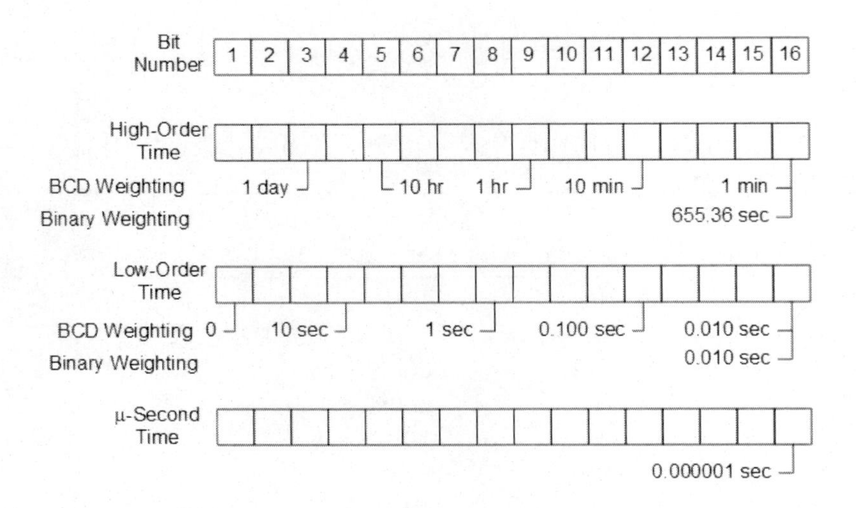

Figure 8.7: IRIG PCM timing word formats with three timing resolutions.

timing information as in the full-frame formats above. Rather, the timing information is included as a datum along with the other measurements in the frame. Figure 8.7 illustrates that the PCM timing word format takes on one of three variations, depending upon the level of timing resolution used. The timing is expressed in either a BCD or binary format over the same timing word length. The standard recommends that the timing information be the initial data in the minor frame. The standard requires that the order consist of three timing words: high-order time word, low-order time word, and microsecond time word.

The BCD format encodes the high-order time as days (3 bits), 10 hours (2 bits), 1 hour (4 bits), 10 minutes (3 bits), and 1 minute (4 bits), seconds, and fractions of a second. The binary weighting format encodes the high-order time as multiples of 655.36 s. The BCD format encodes the low-order time in terms of 10 s (3 bits), 1 s (4 bits), 0.100 s (4 bits), and 0.010 s (4 bits). The binary encoding of low-order time encodes it in 16 bits with a resolution of 0.010 s. The microsecond time word encodes in multiples of 1 μs until 10 ms have elapsed, at which time the microsecond counter is reset. There is no BCD encoding of the microsecond time word.

One concern with 16-bit word formats is how the designer places these data values in smaller-format telemetry frame words, for example, 12-bit words. The recommended manner in the standard is to use two adjacent telemetry frame words and have the 16-bit word span the smaller frame

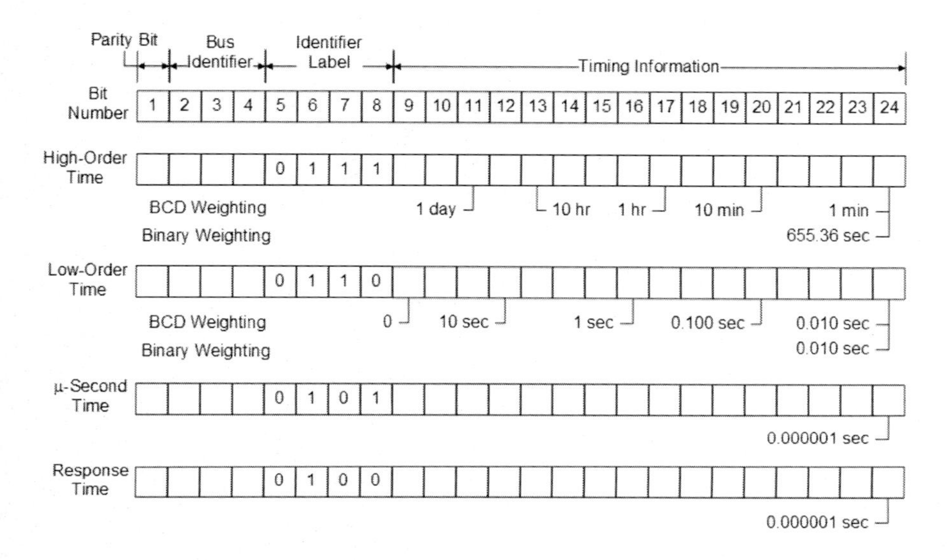

Figure 8.8: MIL-STD-1553 timing information packaged for use in standard telemetry frame structures [IRIG106].

words [IRIG106]. Designers use zero padding at the end of the 16-bit word to fill the entire two-word sequence. The standard allocates 24 bits to hold the timing information. Of the 24 bits, 16 bits contain actual timing information and 8 bits contain the zero padding bits.

8.4.2.3 Inter-Range Instrumentation Group MIL-STD-1553 Time Formats

Figure 8.8 illustrates the IRIG 106 standard MIL-STD-1553 packet time code formats [IRIG106]. If one compares Figures 8.7 and 8.8, the MIL-STD-1553 formats are an extension of the PCM format described above with the 1553 bus identification information and the other additional information used to place 1553 information in a standard IRIG PCM telemetry frame format. The timing resolution and use is the same as the PCM timing word usage.

8.4.3 National Institute of Standards and Technology

National Institute of Standards and Technology (NIST) provides time code services using voice and digital radio signals, computer-based utilities for use over the Internet, and computer dial-up services [Lomb02; Lowe13; Nels05]. The radio signals come from both terrestrial radio towers in

Colorado and Hawaii. Table 8.8 summarizes the characteristics of the voice/tone and BCD time code terrestrial services on stations WWV and WWVH. Station WWVB transmits only the BCD time code.

Figure 8.9 illustrates the hourly and minute WWV voice/tone service broadcast format. The hourly broadcast schedule shows which minute format is used for a given minute within the hour by the shading used to display the minute in part (a) of Figure 8.9. The hourly broadcast also contains various alert and report services as an audio message according to the schedule. The format for the WWVH station is similar. As part (b) of Figure 8.9 shows, WWV uses three formats, consisting of a 500-Hz tone, a 600-Hz tone, and no audio tone, for each minute within the hour. The tone, if present, lasts from seconds 0 through seconds 45. The period from seconds 45 to seconds 52.5 is silent except for the ticks. The period from seconds 52.5 until seconds 60 contains the audio announcement for the next minute of UTC.

NIST ties the time services to the atomic clocks to maintain their intrinsic accuracy. However, due to variable propagation delays and effects through the atmosphere, the usable timing is at best ± 1 ms after best calibrations. The voice/tone service broadcasts are in UTC seconds, which NIST keeps to within 0.9 s of UT1 by the addition or deletion of leap seconds. If UT1 is required to higher accuracy, the user applies a translation based on the number of double ticks heard at the start of each minute. A +0.1-s increment is added for each double tick in the first 8 s, while a −0.1-s decrement is added for each double-tick in the second 8 s.

The WWVB station transmits a digital time encoding using PWM and Phase Modulation (PM) formats. Table 8.9 provides the data mapping to the specific bits in the 60-s frame format for each modulation type. The PWM format uses a BCD time code in a modified IRIG H format. WWVB transmits the time code at 1 bps and it contains the date encoded as the last two digits of the year (Yr80 through Yr1), day number of the date (Day200 through Day1), and hours (Hr20 through Hr1) and minutes (Min40 through Min1) of time. The timing frame also includes markers for leap year indication (LYI), leap second warning (LSW), Daylight Savings Time (DST) (DST0 and DST1), and the correction between UT1 and UTC (UT1c0.8 through UT1c0.1).

The PM is also transmitted at 1 bps but the time is encoded as a 26-bit minute counter (Tim25 through Tim0) relative to the start of each century. Each frame begins with a 13-bit synchronization word (Sync12 through Synch0). The time encoding carries a 5-bit parity check code

Table 8.8: NIST Terrestrial Time Services [Lomb02]

Characteristics and Services	WWV	WWH	WWVB
Geographical Coordinates	40°40′53.1″ N 105°2′228.5″ W	21°59′15.3″ N 159°45′50.0″ W	40°40′28.3″ N 105°2′39.5″ W
Carrier Frequencies	2.5 and 20 MHz; 5, 10 and 15 MHz	2.5 MHz; 5, 10 and 15 MHz	60 kHz
Power (W)	2500 / 10 000	5000 / 10 000	13 000
Standard Audio Tone Frequencies	440, 500 and 600 Hz		None
Time Intervals	1 pulse/second; minute mark; hour mark		Sec and min
Voice Time Signals	Once per minute		None
Coded Time Signals	BCD code on 100 Hz subcarrier at 1 pulse per second		BCD code
UT1 Corrections	UT1 corrections broadcast with an accuracy of ±0.1 s		
Special Announcements	Geo-alerts, Marine storm warnings, Global Positioning System status reports		None

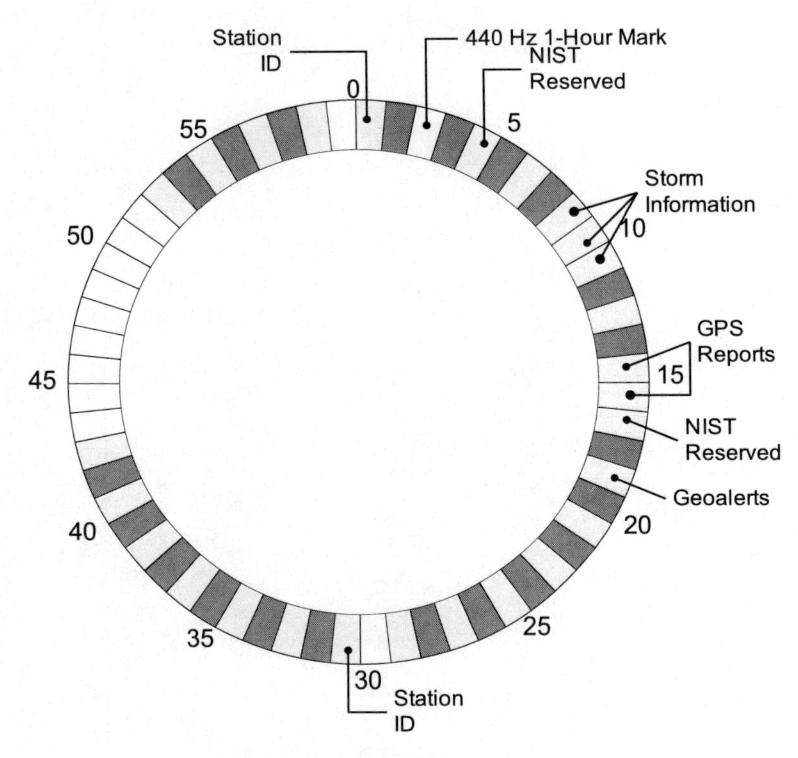

(a) WWV hourly broadcast schedule

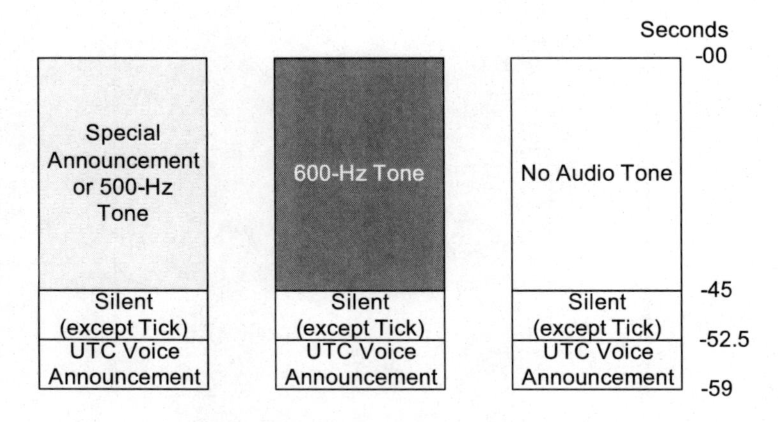

(b) WWV minute broadcast schedule

Figure 8.9: The WWV hourly and minute broadcast schedule [Lomb02].

Table 8.9: Timing Frame Used in WWVB for the Legacy Modulation and Phase Modulation Formats [Lowe13]

Second	**0**	**1**	**2**	**3**	**4**	**5**	**6**	**7**	**8**	**9**
PWM	Mark	Min40	Min20	Min10	0	Min8	Min4	Min2	Min1	P1 Mark
PM	Sync12	Sync11	Sync10	Sync9	Sync8	Sync7	Sync6	Sync5	Sync4	Sync3
Second	10	11	12	13	14	15	16	17	18	19
PWM	0	0	Hr20	Hr10	0	Hr8	Hr4	Hr2	Hr1	P2 Mark
PM	Sync2	Sync1	Sync0	Par4	Par3	Par2	Par1	Par0	Tim25	Tim0
Second	20	21	22	23	24	25	26	27	28	29
PWM	0	0	Day200	Day100	0	Day80	Day40	Day20	Day10	P3 Mark
PM	Tim24	Tim23	Tim22	Tim21	Tim20	Tim19	Tim18	Tim17	Tim16	Rsrvd
Second	30	31	32	33	34	35	36	37	38	39
PWM	Day8	Day4	Day2	Day1	0	0	UT1Sgn1	UT1Sgn2	UT1Sgn3	P4 Mark
PM	Tim15	Tim14	Tim13	Tim12	Tim11	Tim10	Tim9	Tim8	Time7	Rsrvd
Second	40	41	42	43	44	45	46	47	48	49
PWM	UT1c0.8	UT1c0.4	UT1c0.2	UT1c0.1	0	Yr80	Yr40	Yr20	Yr10	P5 Mark
PM	Tim6	Tim5	Tim4	Tim3	Tim2	Tim1	Tim0	DST-LS4	DST-LS3	Note
Second	50	51	52	53	54	55	56	57	58	59
PWM	Yr8	Yr4	Yr2	Yr1	0	LYI	LSW	DST1	DST0	P0 Mark
PM	DST-LS2	DST-LS1	DST-LS0	DSTNxt5	DSTNxt4	DSTNxt3	DSTNxt2	DSTNxt1	DSTNxt0	0

(Par4 through Par0). The DST and leap second logic is encoded in 5 bits (DST-LS4 through DST-LS0). Because the DST start date and end date vary from year to year, the changeover is marked with the six Next DST transition bits (DSTNxt5 through DSTNxt5). See [Lowe13] for further details on the other fields and the alternate message frame format that can be transmitted in lieu of a time frame.

To assist slaving a computer clock to the time standard, NIST provides a utility, `nistime-32.exe`, that runs in background mode as an alternative to the time synchronization utilities available in the current Windows operating system. This keeps the computer clock to within less than one second of the NIST clocks. The exact variation depends upon the computer-network delay characteristics. The utility is available at the NIST Web site `http://www.nist.gov/pml/div688/grp40/its.cfm`. This program uses the standard Internet Network Time protocol for exchanging timing information.

8.4.4 Consultative Committee for Space Data Systems

The international space community, under the auspices of the Consultative Committee for Space Data Systems (CCSDS), has developed time code standards [CCSDS3010] to promote interoperability. The time codes fall into two major divisions: binary data format and American Standard Code for Information Interchange (ASCII) character format.

The CCSDS binary data format is one that is included easily in a packet data system. The CCSDS time code family allows the user three variations in the code's precision level. One variation is a nonsegmented code, while the other two variations are segmented codes. Only the segmented codes are covered here.

Each binary time code starts with a preamble that specifies the time code type, the timing epoch, the time format, and the timing resolution. Table 8.10 shows the preamble encoding. The formatted CCSDS time codes use two encoding methods. The first counts integer days from a standard epoch, and the second gives the calendar information for the day in the current year. The standard epoch used is the TAI zero epoch of 1 January 1958 without leap second adjustments. Figure 8.10 illustrates the CCSDS time codes. For the day-segmented time code format, each segment is composed of the output from a right-adjusted binary counter. For calendar-segmented time codes, each segment has BCD encoded decimal digits with the length of the segment specifying the number of digits contained in the segment.

Table 8.10: CCSDS Segmented Time Code Preamble [CCSDS3010]

Field	**Day Segment Time Code**	**Calendar Segment Time Code**
ID (Bits 1-3)	100	101
Time Mode (Bit 4)	0 - TAI Epoch	0 - month and day of year
	1 - Agency-Defined Epoch	1 - day of year
Resolution	Bit 5 = 0/1 - 16/24-bit day	Bits 5-7 = 000 - 1 second
(Bits 5-7)	Bits 6-7 = 00 - milliseconds	= 001 - 0.01 s
	= 01 - microseconds	= 010 - 0.1 ms
	= 10 - picosecond	= 011 - 1 μs
	= 11 - reserved for	= 100 - 10 ns
	future use	= 101 - 0.1 ns
		= 110 - 1 ps
		= 111 - not used

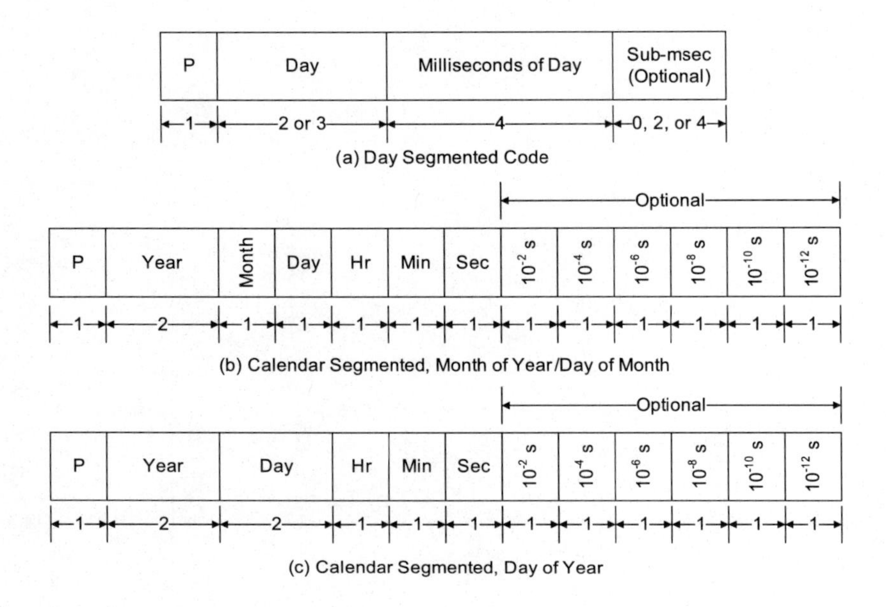

Figure 8.10: CCSDS segmented time codes. The width of each segment in octets (bytes) is also indicated [CCSDS3010].

The ASCII character format is based on the ISO 8601 extended date and time encoding we examined earlier in this chapter. A major difference from the binary formats is that the character time encoding is based on UTC with leap second adjustments. The standard permits the designer to use either of two formats: calendar date with UTC time, and ordinal date with UTC time. The standard allows the designer to specify the number of digits in the fractional seconds according to the system's need. Also, the "Z" character is not required at the end of the time field.

8.5 GLOBAL POSITIONING SYSTEM TIME AND POSITION

The Global Positioning System (GPS) has done much to revolutionize how people think of measuring time and distance [Lewa91; Park96; Spil96a; Spil96b]. The great number of applications for GPS and the dramatic changes in the size and ease of use of electronics have made using the system ubiquitous, like using cellular telephones. In this section, we describe the basics of the GPS and explain how positions and time are derived from the satellite constellation. Developers have developed many ways to use, and expand on, this information.

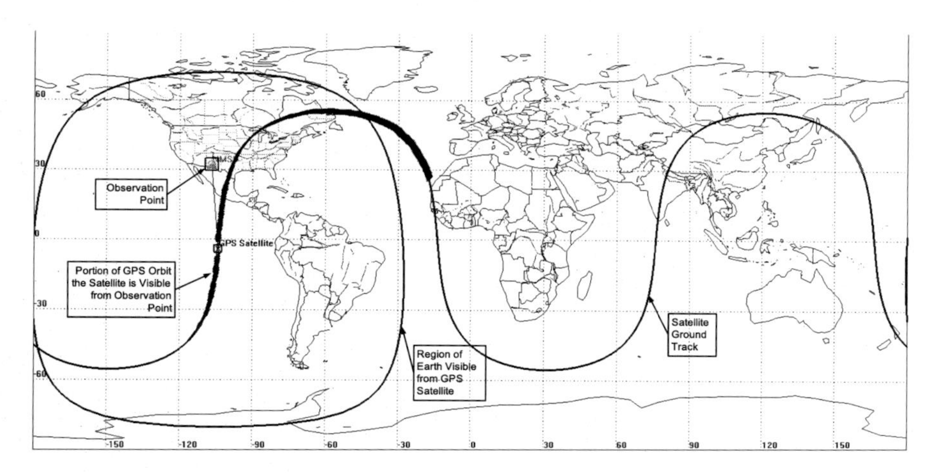

Figure 8.11: GPS ground track over a 24 h period. Also indicated is the portion of the orbit visible to a user and the region of the Earth visible from the satellite at a moment during the orbit.

8.5.1 Global Positioning System Definition

The GPS depends on a constellation of spacecraft circling the Earth approximately every 12 h (718 min). The satellites are in orbit with a nominal 55° inclination angle (some older satellites are at a 63° inclination angle) relative to the equator to provide useful coverage over the Earth. Each satellite orbits at an average altitude of 20 000 km, which yields a 71° angular radius satellite footprint (assuming a minimum 5° elevation angle). Figure 8.11 illustrates a satellite orbital ground track for a 24-h period. The figure also highlights that part of the orbit visible to a user located in the southwestern United States.

A "typical" user sees the satellite over a large part of its orbit. The GPS constellation has at least 24 active satellites with on-orbit spares (see `http://www.gps.gov/systems/gps/space/` for current constellation status), with the orbits divided into six planes uniformly distributed around the Earth and four satellites uniformly spaced within each orbital plane. This satellite constellation plan allows several satellites to orbit above a user's horizon at all times.

The GPS satellites broadcast at two L-Band frequencies: 1575.42 MHz and 1227.60 MHz. These frequencies are slaved to the satellite's on-board system clock operating at 10.23 MHz. The actual clock rate is set at 10.229 999 995 45 MHz to allow for relativistic effects as the signals ap-

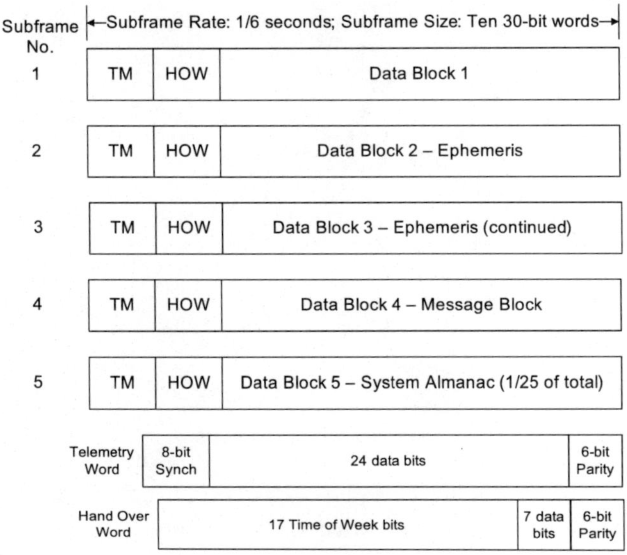

Subframe No. | ←Subframe Rate: 1/6 seconds; Subframe Size: Ten 30-bit words→

1 TM HOW Data Block 1

2 TM HOW Data Block 2 – Ephemeris

3 TM HOW Data Block 3 – Ephemeris (continued)

4 TM HOW Data Block 4 – Message Block

5 TM HOW Data Block 5 – System Almanac (1/25 of total)

Telemetry Word: 8-bit Synch | 24 data bits | 6-bit Parity

Hand Over Word: 17 Time of Week bits | 7 data bits | 6-bit Parity

Figure 8.12: GPS navigation messages.

proach the Earth. The clock rate is based on an atomic clock in the spacecraft that is monitored for each rate by the GPS system control center. Figure 8.12 illustrates how the user derives the satellite timing information in real time by determining the exact phase of two pseudo-random access code sequences broadcast by the satellite and a counter called the Hand Over Word (HOW) broadcast as part of the satellite navigation message.

The shorter of the two codes is called the C/A, or clear access code, and has a bit rate of 1.023 Mbps that produces a code repetition period of 1 ms. The second code is the P (precise or protected) code and GPS transmits it at 10.23 Mbps. This code is really only intended for authorized GPS users, while unauthorized users may only be able to achieve the accuracy fixed by the longer bit period of the C/A code. The full P code has a repetition period of 267 d. However, the code is used in segments of seven days with a unique segment for each operating GPS satellite and reset within the satellite once each week. GPS performs a direct sequence spread spectrum encoding of the navigation message as broadcast on the two L-Band carriers with these codes.

GPS broadcasts the navigation message in a major-frame format with five minor frames per major frame [Spil96b]. Each minor frame is ten words long transmitted at 50 bps. Within the minor frame are Telemetry (TM) and HOW words at the start followed by subcommutated data blocks dealing with satellite and system parameters. The first minor frame data block contains correction parameters for the on-board clock, the tag word for the age of the data, and propagation model coefficients. The second and third data blocks contain the satellite ephemeris information. The fourth minor frame contains space for twenty-three 8-bit ASCII characters for messages from the system control center. The fifth minor frame contains system almanac information. The almanac information for each spacecraft is a truncated version of the information supplied in the previous minor frames in a time-shared manner. GPS receivers use the almanac information to determine the visible satellites initially and aid basic signal acquisition. Twenty-five major frames are required to receive all almanac information. Modern receivers often store the necessary data in memory and have multiple-channel receive capabilities. These features allow the receiver to synchronize quickly to the system on power-up.

The L_1 carrier at 1575.42 MHz is an unbalanced in-phase and quadrature phase carrier, as follows [Spil96a]:

$$S_{L1}(t) = A_P P_I(t) D_I(t) \cos(\omega_{L1} t + \varphi) + A_C G_I(t) D_I(t) \sin(\omega_{L1} t + \varphi) \tag{8.3}$$

where $D_I(t)$ represents contents of the navigation message, $P_I(t)$ is the P-code's Pseudorandom Noise (PN) code, $G_I(t)$ is the C/A-code's PN code, and A_P and A_C are the respective amplitude weights. Typically, the C/A phase is 3 to 6 dB stronger than the P phase. The phase offset, φ, is due to internal phase noise and oscillator drift.

The L_2 carrier is capable of being modulated by either the P-code or the C/A-code as selected by the GPS system control center. The data are bi-phase encoded for modulation and the carrier is expressed as [Spil96a]

$$S_{L2}(t) = B C_I(t) D_I(t) \cos(\omega_{L2} t + \varphi) \tag{8.4}$$

where B is the signal amplitude, $C_I(t)$ is the PN code being used (either P or C/A) and $D_I(t)$ is the data.

8.5.2 Time and Position Determination

The GPS satellites broadcast a known signal format on a periodic basis. If a user observes three satellites simultaneously then, in principle, the user can determine the relative delays and derive a position in three-dimensional space. The determination algorithm does not confine the position to the Earth's surface but the position can be in orbit around the Earth too. In practice, one must have a fourth satellite visible because there are timing unknowns in the system unless one has a calibrated standard clock available. The problem is the solution of four equations in four unknowns. The three spatial equations form intersecting arcs of constant delay from each satellite with the user located outside of the intersection due to the relative timing offset. The time equation brings the intersection of the distance arcs to lie over the user. The user's receiver then solves for the required quantities by choosing four satellites in real-time and listening to their broadcast signals.

Figure 8.13 illustrates a snapshot of the GPS satellites visible during a position measurement. The navigation solution looks for at least four, well-separated satellites. Of the available satellites above the horizon, the receiver should choose those with the greatest relative separation as seen from the center of the Earth up through the user's position. That is, it is best not to choose all of the satellites in the same relative portion of the sky. The desired configuration is to have one satellite passing near the zenith (the "Z" in the figure) and the other three separated as far as possible, yet maintaining sufficient elevation above the horizon for good signal reception. If this occurs, system analysis shows that the precision of the position determination is inversely proportional to the volume of the solid described by the position vectors between the user and the satellites. In modern user support software, the program uses the entire visible field of satellites in determining the position.

Prior to obtaining a full solution, the observer obtains a *pseudorange* from the satellite. Figure 8.14 illustrates that a pseudorange is the true range modified by the system unknowns other than the user's position. The pseudorange, R_P, is related to the true range, R_T, by

$$R_P = R_T + c\Delta t_A + c\left(\Delta T_u - \Delta t_S\right) \tag{8.5}$$

where c is the speed of light, Δt_A is the atmospheric propagation delay, Δt_U is the user clock offset from the GPS time, and Δt_S is the satellite clock offset from GPS time. The GPS receiver processes these pseudoranges to find location and time by using them in four equations with the unknowns being the position and the clock offset.

GPS ground controllers have the ability to reduce the precision of the receiver's position determination. When reduced precision is in effect, the controllers modify the clocking in the signal in such a way as to add small-scale variability to the navigation solution. This process is called the Selective Availability (SA) mode. In mid-2001, the GPS controllers

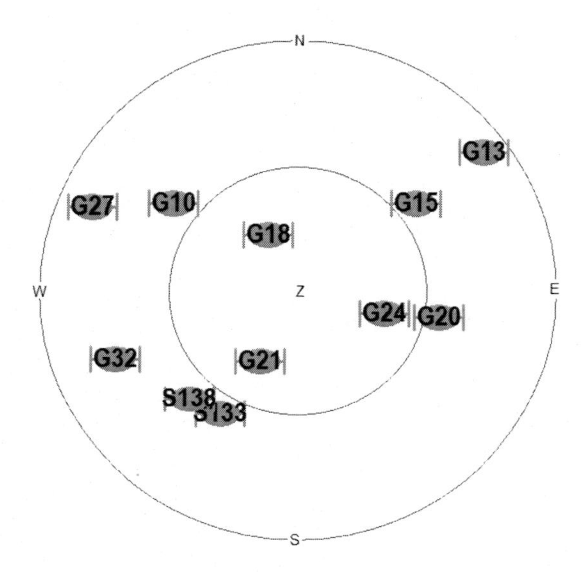

Figure 8.13: An example of GPS constellation satellites visible during a position determination. The plot shows the satellites' azimuthal and elevation pointing information. The outer ring is the horizon and the "Z" is the local zenith.

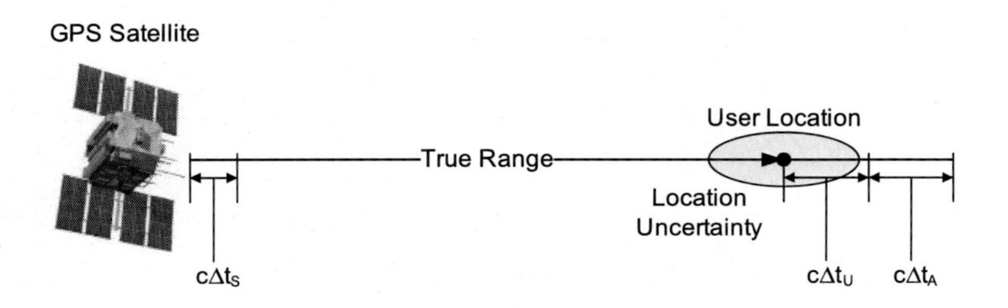

Figure 8.14: GPS pseudorange components.

turned off the SA mode; however, the government could decide to return to this mode of operation.

GPS time uses atomic clocks; it keeps time like TAI. The zero point for the GPS system time is 0^h UT on 6 January 1980. The GPS time reported is given by the system's composite clock that has its zero point at 0^h UT on 17 June 1990. GPS time has no leap seconds; hence it differs from UTC by the number of leap seconds since January 1980. As of 1 June 2017, this difference is 18 s in the sense that GPS time lead UTC by that amount. A listing of the leap second additions is provided on the NIST Web site `http://www.nist.gov/pml/div688/grp50/leapsecond.cfm`.

Each GPS satellite broadcasts its own clock correction as determined by the system control center. This correction is in the form of two coefficients: a_0, and a_1. This allows correction to the United States Naval Observatory Master Clock. One of the peculiarities of the GPS system is that the system keeps its dating internally as a number of weeks in the system clock. This week counter has a length of 1024 and it began 5 January 1980, so that the system has rollover weeks periodically. The first of these occurred in August 1999. The GPS receiver unit's software must account for these rollover corrections and make them prior to providing data to the user. Typically, these corrections are invisible to the user and the correct value appears in the navigation sentence that is output from the unit.

Commercial vendors have developed chip sets and special processing boards that take the system parameters as broadcast by the satellites and the received pseudoranges to compute a solution to the current time and position on the Earth for the receiving station. The user typically views the processed results in the form of navigation sentences (described in the next section) — not the raw navigation messages as broadcast by the satellites. Only the users' vision seems to limit the number of ways users add value to all of this information. For example, the designers of a telemetry system remove the time and position information from a navigation sentence and inset the data in a telemetry frame to tell the ground segment the current position of the payload and time-tag the data. Ground segments use the GPS time to coordinate activities, keep system clocks consistent, etc.

Table 8.11: Standard GPS Message Types

Identifier	Name
GGA	GPS Fix Data
GLL	GPS Geographic Position Longitude and Latitude
GSA	GPS Satellites Active
GSV	GPS Satellites in View
RMC	Recommended Minimum Navigation Information
VTG	Velocity Track Good and Ground Speed
ZDA	Time and Date

8.5.3 National Marine Electronics Association Navigation Sentences

GPS users typically receive time and position information formatted in National Marine Electronics Association (NMEA) standard navigation sentences over a serial interface or using a direct plug-in card. This NMEA format is a structured text string where commas separate each field in the message, for example

$$\$GPidentifier, field_1, field_2, field_3, \ldots, field_N, *hh \tag{8.6}$$

Here, *identifier* is the name of the message type and the message-specific parameters appear in $field_1$ through $field_N$ (each message has a different parameter set). Finally, the message ends with **hh* composed of an asterisk followed by a two digit checksum, *hh*. Table 8.11 lists messages specifically for time and position determination. The exact field order and contents are specific to the individual message types. For example, Table 8.12 lists the fields for the GGA, GLL, and RMC messages. Figure 8.15 illustrates an example of the message output from a commercial GPS receiver.

Navigation messages, other than those mentioned in Table 8.11, give more information about the system but are not as useful for specific time and position determination. The time and position data in the fields have the following formats:

- Longitudes are expressed in degree, minute, and fractional minute format; e.g., 100°12.345 67′ is encoded as 10 012.345 67. A second field for east or west is encoded as `E` or `W`.

- Latitudes are expressed in degree, minute, and fractional minute

Table 8.12: NMEA GPS Message Formats

Field	Description	Field	Description	Field	Description
GGA	Identifier	GLL	Identifier	RMC	Identifier
hhmmss.ss	UTC	ddmm.mmmm	Latitude	hhmmss.ss	UTC
ddmm.mmmm	Latitude	N/S	North or south	A/V	status (valid/invalid)
N/S	North or south	dddmm.mmmm	Longitude	ddmm.mmmm	Latitude
dddmm.mmmm	Longitude	E/W	east or west	N/S	north or south
E/W	East or west	hhmmss.ss	UTC	dddmm.mmmm	Longitude
i	Quality	A/V	status (valid/invalid)	E/W	East or west
ii	no. of satellites	*		speed	knots
d.d	HDOP	hh	checksum	course	degrees
iiiii	altitude			ddmmyy	Date
M	meters			dd.	magnetic variation
	not used			E/W	variation sense
	not used			*	
	not used			hh	checksum
	not used				
*					
hh	checksum				

```
$GPRMC,204329.00,A,3707.81446,N,07622.02965,W,0.003,,041116,,,D*6C
$GPVTG,,T,,M,0.003,N,0.006,K,D*23
$GPGGA,204329.00,3707.81446,N,07622.02965,W,2,10,0.99,7.1,M,-35.8,M,,0000*62
$GPGSA,A,3,46,21,18,51,24,20,32,10,13,15,,,1.78,0.99,1.48*0F
$GPGSV,3,1,12,10,28,300,32,13,16,049,37,15,47,047,41,18,60,320,48*7E
$GPGSV,3,2,12,20,48,089,44,21,70,228,47,24,55,124,46,27,12,303,*7E
$GPGSV,3,3,12,29,01,193,,32,15,242,21,46,38,211,36,51,36,224,45*73
$GPGLL,3707.81446,N,07622.02965,W,204329.00,A,D*75
$GPRMC,204330.00,A,3707.81447,N,07622.02960,W,0.010,,041116,,,D*62
$GPVTG,,T,,M,0.010,N,0.019,K,D*2F
$GPGGA,204330.00,3707.81447,N,07622.02960,W,2,11,0.99,7.1,M,-35.8,M,,0000*6F
$GPGSA,A,3,46,21,18,51,24,20,27,32,10,13,15,,1.78,0.99,1.48*0A
$GPGSV,3,1,12,10,28,300,28,13,16,049,33,15,47,047,42,18,60,320,47*7D
$GPGSV,3,2,12,20,48,089,45,21,70,228,47,24,55,124,46,27,12,303,27*7A
$GPGSV,3,3,12,29,01,193,,32,15,242,18,46,38,211,37,51,36,224,45*78
$GPGLL,3707.81447,N,07622.02960,W,204330.00,A,D*79
```

Figure 8.15: Example of GPS navigation information captured over a two-second interval.

format; e.g., 10°23.45678′ is encoded as 1023.456 78. A second field for north or south is encoded as `N` or `S`.

- UTC time is encoded as hours, minutes, seconds, and fractional seconds as `hhmmss.ss`.
- Message status is encoded with an `A` to indicate a valid navigational message, while a `V` is used to indicate an invalid navigational solution.
- The date is encoded as day, month, and last two digits of the year; e.g., the ISO 8601-formatted date of 2014-08-28 is encoded as 280 814.

Figure 8.16 shows two options for the serial interface to common serial interface GPS units. The first interface option connects the unit to the computer's serial port with a TIA-232 serial cable operating at 4800 bps using 8 data bits, no parity, and one stop bit (8N1). Typically, the serial connection does not need flow control. Figure 8.17 gives a LabVIEW® Virtual Instrument (VI) to decode a navigation sentence using this interface. The second option is to use a Universal Serial Bus (USB) serial interface. This option has the advantage of also powering the GPS unit. With many manufacturers, their application software allows the user to access and control the GPS unit via calls to a serial port so that it looks like a TIA-232 interface and not a USB port. With this configuration, the user can still utilize software written for the traditional serial port method.

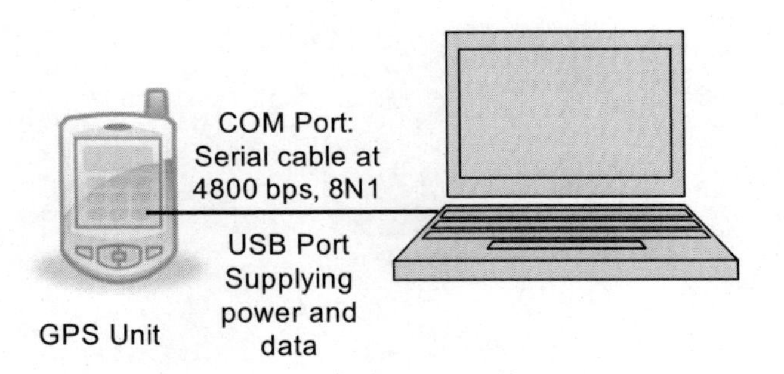

Figure 8.16: GPS serial interface. Two common options are shown: using a TIA-232 serial port or a USB port.

8.6 REFERENCES

[CCSDS3010] *Time Code Formats.* CCSDS 301.0-B-4. Consultative Committee for Space Data Systems. Washington, D.C., Nov. 2010. URL: https://public.ccsds.org/Pubs/301x0b4e1.pdf.

[Hora84] S. Horan. "High Speed Time to Digital Conversion." PhD thesis. Las Cruces: New Mexico State Univ., May 1984.

[IRIG106] Telemetry Group. *Telemetry Standards.* 106-15. Secretariat, Range Commanders Council. White Sands, NM, July 2015. URL: http://www.wsmr.army.mil/RCCsite/Documents/Forms/AllItems.aspx.

[IRIG200] Telecommunications Timing Committee and Timing Group. *IRIG Serial Time Code Formats.* 200-04. Secretariat, Range Commanders Council. White Sands, NM, Sept. 2004. URL: http://www.wsmr.army.mil/RCCsite/Pages/Publications.aspx.

[ISO8601] *Data elements and interchange formats – Information interchange - Representation of dates and times.* 3rd. ISO 8601:2004(E). International Organization for Standardization. Geneva, Switzerland, Dec. 2004.

[Lewa91] W. Lewandowski and C. Thomas. "GPS Time Transfer." In: *Proceedings of the IEEE* 79.7 (1991), pp. 991 –1000. DOI: 10.1109/5.84976.

[Lomb02] M. A. Lombardi. *NIST Time and Frequency Services.* NIST Special Publication 432. National Institute of Standards and Technology. Washington, D.C.: US Government Printing Office, 2002. 76 pp. URL: http://tf.nist.gov/timefreq/general/pdf/1383.pdf.

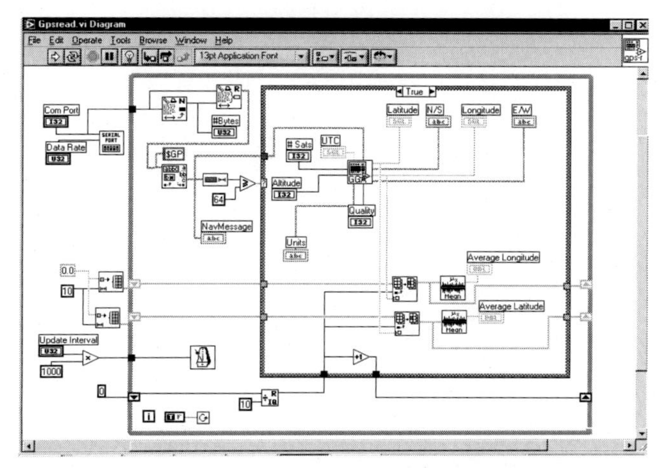

(a) LabVIEW VI for reading a GPS receiver attached to a computer's serial port.

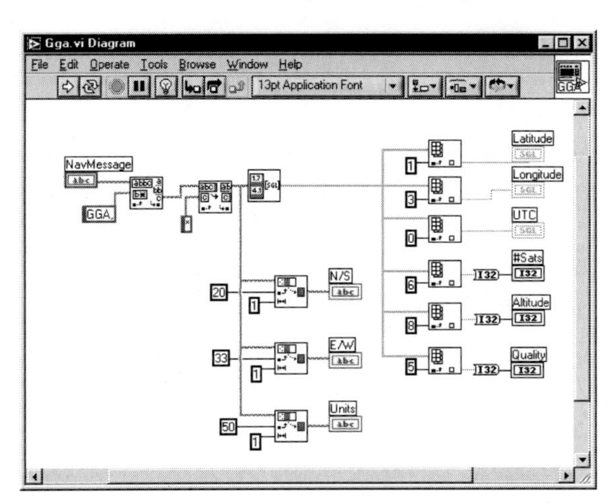

(b) LabVIEW VI for parsing the GGA navigation message.

Figure 8.17: VIs to read and parse NMEA navigation sentences for a GPS receiver.

[Lowe13] J. Lowe. *Enhanced WWVB Broadcast Format.* Version 1.01. National Institute of Standards and Technology. Nov. 2013. URL: http://www.nist.gov/pml/div688/grp40/upload/NIST-Enhanced-WWVB-Broadcast-Format-1_01-2013-11-06.pdf.

[Mcca91] D. D. McCarthy. "Astronomical Time." In: *Proceedings of the IEEE* 79.7 (1991), pp. 915 –920. DOI: 10.1109/5.84967.

[Meeu88] J. Meeus. *Astronomical Formulae for Calculators.* 4th Edition. Richmond, VA: Willmann-Bell, 1988, pp. 23 –25. ISBN: 0943396220.

[MICA05] U.S. Naval Observatory. *Multiyear Interactive Computer Almanac 1800 - 2050.* Richmond, VA: Willmann-Bell, Inc., 2005.

[Nels05] G. K. Nelson, M. A. Lombardi, and D. T. Okayama. *NIST Time and Frequency Radio Stations: WWV, WWVH, and WWVB.* NIST Special Publication 250-67. National Institute of Standards and Technology. Washington, D.C., Jan. 2005. 160 pp. URL: http://tf.nist.gov/general/pdf/1969.pdf.

[Park96] B. P. Parkinson. "Introduction and Heritage of NAVSTAR, the Global Positioning System." In: *Global Positioning System: Theory and Applications.* Ed. by B. W. Parkinson and J. J. Spilker. Vol. I. Washington, D.C.: American Institute of Aeronautics and Astronautics, 1996. ISBN: 978-1-56347-106-3.

[Quin91] Terry J. Quinn. "The BIPM and the Accurate Measurement of Time." In: *Proceedings of the IEEE* 79.7 (1991), pp. 894 –905. DOI: 10.1109/5.84965.

[Sobe95] D. Sobel. *Longitude: The Story of a Lone Genius Who Solved the Greatest Scientific Problem of His Time.* New York, NY: Walker Publishing Co., 1995. ISBN: 0-8027-9967-1.

[Spil96a] J. J. Spilker. "GPS Signal Structure and Theoretical Performance." In: *Global Positioning System: Theory and Applications.* Ed. by B.W. Parkinson and J.J. Spilker. Vol. I. Washington, D.C.: American Institute of Aeronautics and Astronautics, 1996. ISBN: 978-1-56347-106-3.

[Spil96b] J. J. Spilker. "GPS Navigation Data." In: *Global Positioning System: Theory and Applications.* Ed. by B.W. Parkinson and J.J. Spilker. Vol. I. Washington, D.C.: American Institute of Aeronautics and Astronautics, 1996. ISBN: 978-1-56347-106-3.

8.7 PROBLEMS

1. Design a counter-based timer to measure elapsed time to meet the following specifications: ± 1 count resolution of 40 ns and a maximum range of 2.5 ms.

2. If the crystal used to drive the counter in Problem 1 has a stability of $\pm 1 \times 10^{-6}$ and a drift due to aging of 10^{-7}/year, estimate the range of counts for a timing measurement of 1 ms at initial operation and after 1, 5, and 10 years. For the drift, assume a linearly increasing change in frequency with time.

3. Devise a use for control signals in the IRIG time code format.

4. Encode today's date plus a time to the maximum resolution possible using the IRIG B time code format.

5. Illustrate the IRIG 16-bit time code placement in a series of 12-bit telemetry frame words.

6. Encode today's date plus a time to the maximum resolution possible using the IRIG MIL-STD-1553 time code format.

7. The CCSDS binary time format does not track leap seconds. Why is this a reasonable tactic for time encoding?

8. Encode today's date plus a time to the maximum resolution possible using all three variations of the CCSDS formatted time codes.

9. Encode today's date plus a time using both the BCD time code format of the PWM modulation format and the digital PM modulation format for WWVB. Refer to [Lowe13] for the full details of the specification.

10. Investigate the propagation profiles through the atmosphere at 2 to 20 MHz. How long does a radio hop from WWV or WWVH take to propagate to your longitude and latitude?

11. Use basic trigonometry and derive an expression for the slant path between a satellite and a receiving station. Use this to derive the average time code propagation delay from a GPS satellite to your location.

12. If you were permitted to use only four of the GPS satellites in Figure 8.13, which four would you choose? Justify your answer.

13. Determine the time, date, and location from the GPS navigation data presented in Figure 8.15.

14. Show how to place time information from the NMEA format used in GPS to IRIG PCM word format, IRIG frame format, and CCSDS segmented format. Which options do these formats not permit?

CHAPTER 9

TELECOMMAND TRANSMISSION SYSTEMS

9.1 INTRODUCTION

As we saw in the discussion of command user interfaces in Section 4.4.4, telecommand (or command for short) systems play important parts in the overall payload/operator system. The system's command component has become more important over the years because payloads are not necessarily passive devices. Rather, they interact with operators when changing configurations, performing calibrations, sending data, or engaging in real-time operations. It is typical for remote payloads, e.g., a remote weather station, to include communications interfaces to allow users to interact with the payload over a radio link, a wireless telephone, or even an Internet link to accomplish these tasks.

As the highlighted blocks in Figure 9.1 show, both the user interface and the payload are involved in the command processes. The user interface provides the means to translate between ways of encoding the command data so that both the user and the payload understand the instructions. Additionally, the telemetry stream becomes a feedback loop to the operators to indicate whether the payload has received the commands properly.

Certain system characteristics distinguish the telecommand process from the simple links between a computer and its peripheral devices. While many peripheral devices may have similar command data structures,

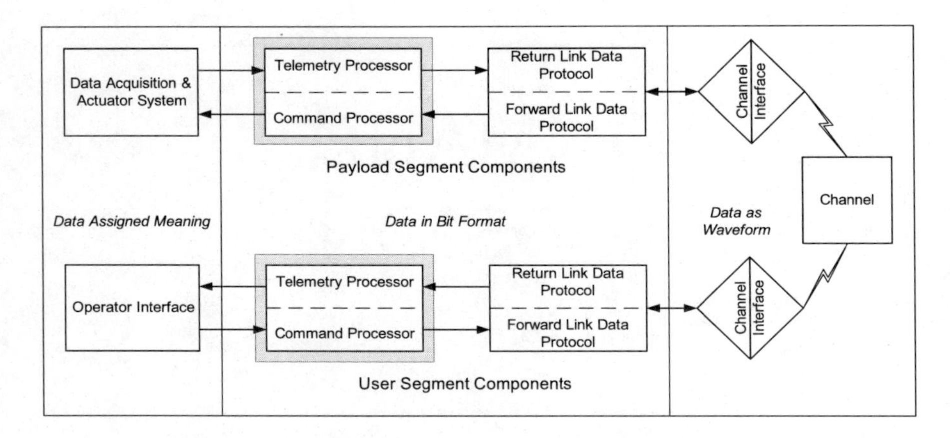

Figure 9.1: The highlighted blocks correspond with the command subsystem components in the payload and user interface as part of the overall system.

for example the standard configuration commands used in telecommunications equipment, command system designers have no industry-standard command sets on which to base the payload command system design. Rather, payload designers have the option to design their own unique syntax. Another difference, which we saw in Section 4.4.4, is that payload commands may be time-tagged for later execution, while peripheral commands are designed for immediate execution. It is typical for system designers to send *no operation* or *no-op* commands to remote payloads to maintain command link integrity when using radio links. Peripheral command sets do not include no-ops because they link directly to the host or over short-distance radio links like Bluetooth®. In most applications, the command transmission rate is significantly smaller than the telemetry transmission rate. This link imbalance actually may cause performance degradation across networks such as the Internet. Finally, command systems usually include security mechanisms to keep unauthorized users from spoofing the payload with improper commands. The designer can employ a digital encryption protocol or an authorization and validation check.

This chapter complements Chapter 4 because here we discuss command composition and command strategies for actual transmission over a radio link, the problems encountered in transmitting commands, and the impacts of commanding that leads to corruption by the links. The emphasis is on the payload here, while the emphasis was on the user interface in Chapter 4. The reader should note that the techniques described for configuring the payload are similar to the techniques used in configuring the user segment equipment. From a communications view, the two commanding environments are the same — one set simply has to travel a bit further.

9.2 OBJECTIVES

Commanding gives us the opportunity to make payloads adaptive to changes in the data acquisition environment. Our objectives for this chapter are to enable the reader to

- Describe three strategies for transmitting command data to a remote device over a radio link.
- Describe the fields found in a transmitted command and give reasons for their usage.
- Compute the probabilities for a command to be corrupted by a transmission error.
- Describe operational impacts from transmission artifacts and considerations in the command system's design.

These telecommand operations will also have a connection to telemetry operations.

9.3 COMMAND COMPOSITION

In any system, the command's function is to transmit information from a master device to a slave device. As noted above, the master device can be the user segment controlling the payload in a telecommand system or a computer controlling an attached device to configure the device for operation. As we saw in the *command dictionary* example for the balloon payload in Table 4.4, the user's command input format usually starts with a mnemonic for the function the payload performs and an associated datum. The datum can originate from the user interface or from a database that contains values for the desired state. The user interface system must package the command information for reliable and secure transmission from the user interface to the payload over the channel.

For example, Table 9.1 gives the command dictionary entries for the

Table 9.1: GPS Command Dictionary

Command Name	Data
GPON	Power-up GPS
GPOF	Power-down GPS
GPIN	Initialize GPS Unit
GPST	Report GPS unit status
GPCDGGA	Acquire GPS GGA message
GPCDZDA	Acquire GPS ZDA message
GPCDGLL	Acquire GPS GLL message
GPCDRMC	Acquire GPS RMC message
GPCDGSA	Acquire GPS GSA message
GPCDGSV	Acquire GPS GSV message
GPCDVTG	Acquire GPS VTG message

Global Positioning System (GPS) receiver used in the balloon payload in Chapter 4. Figure 9.2 shows the LabVIEW® Virtual Instrument (VI) that specifically requests one of the available GPS navigation messages based on the user's specification. The GPS receiver connects to the payload computer as a peripheral using the TIA-232 protocol. The reader should be able to anticipate that designers can develop similar commands in a user interface, which is responding to that data input from the user, and the interface sends the commands over a remote data link to accomplish the same task. The next step is to see how the designer can configure this basic format to allow transmission over a communications link, have the payload synchronize to the data, validate the data, and then execute the commands.

To generalize the command syntax from the example above, it is helpful to think of the transmitted command as being a short packet or frame of information. This packet may contain additional or information encoded differently from what the originator generated. Since the payload controller often sends commands individually, and not in a continuous stream as the payload sends telemetry-frame data, we must build synchronization and timing into the overall structure. Figure 9.3 illustrates several standard fields that typically define commands. These fields are independent of the modulation format and designers do not use some of the fields if the controller sends the commands in a baseband format. Regardless of the number of fields or bits, we refer to the whole

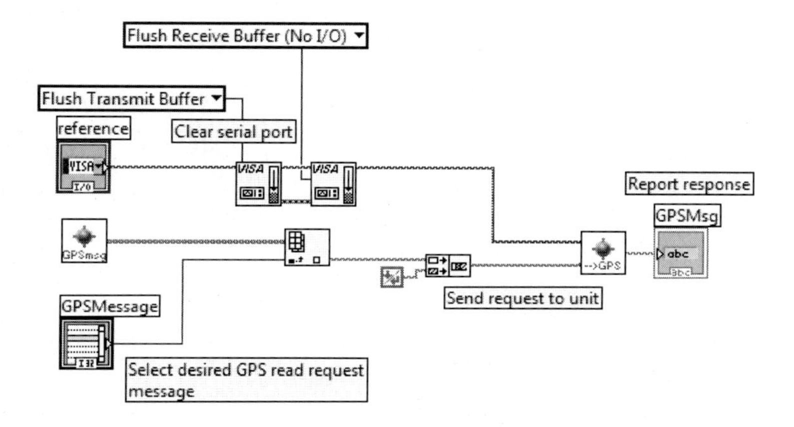

Figure 9.2: VI to send the text string to read a specific message from the GPS unit.

Initial Timing Mark	Synchronization Word	Internal Address	Command Data	Error Check	Final Timing Mark

Figure 9.3: Typical fields in a transmitted command word.

of the command as a *command word.* Typical Command (CMD) fields are as follows:

Initial and Final Timing Marks are bit fields used to cleanly delineate the start and stop of the command word, and, in the case of the initial timing marks, provide for locking the payload's electronics onto the command channel's data stream. These fields are important when the controller sends commands individually via radio or other non-baseband formats with time gaps between commands. Designers do not use these fields with baseband transmission techniques or when the command radio keeps the channel filled.

Word Synchronization is a unique pattern (usually not repeated in the command address or data) to mark the start of the command word's information content. This is the synchronization word as used in packet or frame telemetry and serves the same function. If the designer chooses packet commanding, the packet header serves this purpose. This field is used both in baseband and band-pass command transmission systems.

Internal Address is the payload address of the subsystem within the payload to take an action represented by the command data (similar to a computer memory or peripheral address). This is a repeat of the user interface information or a convenient mapping to fewer bits.

Command Data is the payload subsystem instruction to be performed (similar to a computer's Central Processing Unit (CPU) or peripheral instruction code). This also is a repeat of the user interface information or a convenient mapping to fewer bits.

Error Detection is a parity field or other error detection algorithm used to detect transmission or formatting errors.

All of these fields comprise the entire transmitted command word. The system designer must determine the method to map the user's command input like that found in Table 9.1 into the transmission format found in Figure 9.3. One simple solution is to directly place the text into the structure and transmit the command without modification. Figure 9.4 shows how the system designer can use a structured mapping between the user command input and a binary code to describe the relevant subsystem

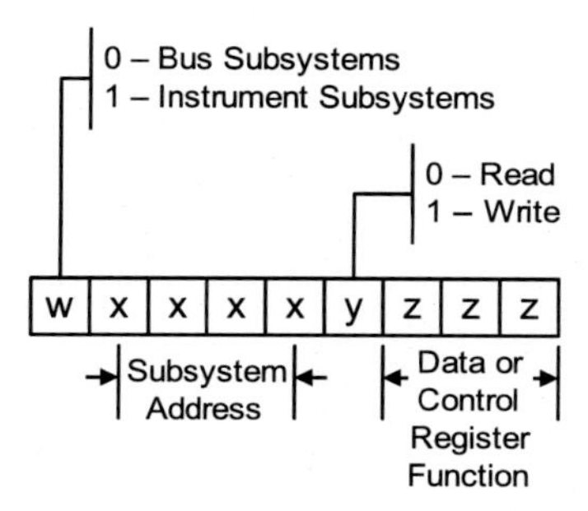

Figure 9.4: Example of a command code mapping to subsystems and functions.

address and command function. The initial bit determines if this command addresses the system's support systems or the instrument systems, the next several bits specify the internal address, the next bit specifies if the command is for a read or write function, and the remaining bits specify the specific data or control registers for the command operation. The user interface appends any specific data for the command after this bit pattern.

As we saw in Section 4.5.2, the payload has an on-board command processor to evaluate the fields and determine whether the command is valid. The command processor cannot determine whether the command is valid for the current payload state — only the payload operator can do that. The payload command processor only ensures that the command has syntactically valid fields. As we saw in the command dictionary in Table 4.4, the payload command processor has a means of reporting back the received command validity through either an error code and/or a copy of the received command so that the operator knows the status.

Figure 9.5 shows how we use a timing diagram to represent *command transmission timing* in a telecommand/telemetry system. It is similar to that used to describe computer-to-computer communications [Stal97]. The timing diagram proceeds from top to bottom with time increasing as one follows the diagram down. Command operations originate on the left-hand side of the diagram and payload operations are performed on the right. Commands flow from left to right and telemetry flows from right to left. The width of the command or telemetry data illustrates the time necessary to transmit that data. For example, a command of 128 bits transmitted at 1024 bits per second takes 125 ms to transmit.

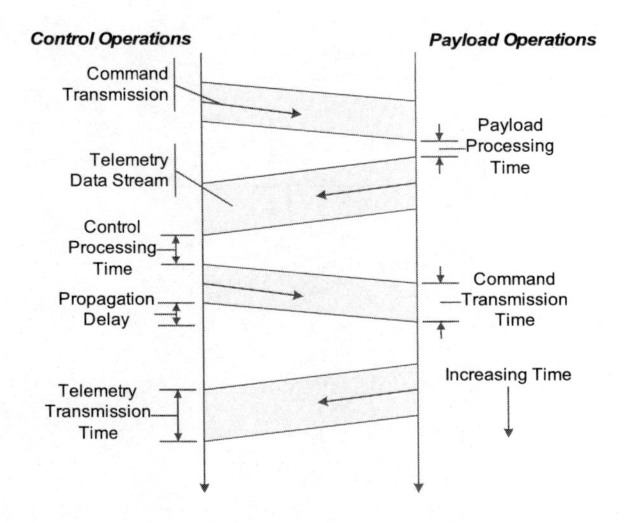

Figure 9.5: Generic command-timing diagram.

The slope of the transmission is proportional to the propagation rate and the total time from the user originating the command to the payload is the total propagation delay. Gaps between incoming and outgoing data are processing delays.

Figure 9.5 shows several command transmissions. We assume that the telemetry data echoes and/or provides a status indicator of the payload's resulting command action. Note: in this type of handshake, the operator cannot send more than one command per telemetry frame or packet or the operator may lose command processing information. In the following section, we use this diagram to illustrate telecommand transmission protocols.

9.4 COMMAND TRANSMISSION STRATEGIES

This section examines three command transmission strategies that the designer can use for a variety of payload classes. The three basic philosophies are

- Repeat and execute
- Command, verify, and then execute
- Execute immediately

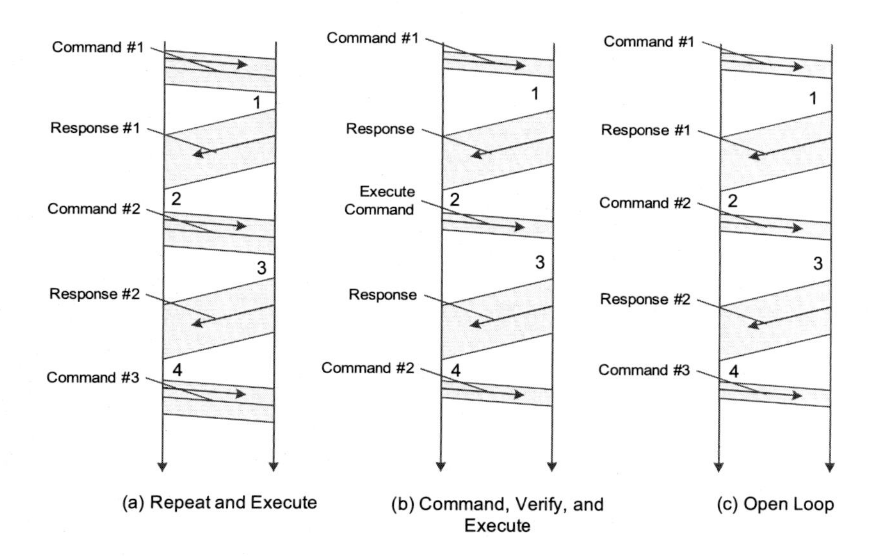

Figure 9.6: Timing diagram for Repeat-and-Execute, Command-Verify-Execute, and Open-Loop command modes.

Figure 9.6 is a data flow diagram for the three protocols based on the general timing diagram given in Figure 9.5. Factors such as command criticality, link delay, link quality, and the sophistication of the user segment and/or payload segment dictate the designer's method of choice.

In command protocols, we sometimes refer to burst commanding and continuous commanding. The *burst commanding* mode sends the commands with some arbitrary time gap between command groups. Designers commonly use burst commanding in low-command-rate systems. *Continuous commanding* means that command words arrive back-to-back without any intermediate gaps. Designers commonly use continuous commanding in high command-rate systems. We also have a combination where the user makes periodic configuration updates to the payload. Here, a command burst may be many commands long, with no time gap between commands within the burst and an indeterminate time gap between command bursts.

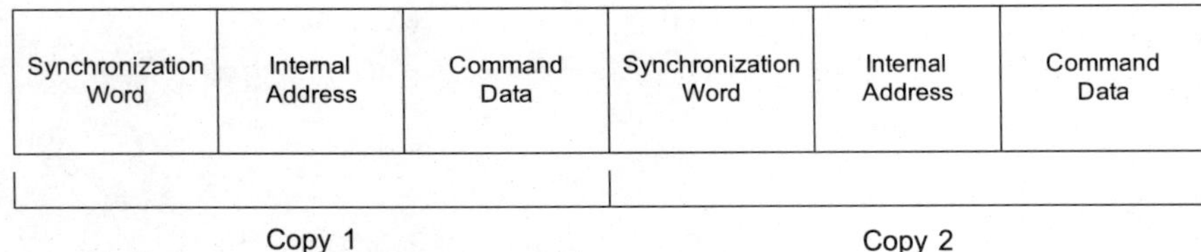

Figure 9.7: Repeat command format.

9.4.1 Repeat-and-Execute Command Protocol

Figure 9.6 illustrates the first case, the repeat-and-execute command protocol. The hardware required to support this strategy is relatively simple and inexpensive. Figure 9.7 shows that the control point transmits the command twice, without any gap in time (and shown as double the width of the other protocols in Figure 9.6). The payload receives and processes the double-width command at Time 1 in the diagram. If both copies of the command are identical when received, the payload executes the command. If copies differ, the payload does not process the command and returns an appropriate error code to the user.

The payload typically echoes the received commands with the telemetry data to show the user the received information; the user receives the notification at Time 2. As we show in the command dictionary, the command echo also has an error code with it to indicate valid or invalid command reception. In the command dictionary example, an error code of "0" signifies "no error," while a nonzero value indicates the error type the payload processor detected. The nonzero codes warn the user that corrective action is required. This "bad command" notification is also reported via the telemetry stream and arrives at Time 2. At Time 3, the payload receives either a repeat of the previous command or a new one, depending upon the status of the previous command. This process continues with subsequent commands.

The designer can choose repeat-and-execute command protocol for both burst and continuous commanding. In both cases, the system designer assumes that the transmission channel is mostly reliable; that is, the commands arrive intact most of the time and the user does not need to resend commands frequently. In a continuous commanding mode, the controller sends the commands in dual format stacked one after the other. High altitude balloon payloads tend to fall in this category. System designers also use the technique with long-propagation-delay command links where real-time feedback is not possible.

9.4.2 Verify-and-Execute Command Protocol

The verify-and-execute protocol is the second timing diagram that Figure 9.6 illustrates. This also is a relatively simple and inexpensive protocol. The user transmits a command and the payload decodes it at Time 1. The payload echoes the received command with the telemetry data and the user receives the information at Time 2. If the user agrees that the payload received the command correctly, the user then transmits a special execute command. If the execute command is correctly received at Time 3, the payload takes desired action. The user should also verify that the payload executed the command by inspecting the telemetry data, which the user subsequently receives at Time 4.

System designers typically use this command protocol in a burst command system with a low command rate due to having the user's involvement in the feedback loop. This protocol may be cumbersome when users are correcting payload anomalies or attempting to command two payload subsystems simultaneously. The latter is a problem because the payload's command decoder needs to "lock out" other commands from the command decoder until the payload receives the execute command. The system also needs a watchdog timer that clears the command decoder to receive a new command if the execute command is not received within a certain timing window. The payload must report flushing the received command buffer due to a timeout with the telemetry data so that the user knows the payload's state and takes corrective action.

9.4.3 Open-Loop Command Protocol

The open-loop command protocol is the third case that Figure 9.6 shows. This strategy is simple for the ground terminal and payload to realize and the simplest from a hardware and software view. However, it also is the most dangerous because it is an open-loop control system. The previous example was a closed loop system with built-in user verification. The repeat-and-execute protocol has strong error checking to prevent execution of damaged commands. In this open-loop protocol, the user transmits the command and the payload receives it at Time 1 when the payload performs a validity check. If the command passes the check, the payload immediately executes it. If the telecommand protocol does not include intrinsic error checking, this functional-only checking is not as rigorous as the previous modes because the data part of the command (only a number) may have a large valid range. A single bit error can permute a valid command into an invalid one and the payload's syntax-checking algorithm may not be able to detect it unless some form of explicit error detection encoding strategy is used. The payload should still echo the command and any associated error codes with the telemetry data at Time 2.

Obviously, the commands in this protocol need a strong validity and error check to ensure that the payload does not execute a damaged command. A simple parity bit may be all that is required on low-error-rate links. Designers may add explicit forward error correcting codes, like those discussed in Section 7.7.3, on less reliable links. This commanding technique is very useful in high command rate environments, in environments with long propagation delays, and when payload anomalies need correcting but cannot wait for the operator to check each transmitted command individually for correct reception.

9.4.4 Command Packaging Examples

This section discusses two command examples. The first is a LabVIEW® VI and the second is from a satellite command system.

Figure 9.8 illustrates the LabVIEW® command generation example for the balloon payload application. Figure 9.8(a) shows the user interface based on pull-down menus and the *Transmit* button to set the command state. Figure 9.8(b) shows the associated processing. When the user presses *Transmit*, the software builds the command from the user's menu selection and any associated data. The command synchronization word is *$$* and the command end delimiter is *. The software uses a string append operator to place the operator's command entry data between these markers. After assembling the command text string, the software transmits the command word to the payload over a TIA-232 serial link connected to the computer's COM port. The software also appends a two-byte checksum at the end of the command for error detection purposes (no error correction).

The user interface design allows the user to see the transmitted command word and the command echo from the telemetry data. The payload command processing software checks for valid command subsystems and functions. If the payload processing determines that it received a valid command word, the software executes the command. If the payload re-

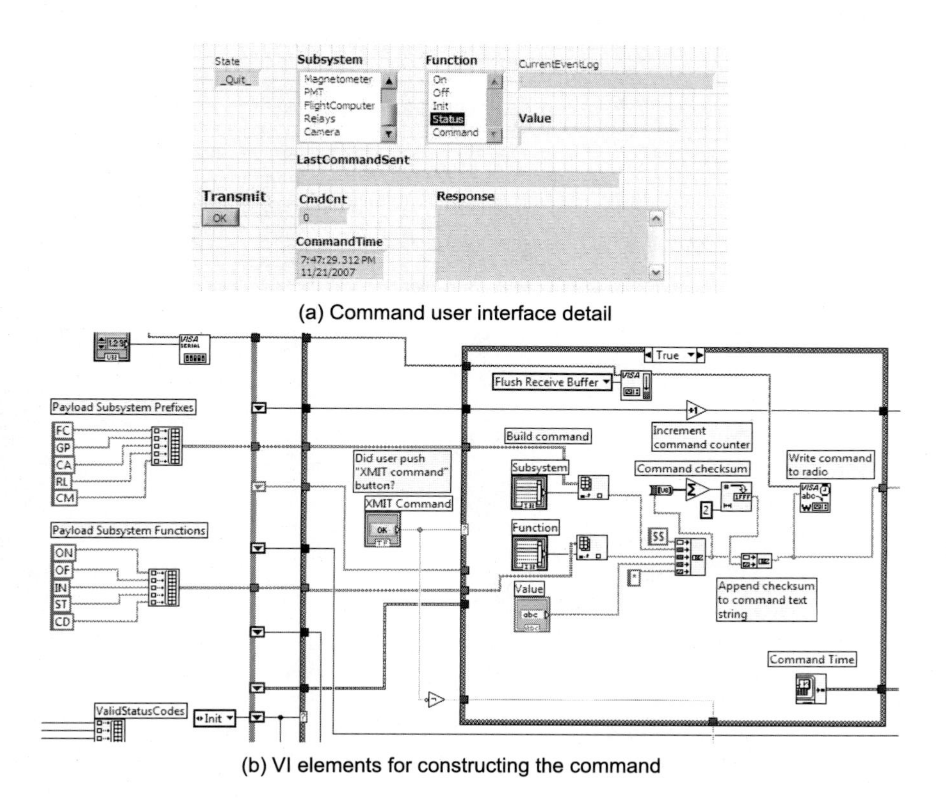

Figure 9.8: LabVIEW VI elements to support the balloon payload example from Chapter 4. Part (a) shows the detail for the user interface including the command menus and field for the payload's response. Part (b) shows the related code in the user interface for packaging the command and writing it to the radio transmitter and sending it to the payload.

ceived an invalid command word, the payload transmits the *command error code* with the telemetry data to warn the user. This example uses an open loop command strategy.

The second command example is from the satellite control industry and is based upon commercially available command processors [NASA85]. Figure 9.9 illustrates a 64-bit length command with an 8-bit Inter-Range Instrumentation Group (IRIG) synchronization code (see Table 6.3) to start the command. Addressing uses twelve bits. Five are reserved to identify the payload and the payload processor and the remaining seven

Synch Code	Payload & Command Processor Identifier	Internal Subsystem Address	Parity	Command Message	Post-amble
8 bits	5 bits	7 bits	1 bit	lsb 16 bits msb	3 bits

Figure 9.9: Example of an open-loop command word.

are for internal subsystem addresses. Error detection uses a single parity bit. The postamble brings the total command word to standard length.

9.5 OPERATIONAL CONSIDERATIONS

The system designer's command system choice is not a simple one. This section covers operational considerations that can influence the designer's choice. We explore synchronization issues, command verification, subsystem constraints, pre-event commanding, and command security.

9.5.1 Command Synchronization

Synchronizing the payload to the command data stream is analogous to the user segment synchronizing to the telemetry data stream. The expressions for computing the probabilities for command word transmission error, for missing a synchronization marker, and for a synchronization marker looking like a data field are the same as for the telemetry probabilities with two main differences. The command word is usually short compared to a telemetry frame or packet, and the payload must receive the command correctly the first time. This implies that the designer may need to prefix the first command after a transmission gap to enable the receiver to lock onto the command data stream.

Since initial bit synchronization across the radio channel is a random process, the designer is never certain when the receiver finally locks onto the data. To avoid ambiguity between the command word and the transmission's synchronization parts, the designer may require that each field within the command word not appear to have the same pattern as the synchronization marker, thereby helping to avoid that error case.

Figure 9.10 illustrates the continuous and burst command transmission process. The designer uses the continuous commanding shown in Figure 9.10(a) when sending a group of commands without any gaps between command words. The burst commanding of Figure 9.10(b) ap-

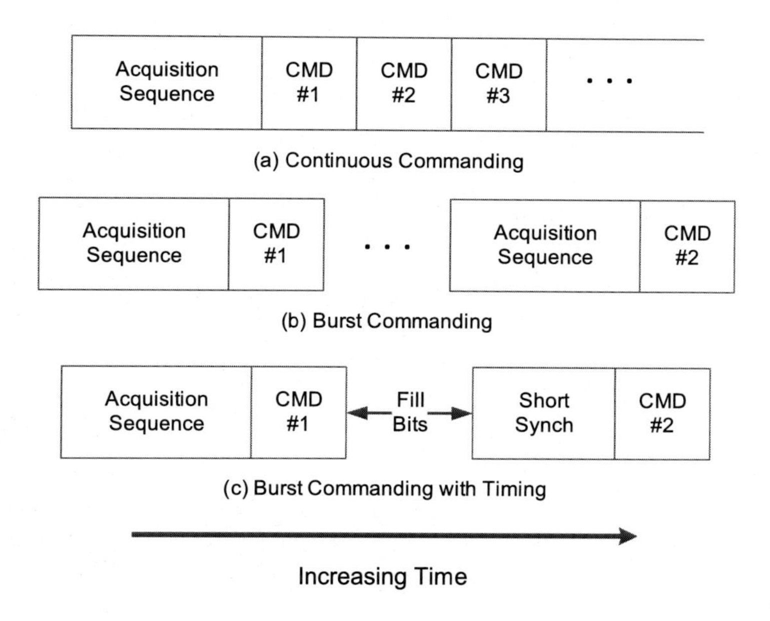

Figure 9.10: Command synchronization sequence examples.

plies when sending each command as an isolated transmission. The burst commanding with timing information sequence occurs when sending commands as isolated data and the transmitter keeps the link active between command words. In all cases, the figure's acquisition sequence part is the bit pattern used to lock the payload's receiver to the incoming command data stream.

Individual command lengths typically are 40 to 64 bits. However, if the user performs commanding in bursts of a few commands at a time, the payload's command receiver has difficulty locking onto the incoming command stream and decoding the command data before the burst is over. Reliably locking onto a bit stream from an initial "noise" input may take over 100 bits, depending upon the link signal-to-noise ratio. For this reason, both burst and continuous commanding protocols need an acquisition sequence [Buro82].

For continuous commanding, the user interface sends this acquisition burst once each time the command link is established and it does not represent a major overhead. For burst commanding, the acquisition sequence may be of approximately the same length as the command burst and represents a major cost. While the exact details of establishing the command synchronization are functions of the command processor hardware, one method that designers use to establish a reliable link is to send an acquisition sequence of several no-op commands before transmitting the desired command sequence. Alternatively, the designer can use the alternating 0s and 1s ($0x55$ or $0xAA$) as in Section 7.6.2.

Figure 9.10(c) shows a strategy for maintaining the synchronization. The transmitter keeps the receiver in the payload synchronized by sending clock pulses (alternating 0/1 pattern) between the actual command word transmissions. At first thought, this technique may seem to waste transmission power. However, the improved synchronization is worth the effort and makes the link more reliable.

9.5.2 Command Verification

The system designer must ensure command validity prior to the payload's execution. If the command is not correct, the payload may execute some command that is inappropriate for its current state, may be dangerous to its environment, or may even cause a catastrophic failure in the payload. Both transmission errors and operations errors cause command errors. Proper training and user-segment software syntax checking minimize operational errors. Transmission errors are unavoidable, at least at some level. Therefore, an error checking strategy is important to determine if the payload received the command correctly.

The previous open-loop command example showed using a single parity bit to check for errors. A single parity bit detects all odd-number bit errors but no even-number bit errors. The checksum used in the balloon payload example detects more than a single error but does not correct those errors. Designers select a more powerful forward error correcting code for error checking that is more extensive and even provides correction at the cost of hardware complexity and additional transmission overhead. For example, the system designer can apply the Hamming error correcting code discussed earlier to command data as well as telemetry data. The designer must determine both the expected chance that a transmission error occurs and the expected chance for the error detection strategy to fail. For example, in the single-parity-bit case, the system designer must determine the probability of a single-bit error versus a two-bit error anywhere in the command word and the probability that two bits are in error being insignificant in the channel.

A second component to command verification is checking the individual *command field validity*. In general, there is a correctness check of the

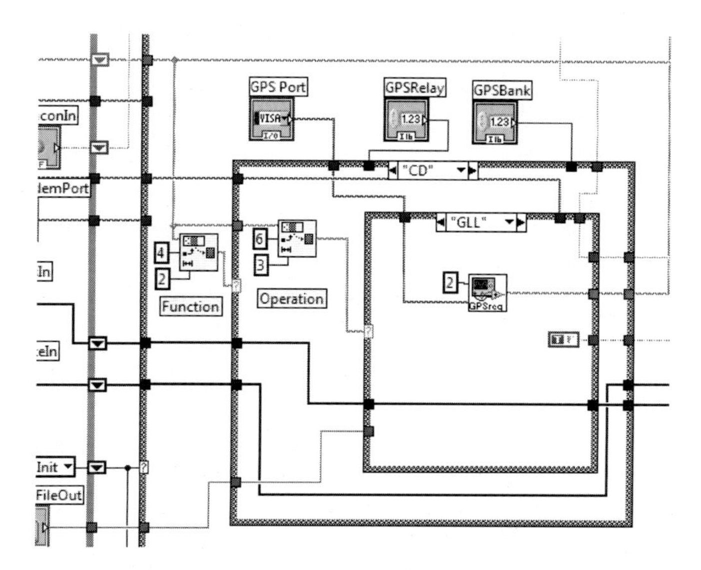

Figure 9.11: A section of the LabVIEW VI to parse command input in the balloon payload example.

initial synchronization code because that tells the payload receiver where to begin processing the command word data. Additional field checking depends on the payload's command-processor complexity level. Figure 9.11 shows a section of the payload's command processing software that checks the validity of the subfields by checking for correct functions and options; this is after the payload software has already checked the subsystem identifier. A command processor cannot verify that the transmission process has not transformed one correct command into a different, yet syntactically valid, different command. This is why system designers use checksums, parity bits, repeats, and operators in the loop in the overall command strategy.

The designer's final form of validity checking is to ensure that the command originates with a *valid user*. This is a concern where the payload receiver is open to transmissions from multiple operator stations. This can occur when the system has multiple operators minding individual payload subsystems and all operators can issue commands to the payload.

9.5.3 Subsystem Command Rates

When the operator commands a payload in continuous command mode, commands to the same payload subsystem may come too rapidly for that subsystem to process and execute them. This is a function of the subsystem design and the manufacturer should specify the maximum command rate. For example, stepping motors have maximum step rates and each command to the stepping motor causes it to advance by one step. The command transmission rate must be consistent with the maximum stepping motor advancement rate.

When this is an issue in the payload, the designer must have a database in the user interface that contains the maximum commanding rate for the payload, and the command processor must take this into account when constructing the command data stream. The user segment software then constructs a command history table for the command data stream. If a timing violation appears imminent, the user interface may insert no-op commands into the command stream to prevent a violation of the subsystem command rate. This adds complexity to the user segment, but this concern arises only in open-loop command situations.

9.5.4 Pre-event Commanding

During payload operations, there are times when the link between the user and the payload is not possible; for example, the payload may not have an active link due to temporarily being out of position, the communications link may have failed for some environmental reason, or the payload may have undergone a fault. For event sequences when it is impractical to command from the user segment, the designer may pre-load commands into the payload to enable the payload to sequence itself.

A prime example of this was the 2015 *New Horizons* spacecraft encounter with Pluto where the link had transmission delays of several hours between the Earth and the spacecraft. The spacecraft controllers cannot actively manipulate the spacecraft's experiments in real time during planetary encounters. Therefore, the command sequences were loaded and verified before the encounters. *New Horizons* flew on autopilot using these command sequences during the encounter. This required that the spacecraft controllers extensively plan the payload sequencing and perform all transmissions and checkouts before the event.

Another example is the payload performing a Built-In Self-Test (BIST) upon power-up or after a power failure. In this case, the designer placed

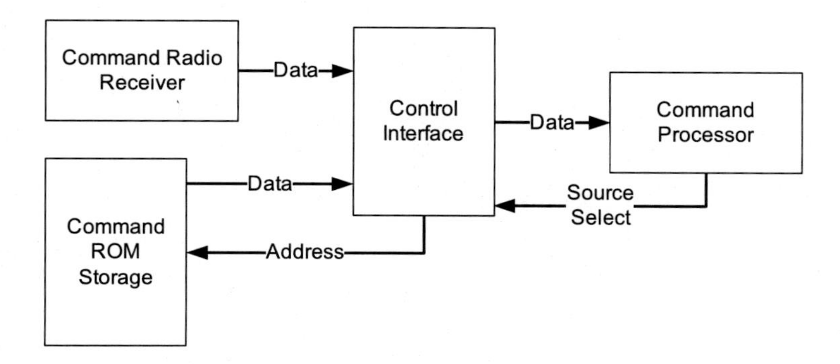

Figure 9.12: Payload configuration allowing both externally received commands and commands stored in ROM to be processed.

the command sequence in memory and the payload executes it as if it received the commands directly from the user. Figure 9.12 uses a block diagram to show a payload configuration for this design. Both a radio and an on-board Read Only Memory (ROM) can be used as the command source. The payload's command processor selects the command source based on the payload's state.

9.5.5 Command Counters

Figure 9.13 shows command data echoing in the telemetry stream. Designers frequently include a *command counter* that tracks how many valid commands the command processor has received in the telemetry stream, along with the command echo. The LabVIEW® interface example in Figure 9.8(a) shows both of these features with the *LastCommandSent* and *CmdCnt* fields in the display. If the history of the commands sent to the payload is a major concern, the user segment keeps a matching command counter to verify that the payload has processed the right number of commands.

Two problem areas occur in this accounting of the command history: lost commands and disagreements between the control station and payload. Lost commands usually occur when the operator believes that the control station transmitted the correct command syntax for each command. If the acquisition process takes too long in the payload, the command receiver does not lock onto the command data stream before the acquisition sequence ends. In that case, the command processor drops commands because it does not find a complete command after the syn-

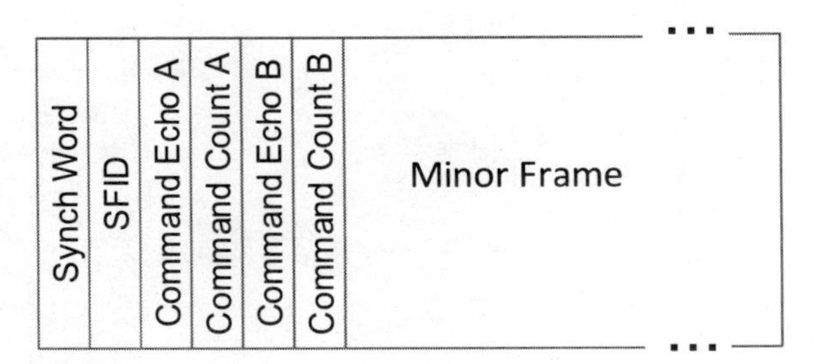

Figure 9.13: Command echo and placement of the command counter in the telemetry minor frame.

chronization sequence. The command processor also drops commands in the event of a transmission bit error so that commands that left the user segment as valid commands are no longer valid upon reception. In some payload processors, not only is the once-valid command lost, but the processor also loses several subsequent commands, while it resynchronizes. This causes a difference in the payload and the ground station command counters. The controller uses this difference to diagnose transmission problems.

Disagreement occurs due to timing and buffering issues with the payload command and telemetry processing design. Consider the case where the payload echoes the latest command once per minor frame. During continuous commanding, commands are processed correctly, as evidenced by a correctly incrementing command counter, but not every command is echoed in the telemetry data. This occurs if commands overfill the payload's command buffer due to their arriving faster than the telemetry processor empties the buffer. This can also occur when the command and telemetry rates are equal (the command rate is equivalent to once per minor frame) and slight timing differences cause commands to appear to become lost due simply to sampling problems. Due to the slight timing differences between the base station and the payload, the command buffer in the payload eventually needs to hold two commands when it can only hold one. The hardware's timing slip causes the dropped command. The command counter verifies this, but it can confuse the operator.

9.5.6 Command Files

So far, we have treated command words as single entities. However, the designer may also package sets of commands into related *command files* (macros) and transmit them as a single entity. This is especially useful when transmitting command sets over commercial networks such as the Internet using services such as File Transfer Protocol (ftp). Command files are useful for tasks such as

- Updating payload constants
- Updating payload software
- Moving structures such as antennas
- Recovering from errors
- Loading operational sequences

As we saw earlier with the balloon payload example, Table 4.3 illustrated such a sequence file for initializing the payload. Each entry contained a time to execute the command, and the actual command value.

9.5.7 Command Error Rates

The payload must validate each command's reception. Usually, the first check is for command transmission errors. In this section, we examine the command transmission's error probability, which is a function of the channel's bit error probability, p. This is the channel's Bit Error Rate (BER). The error probability is also a function of the command block length, N. The command block is either a single command word or a command file.

9.5.7.1 Command Reception Error Probability

To compute the probability that a command is received correctly, we first assume that bit errors are due to random noise in the channel — the standard Additive White Gaussian Noise (AWGN) assumption. In the AWGN case, the probability that the command of length N bits is received correctly, P_0, when the channel BER is p, with $p < 1$, is

$$P_0 = (1 - p)^N \tag{9.1}$$

Related to this is the probability that the command is received with some form of error, P_e. This means that one, two, or more transmission

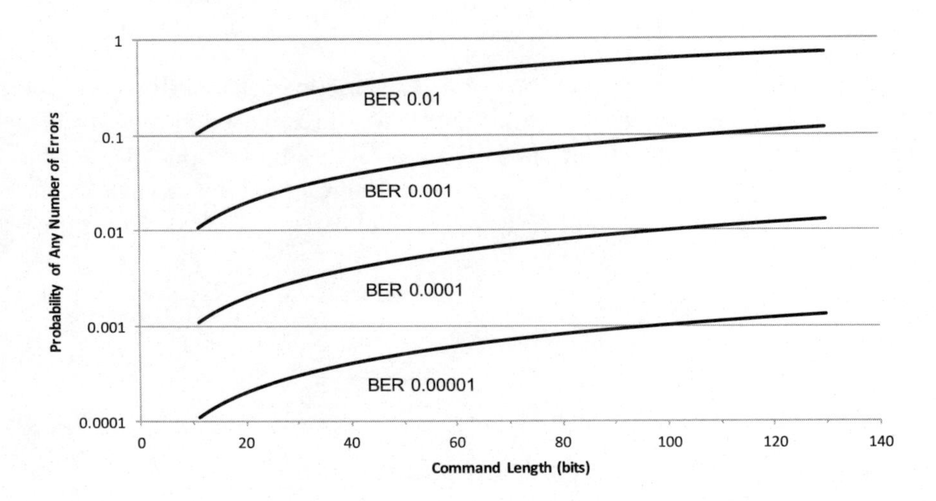

Figure 9.14: Command error probability as a function of channel BER and command length.

errors may have occurred. This error probability is the complement of Equation (9.1) and it is

$$P_e = 1 - P_0 = 1 - (1 - p)^N \tag{9.2}$$

Figure 9.14 illustrates the probability of receiving commands in error for channel BERs of 0.01, 0.001, 0.0001, and 0.000 01, and various command lengths. When we say that the command is in error, we mean that any number of bits in the command block is in error. That is, any error case except for the case of zero errors. As expected, channels with high error rates have a high probability of receiving a command in error. Additionally, as the command block length increases, the probability of an error increases as well.

Next, we must determine a reasonable estimate for the number of transmission errors we expect in the command block because this tells us how powerful the error detection method should be. For example, there is no need to detect two bit errors in a command if the probability of a two-bit error occurrence is once per year and the payload accepts commands only for one week. For a command word of length N, the probability, P_k, of k or more errors with a channel bit error rate, p, is computed using the same relationship used for computing the probabilities of random errors for any block of data and is given by

$$P_k = \sum_{j=k}^{N} \binom{N}{j} p^j (1-p)^{N-j} \tag{9.3}$$

Quantity () is the binomial coefficient defined as

$$\binom{N}{j} = \frac{N!}{j!(N-j)!} = \frac{N(N-1)(N-2)\cdots(N-j+1)}{j!} \tag{9.4}$$

(The right-hand form is often more efficient for computational purposes). We use these results to compute the probability of reception error for several strategies. If the assumption of AWGN is not valid, then these relationships no longer hold. Usually, AWGN is violated during times of burst errors due to atmospheric or other transmission and environmental disturbances.

9.5.7.2 Parity Error Detection Strategies

We determine the probability of all detectable errors and all nondetectable errors using the relationships above for the case of using a single parity bit in the command block as the error detection method. Since parity error detection finds all odd number errors, the probability of receiving a detectable error is

$$P_{1,3,5,\ldots} = \sum_{j=1,3,5,\ldots}^{N} \binom{N}{j} p^j (1-p)^{N-j} \tag{9.5}$$

For most channels, the first term approximates that summation. Similarly, nondetectable errors come from all even-numbered error events. The probability of receiving a nondetectable error is

$$P_{2,4,6,\ldots} = \sum_{j=2,4,6,\ldots}^{N} \binom{N}{j} p^j (1-p)^{N-j} \tag{9.6}$$

Figure 9.15 illustrates the ratio of the probabilities in Equation (9.5) to Equation (9.6) for various command lengths. For channels with low BERs, the ratio of detectable to nondetectable errors exceeds a factor of 1000. For channels with high BERs, the ratio is approximately 10 and this ratio is usually unacceptable. After determining this ratio, the system designer determines whether an error detection scheme more powerful than a single parity error bit is necessary.

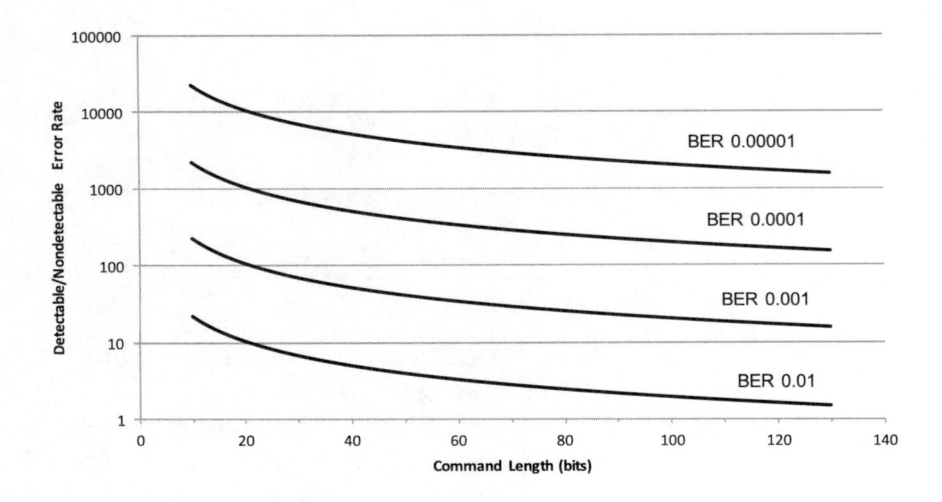

Figure 9.15: Ratio of detectable to nondetectable parity errors for command blocks as a function of channel bit error rate.

9.5.7.3 Repeat Command Strategies

For repeat commanding, the determination of the probability executing a bad command is a bit more complicated. Consider the case in which the full command consists of a full repeat of the desired command, as Figure 9.6 showed. The command processor compares each matching half and, if they are identical, the payload assumes command is valid. Both halves are identical if the command processor receives the command correctly or the two halves underwent identical transmission errors. In this latter case, the command has undergone a nondetectable error. Therefore, the probability that the payload receives what it considers to be a good command, P_g, and its execution without application of any form of syntax checking is given by

$$P_g = P_{0e} + P_{nde} \tag{9.7}$$

Here, P_{0e} is the probability of receiving the command with zero transmission errors and P_{nde} is the probability of an undetected error. We compute these probabilities using the relationships above as a function of the channel bit error rate, p, and the length of the command half, m, in bits (that is, the whole length of the command is $2m$ bits). The probability of zero errors in both halves of the command is the probability that all bits were received correctly, given by

$$P_{0e} = (1 - p)^{2m} \tag{9.8}$$

Synchronization Mark (16 bits)
Subsystem Address (8 bits)
Command Number (8 bits)
Command Data (16 bits)
Timing Mark (4 bits)
Subsystem Address (8 bits)
Command Number (8 bits)
Command Data (16 bits)

Figure 9.16: Example of a modified repeat command structure without repeating the synchronization word.

We compute the nondetectable error probability in the command word with the sum over m bit errors as

$$P_{nde} = P_{1nde} + P_{2nde} + P_{3nde} + \cdots + P_{mnde} \tag{9.9}$$

P_{inde} is the probability that a number i nondetectable errors occurred. We approximate the total nondetectable error probability by the first term only. This is

$$P_{1nde} = \binom{m}{1} p(1-p)^{m-1} p(1-p)^{m-1} \tag{9.10}$$

Similarly, we compute the higher number of error terms above P_{1nde}, if needed.

Figure 9.16 gives a more realistic repeat command structure. The structure repeats only the address and command data, while the synchronization field is not. Using the lengths of the fields in Figure 9.16 as illustrative of the process, the probability that the entire 84-bit long command is received correctly when the channel BER is p is

$$P_0 = (1-p)^{84} \tag{9.11}$$

Assuming that the receiver explicitly checks the synchronizing and timing markers, the probability of a single, nondetectable error is

$$P_{1nde} = (1-p)^{20} \binom{32}{1} \left[p(1-p)^{32-1}\right]^2 \qquad (9.12)$$

This result is similar to the earlier one. We leave the differences as an exercise at the end of the chapter.

9.5.8 Command Security

The security concerns we examined in Section 4.6 apply to command links as well as to telemetry links. When the operator sends commands across a closed link that is part of the overall telecommand and telemetry system, command security may not be an issue if the communications links are well secured from outside entities. If the operator is sending commands over an open, public link, then the designers will frequently include a form of command encryption to prevent others from both intercepting the command data and sending false commands to the payload. In addition to using encryption to mask the command contents, designers can also use it as part of the means to prevent external parties from "spoofing" the payload.

Designers of secure telecommand links must consider the following attributes in the design of the overall system, the data link, and the selection of the data transmission protocols [CCSDS3500]:

Access Controls are the system's formal process(es) for granting access to system resources, processes, or data only to authorized users, programs, processes, or other systems. Access controls may be resource level-specific to grant users only specific access rights, e.g., read-only data access.

Authentication is the system's ability to verify the identity of a user, process, device, or other entity before allowing access to resources or data in the system. Authentication also ensures that the telecommand data actually originates from the claimed source and that the telecommand data has not been duplicated or modified.

Availability is the system's ability to ensure that system resources and data are usable when needed, including the assurance that the system will not be taken down by denial-of-service attacks or other malicious access attempts. This also implies that the system managers have proper system back-ups and redundancy to preserve operations.

Confidentiality is the system's ability to ensure that the telecommand information is not available or disclosed to unauthorized entities or processes; the telecommand data is provided only to those entities or processes that are authorized to access it.

Data Integrity is the system's ability to ensure that the transmitted telecommand data is unchanged and that the system has the ability to detect any unauthorized modifications, including data erasure.

Even systems with minimal security concerns usually need to consider telecommand authentication, data integrity, and confidentiality attributes. If the designers encrypt the commands, they will also need to develop a protocol to manage the encryption keys during operations. If the encrypted commands are sent over a link with a relatively high error rate, then the designer may also need to include error correcting codes to protect the encrypted data.

9.6 PACKET COMMAND SYSTEMS

We saw in Section 6.7 how packet communications standards are used to provide the data transport for the telemetry side of the overall system. We discussed the networking communications protocols in Section 6.10 and these protocols are applicable to packet commands as well. It should come as no surprise if the designer also configures the system to use packet standards for the command link. The two behave similarly. The major differences are as follows:

- The command packets usually are smaller than the telemetry packets.
- The command packet rate usually is lower than the telemetry packet rate.
- The telemetry packets appear to be more continuous; the command packets appear as bursts of packets.
- The command link is more concerned with delivery order than is the telemetry link.

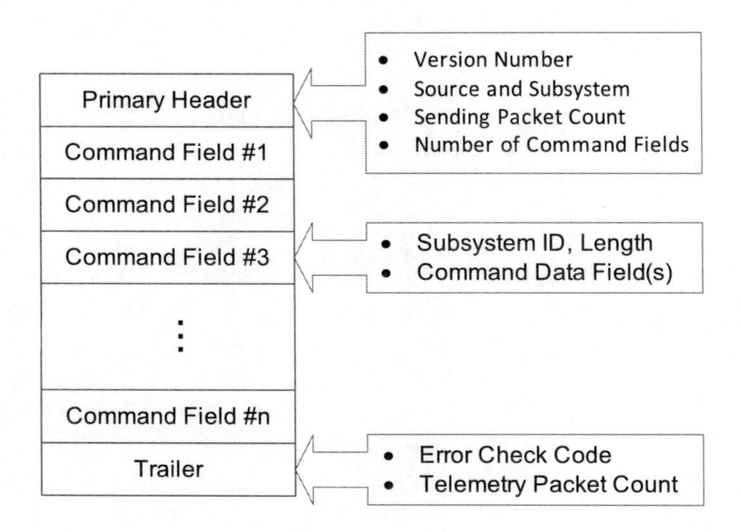

Figure 9.17: Example of a command file arranged in a packet format.

Figure 9.17 illustrates a command packet intended to complement a corresponding telemetry packet, hence the packet counters for the command and telemetry links. The command fields are generic and assume that the individual subsystems know the format of the data fields. This format fits easily into a standard Transmission Control Protocol/Internet Protocol (TCP/IP) or User Datagram Protocol/Internet Protocol (UDP/IP) format. Packet processors use simple counters to delineate the command field lengths and tell the payload how many commands are in each packet. The system designer chooses either the connection-oriented mode or the connectionless mode depending upon metrics for link establishment time, link latency, command frequency, and command structure (burst commands, continuous commands, or command files).

The move to a packet system from a command word system does not mean that the command functions or processing has radically changed. Using packets still requires that the command receiver lock onto the packet structure as it did for the command word examples. This includes sending an acquisition sequence if the command source does not occupy the communications channel fully with packet transfers or if the transmitter does not send a clocking signal from the ground station to the payload between packet transmissions. We proceed as before in computing the packet's error probability, missing the command synchronization marker, or mistaking the command synchronization marker with valid data, but with the appropriate packet parameters. The Consultative Committee for Space Data Systems (CCSDS) standard provides an example of a fully complementary command and telemetry packet system [CCSDS1300].

9.7 REFERENCES

[Buro82] N. A. Burow and M. K. Tam. "Command System." In: *Deep Space Telecommunications Systems Engineering*. Ed. by J. H. Yuen. JPL Publication 82-76. Pasadena, CA: Jet Propulsion Laboratory, California Institute of Technology, National Aeronautics, and Space Administration, 1982. Chap. 6, pp. 343–382.

[CCSDS1300] *Overview of Space Communications Protocols*. CCSDS 130.0-G-3. Consultative Committee for Space Data Systems. Washington, D.C., July 2014. URL: `https://public.ccsds.org/Pubs/130x0g3.pdf`.

[CCSDS3500] *The Application of CCSDS Protocols to Secure Systems*. CCSDS 350.0-G-2. Consultative Committee for Space Data Systems. Washington, D.C., Jan. 2006. URL: `https://public.ccsds.org/Pubs/350x0g2.pdf`.

[NASA85] *Space Network TDRSS Data Book*. TDRSS Program Division, Code TX. Washington, D.C.: National Aeronautics and Space Administration, 1985.

[Stal97] W. Stallings. *Data and Computer Communications*. 5th Edition. Upper Saddle River, NJ: Prentice Hall, 1997. ISBN: 0-02-415425-3.

9.8 PROBLEMS

1. What is a potential advantage of using the binary mapping that is illustrated in Figure 9.4 over using a direct transmission of the textual command format illustrated in Table 9.1 when transmitting the command over a channel?

2. If a link has an error probability of 10^{-5}, what is the probability that the command is received in an error-free state under the strategies of (a) open-loop commanding with a command length of 48 bits, (b) dual commands sent back-to-back with each command is 20 bits long. What is the probability of a nondetected error for these command strategies?

3. If a command uses a single parity bit to detect errors, what is the probability of receiving a command with (a) zero errors, (b) one error, (c) two errors, and (d) more than two errors. You are to assume that the parity is computed over 32 bits. Consider both cases of the link error probability is 10^{-3} and 10^{-5}.

4. Consider a 32-bit data block transmitted over a channel with a bit error rate of 0.0001. Make each of the following estimates:

   (a) What is the probability that the data block is received with no transmission errors?

   (b) If a single parity bit is included in the data block and used as the error detection strategy, what is a reasonable estimate for ratio of detectable to nondetectable transmission errors?

   (c) If this data block were to be used in an open-loop command protocol (1) indicate your pick for an error detection methodology and why (it does not have to be the same as question (b) but it could be), (2) sketch the timing for transmitting the command when there is a transmission error, and (3) indicate in the timing diagram when the command is executed and when the operator knows that it is executed.

5. Two command links have respective propagation delays of 0.125 s and 0.000 125 s. Compute an expected time to execute a command for both command-verify-and-execute commanding and execute-immediately commanding when the system parameters are as follows:

   (a) The command word length is 64 bits

   (b) The command channel transmits at 38.4 kbps

   (c) Commands require 64 bits for synchronization (prior to 64-bit command word)

   (d) The payload processor takes 0.5 s to place a received command in the outgoing telemetry stream

   (e) The telemetry frames are 2048 bits long and are sent at 115.2 kbps

   (f) Ground computers require 0.001 s to generate the command and 0.005 s for any subsequent processing steps

6. Develop an expression to compute P_{2nde} when the full command is repeated in the repeat commanding protocol and compare the result with the computation for P_{1nde}. Can you justify the assertion that you can ignore P_{2nde}?

7. Derive the expression for the cases of two and three nondetectable errors for the command protocol illustrated in Figure 9.16 .

8. Develop a hardware circuit to verify the repeat command structure given in Figure 9.16.

9. Design challenge problem: use a product such as LabVIEW® to develop a basic command system to use repeat commanding, validate the received command, and inform the user of the command state.

10. How might you apply digital signatures to command verification? What are the lengths of commercially available digital signatures? How much processing do you need to validate the signature? How susceptible is the digital signature to transmission errors?

11. If a (7, 4) Hamming code is used to encode a 64-bit command data word for a total of 128 bits and the channel BER for uncoded data is 10^{-3}, how much does the application of the FEC affect the performance? Suppose the channel BER is 10^{-6}.

III

DATA TRANSMISSION TECHNIQUES

Here we look at the techniques used in transmitting telemetry and telecommand data. While we can use the Internet and other network-based communications systems for data transmission, we concentrate on radio-based systems because the majority of traditional telemetry systems use radio systems. Other systems have better defined channels and standards to support the transmission technology. The standards often use the same transmission techniques as used in radio channels but with protocol-defined methods. In Part III, we investigate the following topics:

Modulation Techniques: Techniques for analog and digital modulation and demodulation for sending data, and introduction to issues related to Radio Frequency (RF) spectrum planning

Microwave Transmission: An introduction to RF transmission on the microwave bands, components used for data transmission and reception, and propagation effects on the channel that affect the quality of the data transmission

The topics also cover the parameters designers use in the system design and system performance measures. The relationships between these parameters are illustrated in the following concept map. The link characterization investigates noise found on the RF link. This section includes a radio-propagation-link analysis tool to assist the link designer in determining if the system components meet the desired performance specifications.

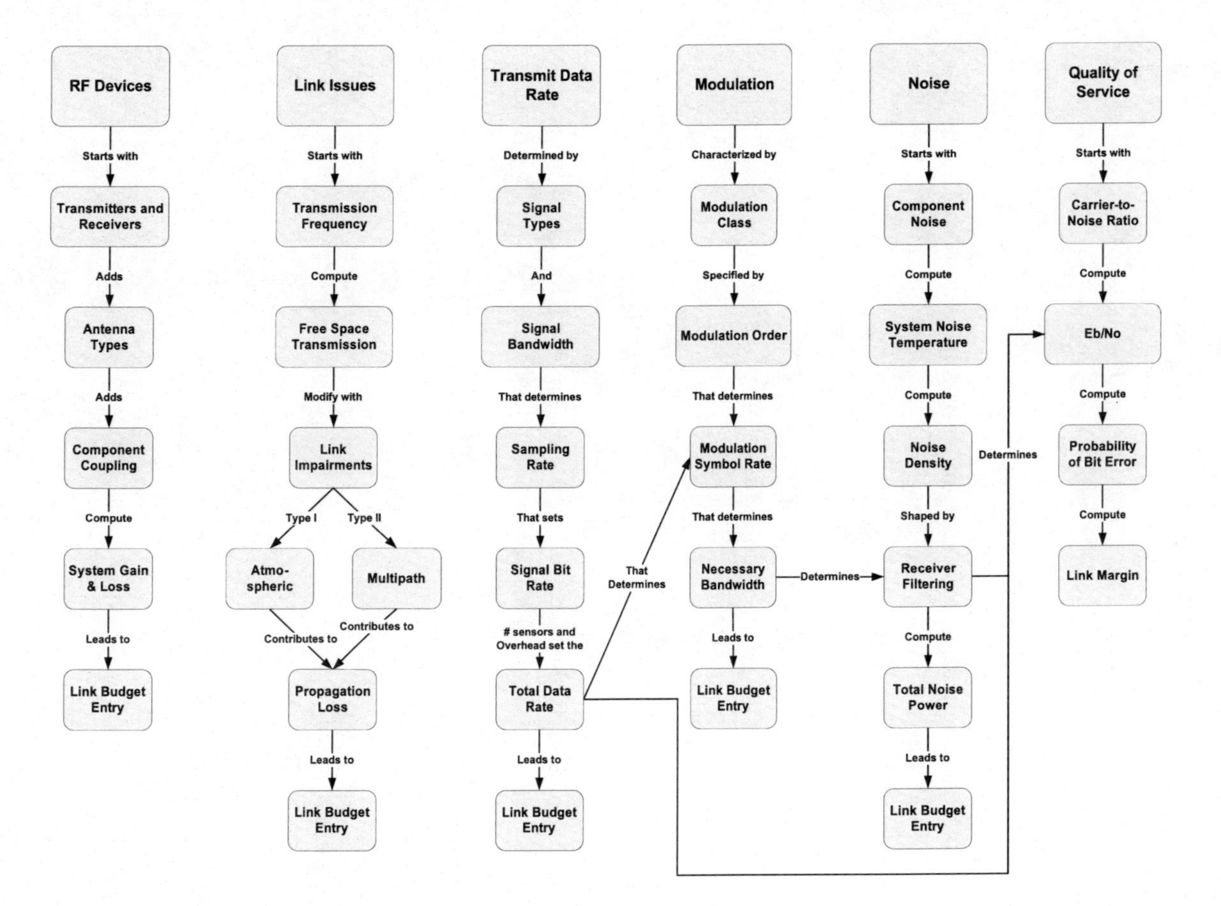

The system's data transmission architecture and how the architecture supports the overall link analysis.

CHAPTER 10

MODULATION TECHNIQUES

10.1 INTRODUCTION

Both the telemetry and the telecommand link require channels to carry their respective data transmissions. It is typical for these links to be on entirely separate channels with different data rates to match the needs of the respective sources. In this section, we assume that the system designer has opted to use a radio system to sustain the data transmission needs.

The highlighted blocks of Figure 10.1 illustrate that the payload is not the only active component involved in transmission. The user segment is also active both as a data source and data sink. We are interested in transmitting the data via a standard carrier signal. At the physical layer, the radio channel does not distinguish between telemetry and telecommand data. At that level, we are simply dealing with data waveforms that must be transmitted.

Modulation is the process of modifying the *carrier* by some well-defined function of the transmitted data. The simplest radio carrier signal is a sinusoidal wave whose amplitude, frequency, phase, or some combination of these quantities, varies in time due to the data being transmitted. The mathematical form for the general Radio Frequency (RF) carrier is a voltage level as a function of time, $s(t)$, represented by

$$s(t) = A(t)\cos\left(\omega(t)t + \theta(t)\right) \tag{10.1}$$

The parameters, which can be modified by the data, are the amplitude, $A(t)$, the frequency, $\omega(t)$, and/or the phase, $\theta(t)$. Generally, the

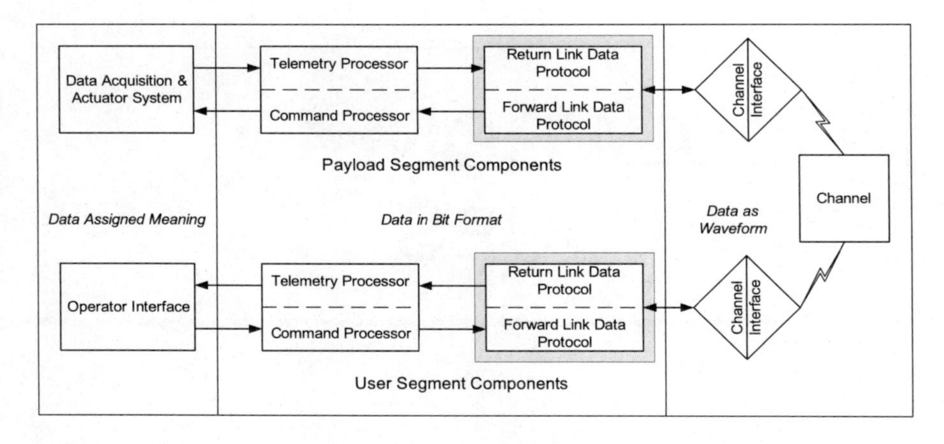

Figure 10.1: The highlighted blocks correspond with the modulation and demodulation component locations in the overall telemetry and command system.

modulator holds at least one of these parameters constant and the ones that vary with the data are the only ones shown as functions of time. However, by the time the receiver acquires the signal, the presumed constant parameter may be time variable due to the channel conditions and the receiver must track the variation.

Engineers have devised clever names for the modulation types used in radio systems. If the transmitter modifies the carrier's amplitude in response to the data, the modulator produces Amplitude Modulation (AM). Likewise, if the data changes the modulator's frequency, then the modulator produces Frequency Modulation (FM). Phase Modulation (PM) occurs when the data modifies the carrier's phase. A combination example is the data modifying both the carrier's amplitude and phase to produce Quadrature Amplitude Modulation (QAM).

The most commonly used modulation formats for transmitting telemetry and telecommand data are FM and PM. When the modulator sends analog data using FM, the modulation format is not qualified further. However, when a FM modulator sends digital data, we qualify the modulation and call it Frequency Shift Keying (FSK) because the carrier frequency only takes on discrete values. Similarly, when a PM modulator sends digital data, the process is called Phase Shift Keying (PSK) because the carrier phase takes on only discrete values. This chapter covers FM and PM systems for transmitting analog data and PSK and FSK systems for transmitting digital data. We also examine the trade-offs involved in choosing a modulation approach.

10.2 OBJECTIVES

This chapter's contents provide the basics of analog and digital transmission techniques that designers commonly use in telemetry and telecommand systems. Our objectives for this chapter are to enable the reader to perform the following tasks:

- Mathematically describe the functional forms of the output signals for analog frequency modulation, analog phase modulation, digital phase modulation, and digital frequency modulation.
- Estimate the required bandwidth for the modulation format.
- Estimate the probability of a bit error for digital modulation with a given carrier-to-noise ratio.
- Draw typical modulator and demodulator hardware configurations for the different formats.
- Estimate the required bandwidth necessary to transmit digital data with different formats.
- Explain how the radio frequency spectrum is divided in a general sense by International Telecommunications Union (ITU) region and usage.
- Estimate the allowed bandwidth for data transmission under National Telecommunications and Information Administration (NTIA) and Inter-Range Instrumentation Group (IRIG) standards.

10.3 ANALOG MODULATION

We start by considering three possible analog modulation forms when building a telemetry and telecommand system: AM, FM, and PM. In practice, designers generally do not use analog AM-based telemetering systems because FM and PM systems have superior performance in transmitting data over an analog carrier. However, understanding AM gives insight into how signals are represented in the frequency domain. Also, all three analog techniques form the basis for digital modulation techniques, which are restricted versions of the analog format for sending data as a digital signal.

In the U.S., engineers at the military test ranges, under the coordination of the IRIG, standardized the characteristics of analog modulation systems for use in telemetering applications [IRIG106; RCC119]. The interested reader should consult these standards for application details. This section reviews the FM and PM concepts and the transmitted signal characteristics. The initial references on analog modulation developed these concepts in detail [Arms36; Cars22; Cars37; Ever40; Rode31].

10.3.1 Phase and Frequency Definition

For analog phase and frequency modulation, we treat the amplitude as a constant, i.e., $A(t) = A$. We then write the RF carrier as

$$s(t) = A\cos\left(\Theta(t)\right) \tag{10.2}$$

We define $\Theta(t)$ as the carrier's *total instantaneous phase* in the units of *radians*. The total phase is composed of phase contributions from both the carrier frequency and the transmitted data. How the data modifies the total phase depends on the specific modulation type used. We compute the total instantaneous phase using

$$\Theta(t) = \omega_c t + \theta(t) \tag{10.3}$$

The radian carrier frequency, ω_c, is related to the tuning frequency on the receiver or the cyclic carrier frequency, f_c, by

$$\omega_c = 2\pi f_c \tag{10.4}$$

The units on ω_c are *radians/second*, while the units on f_c are expressed as Hz. The quantity $\theta(t)$ is the *instantaneous phase deviation* from the carrier phase in *radian* units. The instantaneous phase deviation is composed of two parts: the initial phase of the carrier, θ_0, plus a function of the data, $F\left[m(t)\right]$. This is written as

$$\theta(t) = \theta_0 + F\left[m(t)\right] \tag{10.5}$$

The *instantaneous frequency*, $\omega_i(t)$ (or $f_i(t)$), is related to the total instantaneous phase via

$$\begin{aligned}
\omega_i(t) &= \frac{d}{dt}\left[\Theta(t)\right] \\
\omega_i(t) &= \frac{d}{dt}\left[\omega_c t + \theta(t)\right] \\
f_i(t) &= \frac{1}{2\pi}\frac{d}{dt}\left[2\pi f_c t + \theta(t)\right]
\end{aligned} \tag{10.6}$$

We use these definitions to define the carrier waveform for FM and PM that the data waveform, $m(t)$, modulates.

10.3.2 Frequency Modulation

We define FM by making the RF carrier's *instantaneous frequency* proportional to the data, $m(t)$. That is

$$f_i(t) = f_c + k_f m(t) \tag{10.7}$$

The scale factor, k_f, is the frequency modulation *deviation sensitivity.* Typical values for k_f are in the range of 100 to 500 Hz/V for commercial telemetry systems. By using Equations (10.6) and (10.7) the carrier's total instantaneous phase becomes

$$\Theta(t) = 2\pi f_c t + 2\pi \int_0^t k_f m(\tau)\mathrm{d}\tau + \theta_0 \tag{10.8}$$

The receiver tracks out any initial phase offset so we do not lose any generality if we assume that $\theta_0 = 0$. By comparing Equation (10.8) with Equation (10.5), we can see that the integral of the data is the $F[m(t)]$ in the instantaneous phase deviation.

One special case for $m(t)$ gives us a closed-form analysis to see how FM operates. This special case is the single-tone source with the data signal being

$$m(t) = A_m \cos(2\pi f_m t) \tag{10.9}$$

where A_m is the signal's amplitude and f_m is the signal's frequency (Hz). If we apply this signal to Equation (10.8), we obtain a total instantaneous phase of

$$\Theta(t) = 2\pi f_c t + \beta_{FM} \sin(2\pi f_m t) \tag{10.10}$$

The constant β_{FM} is the *modulation index.* The index is a unitless quantity that provides a relative indication of how much frequency deviation is present in the carrier from its rest frequency, f_c. If we define the peak frequency deviation, Δf, as the maximum frequency shift that the carrier receives from the rest value due to the modulating data, then the modulation index is computed from

$$\beta_{FM} = \frac{\Delta f}{f_m} = \frac{k_f A_m}{f_m} \tag{10.11}$$

The modulation index is a function of the data tone frequency, f_m, and not the carrier frequency, f_c. For example, consider a system with

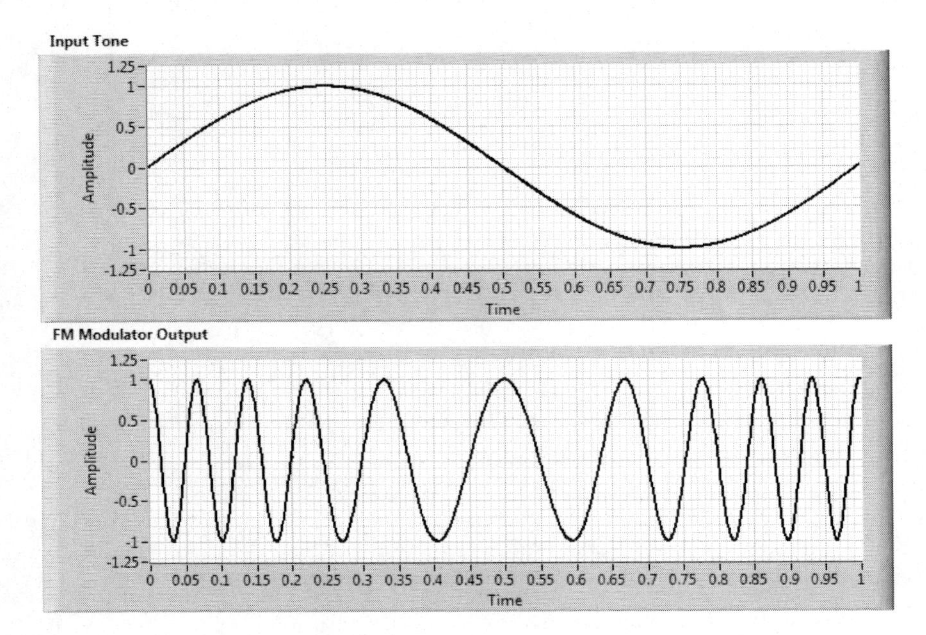

Figure 10.2: A simulated single-frequency modulating tone for FM and the resulting carrier output over a 1-s interval.

a carrier frequency at 40 kHz and a permitted total carrier deviation of 6 kHz. This means that the maximum carrier frequency is 43 kHz and the minimum carrier frequency is 37 kHz. The maximum frequency deviation is $\Delta f = 3\,\text{kHz}$. A representative data source has a frequency of 600 Hz. This yields a modulation index of $\beta_{FM} = 5$.

Figure 10.2 depicts an example of how the transmitted FM signal looks for a single-tone source based on a LabVIEW® simulation output. This is a normalized simulation plot with the tone frequency, f_m, set to 1 Hz, the FM modulator carrier frequency, f_c, set to 10 Hz and the modulation index, β_{FM}, set to 1. This example shows that the carrier frequency is not exactly constant but does change in response to the input tone frequency. The carrier frequency is higher between 0.0 to 0.15 s, evidenced by the proximity of the peaks. This corresponds to the time when the input is positive. Between 0.4 and 0.6 s, the carrier frequency decreases as the peaks become further apart. This corresponds to the time when the input is negative. Notice that the carrier amplitude remains constant.

Engineers classify FM systems by the modulation index. Narrowband Frequency Modulation (NBFM) occurs when $\beta_{FM} << 1$, while Wideband

Frequency Modulation (WBFM) occurs when $\beta_{FM} > 1$. These names arise from the carrier's Power Spectral Density (PSD) width, as Figure 10.3 illustrates. We use the same simulation in Figure 10.2, but the modulation index takes on the values of 0.1, 1.0, and 5.0, respectively. The two biggest influences on the PSD characteristics are the input signal's tone frequency and the modulation index. The input signal's tone frequency controls the spacing of the harmonics around the carrier frequency.

Inspecting the cases with $\beta_{FM} = 1$ and 5, we see that the harmonics are spaced at 1-Hz intervals around the carrier as dictated by the modulating tone's frequency. The modulation index controls the number of harmonics in the PSD and their relative height. For $\beta_{FM} = 0.1$, there is only one major term in the PSD at the carrier frequency of 10 Hz. The first two harmonics are very small and barely above the sampling limit from the simulation. In the middle case where $\beta_{FM} = 1$, we see that there are more harmonics around the carrier frequency harmonic, but these side bands decay quickly. Finally, for $\beta_{FM} = 5$, there are a great number of sidebands around the carrier frequency; the carrier frequency is no longer the dominant term. As the modulation index increases, the channel requires a larger bandwidth to hold the transmitted signal. Changing the modulation index does not change the total power in the signal. Rather, the available power is spread over more frequency components as the modulation index increases. If the designer makes the modulation index too large, the carrier signal can become lost in background noise. Typical telemetry systems operate with β_{FM} in the range $1 < \beta_{FM} < 10$.

The reason the PSD for single-tone FM systems appears as a line spectrum becomes apparent when we analyze the carrier's functional form. As indicated above, the carrier, $s(t)$, is given by

$$s(t) = A\cos\left(\omega_c t + \beta_{FM}\sin\left(\omega_m t\right)\right) \tag{10.12}$$

We can use an infinite Fourier series to represent the FM instantaneous phase deviation of the total instantaneous phase of the carrier. This infinite series has Bessel functions for the Fourier coefficients used the relationship

$$\mathrm{e}^{j\beta_{FM}\sin\omega t} = \sum_{n=-\infty}^{\infty} \mathrm{J}_n\left(\beta_{FM}\right)\mathrm{e}^{jn\omega t} \tag{10.13}$$

The quantity $\mathrm{J}_n(\beta_{FM})$ is the Bessel function of the first kind, of the nth order. By applying the expansion, we obtain the following series

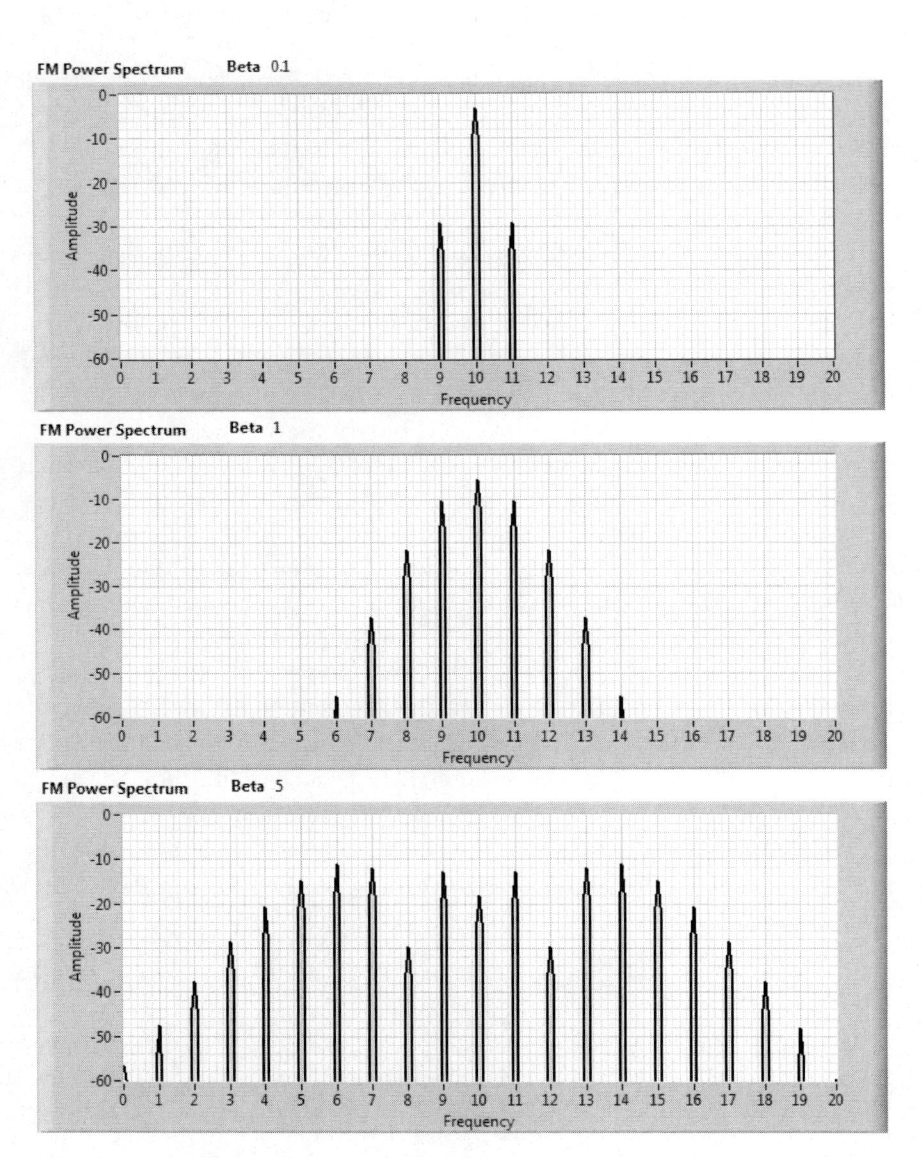

Figure 10.3: Simulated power spectra for FM modulator output with single-tone modulating signals. Carrier modulation index varied from 0.1 to 5.

representation for the carrier signal:

$$\begin{aligned} s(t) &= \Re\left\{A\,\mathrm{e}^{j\omega_c t} \sum_{n=-\infty}^{\infty} \mathrm{J}_n\left(\beta_{FM}\right)\mathrm{e}^{jn\omega_m t}\right\} \\ &= A \sum_{n=-\infty}^{\infty} \mathrm{J}_n\left(\beta_{FM}\right)\cos\left[\left(\omega_c + n\omega_m\right)t\right] \end{aligned} \tag{10.14}$$

When we compute the Fourier transform of Equation (10.14), the series expansion yields an infinite sum of single frequencies displaced from the carrier frequency by integer multiples of the modulating tone frequency. The plots in Figure 10.3 only show a finite number of terms because the Bessel function that becomes the amplitude of the term goes to zero as n and β_{FM} increase.

While the analysis so far has dealt with single-tone modulation, that signal type is not very common in telemetry systems. Generally, sensor output is a multiharmonic signal so we must modify some of the definitions to accommodate the characteristics of useful sensor signals. The concept of a modulation index is only valid for single-tone systems. For wideband sensor signals, designers use the concept of a deviation ratio in place of the modulation index. The *deviation ratio*, D, is defined in terms of the highest significant frequency, f_H, in the input signal and the maximum carrier frequency deviation, Δf, using

$$D = \frac{\Delta f}{f_H} \tag{10.15}$$

Figure 10.4 presents simulation results for a more complicated input signal with components at 1 Hz, 2.35 Hz, 4.99 Hz, 5.77 Hz, and 7.32 Hz. In this simulation, the carrier is at 100 Hz. The first plot in the figure shows the time-domain input signal. The second plot in the figure shows the PSD for the input signal, while the third plot shows the PSD for the output. This output spectrum is a harmonic structure that is more complicated than the input, and the input does not appear directly in the output signal, as is the case with an AM transmission system.

10.3.3 Phase Modulation

PM and FM work in a similar manner. The total *instantaneous phase* for the PM carrier is

$$\Theta(t) = 2\pi f_c t + k_p m(t) \tag{10.16}$$

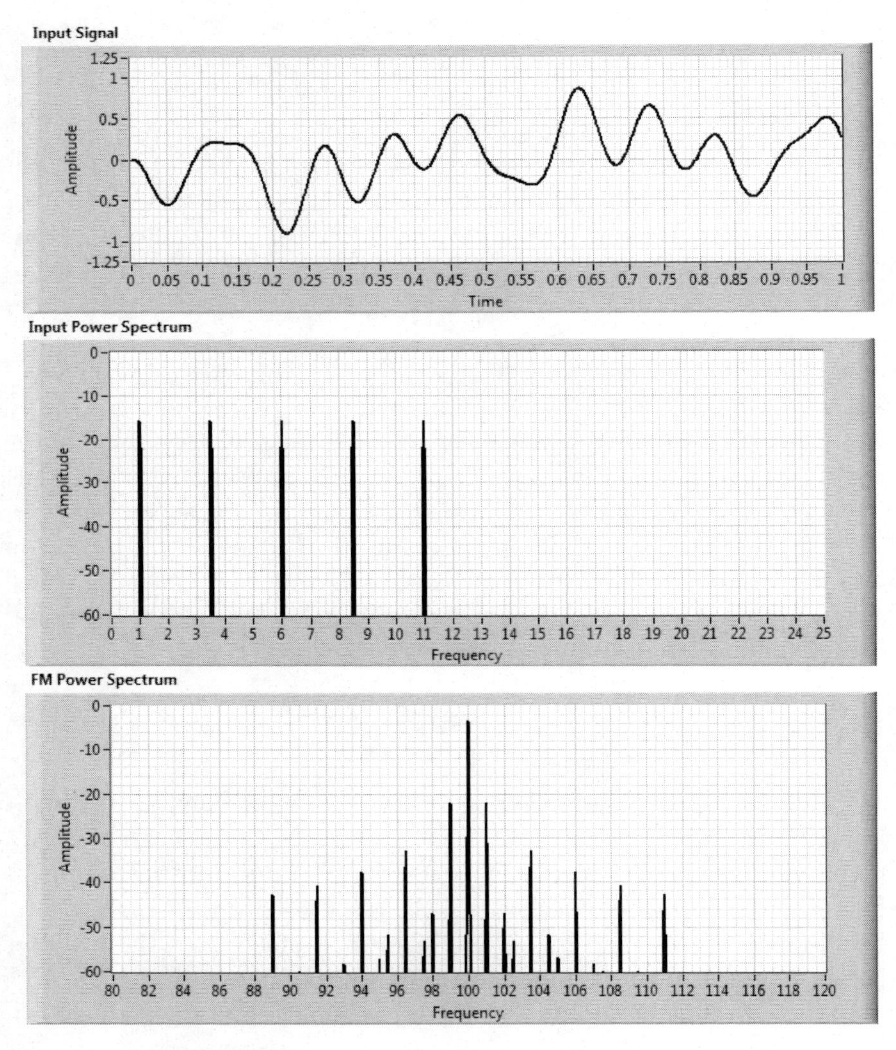

Figure 10.4: Simulated results for a wideband input signal to a FM modulator. The first plot is the time-domain input, the second plot is the input spectrum, and the third plot is the FM output spectrum.

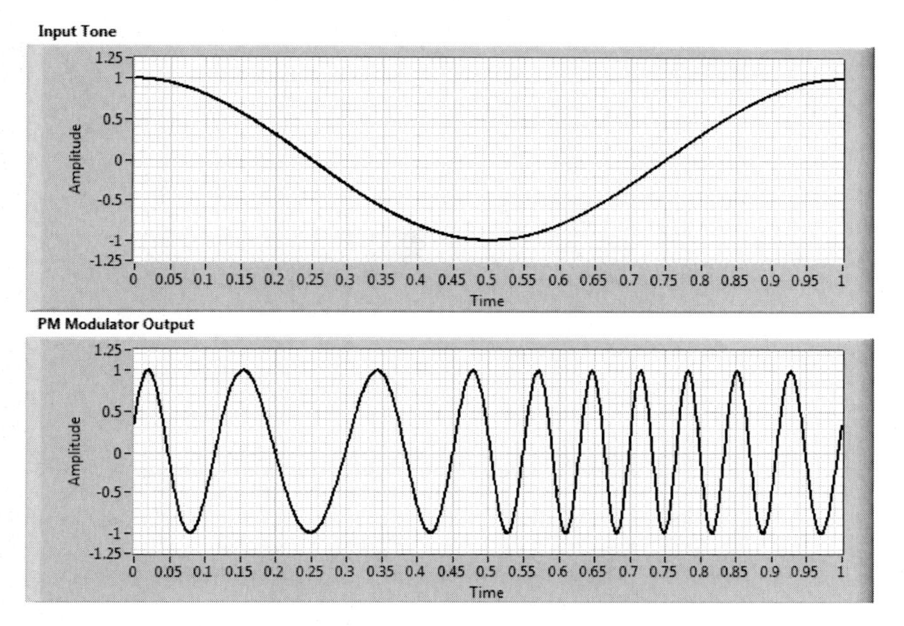

Figure 10.5: A simulated single-frequency modulating tone for PM and the resulting carrier output over a 1-s interval.

Notice that we are omitting the initial phase offset, θ_0, as we did with FM. The scaling constant, k_p, is called the *phase deviation sensitivity* and it has the units of *radians/volt*. PM does not integrate the input message signal, $m(t)$, as FM did so the $F[m(t)]$ of Equation (10.5) is $k_p m(t)$. For the single-tone message signal case, as in Equation (10.9), driving the PM modulator, we define the PM *modulation index* as

$$\beta_{PM} = k_p A_m \tag{10.17}$$

Notice that this does not contain the tone frequency, as did the FM modulation index. Using Equations (10.9), (10.16), and (10.17), we rewrite the transmitted carrier signal, $s(t)$, as

$$s(t) = \cos(2\pi f_c t + \beta_{PM} \cos(2\pi f_m t)) \tag{10.18}$$

Again, this is very similar (but not identical) to the FM form. The simulation results appearing in Figure 10.5 show a 1-Hz tone modulating the PM system. The PM output is very similar to that for the FM output. Again, an apparent frequency difference occurs between 0.0 to 0.4 s and between 0.55 to 0.85 s that corresponds to the periods when the input is positive and negative. The PM PSD also behaves similarly to FM PSD, as Figure 10.6 illustrates. The modulation indices vary from 0.1 to 5 and the output signal has the same harmonic characteristics as the FM output.

10.3.4 Signal-to-Noise Performance

The Signal-to-Noise Ratio (SNR) is an important criterion engineers use to determine the quality and performance of analog transmission systems. A normal way to express the metric is the ratio of the SNR on the demodulator's output to the SNR on the demodulator's input. For properly operating systems, this ratio is greater than unity, which implies that the demodulation process improves the received signal's quality.

Engineers express the output SNR for PM systems in terms of the phase deviation sensitivity, k_p, and the message signal's mean-square value, $\overline{m^2}$, using [Lath98]

$$SNR_{out} = k_p^2 \overline{m^2} SNR_{in} \tag{10.19}$$

The output SNR for FM systems has two components: one that is due to channel noise and one that causes output signal discontinuities in the demodulator due to phase errors. Single-tone modulation is the basis for the theoretical estimates. Analysts use the modulation index, β_{FM}, in the total output SNR computation along with the transmission bandwidth, W, the signal baseband bandwidth, B, and the square of the message signal's peak value, m_p^2, in the equation [Lath98] (see also [Couc01])

$$SNR_{out} = \frac{3\beta_{FM}^2 SNR_{in}\left(\overline{m^2}/m_p^2\right)}{1 + 4\sqrt{3}\left(\beta_{FM}+1\right) SNR_{in} Q\left[\sqrt{SNR_{in}/\left(\beta_{FM}+1\right)}\right]} \tag{10.20}$$

Figure 10.7 plots the output SNR relative to the input SNR for FM systems. Notice the following points:

- In the flat parts of the curves, the output SNR is larger than the input SNR. This implies a signal quality improvement. This is the region where the demodulator should be operating and where the receiver locks to the signal.
- The improvement is larger for larger modulation indices than for smaller modulation indices.

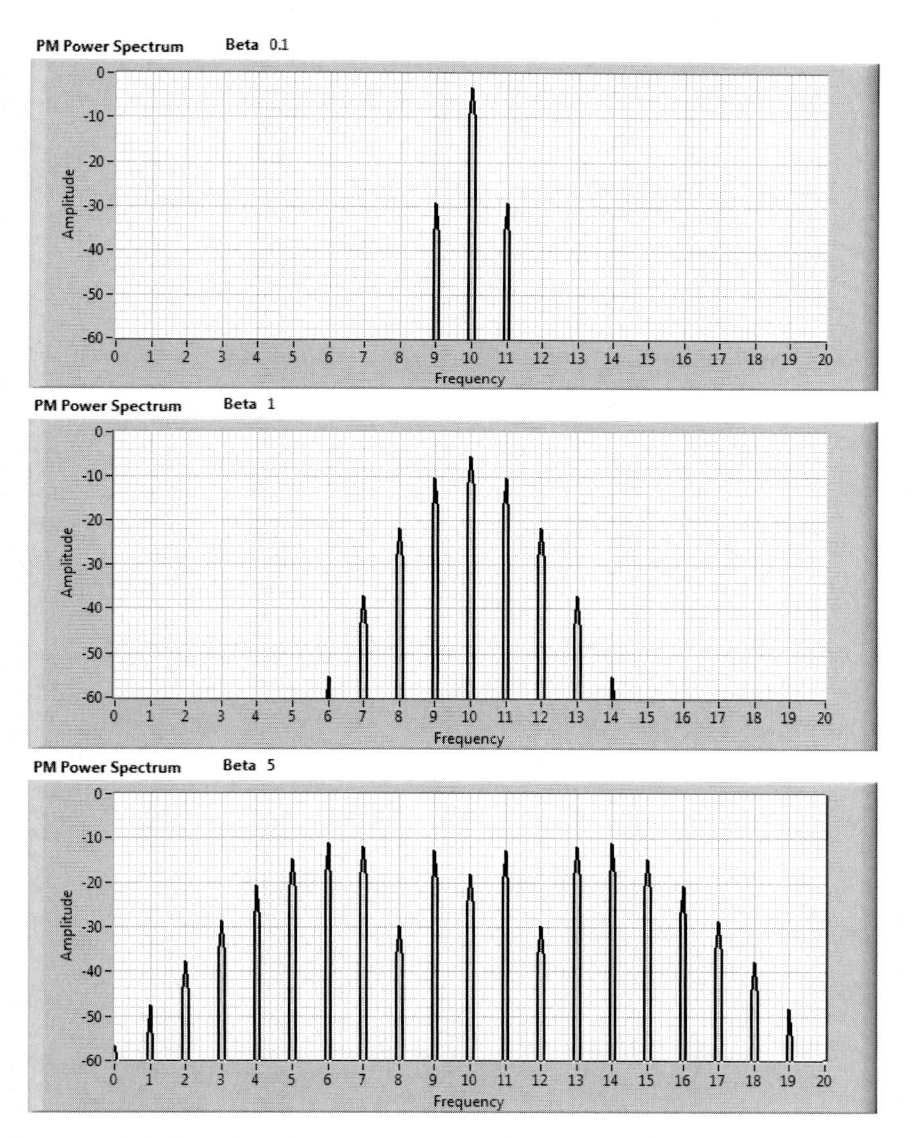

Figure 10.6: Simulated power spectra for PM modulator output with single-tone modulating signals. Carrier modulation index was varied from 0.1 to 5.

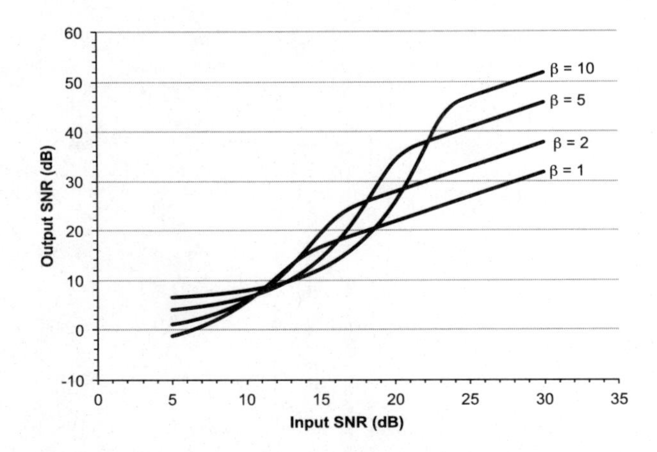

Figure 10.7: Output SNR versus input SNR for FM systems.

- At the low end of the curves, the output SNR is actually smaller than the input SNR, which implies that the performance is worse.

For each of the curves given in Figure 10.7, the curve's knee is the FM *threshold* region. This is the transition point between where the receiver locks onto the signal and gives an improved output and the region where the receiver has many phase errors. Figure 10.8 shows the threshold level as a function of input SNR; engineers estimate the knee point using [Lath98; Couc01]

$$SNR_{threshold} = 20\,(\beta_{FM} + 2) \tag{10.21}$$

Once the FM receiver is locked, engineers approximate the FM improvement using [Lath98; Couc01]

$$SNR_{out} \approx \frac{3}{2}\beta_{FM}^2 SNR_{in} \tag{10.22}$$

This arises from Equation (10.20) by considering only single-tone modulation and noting that the second term in the equation's denominator is approximately 1 with a high SNR, as is the case when the FM receiver is locked to the carrier.

10.3.5 Relative Performance of FM and PM

While FM and PM are very similar, some of their differences are easy to confuse. Table 10.1 lists the parameter differences between FM and PM.

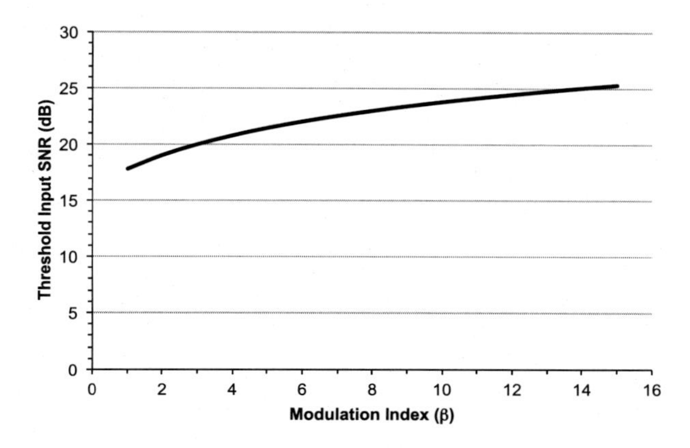

Figure 10.8: FM threshold values as a function of input SNR.

Table 10.1: Parameter Comparison between FM and PM

Parameter	**FM**	**PM**
Data signal processing	Integrate the data	Do not integrate the data
Deviation Sensitivity	k_f (Hz/V)	k_p (rad/V)
Instantaneous Frequency	$\omega_i = \omega_c + 2\pi k_f m(t)$	$\omega_i = \omega_c + k_p m(t)$
Modulation Index (tone modulation)	$\beta_{FM} = \Delta f / f_m = k_f A_m / f_m$	$\beta_{PM} = k_p A_m$
SNR Improvement	$SNR_{out} = \frac{3}{2}\beta_{FM}^2 SNR_{in}$	$SNR_{out} = k_p^2 E[m^2] SNR_{in}$

Note that the units on several parameters are different. One common question designers ask is "which technique performs better?". We look at both techniques in this section.

From detailed studies of FM and PM, engineers have learned that the FM modulation characteristics are more affected by the message signal's amplitude, while PM is more affected by the derivative of the message signal's amplitude. Since the derivative is affected by the signal's harmonic content, PM is more affected by the shape of the message signal's amplitude spectrum. Comparing the output SNR improvement, as a function of the signal's spectrum and transmission characteristics, illustrates this point. Engineers define the *mean-square bandwidth*, in Hz, for a message signal, $m(t)$, in terms of its spectrum, $M(\omega)$, by the equation [Lath98]

$$\overline{B_m^2} = \frac{\int_{-\infty}^{\infty} f^2 M\left(2\pi f\right) \mathrm{d}f}{\int_{-\infty}^{\infty} M\left(2\pi f\right) \mathrm{d}f} \tag{10.23}$$

If we combine this signal bandwidth with the message signal's essential baseband bandwidth, B, and configure the FM and PM modulators to have the same transmission bandwidth, then the ratio of the output SNR improvement for PM and FM is given by [Lath98]

$$\frac{(SNR_{out})_{PM}}{(SNR_{out})_{FM}} = \frac{B^2}{3\overline{B_m^2}} \tag{10.24}$$

This result is only valid if the FM receiver is in lock; i.e., noise discontinuities are relatively infrequent. From Equation (10.24), engineers conclude that FM is superior to PM if the spectrum is more heavily weighted to higher frequencies and about the same if the spectrum is relatively flat. PM is superior if the spectrum has most of its signal at lower frequencies.

10.4 DIGITAL MODULATION

Digital transmission techniques dominate telemetry and telecommand systems. The value-added signal processing available in digital communications networks is also available in telemetry and telecommand systems. Designers frequently realize digital communications techniques as analog modulation with restricted signal sets. However, that is changing as designers utilize modern signal processing design techniques to synthesize the transmitter and receiver functions using software-based techniques and high-speed digital sampling hardware. With digital modulation, designers restrict the variable amplitude, phase, or frequency of Equation (10.1) to specific values and not a continuum of values as is true with analog modulation. Because of the restricted data set, the receiver is only interested in determining the transmitted 0/1 bit values. Consequently, the quality metric becomes the probability of making a bit error and not the faithfulness of recovering the transmitted waveform.

Digital transmission systems frequently transmit more than one bit at a time. Engineers refer to this as *M-ary signaling*. The transmission differentiates between the transmitted symbol rate and the source bit rate. The source data generation rate is R_b bits/second. The *M-ary* modulation technique combines l bits to form each transmitted modulation symbol. The l bits form $M = 2^l$ modulation levels. The *modulation symbol rate*, R_s, is lower than the source data rate by the factor l

$$R_s = \frac{R_b}{l} \tag{10.25}$$

These relationships hold regardless of the modulation type (PSK, FSK, etc.).

We examine the PSK, Minimum Shift Keying (MSK), FSK, and QAM transmission techniques in the subsequent subsections. For further information, refer to standard texts covering digital communications [Ande99; Rice09; Skla01]. In the discussion, we will assume that the logic 0 value is mapped to $+1\,\text{V}$, while the logic 1 value is mapped to the $-1\,\text{V}$ signal levels when we actually transmit the data over the carrier. Alternatively, the system designer can use the reverse mapping as well. In either case, the designer needs to specify the mapping in the system documentation.

10.4.1 Phase Shift Keying

10.4.1.1 Binary Phase Shift Keying and Quadrature Phase Shift Keying

Phase Shift Keying (PSK) results when the carrier's phase changes in response to the data. The transmitted carrier, $s(t)$, is represented by

$$s(t) = A \sin\left[\omega_c t + \theta(t)\right] \tag{10.26}$$

Table 10.2 describes how the data and the modulation level determine the carrier's phase, $\theta(t)$. Figure 10.9 shows the phasor diagram that is the

Table 10.2: PSK Modulation Parameters

Type	M	l	Phase points (rad)
BPSK	2	1	$0, \pi$
QPSK	4	2	$\pi/4, 3\pi/4, 5\pi/4, 7\pi/4$
8-PSK	8	3	$\pi/8, 3\pi/8, 5\pi/8, 7\pi/8, 9\pi/8, 11\pi/8, 13\pi/8, 15\pi/8$
16-PSK	16	4	$\pi/16$, $3\pi/16$, $5\pi/16$, $7\pi/16$, $9\pi/16$, $11\pi/16$, $13\pi/16$, $15\pi/16$, $17\pi/16$, $19\pi/16$, $21\pi/16$, $23\pi/16$, $25\pi/16$, $27\pi/16$, $29\pi/16$, $31\pi/16$

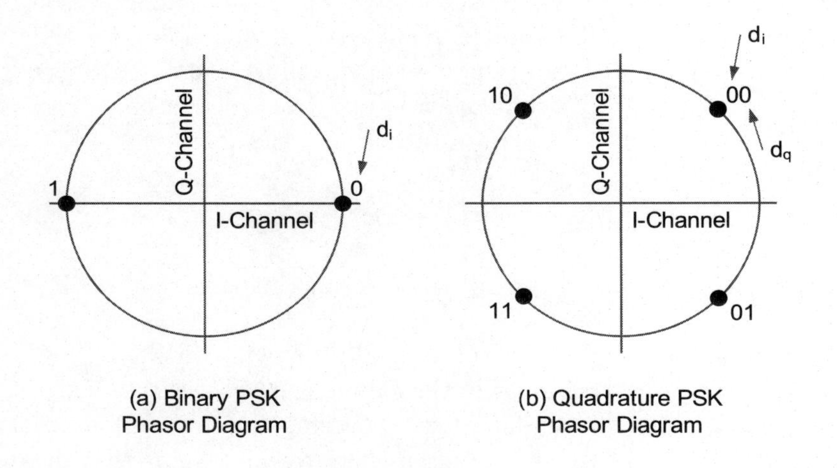

Figure 10.9: Phasor diagram for BPSK and QPSK. The assignment of the logic levels to the phase points is also shown.

major means for tracking the data in Binary Phase Shift Keying (BPSK) and Quadrature Phase Shift Keying (QPSK). The figure shows the mapping of the data values to the phase points. If we use a mapping of a logic 0 to a phase of 0 rad and a logic 1 to a phase of π rad [CCSDS4010] and then using trigonometric identities, we can see that the mapping of the phase points in Figure 10.9 for the BPSK carrier can be used to express the carrier in Equation (10.26) in terms of the data bit value, d_i, as

$$s(t) = d_i \sin(\omega_c t) \tag{10.27}$$

Figure 10.10 illustrates a simulation with random data and a BPSK carrier. Figure 10.11 illustrates the modem structure for BPSK

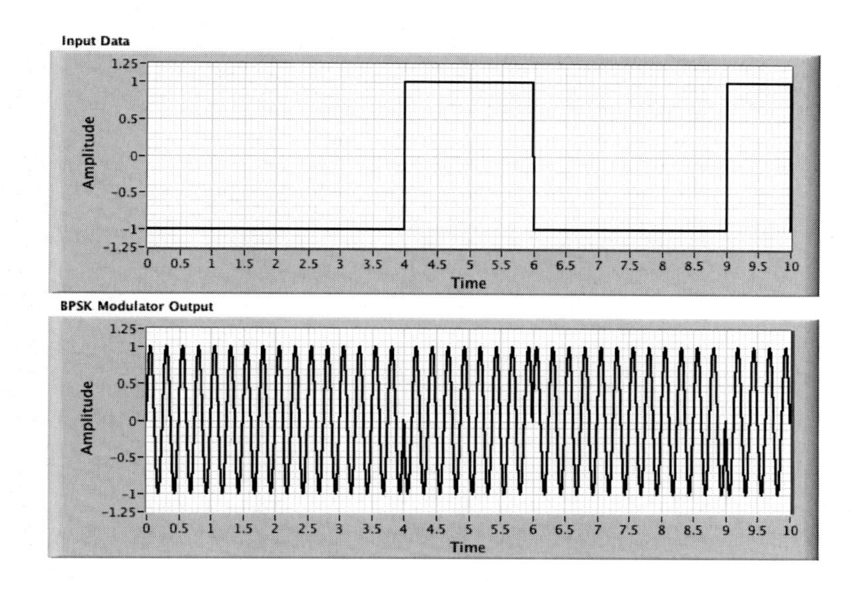

Figure 10.10: Simulated data and BPSK carrier.

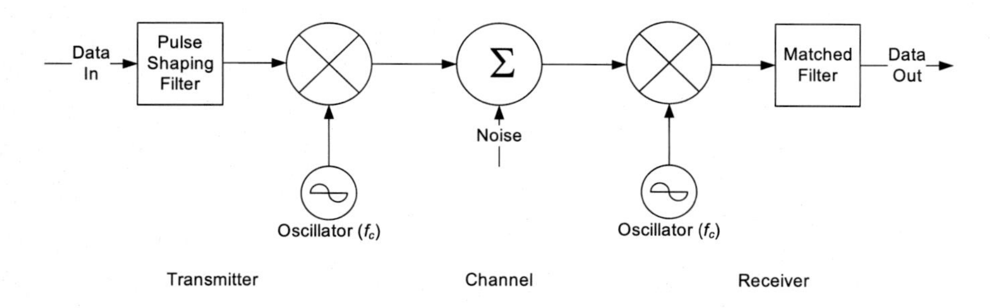

Figure 10.11: BPSK transmission modem structure.

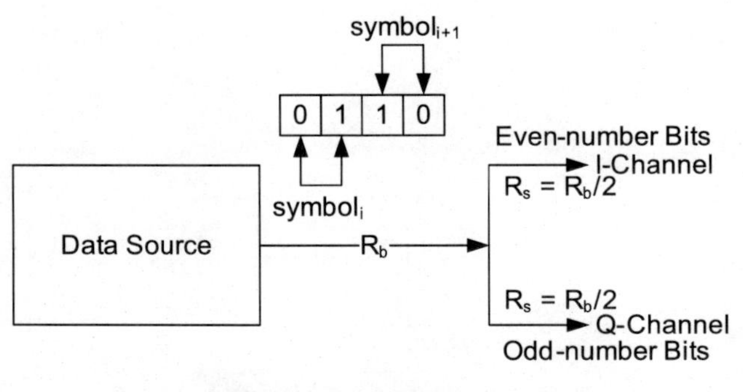

(a) Single source QPSK generation

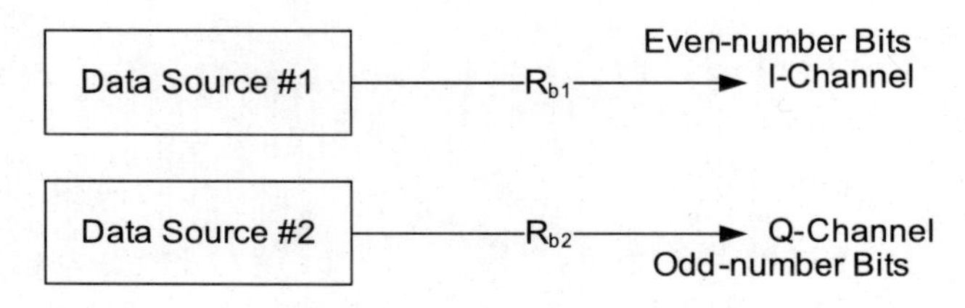

(b) Dual source QPSK generation

Figure 10.12: Bit streams for QPSK transmission. (a) Bits grouped as symbols for QPSK transmission; (b) dual data sources as independent channels.

transmission. The transmitted carrier has constant amplitude and a phase discontinuity occurs at the time of every data transition level. This behavior holds for all orders of PSK. Extending the BPSK results to two bits at a time and using trigonometric identities along with the phase points given in Table 10.2, we find the QPSK carrier is written in terms of the two data bits, d_i and d_q, forming the transmitted symbol using

$$s(t) = d_q \sin(\omega_c t) + d_i \cos(\omega_c t) \tag{10.28}$$

Figure 10.12 illustrates the two bits grouped to form the channel symbol. The symbol can come from either the even- or odd-numbered bits from a single source or two bits from different sources. From Equation (10.28), we see that QPSK is two orthogonal BPSK carriers transmitted at the same carrier frequency. They do not interfere with each other due to that orthogonality. For higher modulation orders, we still represent the carrier signal in a quadrature form similar to Equation (10.28). However,

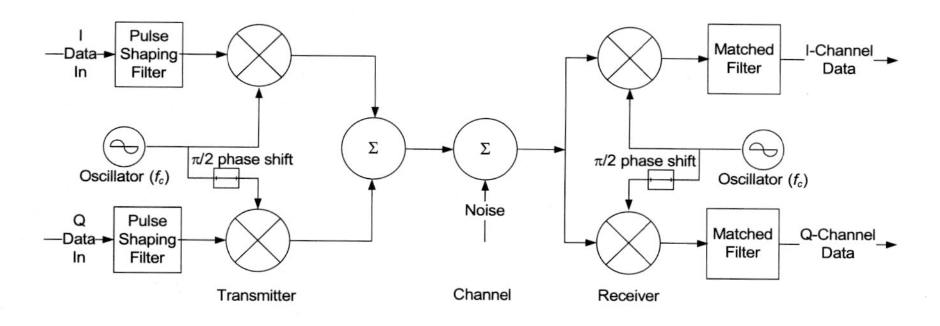

Figure 10.13: QPSK transmission modem structure.

the data symbols on the I- and Q-channels are mixtures of multiple baseband bits and are no longer independent.

Figures 10.11 and 10.13 illustrate how Equations (10.27) and (10.28) form the basis for the BPSK and QPSK modems and that PSK is an amplitude modulation process. The PSK modem in Figure 10.13 takes two bits at a time and passes one to the I-channel and one to the Q-channel. Because the PSK transmission is an AM process, the receiver recovers the data by mixing the received carrier with an in-phase sine or cosine waveform, as appropriate, and then low-pass filtering the result. This receiver structure is an example of a coherent demodulator since the receiver uses a local oscillator that is in-phase with the incoming carrier.

Although Figure 10.13 shows the QPSK modulator taking the two bits from the data source, this is not required as Figure 10.12 illustrates. The data bits can actually come from two different sources and have different transmission rates.

10.4.1.2 Offset Quadrature Phase Shift Keying or Staggered Quadrature Phase Shift Keying

One common variation on QPSK is to offset or stagger the times at which the bits change states within the channel symbol. This produces Offset Quadrature Phase Shift Keying (OQPSK) or Staggered Quadrature Phase Shift Keying (SQPSK) modulation (both names are commonly used). In regular QPSK, two bits are taken at a time from the input data and both change to their new states at the start of each symbol time. In OQPSK, one of the two bits changes at the start of the symbol period and the other changes at the mid-point. This is represented in Figure

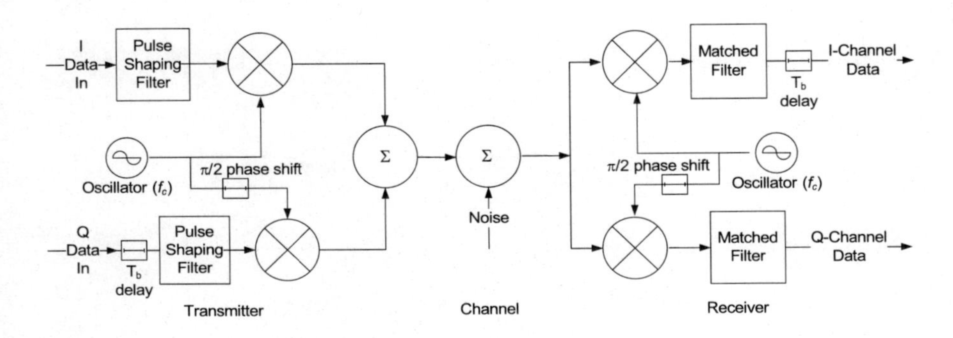

Figure 10.14: OQPSK transmission structure. The block labeled T_b represents a delay of one bit period.

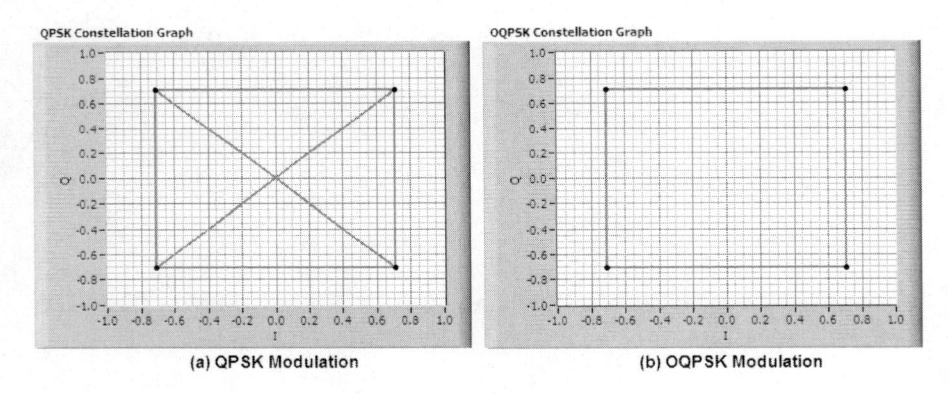

Figure 10.15: I/Q diagrams for QPSK and OQPSK modulation.

10.14 by adding a delay element to one channel of the modulator and a corresponding delay element to the other channel in the receiver.

In a linear system, this timing offset is not required and does not change the system performance. However, nonlinear systems, such as that found in certain types of high-power amplifiers, are often sensitive to phase changes that pass through the origin. This can cause unwanted harmonics that distort the output signal and cause interference. In these cases OQPSK improves the system's performance and brings it closer to the linear system case.

Figure 10.15 shows the I/Q diagrams for QPSK and OQPSK modulation and the output of the modulator's I- and Q-channels plotted against each other. In QPSK, many of the symbols cause the carrier's phase to pass through the origin, while the OQPSK carrier does not generally do so. This is the carrier characteristic that causes undesirable behavior in QPSK with nonlinear amplifiers.

10.4.2 Frequency Shift Keying

If we use a digital signal to change the carrier's frequency, we have Frequency Shift Keying (FSK). The modulator generates FSK either by having a single source generate the multiple frequencies or by having individual tuned sources that are multiplexed to send each symbol. We write the mathematical form of the FSK carrier, $s(t)$, in terms of the *modulation index*, h, and the symbol period, T_s, as [Osbo74]

$$s(t) = \cos\left(\omega_0 t + \frac{a_n h\pi\,(t-n)}{T_s} + \phi_n\right) \tag{10.29}$$

In Equation (10.29), the phase term, ϕ_n, is used to control phase continuity from symbol to symbol, if desired. The value for a_n is related to the number of bits per symbol. For example, Binary Frequency Shift Keying (BFSK) uses a_n from -1, +1 and 4-ary FSK uses a_n from -3, -1, +1, +3. Using Equation (10.29), the carrier's instantaneous frequency becomes

$$\omega_i = \omega_0 + \frac{a_n h\pi}{T_s} \tag{10.30}$$

As with analog FM, FSK is divided into wideband and narrowband versions. Wideband FSK has $h >> 1$ and $\phi_n = 0$. Designers typically choose frequency separations to make the frequencies orthogonal. Figure 10.16 illustrates a modem structure for wideband BFSK and Figure 10.17 provides the results of a simulated BFSK modulator output. We see the carrier's frequency changes at the bit transition times. This simulation exaggerates the ratio between the carrier frequency and the data frequency compared with the ratio used in practice so that we see the effect on the carrier. We also configured this particular simulation to have phase continuity across the bit boundaries. We configured the receiver in Figure 10.16 to use energy detectors to determine which frequency was sent. This is a form of noncoherent detection. Time is usually insufficient to have the receiver generate an in-phase local carrier reference to mix with the incoming data carrier.

Narrowband FSK has a modulation index $0 < h < 1$ and the phase term, ϕ_n, chosen for the desired phase continuity. Discontinuous narrowband FSK has $\phi_n = 0$. A more interesting case is narrowband FSK with continuous phase or Continuous Phase Frequency Shift Keying (CPFSK).

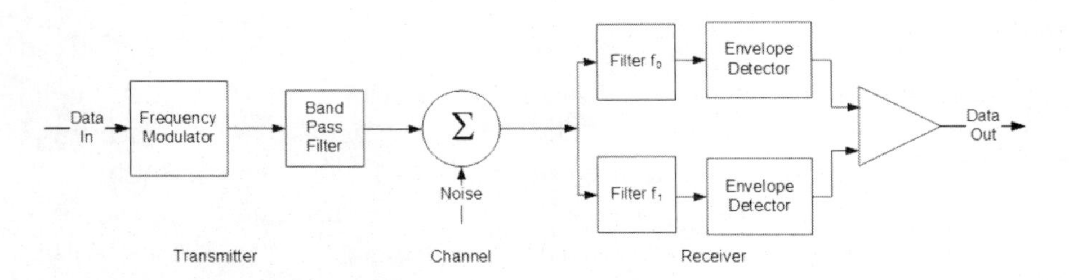

Figure 10.16: Wideband FSK modem structure for binary FSK.

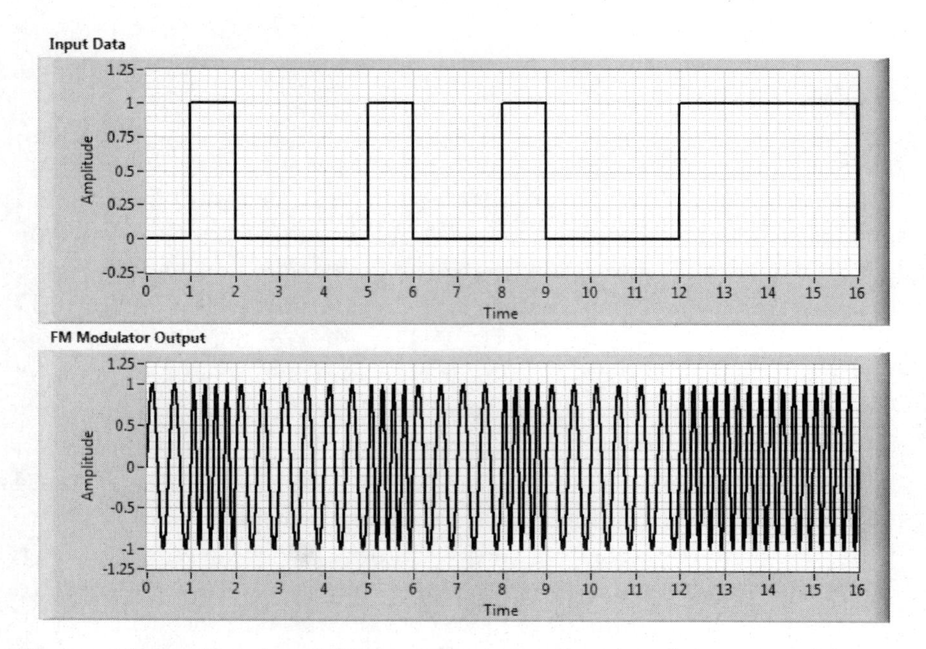

Figure 10.17: Simulated binary data waveform and associated binary CPFSK modulator output.

In this case, the designer selects the phase term to preserve phase continuity by making the phase for each symbol

$$\phi_n = \pi h \sum_{i<n} a_i \tag{10.31}$$

One advantage of CPFSK modulation is that it has a more compact frequency structure than wideband FSK and the phase continuity produces good transmission techniques. The receiver is, however, more complicated.

10.4.3 Pulse-Shaped Modulation Techniques

Figure 10.14 showed that pulse shaping filters can also be added to the time delay element of the QPSK modem. This pulse shaping is added to give the transmitted signal a narrower channel bandwidth [Elze72; Simo76]. In this section, we will look at the MSK, Gaussian Minimum Shift Keying (GMSK) and Shaped Offset Quadrature Phase Shift Keying (SOQPSK) formats as examples of pulse-shaped modulation techniques.

10.4.3.1 Minimum Shift Keying and Gaussian Minimum Shift Keying

Minimum Shift Keying (MSK) is an interesting modulation format. From one point of view it is a form of OQPSK, while from another point of view, it is a form of FSK with a modulation index of 0.5. MSK techniques are widely accepted in the wireless telecommunications world for transmitting data. As we will see in a later section, the MSK waveforms have definite advantages when one considers the bandwidth required for transmitting the signal [Elze72].

The CPFSK realization of MSK uses a carrier that takes the form [Gron76]

$$s(t) = \cos\left[2\pi\left(f_0 + \frac{d_k}{4T_b}\right)t + \phi_k\right] \tag{10.32}$$

where d_k is the data bit, f_0 is the carrier rest frequency, and T_b is the bit period. The CPFSK phase is

$$\phi_k = \left[\phi_{k-1} + \frac{\pi k}{2}\left(d_{k-1} - d_k\right)\right] \mod 2\pi \tag{10.33}$$

The OQPSK signal represents the MSK carrier by [Math74]

$$s(t) = d_I \cos\left(\frac{\pi t}{T_b}\right)\cos\left(\omega_0 t\right) + d_Q \sin\left(\frac{\pi t}{T_b}\right)\sin\left(\omega_0 t\right) \tag{10.34}$$

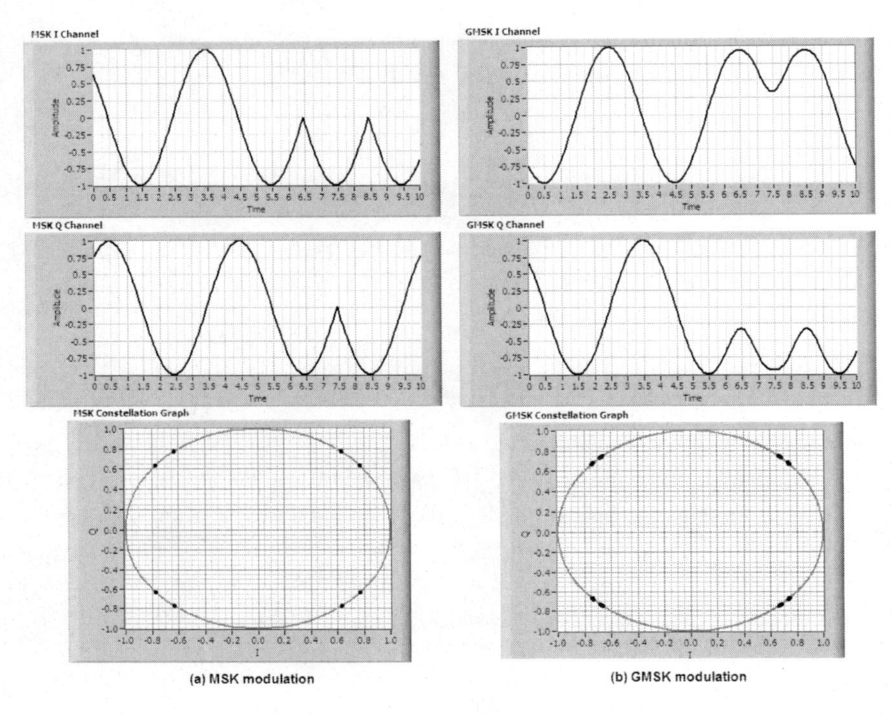

Figure 10.18: Simulated I/Q diagram and phasor diagram for (a) MSK and (b) GMSK transmission.

Equation (10.34) has the same form as Equation (10.28) with the exception that sinusoidal pulse shapes filter the data bits. Because the bits are rounded pulses and are no longer square waves, the bandwidth needed to transmit the data is less than that for square waves.

Figure 10.18(a) gives an example of the pulse shaping for MSK where the modulator applies the sinusoidal MSK pulse shape to the I- and Q-channels. One interesting feature of the MSK transmission is the fact that an OQPSK receiver can be used to demodulate that data and the transmission has the same theoretical error performance as OQPSK as well.

Another form of MSK transmission is to use Gaussian pulse shaping filters, instead of sinusoidal pulse shaping filters, to produce GMSK [Muro81]. Typical filter bandwidths have a Bandwidth-Bit-Period (BT) product of 0.25 to 0.30. GMSK modem chips are currently available at relatively inexpensive prices due to their high demand in the wireless industry. The GMSK modem chips are noncoherent detectors and operate

Table 10.3: Standard Telemetry Group Parameters for SOQPSK Modulation [IRIG106]

Rolloff, α	**Bandwidth Expansion,** B	T_1	T_2
0.7	1.25	1.5	0.5

over several symbol times. Figure 10.18(b) gives an example of the pulse shaping for GMSK where the modulator applies the GMSK pulse shape to the I and Q data channels.

10.4.3.2 Shaped Offset Quadrature Phase Shift Keying

Shaped Offset Quadrature Phase Shift Keying (SOQPSK) makes the variation to OQPSK by adding Nyquist pulse shaping to the bits [Hill00]. The shape of the Nyquist pulse is determined by the rolloff factor, α, the symbol period, T_s, the waveform timing marks T_1 and T_2 under the constraint $T_s = T_1 + T_2$, and the bandwidth expansion factor, B. The resulting SOQPSK pulse, $g(t)$, is given by

$$g(t) = n(t)w(t) \tag{10.35}$$

where $n(t)$ is the Nyquist pulse and $w(t)$ is a window function to time-limit the pulse duration since $n(t)$ extends over all time. The Nyquist pulse is defined by

$$n(t) = \left[\frac{A\cos(\pi\theta_1 t)}{1-4(\theta_1 t)^2}\right]\left[\frac{\sin(\theta_2 t)}{\theta_2 t}\right] \tag{10.36}$$

with $\theta_1 = \dfrac{\alpha B}{T_s}$ and $\theta_2 = \dfrac{\pi B}{T_s}$. The window function is defined by

$$w(t) = \begin{cases} 1 & |\frac{t}{T_s}| \leqslant T_1 \\ \frac{1}{2}\left[1+\cos\left(\frac{\pi\left(|\frac{t}{T_s}|-T_1\right)}{T_2}\right)\right] & T_1 < |\frac{t}{T_s}| \leqslant T_1 + T_2 \\ 0 & |\frac{t}{T_s}| > T_1 + T_2 \end{cases} \tag{10.37}$$

The IRIG Telemetry Group standard has specified key parameters for $g(t)$ as Table 10.3 lists. The standard constrains the amplitude, A in

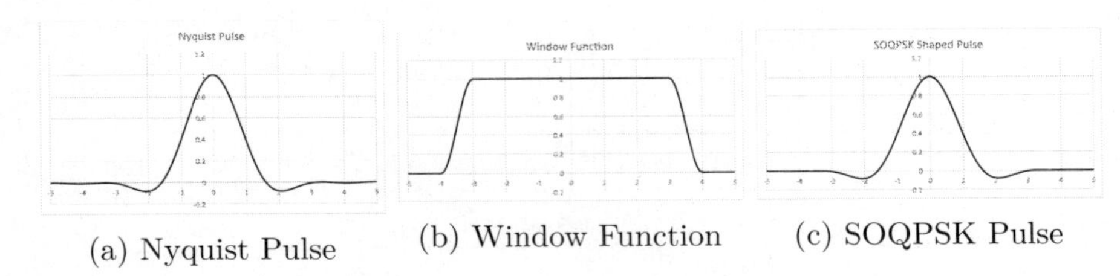

(a) Nyquist Pulse (b) Window Function (c) SOQPSK Pulse

Figure 10.19: Nyquist pulse and window functions generating the SOQPSK pulse function using the IRIG recommended parameters.

$g(t)$ to meet

$$\int_{-T_s}^{T_s} g(t)\mathrm{d}t = \frac{\pi}{2} \tag{10.38}$$

Figure 10.19 illustrates the resulting Nyquist pulse, window function, and SOQPSK pulse for the IRIG parameters. Figure 10.20 illustrates the resulting waveform shapes and the I/Q diagram for the SOQPSK modulation with the IRIG parameters.

10.4.4 Quadrature Amplitude Modulation

One characteristic of FM and PM modulation systems is that the carrier has a constant transmission envelope because the amplitude remains constant, while frequencies or phases have been manipulated. With QAM, this constant envelope restriction is removed and the I- and Q-channel signal levels take on multiple values. Figure 10.21 shows a 16-QAM signal. This multiple-level channel representation represents carrier amplitude and phase as functions of the data symbol value. Actually, we have already seen forms of QAM since BPSK is binary QAM and QPSK is quaternary QAM.

Designers often consider higher modulation orders because the transmitter sends more bits per symbol. Figure 10.22 shows that the QAM constellation can be rectangular or nonrectangular in shape. The figure shows that we consider BPSK and QPSK to be QAM formats in addition to the various 16-ary configurations. If we modify the levels in Equation (10.28) to use nonbinary values determined by the bit pattern, we produce each of the constellation patterns. The amplitudes in Figure 10.22 are not the same but are convenient levels for plotting. For a given application, industry standards often specify the constellation selection. The stan-

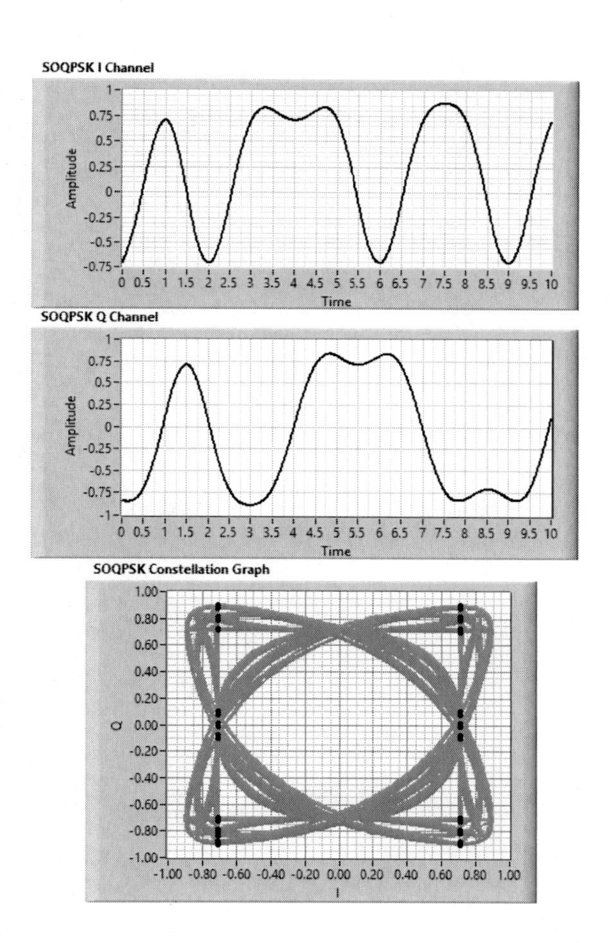

Figure 10.20: Simulated waveform and I/Q diagram for SOQPSK.

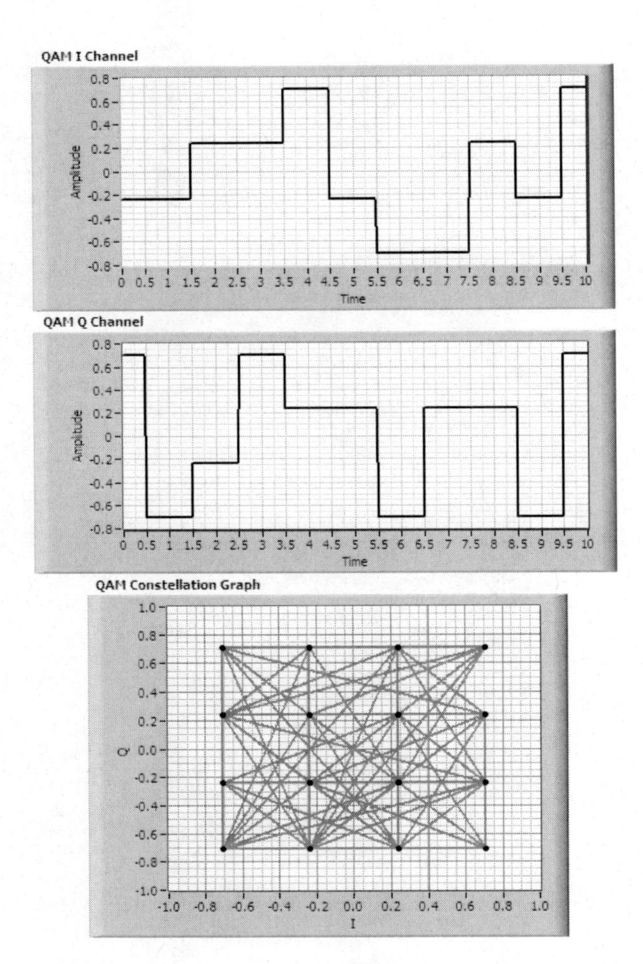

Figure 10.21: Simulated random data with the resulting I/Q diagram for 16-QAM.

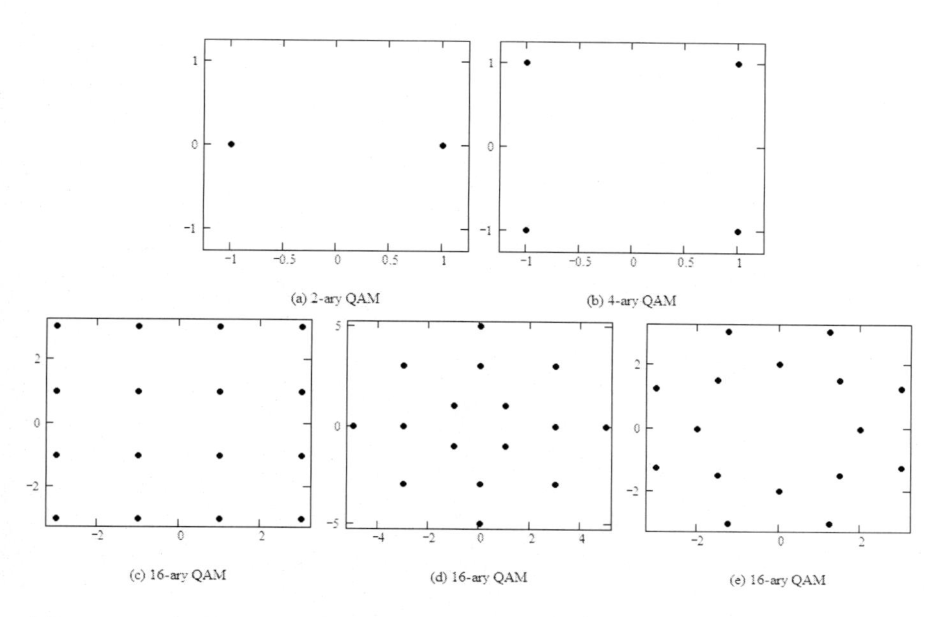

Figure 10.22: Rectangular and nonrectangular 16-QAM constellations.

dards dictate the choice based on requirements related to synchronization, average signal energy, and assumed channel characteristics.

10.4.5 Subcarrier Modulation

Communications engineers for satellites used in space applications such as those with National Aeronautics and Space Administration (NASA)'s Deep Space Network often use a mixed digital/analog transmission method. This method goes by names such as PCM/PSK/PM or subcarrier modulation. The Pulse Code Modulation (PCM) means that the data source comes from a digital PCM source. The PSK means that the PCM data modulates the subcarrier which is a square wave in this case. The PM means that the PCM/PSK signal is transmitted via analog Phase Modulation (PM). We express the transmitted signal, $S_{sc}(t)$, in terms of the transmission power, P, the data, $d(t)$, and the PM modulation index, θ, using [Yuen82b]

$$S_{sc}(t) = \sqrt{2P}\sin\left[\omega_c t + \theta d(t)\sin_{sw}(\omega_{sc}t)\right] \tag{10.39}$$

Here, ω_c is the radian carrier frequency and ω_{sc} is the radian subcarrier frequency. The function $\sin_{sw}(x)$ is a normal sinusoid that has been passed through a limiter so that the output amplitude is ± 1. This pro-

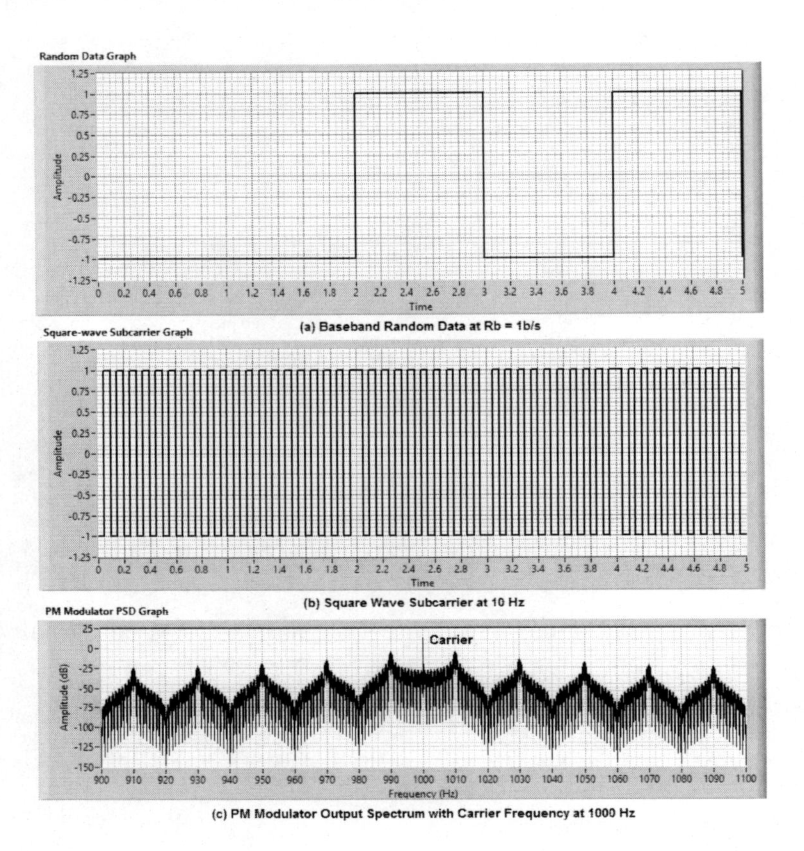

Figure 10.23: Simulated digital subcarrier modulation technique showing the input data, the PSK-modulated digital subcarrier, and the output PM spectrum.

duces a PSK modulated square wave that is used as the input to the PM modulator.

Figure 10.23 illustrates this process. The first panel shows the baseband data and the second panel shows the PSK-modulated square wave from this data. The third panel shows the PM modulator output spectrum. This simulation uses a data bit rate of 1 bps, a subcarrier frequency of 10 Hz, a transmission carrier frequency of 1000 Hz, and a PM modulation index of $\pi/4$.

As we see from the simulation, using a square wave subcarrier with analog PM produces a clean harmonic at the carrier frequency in the transmitted signal's power spectrum. This carrier assists the receiver's

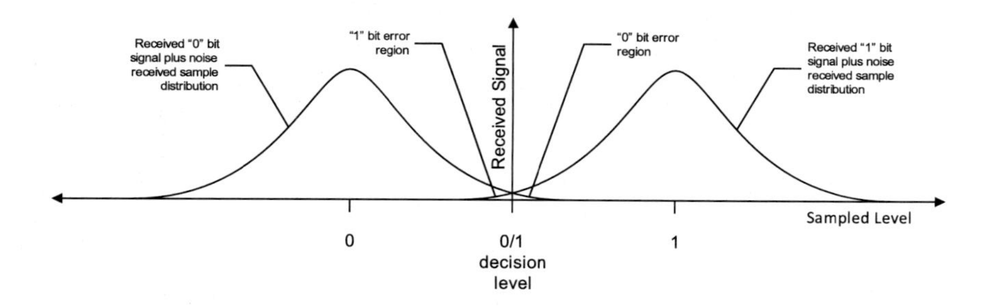

Figure 10.24: BPSK example of adding noise to the received signal points producing the Gaussian received signal distribution and the resulting probability of error regions relative to the 0/1 decision threshold.

tracking loops in locking onto the signal. This is important for weak, noisy signals like those found in deep space communications.

10.4.6 Bit Error Rate Performance

The signal constellations represent the way the modulator encodes the symbols on the carrier. However, actual transmission channels do not pass the carrier without modification. The addition of channel and component noise causes the constellation phase points to grow from dots to regions. Figures 10.11, 10.13, 10.14, and 10.16 show simple channel models where Additive White Gaussian Noise (AWGN) is added to the signal. Figure 10.24 illustrates the effect of the noise on bits received in a BPSK system. The noise causes the received signal level for all of the transmitted bits to form a Gaussian distribution around their normal 0/1 signal voltage levels. The receiver makes a bit error when the noise causes the signal level to go to the other side of the y-axis from the intended side. The signal energy controls the distance between the 0/1 signal level and the y-axis, while the noise power controls the width of the noise distribution.

The important ratio in characterizing the noise is the Energy-per-Bit-to-Noise-Spectral-Density (E_b/N_o) ratio. The E_b/N_o is related to the Carrier-to-Noise Power Ratio (C/N) ratio by

$$\frac{E_b}{N_o} = \frac{C}{N}\frac{W}{R_b} = \frac{C}{N}WT_b \tag{10.40}$$

C is the carrier power and N is the noise power, both in Watts. W is the transmitted signal's channel bandwidth and Bit Rate (R_b) is the bit transmission rate. The data Bit Period (T_b) is the reciprocal of R_b. The receiver's System Noise Temperature (T_{sys}) parameterizes the noise in the communications system. The total *noise power* in the receiver is [Dave87]

$$N = kT_{sys}W \tag{10.41}$$

where k is Boltzmann's constant ($1.380\,648\,52 \times 10^{-23}$ J/K or, in decibel (dB) units, $-228.6\,\text{dBW/K} - \text{Hz}$). The noise spectral density, in W/Hz, is

$$N_o = kT_{sys} = \frac{N}{W} \tag{10.42}$$

Figure 10.25 extends Figure 10.24 to illustrate the effect of AWGN in two dimensions with the QPSK signal constellation. For low values of E_b/N_o, the noise makes the signal phase points spread out from their source transmission values along a two-dimensional Gaussian curve. For high values of the E_b/N_o, the noise still displaces the constellation points from their transmission values but the spread is much smaller. For higher modulation orders, we can imagine this constellation point spreading taking place on the I/Q diagram with the signal constellation points packed closer together. In this section, we provide the equations for the relative performances of the modulation types based on their Bit Error Rate (BER). For derivations of these results, see references such as [Ande99; Skla01]. In these derivations, the transmission bandwidth is infinite. In practice, the bandwidth is less and the theoretical BER results overestimate the performance quality.

Both BPSK and QPSK have the same BER or probability of error (p_e), which is estimated by

$$p_e = Q\left(\sqrt{\frac{2E_b}{N_o}}\right) \tag{10.43}$$

Figure 10.26 plots this relationship, assuming that the receiver is operating in a coherent mode. Operating in a noncoherent mode decreases performance by about 3 dB (the transmission needs about 3 dB more power to achieve the same BER level). Equation (10.43) holds for OQPSK and MSK as well since they are other forms of QPSK. If GMSK is properly detected with multisymbol detection, it also has the same BER as QPSK, while for narrow-filtering cases, the BER depends on the BT product. The GMSK BER is estimated via [Muro81; Rapp96]

$$p_e = Q\sqrt{\frac{2\alpha E_b}{N_o}} \tag{10.44}$$

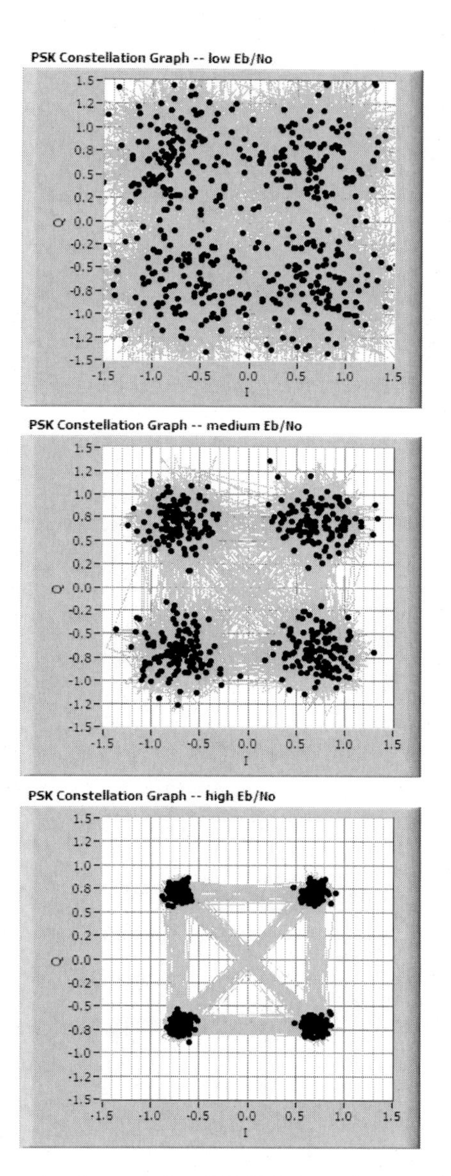

Figure 10.25: Effect of noise on the QPSK signal constellation for different signal-to-noise levels.

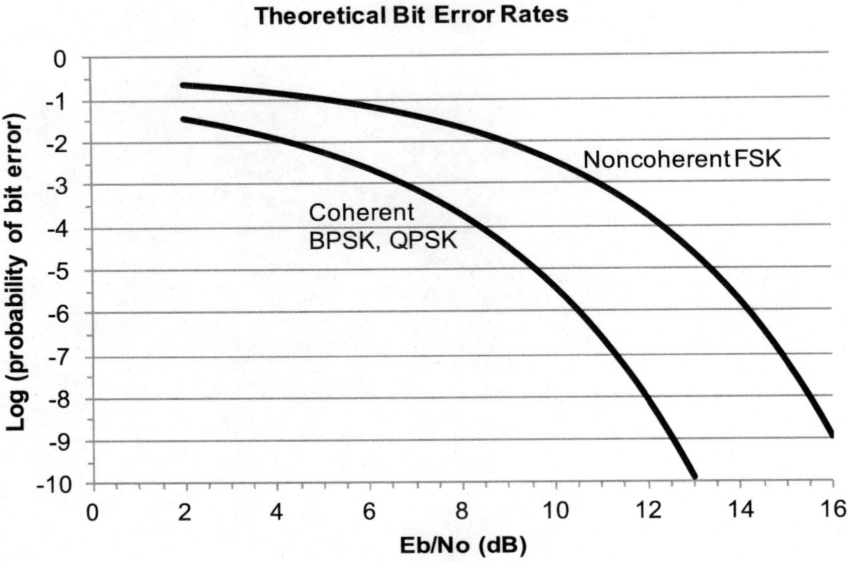

Figure 10.26: Estimated BER for coherent BPSK and QPSK, and noncoherent BFSK.

where α is 0.68 for a BT of 0.25 and 0.85 for an infinite BT.

For higher (M-ary) PSK modulation orders, the BER is estimated using the symbol energy ($E_s = E_b \log_2(M)$) instead of the energy per bit

$$p_e = 2Q\left[\sqrt{\frac{2E_s}{N_o}}\sin\frac{\pi}{M}\right] \tag{10.45}$$

The BER for QAM is more difficult to determine because it is a function of the arrangement of the constellation points. For evenly spaced rectangular constellations, the QAM BER is estimated using the equation [Skla01] (see also [Ande99; Rapp96])

$$p_e = \frac{2\left(1 - L^{-1}\right)}{\log_2 L} Q\left[\sqrt{\left(\frac{3\log_2 L}{L^2 - 1}\right)\frac{2E_b}{N_o}}\right] \tag{10.46}$$

The BER is a function of the number of levels in one dimension, L.

BFSK transmitters normally generate orthogonal signals and the receiver uses a noncoherent detection method. In this case, the BER is estimated using

$$p_e = \frac{1}{2}\,\mathrm{e}^{-\frac{1}{2}\frac{E_b}{N_o}} \tag{10.47}$$

Figure 10.26 also plots Equation (10.47).

10.5 BANDWIDTH ESTIMATES

If we wish to estimate the required bandwidth for an unfiltered analog or digital FM or PM transmission system, the task is relatively easy because it is mathematically infinite. One of the most difficult estimates is the filtered system's practical bandwidth because of the number of different types of bandwidth definitions in existence. Regulators typically use 99% of the total signal strength as the required bandwidth [NTIA13]. Other bandwidth estimates use spectral features or relative performance. This section discusses common bandwidth estimates.

10.5.1 Analog Bandwidth

Designers use the same relationship to estimate the required bandwidth to transmit both analog FM and PM signals [Lath98]. As we saw earlier, the analog modulation spectrum is composed of an infinite number of discrete spectral lines that have their greatest amplitude near the carrier (although the maximum amplitude is not necessarily exactly at the carrier frequency). The series of spectral lines becomes lower in amplitude as they progress towards infinity but never go exactly to zero and remain there. Engineers use *Carson's rule* to estimate the FM bandwidth containing the significant spectral lines and it expresses the transmitted bandwidth, W, in terms of the modulating signal's baseband bandwidth, B, and the peak frequency deviation of the carrier, Δf. Carson's rule is

$$W = 2\left(\Delta f + B\right) \tag{10.48}$$

For single-tone FM, we substitute the modulation index into Equation (10.48) and the signal's bandwidth is the modulating tone frequency, f_m. For non-tone messages, we use the deviation ratio, D, and still have good approximations of the required FM bandwidth

$$\begin{aligned} W &= 2\left(\beta_{FM} + 1\right) f_m \\ W &= 2\left(D + 1\right) B \end{aligned} \tag{10.49}$$

For example, the FM commercial-broadcast standard has a maximum message signal bandwidth of 15 kHz and a deviation ratio of 5. The broadcast FM bandwidth is estimated as $W = 2(5+1)15$ kHz $= 180$ kHz. This compares well with the channel spacing between commercial FM stations, which is 200 kHz.

To apply Carson's rule to PM systems, we first need to compute the peak frequency deviation from the peak instantaneous frequency by taking the first derivative of the carrier's total phase

$$\begin{aligned}\omega_i &= \frac{d\Theta(t)}{dt} = \frac{d}{dt}\left[\omega_c t + k_p m(t)\right] \\ &= \omega_c + k_p \dot{m}(t)\end{aligned} \tag{10.50}$$

The peak frequency deviation occurs at the peak magnitude of the message signal's derivative

$$\dot{m}_{peak} = \left|\left[\frac{dm(t)}{dt}\right]_{max}\right| \tag{10.51}$$

Using these relationships, Carson's rule for the PM bandwidth estimate becomes

$$\begin{aligned}W &= 2\left(\Delta f + B\right) \\ &= 2\left(\frac{k_p \dot{m}_{peak}}{2\pi} + B\right)\end{aligned} \tag{10.52}$$

From this, we deduce that the PM bandwidth is a function of the message signal's amplitude spectrum. All other parameters being equal, a message signal having most of its harmonics concentrated in lower frequencies generally has a lower transmission bandwidth than a signal with many significant high frequency harmonics.

10.5.2 Digital Bandwidth

Two parameters determine the power spectrum's shape for digital communications: the symbol transition rate and the underlying pulse shaping function. As with FM, the digital power spectrum extends to infinity so there is no finite bandwidth representation. However, designers use several common measures to estimate bandwidth. The first null's location in the spectrum (the frequency at which the amplitude spectrum first goes to zero before rising back) is often used because it scales directly with the transmission symbol rate.

Designers and regulators frequently use the magnitude spectrum bandwidth that contains 99% of the total signal spectrum. Any filtering the designers apply to the transmission strongly affects this measure.

First, we examine the unfiltered bandwidth measures. Standard references contain derivations of these results [Ande99; Skla01]. Next, we examine the effects of filtering on the transmitted spectra.

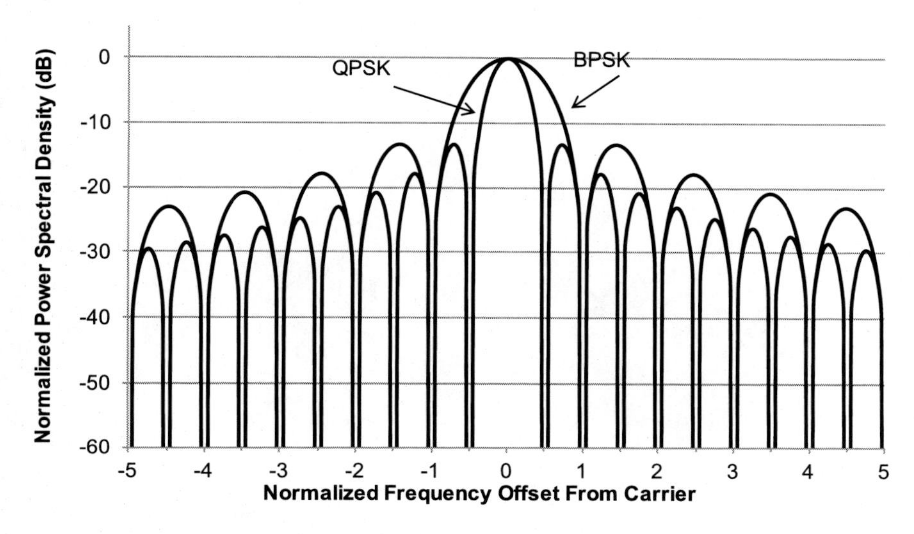

Figure 10.27: Theoretical power spectra for BPSK and QPSK signals relative to the carrier center.

10.5.2.1 Phase Shift Keying

The spectrum shape for unfiltered Phase Shift Keying (PSK) with Non-Return to Zero (NRZ) pulse shaping is determined by the sinc function that describes the NRZ pulse in the frequency domain. The sinc function scales with the bit period, T_b. If we define Δf as the frequency offset from the carrier frequency, then the spectrum, $P(\Delta f)$, for BPSK is

$$P(\Delta f) = T_b \left\{ \frac{\sin(\pi \Delta f T_b)}{\pi \Delta f T_b} \right\}^2 \tag{10.53}$$

Similarly, for QPSK, the spectrum becomes

$$P(\Delta f) = 2T_b \left\{ \frac{\sin(2\pi \Delta f T_b)}{2\pi \Delta f T_b} \right\}^2 \tag{10.54}$$

Figure 10.27 illustrates the normalized, unfiltered transmissions for BPSK and QPSK. In each spectrum, the bit rate 1 bps and the spectrum is relative to a carrier frequency. The figure shows the relative positions of the spectral nulls. Although both transmissions have the same data rate, QPSK has one-half the symbol rate of BPSK so the transmission bandwidth, as measured by the location of the nulls, is half that of BPSK. If we use Bi-Phase (Biϕ) pulse encoding in place of NRZ encoding, the Biϕ pulse spectrum is used in place of the sinc function.

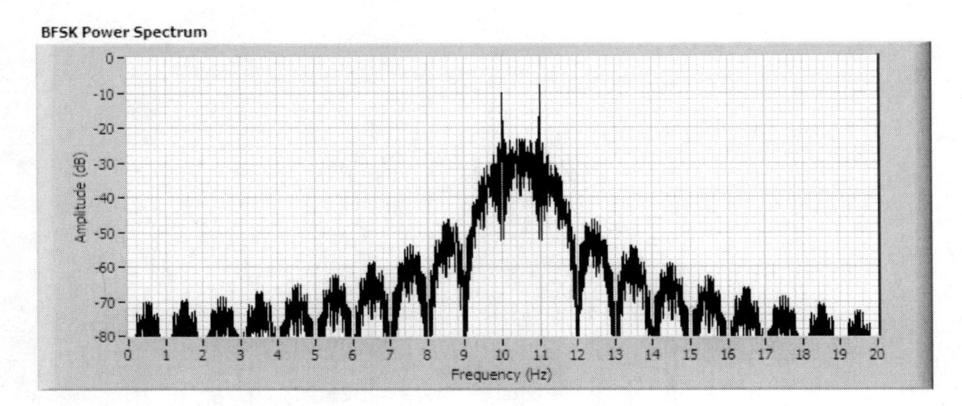

Figure 10.28: Simulated BFSK spectrum with frequency shift equal to R_b.

10.5.2.2 Frequency Shift Keying

The Frequency Shift Keying (FSK) occupied bandwidth is a function of the tone spacing. Figure 10.28 shows a simulated FSK spectrum where the tone spacing is equal to the data rate. Both tone frequencies are in the figure: one at the simulated carrier frequency and one at R_b above the frequency.

We make the transmission bandwidth smaller by using orthogonal frequency spacing, which gives BFSK spectrum as

$$P(\Delta f) = \frac{\sin\left[\pi \Delta f T_b - (\pi/2)\right]}{\pi \Delta f T_b - (\pi/2)} + \frac{\sin\left[\pi \Delta f T_b + (\pi/2)\right]}{\pi \Delta f T_b + (\pi/2)} \tag{10.55}$$

Figure 10.29 plots this equation. The individual tone peaks are no longer in the spectrum but have merged to make a broader central lobe. The first null's location is further from the carrier than the BPSK null.

As we can see in Figures 10.28 and 10.29, the FSK modulation is not band-limited. To remedy this situation, we can apply the pulse shaping techniques that we considered in the PSK modulation techniques to FSK modulation as well. Figure 10.30 illustrates a CPFSK modulation result with Gaussian pulse shaping filtering. As we can see, the Gaussian filtering does greatly reduce the occupied bandwidth with FSK modulation as it does with PSK modulation. This technique is used in commercial and amateur radio low data rate digital modems to control bandwidth on shared channels.

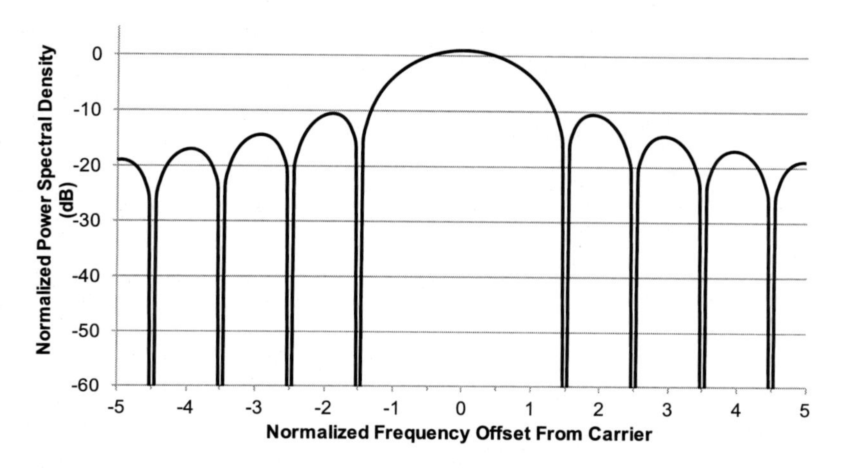

Figure 10.29: Theoretical BFSK spectrum for orthogonal tone spacing.

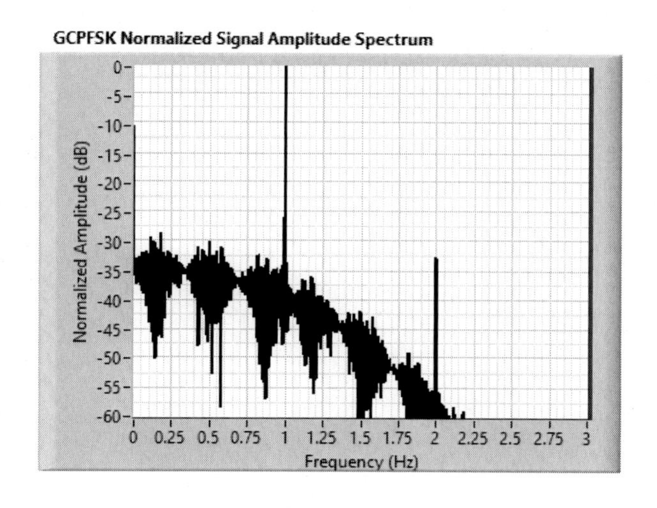

Figure 10.30: Gaussian pulse shaping applied to CPFSK modulation.

10.5.2.3 Minimum Shift Keying, Gaussian Minimum Shift Keying, and Shaped Offset Quadrature Phase Shift Keying

As we saw earlier, we can view the Minimum Shift Keying (MSK), Gaussian Minimum Shift Keying (GMSK), and Shaped Offset Quadrature Phase Shift Keying (SOQPSK) modulation techniques as cases of OQPSK with pulse shaping filters applied before transmission. These filters reduce the side lobes on the QPSK spectrum and make the frequency region containing the 99% total power smaller than the unfiltered 99% total power region. Figure 10.31 illustrates the spectra from simulations of QPSK, MSK, GMSK, and SOQPSK for relative comparison. As we see, the spectra fall off more rapidly than do the unfiltered QPSK spectra.

The theoretical spectrum for MSK is given in terms of the offset relative to the carrier, Δf, using

$$P(\Delta f) = \frac{16}{\pi^2}\left(\frac{\cos 2\pi\Delta f T_s}{1 - 16\Delta f^2 T_s^2}\right)^2 \tag{10.56}$$

The GMSK and SOQPSK spectra are functions of their filter bandwidths. Usually this is measured in normalized fashion with the BT factor where B is the filter bandwidth and T_b is the data bit period. For typical GMSK wireless links, BT ranges from 0.25 to 0.5. For SOQPSK, the standard Raised Cosine roll off factor is 1.0.

10.5.2.4 Quadrature Amplitude Modulation

A typical QAM transmission uses a NRZ pulse shaping on the transmitted data. Because this is the same pulse shaping as used with PSK, the spectrum depends on the symbol transmission rate and not the fact that multiple amplitude levels are used. Figure 10.32 illustrates the output spectrum for a 1-symbol/second QAM transmission. We see that the spectral nulls are in the same location as for a 1 symbol/second BPSK transmissions. However, the 16-QAM transmission is more efficient because it sends four bits per symbol, while BPSK only sends one bit per symbol.

10.5.3 Spectrum Control Issues

Current regulatory and band usage pressures require that transmissions occupy the minimum amount of frequency spectrum possible. See, for example, [IRIG106; RCC120]. The transmission bandwidths of digital and FM signals are not bounded or bandwidth limited, but extend math-

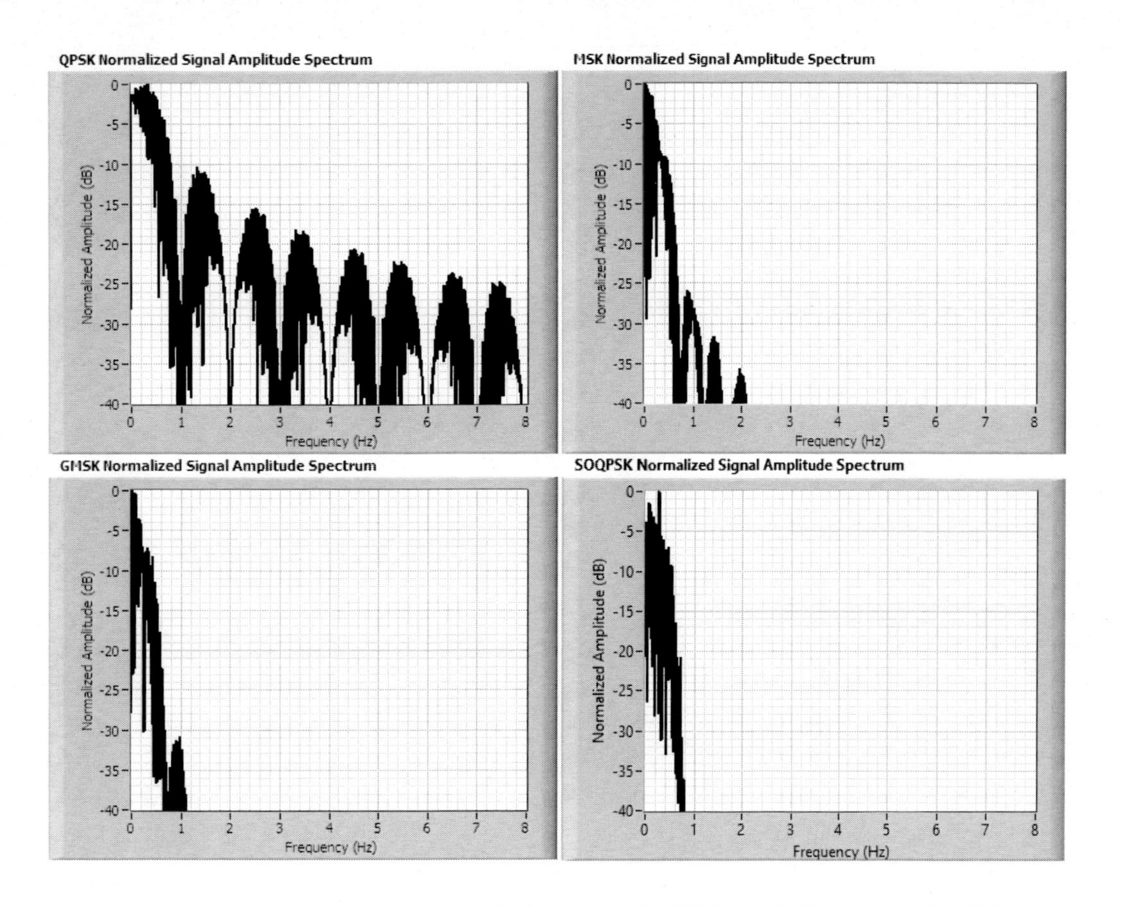

Figure 10.31: Simulated power spectra for QPSK and the related MSK, GMSK, and SOQPSK modulation formats.

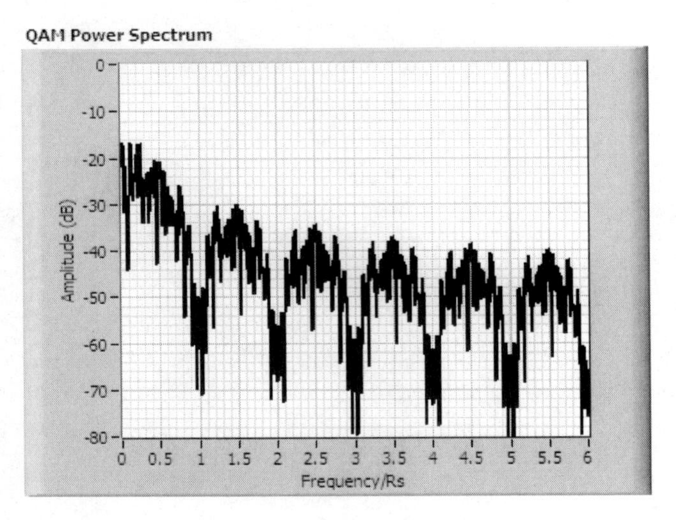

Figure 10.32: Power spectrum for a simulated 16-QAM transmission.

ematically to infinity. In this section, we examine ways to estimate digital transmission bandwidths and how spectral shaping affects the transmission process.

Regulators typically require knowledge of the transmission's integrated spectrum. They usually require knowledge of the 99% total power bandwidth or a similar measure. The question that arises is how to compute this bandwidth. One way is with the Power Out-of-Band (POB) measure used earlier in Section 6.12.6 for baseband signals. The POB is a function of the signal's spectrum, $S(f)$, and is computed using [Math74; Simo76]

$$P_{OB}(f_o) = \frac{\int_{f_o}^{\infty} |S(u)|^2 \, du}{\int_{0}^{\infty} |S(u)|^2 \, du} \tag{10.57}$$

$P_{OB}(f_o)$ measures the amount of the signal that is found from the frequency f_o to infinity. Figure 10.33 shows a comparison of the $P_{OB}(f)$ for QPSK and SOQPSK using the simulations that generated Figure 10.31. For 99% signal power bandwidth, we must determine the frequency yielding a POB of 1% (or a POB of $-20\,$dB). From Figure 10.33, we see that applying the SOQPSK filtering yields a smaller 99% bandwidth (0.5 Hz) than QPSK (8 Hz). Other filtering techniques that designers consider for efficient spectral use are GMSK and Feher's Quadrature

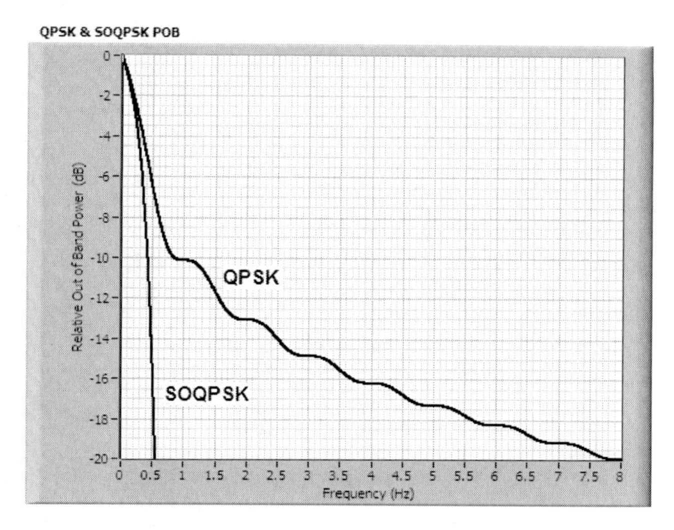

Figure 10.33: The POB computed for QPSK and SOQPSK transmissions.

Phase Shift Keying (FQPSK). GMSK is frequently used in the wireless industry [Rapp96], while FQPSK is part of the range telemetry standards [IRIG106; RCC119]. There are two realizations of FQPSK. The first is using Feher's licensed procedure [Fehe95] and the second is using Simon's approximation [Simo00a; Simo00b; Bora01]. Further research to obtain even greater spectral efficiency than present techniques allow is ongoing.

10.6 SYSTEM PLANNING

The designers and users of telemetry and telecommand systems cannot operate with an arbitrary carrier frequency or bandwidth utilization that suits them. Rather, various international treaties and governmental regulations control communications links for telemetry and telecommand usage, just as the commercial broadcast services are controlled. In this section, we examine the effects of regulatory issues on the design of telemetry and telecommand systems. This section includes a summary of allocations for spectrum usage and a look at emission standards. These issues constrain the designer's selections for the transmission frequency and bandwidth, which affect the designer's hardware choices due to the nature of the available components, the modulation types, bandwidth available for transmission, and other system parameters. The designer's choice of transmission frequency also carries propagation effects that we will discuss in Chapter 11.

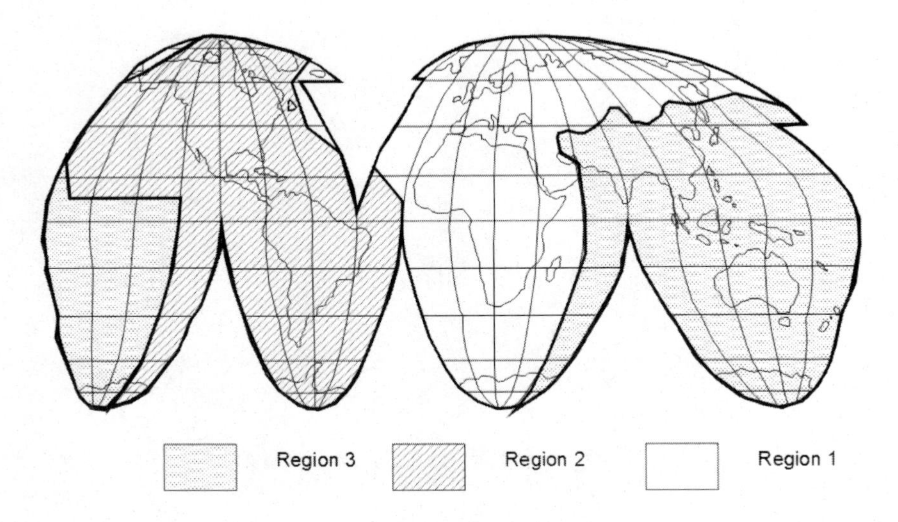

Figure 10.34: The three worldwide ITU regions.

10.6.1 Telemetry Frequency Allocations

International arrangements allocate the radio spectrum in an attempt to provide access to all nations for commercial, military, aerospace, and scientific uses. These allocations are functions of the world region in which the radio emissions emanate and the carriers' frequencies. The *Radio Regulations* from the International Telecommunications Union (ITU) divide the allocations between governmental and nongovernmental users. Figure 10.34 shows the three geographic ITU regions. North and South America comprise Region 2. The allocations in one ITU region may not extend to other ITU regions, so designers need to practice care with designs used internationally.

Can a user transmit on a "telemetry" band without further consideration? The quick answer is "no." Usually, the system designer needs to file a spectrum application request with an appropriate agency or spectrum coordination office. If the application is in order, and after that office has properly coordinated the request with other users or peer coordinators, the agency or office grants an *authorization* to use a particular frequency *assignment* and specify under what conditions the frequency is used. The exact nature of the license or coordination activity depends upon who is using the frequency and for what purpose. The bottom line is that coordination between users is required to ensure equitable access and the user may not start transmitting without the authorization from the appropriate authority.

10.6.1.1 Telemetry Stations and Services

The ITU recognizes certain services specifically for telemetry and telecommand uses. These services are not the only ones that carry telemetry or telecommand data but they represent services specifically named for that purpose. The National Telecommunications and Information Administration (NTIA) *Manual of Regulations and Procedures for Federal Radio Frequency Management* provides the following service definitions [NTIA13] based on the ITU *Radio Regulations*:

Aeronautical Telemetering Land Station: A telemetering land station used in the flight testing of manned or unmanned aircraft, missiles, or major components thereof.

Aeronautical Telemetering Mobile Station: A telemetering mobile station used for transmitting data directly related to the airborne testing of the vehicle (or major components) on which the station is installed.

Earth Exploration-Satellite Service: A radio communication service between Earth stations and one or more space stations, which may include links between space stations, in which:

i. Information relating to the characteristics of the Earth and its natural phenomena is obtained from active sensors or passive sensors on Earth satellites;
ii. Similar information is collected from airborne or Earth-based platforms;
iii. Such information may be distributed to Earth stations within the system concerned;
iv. Platform interrogation may be included.

This service may also include feeder links necessary for its operation.

Flight Telemetering Land Station: A telemetering land station the emissions of which are used for telemetering to a balloon; to a booster or rocket, excluding a booster or rocket in orbit about the Earth or in deep space; or to an aircraft, excluding a station used in the flight testing of an aircraft.

Flight Telemetering Mobile Station: A telemetering mobile station used for transmitting data from an airborne vehicle, excluding data related to airborne testing of the vehicle itself, or major components thereof.

Oceanographic Data Station: A station in the maritime mobile service located on a ship, buoy, or other sensor platform the emissions of which are used for transmitting oceanographic data.

Radiosonde: An automatic radio transmitter in the meteorological aids service usually carried on an aircraft, free balloon, kite, or parachute, and which transmits meteorological data.

Radiotelemetry: Telemetry by means of radio waves.

Space Operation Service: A radiocommunication service concerned exclusively with the operation of spacecraft, in particular space tracking, space telemetry and space telecommand. These functions will normally be provided within the service in which the space station is operating.

Space Telecommand: The use of radiocommunication for the transmission of signals to a space station to initiate, modify, or terminate functions of equipment on an associated space object, including the space station.

Space Telemetry: The use of telemetry for the transmission from a space station of results of measurements made in a spacecraft, including those relating to the functioning of spacecraft.

Surface Telemetering Land Station: A telemetering land station the emission of which are intended to be received on the surface of the Earth.

Surface Telemetering Mobile Station: A telemetering mobile station located on the surface of the Earth and the emissions of which are intended to be received on the surface of the Earth.

Telecommand: The use of telecommunication for the transmission of signals to initiate, modify, or terminate functions of equipment at a distance.

Telecommand Aeronautical Station: A land station in the aeronautical mobile service the emissions of which are used for terrestrial telecommand.

Telecommand Aircraft Station: A mobile station in the aeronautical mobile service the emissions of which are used for terrestrial telecommand.

Telecommand Base Station: A land station in the land mobile service the emissions of which are used for terrestrial telecommand.

Telecommand Coast Station: A land station in the maritime mobile service the emissions of which are used for terrestrial telecommand.

Telecommand Fixed Station: A fixed station in the fixed service the emissions of which are used for terrestrial telecommand.

Telecommand Land Station: A land station in the land mobile service the emissions of which are used for terrestrial telecommand.

Telecommand Land Mobile Station: A mobile station in the land mobile service the emissions of which are used for terrestrial telecommand.

Telecommand Mobile Station: A mobile station in the mobile service the emissions of which are used for terrestrial telecommand.

Telecommand Ship Station: A mobile station in the maritime mobile service the emissions of which are used for terrestrial telecommand.

Telemetering Fixed Station: A fixed station the emissions of which are used for telemetering.

Telemetering Land Station: A land station the emissions of which are used for telemetering.

Telemetering Mobile Station: A mobile station the emissions of which are used for telemetering.

Telemetry: The use of telecommunications for automatically indicating or recording measurements at a distance from the measuring instrument.

Table 10.4: Specific Range Telemetry Bands [IRIG106]

Frequency Range	Usage
1435 to 1525 MHz	Aeronautical telemetry
1525 to 1535 MHz	Mobile service, including aeronautical telemetry, as a secondary allocation
2200 to 2290 MHz	Government fixed, mobile, space research, space operation, and Earth-exploration services
2290 to 2300 MHz	Space research for deep space shared with fixed and mobile (except aeronautical mobile) services
2310 to 2360 MHz	Aeronautical telemetry as a secondary service
2360 to 2395 MHz	Aeronautical telemetry and telecommand for flight testing
4400 to 4940 MHz	Telemetry as part of the Mobile Service
5091 to 5150 MHz	Aeronautical mobile telemetry
5925 to 6700 MHz	Mobile Service

Telemetering and telecommand system designers may use other land, mobile, aeronautical, and space services for various data services. These uses are not explicitly included in the tables as telemetry or telecommand uses but come under general usage and the designer accesses them with proper coordination. Table 10.4 lists frequency bands commonly used in range telemetry [IRIG106], while Table 10.5 lists representative bands from the U.S. frequency allocation table [NTIA13]. The reader is referred to the current U.S. allocation table for other possibilities.

10.6.1.2 Band Sharing

One future trend in band allocation, especially in the Ultra High Frequency (UHF) and Super High Frequency (SHF) bands, is the pressure to re-allocate the bands from their traditional telemetry services. In some cases, spectrum actions to move spectrum allocations from government-usage bands to commercial-usage bands will continue. In other cases, bands will change from exclusive use to shared use where government and commercial users will need to cooperate on band usage. In either case, the link designers will need to track how the spectrum allocations change as time goes forward.

Table 10.5: Other Telemetry and Telecommand Bands [NTIA13]

Frequency Range	Usage
400 to 406 MHz	Radiosonde; space operation; Earth exploration satellite
449.75 to 450.25 MHz	Space operation and space research services
902 to 928 MHz	Industrial, scientific and medical (ISM) band
1400 to 1427 MHz	Earth exploration satellite
1427 to 1435 MHz	Land mobile (telemetering and telecommand); Fixed (telemetering)
1670 to 1700 MHz	Radiosonde
2360 to 2400 MHz	Medical radio operations
2400 to 2500 MHz	Industrial, scientific and medical (ISM) band
5725 to 5875 MHz	Industrial, scientific and medical (ISM) band

10.6.1.3 Matched Bands

When system designers choose frequency bands for telemetry and telecommand, certain bands on the allocation table have predesignated pairs of links. This allows users to design transmitters and receivers that have sufficient carrier frequency separation to avoid interference but are close enough that convenient hardware design allows for commonality in components used to synthesize the oscillators used to make the transmitters and receivers. For example, NASA uses paired S-Band frequencies ($\approx$2.2 GHz) for near-Earth telemetry and telecommand services and paired X-Band frequencies ($\approx$8 GHz) for deep-space operations.

10.6.2 Emission Standards

In addition to restrictions on carrier frequency selection imposed by regulations, the output spectrum is subject to restrictions in addition to the bandwidth restrictions dictated by hardware sensitivity. These restrictions must achieve a compromise between allowing multiple users to share the RF spectrum and not degrading the transmission system's performance. This section covers these regulatory restrictions.

10.6.2.1 Necessary Bandwidth

Regulators use several common rules for estimating the necessary bandwidth, B_n, to transmit data using various modulation techniques. For any emission class, the *necessary bandwidth* is defined as "the width of the frequency band which is just sufficient to ensure the transmission of information at the rate and with the quality required under specified conditions" [NTIA13]. System designers use the following equations from the NTIA *Manual* [NTIA13] to estimate necessary bandwidth. This necessary bandwidth then becomes the basis for the *authorized bandwidth* that the approving authority uses when granting a transmission license or permission to operate.

The BFSK necessary bandwidth, B_n, is computed as a function of the peak frequency deviation, D, and the digital data rate, R, using one of the following equations based on the values of D and R:

$$\begin{aligned} \left(0.03 < \frac{2D}{R} < 1.0\right) &: B_n = 3.86D + 0.27R \\ \left(1.0 < \frac{2D}{R} < 20\right) &: B_n = 2.4D + 1.0R \end{aligned} \tag{10.58}$$

The PSK necessary bandwidth is computed in terms of the number of digital symbols, S, the digital data rate, R, and a transmission factor, K, that includes the effect of signal distortion, using the relationship

$$B_n = \frac{2RK}{\log_2 S} \tag{10.59}$$

QAM uses a similar relationship for computing the necessary bandwidth

$$B_n = \frac{2R}{\log_2 S} \tag{10.60}$$

The MSK necessary bandwidth scales in terms of the digital data rate, R

$$\begin{aligned} 2-ary\ MSK &: \ B_n = 1.18R \\ 4-ary\ MSK &: \ B_n = 2.34R \end{aligned} \tag{10.61}$$

10.6.2.2 Spectral Masks

The necessary bandwidth and POB measurements mentioned above are useful for estimating the 99% power points in the transmission. However, regulators also look at how quickly the transmission spectrum decays and how closely data channels are packed. To do this, system designers compare the transmission PSD with spectral masks to determine if the transmission interferes with other users that may have a carrier frequency close to the user's.

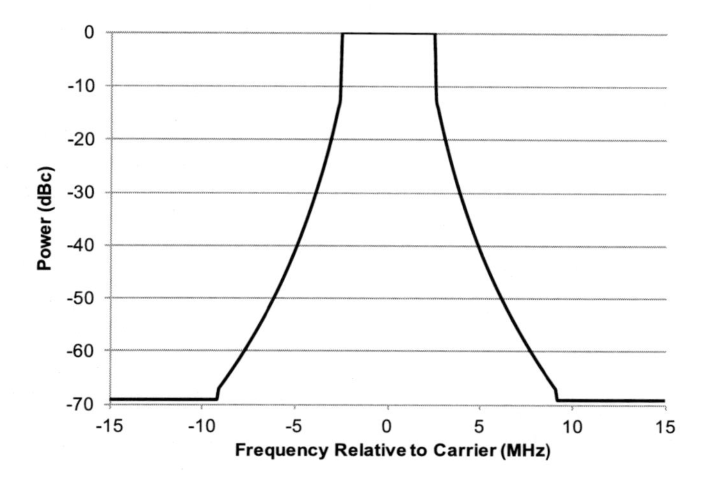

Figure 10.35: Transmission spectral mask example for a 10-Mbps data link using 4-ary modulation.

The telemetry standards [IRIG106] define the range telemetry spectral mask for analog and digital transmitters relative to the carrier power, P. All spectral components that are attenuated less than $(55 + 10log(P))$ dB below the authorized carrier power or are greater than -25 dBm absolute power must be below the mask level. The standard defines the mask function, $M(f)$, for those frequencies, f, in MHz, relative to the carrier frequency, f_c, outside the region $|f - f_c| \geqslant R/m$ where R is the bit rate in Mbps and m is the number of states in the modulation technique. The standard mask is

$$M(f) = K + 90\log(R) - 100\log|f - f_c| \qquad (10.62)$$

K is a scale factor depending upon the modulation technique used. For analog signals, $K = -20$, for binary signals, $K = -28$, and for SOQPSK, $K = -61$. The -25-dBm bandwidth restriction does not apply to frequencies within 1 MHz of the carrier frequency. Figure 10.35 gives the spectral mask for a 10 Mbps transmission using QPSK modulation and a maximum transmission power of 25 W.

The Federal Communications Commission (FCC) regulates the various radio emission types produced by electronic equipment. Its regulations influence the design and performance of actual telemetry systems. The regulations limit the emissions outside the necessary transmission bandwidth that the user is authorized to occupy just as regulations on the

Table 10.6: Spectral Mask for Telemetry Transmission [NTIA13]

% Authorized Bandwidth	Required Attenuation
$f < 50\%$	0 dB
$50\% < f \leqslant 250\%$	The greater of 50 dB or $A = 35 + 0.8\,(W(f)\% - 50\%) + 10\log(W)$ The attenuation need not exceed 80 dB
$f > 250\%$	The lesser of 80 dB or $A = 43 + 10\log(P)$

telemetry ranges do. The exact mask function depends on the frequency band and service in use and the discussion here illustrates a typical example. The spectral emission rules are in terms of the percent of authorized bandwidth which is generally the necessary bandwidth for the data and modulation format. If the user's authorized transmission bandwidth is W, then the percent of authorized bandwidth, $W(f)\%$, as a function of frequency, f, is defined in terms of the carrier frequency, f_c, as

$$W(f)\% = \frac{|f - f_c|}{W} \times 100\% \tag{10.63}$$

Table 10.6 presents a typical spectral mask relative to the average transmitter power, P, for a terrestrial telemetry service. The operation is similar to the one in Equation (10.62), not in terms of spectral density, but as a 4-kHz bandwidth at every frequency sampling point. Mathematically, this is represented for a given frequency point, f_i, and a spectral density, $S(f)$, as

$$M\left(f_i\right) = \int_{f_i - 2000Hz}^{f_i + 2000Hz} S(f)\mathrm{d}f \tag{10.64}$$

The system designer must evaluate this integral at each frequency point across the bandwidth of interest.

10.6.3 Intermodulation Effects

One spectral effect that can cause problems is Intermodulation (IM) interference that occurs when a nonlinear device is used in the signal mixing process. The mixer multiplies the desired signals and produces

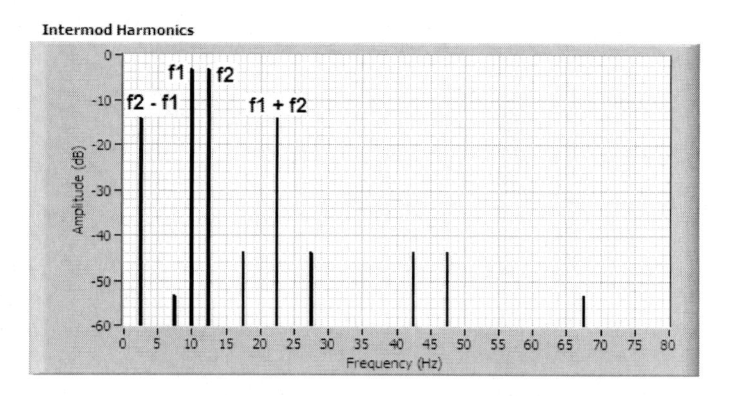

Figure 10.36: Spurious emissions caused by a nonlinear mixing process.

harmonics that can cause interference with signals near the desired carrier frequency. For example, an ideal commercial mixer may take two sinusoidal signals at f_1 and f_2 and produce harmonics at f_1, f_2, $f_1 + f_2$, and $f_2 - f_1$. Depending upon the application, filters isolate the desired components: $f_1 + f_2$ or $f_2 - f_1$. A nonideal mixer might have its output represented by

$$s(t) = a_0 \sin(\omega_1 t) + a_0 \sin(\omega_2 t) + a_1 \left[\sin(\omega_1 t) \sin(\omega_2 t)\right] + a_2 \left[\sin(\omega_1 t) \sin(\omega_2 t)\right]^3 \tag{10.65}$$

The first three terms give the same output as the ideal mixer. The final term causes a signal distortion. The output spectrum in Figure 10.36 shows the IM distortion from the mixing process in Equation (10.65) in the form of excess components in the spectrum. The labeled components in Figure 10.36 are the expected ones. The unlabeled components are spurious harmonics that cause interference with other users if transmitted.

10.6.4 Unequal Data Rates

One characteristic of telemetry and telecommand links is that the link planner does not require the same bandwidth on the telecommand link and telemetry links. For example, the telecommand link at 450 MHz in Table 10.5 has a relatively narrow bandwidth and, consequently, only a low data rate is supported. Since commands generally do not require the same data volume as the telemetry return link to support operations , this is not a problem. The telemetry link, usually in another band, supports

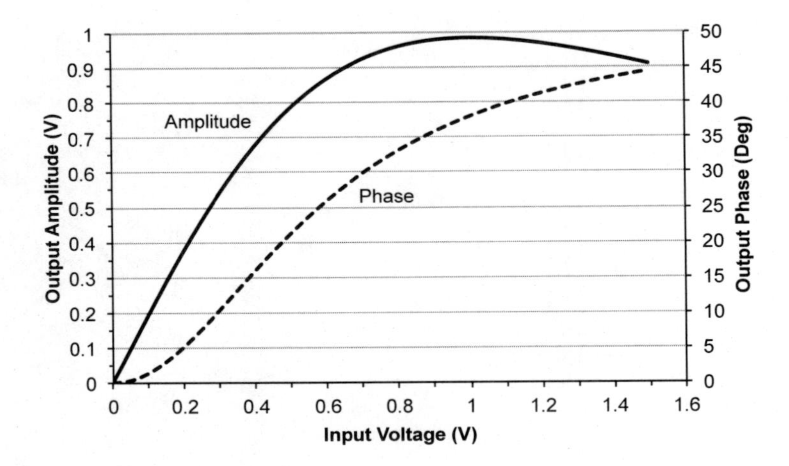

Figure 10.37: Traveling Wave Tube Amplifier amplitude and phase model.

the data bandwidth requirements of that link and not the telecommand link. The one condition where this may affect operations is when the system uses Transmission Control Protocol/Internet Protocol (TCP/IP) based transmissions, as we discussed in Section 6.10.4. The unequal data rates do not allow acknowledgments to flow as quickly as equal-rate channels allow so the throughput may drop.

10.6.5 Spectral Regrowth

When nonlinear devices amplify modulator outputs, the power spectrum for the transmission can grow wider due to the nonlinear effects. This is especially bothersome in closely spaced transmission channels where the spectral growth causes *adjacent channel interference* between users. This spectral regrowth can even occur when the modulator filters the output. To see this effect, we have simulated both a FQPSK transmission and a QPSK transmission through a Traveling Wave Tube Amplifier (TWTA) with the tube operating at saturation (maximum amplitude gain). The model for the TWTA is from [Sale81]. Figure 10.37 shows the amplitude and phase characteristics of the TWTA model. We see that the amplitude and phase do not conform to the ideal filter specifications given in Section 5.7.3.1 so nonlinear behavior is expected.

Figure 10.38 gives the results for FQPSK modulation where the side lobes of the modulation are nearly identical on the input and the output

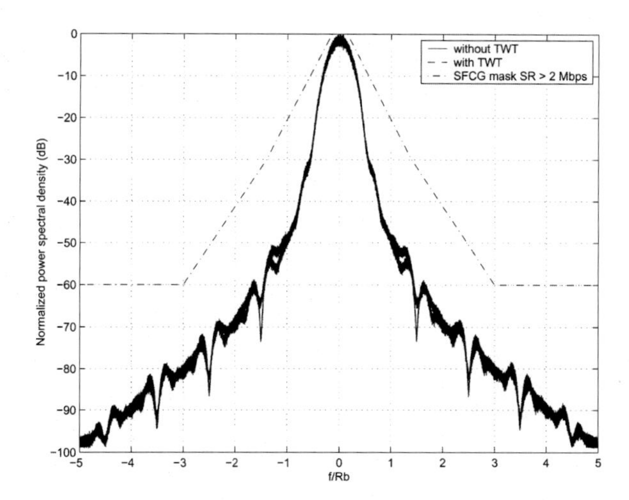

Figure 10.38: Simulated FQPSK spectral regrowth performance before and after the TWTA [Bora01].

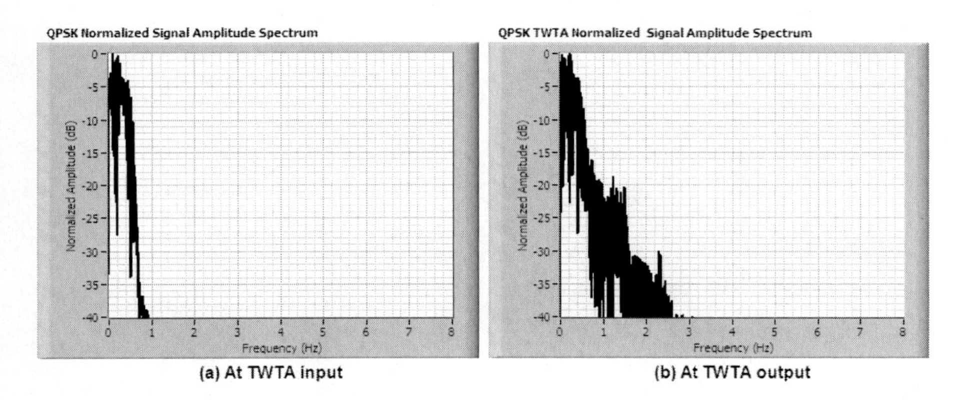

Figure 10.39: Simulated QPSK spectral regrowth performance before and after the TWTA.

Table 10.7: Recommended Settings for Transmitting Digital Data Using Analog Modulators [RCC119]

	Waveform Encoding/Modulation			
Parameter	**NRZ/FM**	**Biϕ/FM**	**NRZ/PM**	**Biϕ/PM**
Peak deviation	$0.35R_b$	$0.65R_b$	90°	90° for -M or -S; 75° for -L
Pre-modulation filtering bandwidth	$\geqslant 0.7R_b$	$\geqslant 1.4R_b$	$\geqslant 1.0R_b$	$\geqslant 2R_b$
Receiver IF bandwidth	$\approx R_b$	$\geqslant 2R_b$	As wide as possible	As wide as possible
Output bandwidth	$0.5R_b$ to Rb	R_b to $2R_b$	$\geqslant 1.0R_b$	$\geqslant 2R_b$
Maximum bit rate in 1-MHz channel	740 kbps	333 kbps	333 kbps	125 kbps (90°); 166 kbps (75°)

of the transmission. This implies that FQPSK has very little spectral regrowth and this performance is very desirable for closely spaced channels. Figure 10.39 illustrates the results for QPSK filtered at the first spectral null. The spectral regrowth here is undesirable and the benefits of the filtering are lost due to the amplifier. The alternative is to back off the amplifier from saturation to its linear region. While this gives better spectral performance, it also reduces the amplifier's gain, which may affect transmission performance.

10.7 DIGITAL OVER ANALOG TRANSMISSION

While this chapter has emphasized digital modulators, system designers can send digital data with analog components. Table 10.7 lists standard experimentally determined parameter settings for NRZ and Biϕ waveform encoding used with FM and PM modulators [RCC119]. In the table, R_b is the data bit rate.

10.8 REFERENCES

[Ande99] J. B. Anderson. *Digital Transmission Engineering.* Piscataway, NJ: IEEE Press, 1999.

[Arms36] E. H. Armstrong. "A Method of Reducing Disturbances in Radio Signaling by a System of Frequency Modulation." In: *Proceedings of the Institute of Radio Engineers* 24.5 (May 1936), pp. 689 –740. DOI: 10.1109/JRPROC.1936.227383.

[Bora01] D. K. Borah and S. Horan. *EFQPSK versus CERN: A Comparative Study.* NMSU-ECE- 01-014. New Mexico State University. Las Cruces, NM, Sept. 2001.

[Cars22] J. R. Carson. "Notes on the Theory of Modulation." In: *Proceedings of the Institute of Radio Engineers* 10.1 (Feb. 1922), pp. 57 –64. DOI: 10.1109/JRPROC.1922.219793.

[Cars37] J. R. Carson and T. C. Fry. "Variable frequency electric circuit theory with application to the theory of frequency-modulation." In: *Bell System Technical Journal* 16.4 (Oct. 1937), pp. 513 –540. DOI: 10.1002/j.1538-7305.1937.tb00766.x.

[CCSDS4010] *Radio Frequency and Modulation Systems – Part 1 Earth Stations and Spacecraft.* CCSDS 401.0-B-26. Consultative Committee for Space Data Systems. Washington, D.C., Oct. 2016. URL: https://public.ccsds.org/Pubs/401x0b26.pdf.

[Couc01] L. W. Couch. *Digital and Analog Communications Systems.* 6th Edition. Upper Saddle River, NJ: Prentice Hall, 2001. ISBN: 0-13-081223-4.

[Dave87] W. B. Davenport and W. L. Root. *An Introduction to the Theory of Random Signals and Noise.* New York, NY: IEEE Press, 1987. ISBN: 0-87942-235-1.

[Elze72] H. C. van den Elzen and P. van der Wurf. "A Simple Method of Calculating; the Characteristics of FSK Signals with Modulation Index 0.5." In: *IEEE Transactions on Communications* COM-20.2 (Apr. 1972), pp. 139 –147. DOI: 10.1109/TCOM.1972.1091143.

[Ever40] W. L. Everitt. "Frequency Modulation." In: *Transactions of the American Institute of Electrical Engineers* 59.11 (Nov. 1940), pp. 613 –625. DOI: 10.1109/T-AIEE.1940.5058020.

[Fehe95] K. Feher. *Wireless Communications: Modulation And Spread Spectrum Applications.* Upper Saddle River, NJ: Prentice Hall PTR, 1995. ISBN: 0130986178.

[Gron76] S. Gronemeyer and A. McBride. "MSK and Offset QPSK Modulation." In: *IEEE Transactions on Communications* COM-24.8 (Aug. 1976), pp. 809 –820. DOI: 10.1109/TCOM.1976.1093392.

[Hill00] T. J. Hill. "An Enhanced, Constant Envelope, Interoperable Shaped Offset QPSK (SOQPSK) Waveform for Improved Spectral Efficiency." In: *Proc. International Telemetering Conf.* Vol. 36. San Diego, CA, Oct. 2000, p. 00.02.5.

[IRIG106] Telemetry Group. *Telemetry Standards.* 106-15. Secretariat, Range Commanders Council. White Sands, NM, July 2015. URL: http://www.wsmr.army.mil/RCCsite/Documents/Forms/AllItems.aspx.

[Lath98] B. P. Lathi. *Modern Digital and Analog Communication Systems.* 3rd Edition. New York, NY: Oxford University Press, 1998.

[Math74] H. Mathwich, J. Balcewicz, and M. Hecht. "The Effect of Tandem Band and Amplitude Limiting on the Eb/No Performance of Minimum (Frequency) Shift Keying (MSK)." In: *IEEE Transactions on Communications* COM-22.10 (Oct. 1974), pp. 1525 –1540. DOI: 10.1109/TCOM.1974.1092108.

[Muro81] K. Murota and K. Hirade. "GMSK Modulation for Digital Mobile Radio Telephony." In: *IEEE Transactions on Communications* COM-29.7 (July 1981), pp. 1044 –1050. DOI: 10.1109/TCOM.1981.1095089.

[NTIA13] *Manual of Regulations and Procedures for Federal Radio Frequency Management.* 2013th ed. Dept. of Commerce, National Telecommunications and Information Administration. Washington, D.C., May 2013. URL: http://www.ntia.doc.gov/files/ntia/publications/redbook/2013/May_2013_Edition_of_the_NTIA_Manual.pdf.

[Osbo74] W. P. Osborne and M. Luntz. "Coherent and Noncoherent Detection of CPFSK." In: *IEEE Transactions on Communications* COM-22.8 (Aug. 1974), pp. 1023 –1036. DOI: 10.1109/TCOM.1974.1092333.

[Rapp96] T. S. Rappaport. *Wireless Communications Principles and Practice.* Upper Saddle River, NJ: Prentice Hall PTR, 1996. ISBN: 0-13-461088-1.

[RCC119] E. Law. *Telemetry Applications Handbook.* 119-06. Secretariat, Range Commanders Council. White Sands, NM, May 2006. URL: http://www.wsmr.army.mil/RCCsite/Pages/Publications.aspx.

[RCC120] Telemetry Group RF Systems Committee. *Telemetry (TM) Systems Radio Frequency (RF) Handbook*. 120-08. Secretariat, Range Commanders Council. White Sands, NM, May 2008. URL: `http://www.wsmr.army.mil/RCCsite/Pages/Publications.aspx`.

[Rice09] M. Rice. *Digital Communications: A Discrete-Time Approach*. Upper Saddle River, NJ: Pearson Prentice Hall, 2009. ISBN: 0-13-030497-2.

[Rode31] H. Roder. "Amplitude, Phase, and Frequency Modulation." In: *Proceedings of the Institute of Radio Engineers* 19.12 (Dec. 1931), pp. 2145 –2176. DOI: `10.1109/JRPROC.1931.222283`.

[Sale81] A. A. M. Saleh. "Frequency-Independent and Frequency-Dependent Nonlinear Models of TWT Amplifiers." In: *IEEE Transactions on Communications* COM-29.11 (Nov. 1981), pp. 1715 –1720. DOI: `10.1109/TCOM.1981.1094911`.

[Simo00a] M. K. Simon and T.-Y. Yan. "Unfiltered Feher-patented quadrature phaseshift-keying (FQPSK): Another interpretation and further enhancements, Part 1: Transmitter implementation and optimum reception." In: *Applied Microwave & Wireless* (Feb. 2000), pp. 76 –96.

[Simo00b] M. K. Simon and T.-Y. Yan. "Unfiltered Feher-patented quadrature phaseshift-keying (FQPSK): Another interpretation and further enhancements, Part 2: Suboptimum reception and performance comparisons." In: *Applied Microwave & Wireless* (Mar. 2000), pp. 100 –105.

[Simo76] M. K. Simon. "A Generalization of Minimum-Shift-Keying (MSK)-Type Signaling Based Upon Input Data Symbol Pulse Shaping." In: *IEEE Transactions on Communications* COM-24.8 (Aug. 1976). DOI: `10.1109/TCOM.1976.1093380`.

[Skla01] B. Sklar. *Digital Communications: Fundamentals and Applications*. 2nd Edition. Upper Saddle River, NJ: Prentice Hall PTR, 2001.

[Yuen82b] J. H. Yuen et al. "Telemetry System." In: *Deep Space Telecommunications Systems Engineering*. Ed. by J. H. Yuen. JPL Publication 82-76. Pasadena, CA: Jet Propulsion Laboratory, California Institute of Technology, National Aeronautics, and Space Administration, 1982. Chap. 5, pp. 179 –341.

10.9 PROBLEMS

1. The signal, $m(t)$, used to generate Figure 10.4 is

$$m(t) = 0.74\sin(2\pi t) + 0.25\sin(4.69134\pi t) - 0.1\sin(9.9752\pi t) + 0.15\sin(11.5308\pi t) - 0.1\sin(14.642\pi t)$$

   Find the instantaneous phase and instantaneous frequency for this signal when FM is used. Assume that $k_f = 100\,\text{Hz/V}$. Plot the values for $m(t)$, the instantaneous phase, and the instantaneous frequency versus time.

2. Repeat Problem 1 for PM. Assume that $k_p = 0.75\,\text{rad/V}$.

3. Using the results from Problems 1 and 2, plot the carrier as a function of time when the carrier frequency is 200 Hz.

4. Plot the instantaneous phase, the instantaneous frequency, and the carrier for an analog FM system when the carrier frequency is 3 Hz and the modulating data signal is digital with a bit period of 1 bps. The pulses have logic 0 as -1 V and logic 1 as $+1$ V. The data bits are 101 011 001 101. The deviation sensitivity is $k_f = 1.0\,\text{Hz/V}$.

5. Repeat Problem 4 for PM with $k_p = \pi/2$ rad/V.

6. A baseband signal's PSD is represented using the same form as a Rayleigh distribution using a parameter $\alpha = 3000$ Hz

$$S(f) = \frac{|f|}{\alpha}\, e^{-f^2/2\alpha^2}$$

   Is this signal better transmitted with FM or PM? Justify in terms of the mean square bandwidth for the signal.

7. Assume that a single tone modulates a FM system with β_{FM}=2 and a tone frequency of 500 Hz. Sketch the Bessel function-based amplitude spectrum. Indicate the expected bandwidth using Carson's rule. How do the amplitudes match with the estimate? What is the total signal power contained in the Carson's rule bandwidth?

8. Estimate the bandwidth for the signal in Problems 1 and 2.

9. Assume that the modulating data signal is an exponential Fourier series representation. How might the carrier signal for a FM system be represented for the signal when computed in terms of the Bessel function expansion?

10. Justify the approximation in Equation (10.21) for the FM threshold point by plotting this threshold point with Equation (10.20) for various values of the modulation index, β_{FM}. Justify that the FM improvement ratio can be approximated by $3/2\beta_{FM}^2$. For what input SNR value does this break down when the modulation index is $\beta_{FM} = 1$, 5, and 10?

11. Show that the "phase modulation" form for the carrier representation in Equation (10.26) is equivalent to the "amplitude modulation" form of BPSK in Equation (10.43) and QPSK in Equation (10.28).

12. Plot the carrier waveform for BPSK and QPSK modulation when the carrier frequency is 3 Hz and the modulating data signal is digital with a bit period of 1 bps. The pulses have logic 0 as +1 V and logic 1 as −1 V. The data bits are 101 011 001 101.

13. Determine the minimum tone spacing for two sinusoids to make them orthogonal. Relate this tone spacing to a binary bit rate so that orthogonal FSK is made.

14. Plot the carrier waveform for BFSK modulation when the rest carrier frequency is 3 Hz and the modulating data signal is digital with a bit period of 1 bps. The pulses have logic 0 as +1 V and logic 1 as −1 V. The data bits are 101 011 001 101. Use the minimum tone spacing found in Problem 13.

15. For SOQPSK pulse shaping, determine the amplitude value, A, to meet the constraint given in Equation (10.38).

16. Develop a Grey code mapping like Figure 10.9 shows for QPSK and apply that to the QAM constellations shown in Figure 10.22.

17. Sketch the transmitted carrier for 16-QAM using the constellations in Figure 10.22.

18. For a required BER of 10^{-5}, an expected input C/N of 10, and the signal is bandlimited to the first null when (a) BFSK, (b) BPSK, and (c) QPSK modulation is used. What is the expected link margin (the difference between the available C/N and the required C/N, in dB units, for the specified BER) for each modulation type?

19. For the hybrid modulation given in Equation (10.39), construct a simulation similar to that in Figure 10.23. What happens in the

simulation when the PM modulation index is $\theta \approx 0$? What happens when $\theta = \pi/2$? What kind of modulation does this represent at that point?

20. Which modulation technique requires more transmission power for achieving a transmission BER of 10^{-6}: 16-QAM or 16-PSK?

21. Compare the necessary bandwidth, B_n, estimates with the POB estimates for QPSK and 4-ary MSK. Assume $K = 1$ for the computation.

22. A telemetry system, with a maximum output power of 25 W, has an operational bandwidth of 2 MHz. If QPSK modulation is used on the system and filtering is performed at 1.5 times the symbol rate, determine the maximum data rate that can be transmitted and still maintain the power level within the IRIG emission profile.

23. Repeat Problem 22 for 4-ary MSK.

24. Repeat Problem 23 using the FCC mask.

25. Use trigonometric identities to expand Equation (10.65) to locate all of the harmonics produced by the mixing process.

26. Use the Saleh model for a TWTA to simulate spectral regrowth with MSK or QPSK. Operate the amplifier both at saturation and in the linear region. Try adding filters at various points to see if the situation improves. Comment on the effects on spectral regrowth.

CHAPTER 11

MICROWAVE TRANSMISSION

11.1 INTRODUCTION

As we mentioned earlier, we assume that the telemetry and telecommand data are transmitted over a microwave radio channel. We need to understand this process as a system composed of the Radio Frequency (RF) electronic components, the data transmission channel, and possible sources of interference from a variety of sources. Figure 11.1 illustrates these elements that we must consider. The system designer must consider the components and the noise and interference sources when formulating the link design.

This chapter covers practical topics for the system designer including radio transmission through the vacuum and the atmosphere, antennas types and performance metrics in link design, major RF components for transmission and reception, and issues in computing the radio link performance. The highlighted channel block of Figure 11.2 illustrates that these topics apply to both the telemetry and telecommand data links. We cover the topics in the following order:

- Antenna basics
- Link power budget
- System noise temperature
- Rain loss
- Atmospheric phenomena

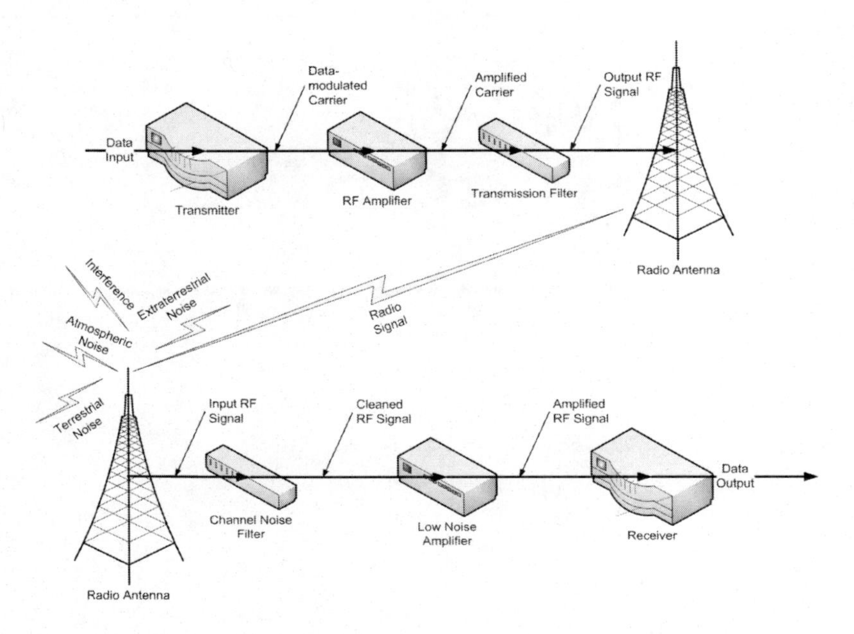

Figure 11.1: Schematic of the microwave channel including the RF components, noise and interference sources, and the transmitted signal.

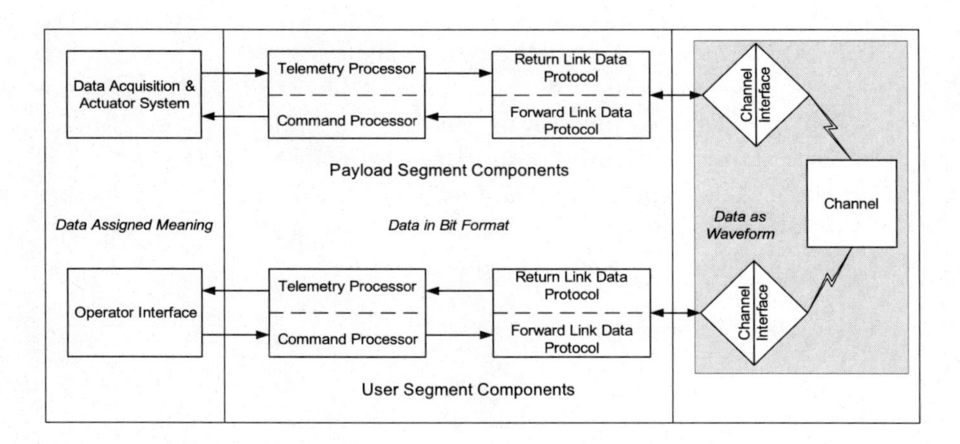

Figure 11.2: Location of the radio channel and associated components in the overall telemetry and telecommand system.

- Radio wave diffraction
- Multipath and mobile propagation issues

These topics give the system designer the tools to evaluate whether the designer's data-transmission component choices perform required tasks with the desired performance metrics

11.2 OBJECTIVES

The system designer's choices for the radio link affect the quality of the transmitted data. It also influences the system designer's choices in other areas, such as the overall system design for weight and power, so the designer's proper understanding of the issues involved is critical to overall success. Our objectives for this chapter are to enable the reader to perform the following tasks:

- Describe how designers use various RF components in configuring transmission and reception hardware.
- Compute parameters for estimating antenna performance.
- Compute a link budget that accounts for path loss, transmission Effective Isotropic Radiated Power (EIRP), data rate, and receiver figure of merit.
- Compute the system noise temperature for a cascaded string of modules using module gain or loss, noise figure, and noise temperature.
- Compute the rain attenuation loss and its effects on link performance.
- Describe other sources of disturbances found in the atmosphere and their effects on propagation through the link.

This chapter gives the designer a set of computational tools to plan the radio link and understand the potential problems. While these tools give the link designer a workable paper solution, the designer must still have detailed performance measurements using the actual equipment and conditions to give precise understanding of link performance.

Table 11.1: Standard Radio Band Designations [NTIA13]

Band Number	Designation	Frequency Range
3	Extremely Low Frequency (ELF)	<3 kHz
4	Very Low Frequency (VLF)	3 to 30 kHz
5	Low Frequency (LF)	30 to 300 kHz
6	Medium Frequency (MF)	300 to 3000 kHz
7	High Frequency (HF)	3 to 30 MHz
8	Very High Frequency (VHF)	30 to 300 MHz
9	Ultra High Frequency (UHF)	300 to 3000 MHz
10	Super High Frequency (SHF)	3 to 30 GHz
11	Extremely High Frequency (EHF)	30 to 300 GHz
12	Sub-millimeter	300 to 3000 GHz

11.3 BACKGROUND

This section describes the structure of the microwave spectrum and the characteristics of the Earth's atmosphere that affect radio propagation. The atmosphere greatly influences transmission quality in ways that system designers do not encounter in a vacuum environment like space. We will see how this influences the designer's choice for band selection, in addition to the regulatory constraints discussed in Section 10.6, for the desired transmission.

11.3.1 Microwave Bands

The *Radio Regulations* from the International Telecommunications Union (ITU) define microwave bands used in transmissions based on the frequency of the carrier. Table 11.1 lists the standard band designations [NTIA13]. A PDF file showing the allocations to specific services within each band is available at `http://www.ntia.doc.gov/files/ntia/publications/spectrum_wall_chart_aug2011.pdf`. Table 11.2 shows how engineers in the communications industry also use a letter designation for the bands [IEEE521]. This naming convention comes from WWII radar designations, which were random designations for security reasons. Radio regulators do not recognize this letter designation for their purposes. The designations show some variability, depending upon the reference source cited.

Table 11.2: Radar Frequency Band Designations [IEEE521]

Band	Frequency Range
VHF	30 to 300 MHz
UHF	300 to 1000 MHz
L	1 to 2 GHz
S	2 to 4 GHz
C	4 to 8 GHz
X	8 to 12 GHz
Ku	12 to 18 GHz
K	18 to 27 GHz
Ka	27 to 40 GHz
V	40 to 75 GHz
W	75 to 110 GHz
mm	110 to 300 GHz

Note: mm is sometimes called G Band

11.3.2 Structure of the Atmosphere

The Earth's atmosphere greatly influences the type and quality of radio propagation as a function of frequency and altitude. Link designers consider two major atmospheric regions for telemetry and telecommand system design: the troposphere and the ionosphere. Figure 11.3 shows the troposphere is the region where we live and where the weather occurs. The ionosphere is above the troposphere and is the region where atmospheric gases are ionized. It is divided into layers or regions where the electron density is relatively high [ARRL05; Hall96]. The ionosphere layers that are of concern for link planning are

- *D* region at approximately 80 km; it is important during the day and fades away at night
- *E* region at approximately 110 km
- *F* region which forms two areas during the day, *F1* at 200 km and *F2* at 300 km, and it fades into a single region at night

The ionosphere influences radio propagation by causing disturbances that are functions of solar activity, time of day, frequency, and the angle at which the radio waves enter. The link designer is concerned with effects such as

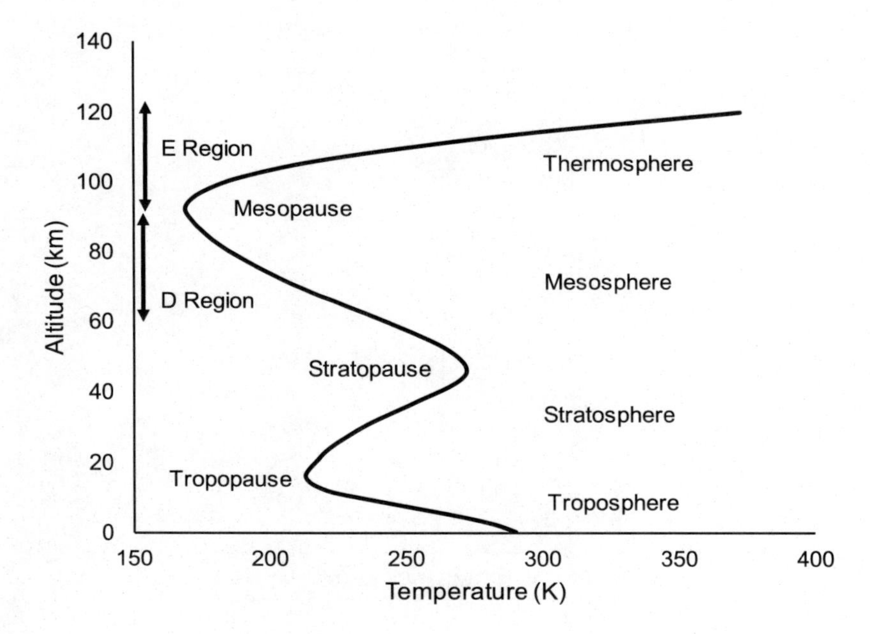

Figure 11.3: Structure of the Earth's atmosphere based on the COSPAR Standard Model Atmosphere. Model is for June at 40° N [COSPAR].

- Group delay or propagation delay in the carrier
- Angle of arrival variations
- Multipath interference effects
- Signal absorption
- Ionospheric scintillation (rapid amplitude, phase, and polarization changes)
- Carrier fading carrier (slow changes in amplitude)

The troposphere influences radio propagation by causing effects such as

- Signal attenuation due to gaseous absorption and rain, cloud, and fog attenuation
- Depolarization of the carrier due to rain and ice, and multipath effects

- Scintillation in the carrier's amplitude and phase due to atmospheric turbulence
- Angle of arrival variations
- Dispersion that changes the frequency and/or phase of the carrier and limits coherence bandwidth of the channel
- Wet surface effects (rain or dew on the antenna and/or feed surfaces)
- Radio noise due to atmospheric gases

Based on the list of possible effects, the link designer might wonder whether any signals ever arrive at their destinations. However, not all of the effects are present at all times and at all frequencies. The job of the link designer is to know which effects are prevalent in the desired channel and influence the transmission. The designer then selects the carrier frequency and transmission mode in accord with the regulatory service constraints. The remainder of this chapter deals with the most prominent effects and the radio bands they most affect.

11.3.3 Radio Propagation Modes

If we are used to thinking of data transmission only over wires, we may be surprised at the number of ways radio waves move between two locations. Figure 11.4 illustrates that radio waves travel between two antennas via four propagation modes:

Sky waves — Radio waves that reflect off the ionosphere

Line of sight — Radio waves that travel along a direct path between the two antennas

Reflected waves — Radio waves that reflect off the surface of the Earth or structures

Surface waves — Radio waves that travel along the Earth's surface

The link designer usually desires to have the data sent along the direct line-of-sight path. However, obstructions in the channel or near the antenna cause reflected waves, for example. Sky waves and surface waves are important at some frequencies; for example, surface waves are important over a distance of 80 km at 1 MHz but only 8 km at 1 GHz. Figure 11.5 illustrates that sky wave propagation is more complicated.

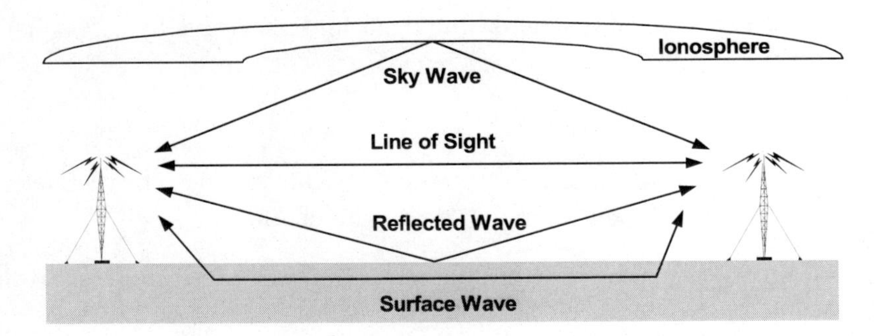

Figure 11.4: Propagation modes for radio propagation.

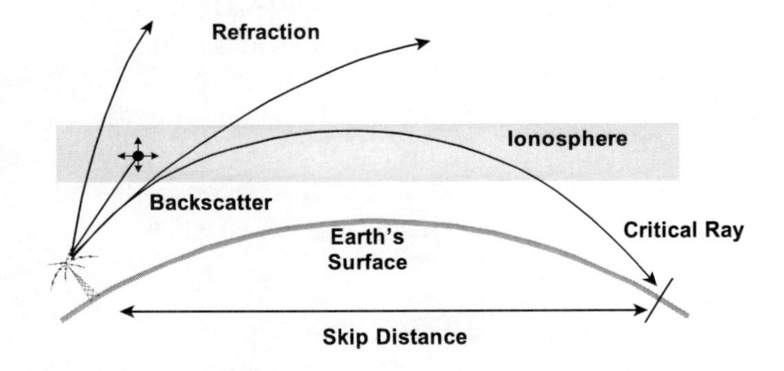

Figure 11.5: Specific propagation paths for sky wave propagation.

Whether a radio wave is scattered, reflected from the ionosphere, or passes directly through it is a function of frequency and angle of arrival.

Sky wave backscatter is a localized reflection of the radio wave back from the ionosphere. Generally, it is not a smooth reflection as from a mirror, but has amplitude and phase variability. The electron density in the ionosphere determines the *critical frequency* and radio transmissions above this frequency can pass through the ionosphere. RF signals below the critical frequency and below a zenith angle known as the *critical angle* will be reflected back to the ground. When a communications link between two points on the Earth is using this reflected path, there is a Maximum Usable Frequency (MUF), which is below the critical frequency. The MUF is a function of time of day, position, and variables affecting the electron density so link connectivity is not assured at all times. Generally, the MUF is in the tens of MHz (HF/VHF) region. Carriers in the UHF

Table 11.3: Propagation Characteristics of Radio Bands

Band	Frequency	Characteristic
HF	3 to 30 MHz	Atmospheric noise, ionospheric reflections, long-distance links; affected by solar flux
VHF	30 to 300 MHz	Some ionospheric reflections, sporadic E, meteor scatter, basically line-of-sight propagation
UHF	300 to 3000 MHz	Basically line-of-sight propagation
SHF	3 to 30 GHz	Line-of-sight propagation; atmospheric absorption at higher frequencies
EHF	30 to 300 GHz	Line-of-sight propagation; subject to atmospheric absorption

and higher bands generally pass through the ionosphere and do not generate sky waves [Hall96].

11.3.4 Band Characteristics

The link designer's choice of carrier frequency greatly influences the magnitude of the atmospheric and ionospheric effects. Table 11.3 shows the dominant propagation characteristics in the different bands. Lower frequencies give the designer longer propagation distances. However, noise, amplitude fluctuations, and other effects greatly affect links over these long distances. Data transport is possible over these bands but at a low bit rate and the designer needs to add error correction techniques to mitigate the effects of the ionosphere. If the designer needs higher data rates, then the designer needs higher carrier frequencies as well. The high-frequency carriers force the designer to use only line-of-sight communications. If the designer also needs long distances with the high data rates, then the system needs some form of relay network.

11.4 RADIO FREQUENCY DEVICES

We examined various forms of modulation and demodulation in Chapter 10. In this chapter, we examine how designers realize these techniques in the transmitter and receiver components. The antennas that provide the interface between the transmission components and the channel connect

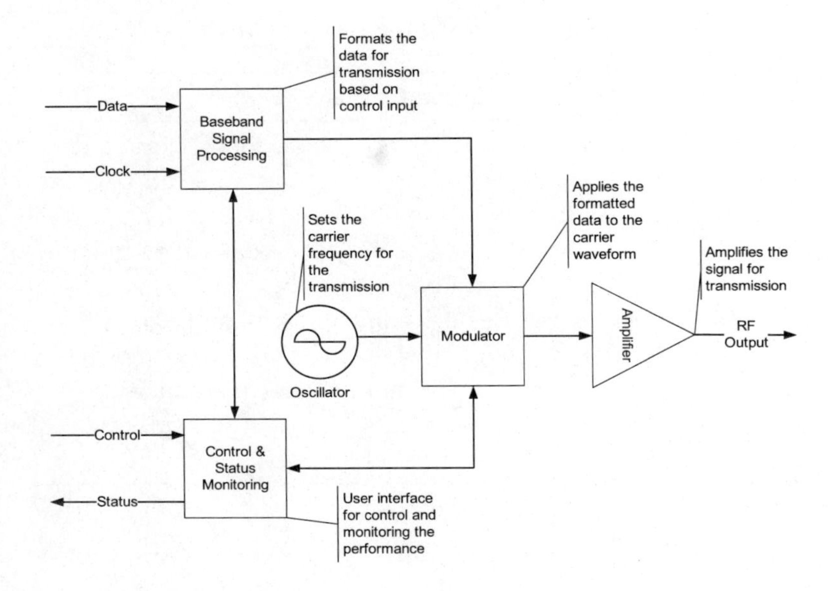

Figure 11.6: Functional components found in typical RF transmitters.

to the transmitters and receivers via specialized wiring. This section provides a brief overview of these RF devices in the system.

11.4.1 Transmitters and Receivers

The transmitters and receivers hold the modulation and demodulation circuitry, respectively, for the RF components illustrated in Figure 11.1. Manufacturers package them as either separate devices or a combination device, which is a *transceiver*.

Figure 11.6 illustrates the major parts of the transmitter. This figure uses characteristics of several commercially available products and it does not represent any specific device. The transmitter converts the data from its baseband encoding into a waveform that flows across the channel. The transmitter does not care if the data represent commands to control the system or telemetry gathered about the environment. It is important for the system designer to remember that whatever the transmitter does must be recoverable on the receiver side. This means that the process needs to be deterministic and synchronized. The functional parts of the transmitter include the following elements:

Baseband Signal Processing to ingest the bits and perform any wave

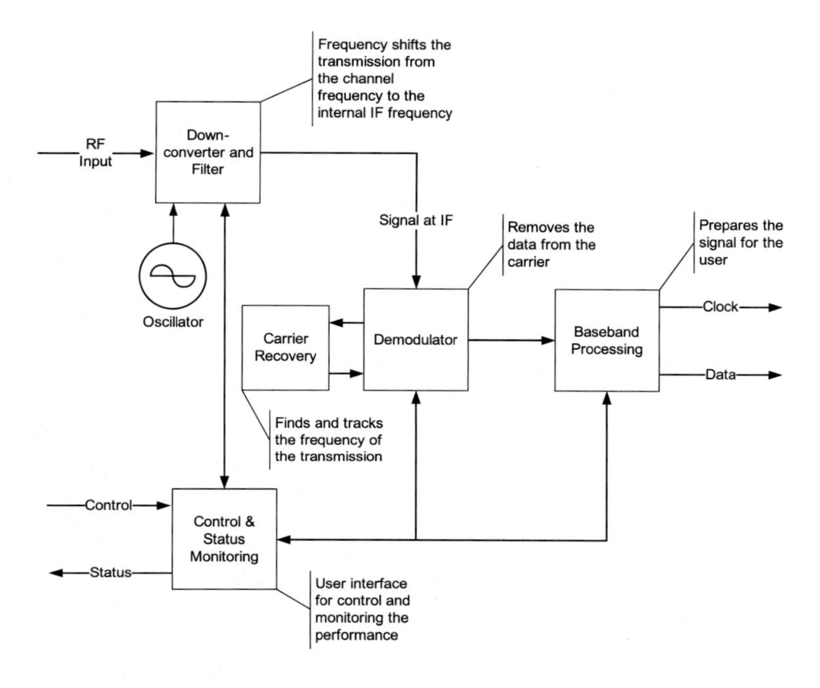

Figure 11.7: Functional components found in typical RF receivers.

shaping necessary; there may be other processing such as error-correcting coding applied as well.

Control and Status Monitoring to provide user control of options and reporting of the transmitter's status.

High Power Amplifier (HPA) to boost the signal strength for transmission with an external HPA that may follow if the transmitter needs an even stronger signals.

Modulator to modify the carrier according to the data where the control block may choose the exact form of the modulation used.

Oscillator to provide the stable carrier frequency.

Figure 11.7 illustrates the major parts of the receiver. As with the transmitter, this figure uses the characteristics of several commercially available products and it does not represent any specific device. The receiver converts the modulated channel signal into its baseband waveform. The receiver must perform this task despite the noise and corruption that the channel imposes on the transmitted signal. The receiver does not care if the data represent commands to control the system or telemetry gathered about the environment. The functional parts of the receiver include the following elements:

Baseband Processing to extract data clock from the data waveform; to sample and cleanly shapes the bits.

Carrier Recovery to find the transmitted carrier signal despite not knowing the current phase of the carrier, any Doppler shifts, and added noise from the channel, etc.

Control and Status Monitoring to provide user control of options and reporting the receiver's status.

Demodulator to remove the data signal from the carrier signal.

Downconverter and Filter to move the transmitted frequency from the selected carrier frequency to a fixed Intermediate Frequency (IF) for further processing with filtering that may be applied to reduce channel noise effects.

In addition to these functions, there are three important receiver characteristics that the designer needs to consider when selecting a receiver for the design:

- Minimum Carrier Power — the minimum signal strength that the receiver electronics detects; the manufacturer usually gives this specification in dBm, e.g., −120 dBm.
- Minimum Signal-to-Noise Ratio — how clean does the signal need to be for the detectors and tracking loops to work properly; this might be from 0 to 5 dB depending upon the quality of the electronics.
- Maximum Carrier Power — an input signal that is too strong may damage the receiver electronics; the manufacturer usually gives this specification in dBm, e.g., −90 dBm.

The link budget spreadsheet that we develop in Section 11.5.5 contains checks for these receiver characteristics as part of the computations.

11.4.2 Radio Frequency Components

The system designer includes at least one RF component between the transmitter or receiver and the antenna: some form of cabling. The designer may also include other RF devices to meet other design drivers. We will look at these types of devices in this section, starting with RF cabling.

When selecting the RF devices, the system designer must also keep the impedance of the devices in mind. For example, the RF port on the transmitter, the RF cabling, and the antenna may have impedances of $50\,\Omega$. If the designer chooses a component with a different impedance than the others in the group, for example $300\,\Omega$ when the others are $50\,\Omega$, it causes an impedance mismatch and degrades performance via power loss and unwanted device heating.

For more information on microwave components, see references such as [Hick07; Poza01; Poza05].

11.4.2.1 Radio Frequency Cabling

Devices do not exchange RF signals using regular electrical hook-up wire. Rather, the system designer uses special RF conductors such as

- Co-axial cables
- Semi-rigid and rigid cables
- Waveguides

Co-axial cable is usually the least expensive but has the lowest performance (greatest loss per unit length as a function of frequency). It also has the best flexibility. Semi-rigid and rigid cables have higher performance (less loss per unit length) but less flexibility than co-ax. Waveguides are custom metal channels with highest performance (little loss per unit length) but the least flexibility. The system designer chooses the cabling based on cost, mass, acceptable loss, and concerns about routing the cabling in the system.

11.4.2.2 Active and Passive Devices

There are several active and passive RF devices designers use to route signals in command and telemetry systems. Designers use these components to provide antenna placement options, redundancy options, or routing flexibility. In this section, we look at

- Circulator and Diplexer
- Coupler and Transfer Switch
- Splitter and Combiner

Circulator and Diplexer Designers use circulators to isolate signals or force signals to flow in a particular direction. Figure 11.8(a) shows the circuit symbol for a circulator where the signal flow goes from port 1 to port 2, port 2 to port 3, and port 3 to port 1. For example, designers insert the circulator between the transmitter and the filter to keep reflected signals from flowing back into the transmitter and damaging it, as in Figure 11.8(b). Designers also use circulators to direct signal flow between the antenna and the transmitter and receiver components, as in Figure 11.8(c).

The diplexer is similar to the directional flow characteristics of the circulator (duplexer). The diplexer is a special filter that permits the antenna to be connected to the transmitter and receiver that are operating at different frequencies. In many applications, the transmitter and receiver are on different frequencies in the same general band and the diplexer lets both devices use the same antenna. Figure 11.9 illustrates this use of the diplexer.

Coupler and Transfer Switch A coupler, also known as a hybrid coupler, is a passive RF device that allows signals to flow in both directions without the need for device control signals, albeit with some attenuation and phase shift. A coupler is a four-port device that designers frequently terminate with a resistor on one port. Signals on the input appear on both outputs. The phase shift is usually either 90° or 180° (user selectable); also known as a 90° or 180° hybrid device. The coupled output is at least 3 dB attenuated, but the user can select coupling ratios from manufacturers. Figure 11.10(a) illustrates the circuit symbol for a coupler also showing the relative phase shifts. Figure 11.10(b) illustrates using the coupler to permit antenna diversity, e.g., antennas facing different directions, from a single transmitter/receiver.

A transfer switch is a RF device that allows for redundant connections and the designer configures it by externally generated control signals. Figure 11.11(a) shows the switch has two inputs and two outputs. In the first position, jack 1 connects to jack 3, while jack 2 connects to jack 4. In the second position, jack 1 connects to jack 2, while jack

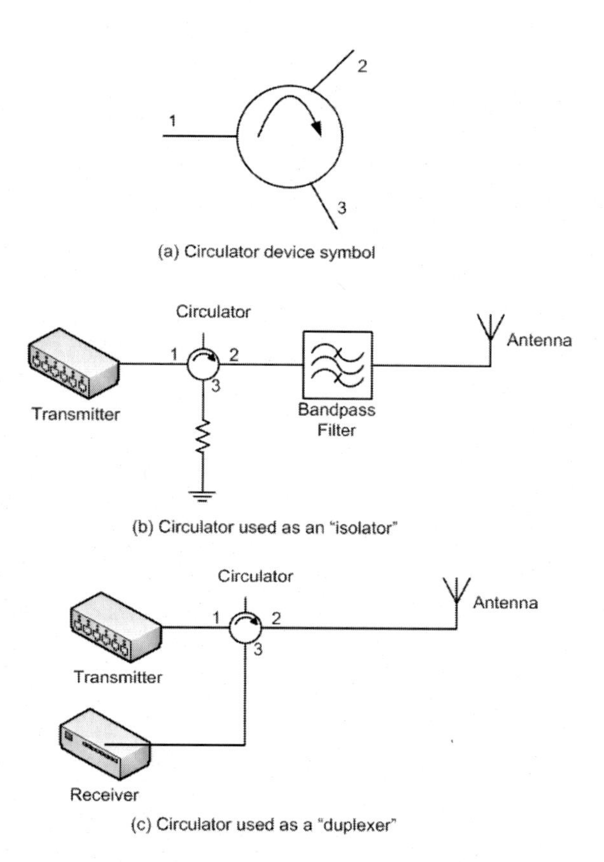

Figure 11.8: Symbol for a circulator and example use in a RF circuit.

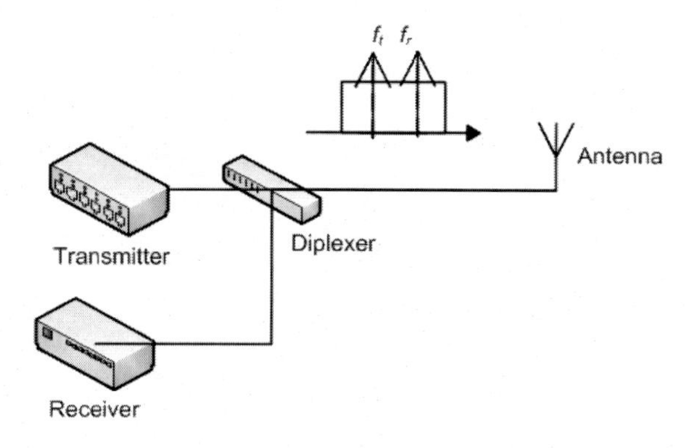

Figure 11.9: Example use for a diplexer in a RF circuit.

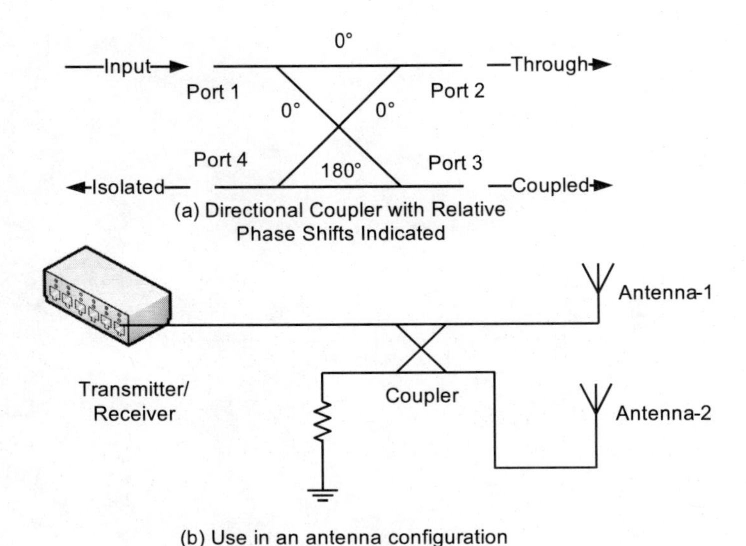

Figure 11.10: Circuit symbols for a hybrid coupler and example use in a RF circuit.

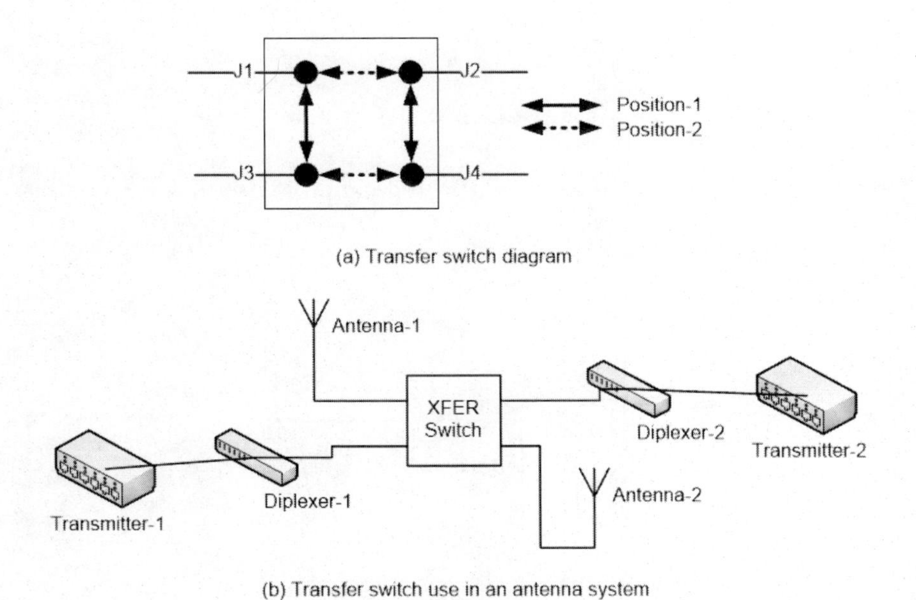

Figure 11.11: Symbol for a transfer switch and example use in a RF circuit.

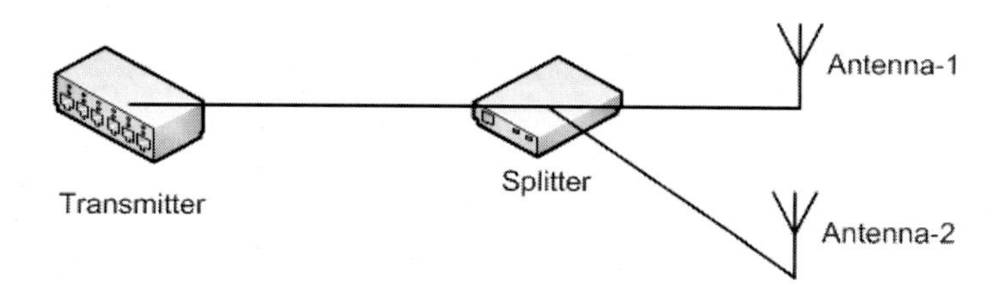

Figure 11.12: Symbol for a splitter and example use in a RF circuit.

3 connects to jack 4. Generally, losses through the device are small. Figure 11.11(b) illustrates how the transfer switch allows the designer to switch connections between devices to design a high-reliability, redundant transmission system with two transmitters and two antennas. This allows devices to be in either active or standby mode in use and switch if there is a fault.

Power Splitter and Combiner A power splitter takes the input signal and routes it to multiple devices. The splitter does not amplify the signal so the power to each individual device is less than the input power level. The splitter is a three-port electronic circuit having one input and at least two outputs. Unlike the coupler, there is no terminated port in the splitter; all outputs connect to devices such as antennas. If a splitter is properly designed, it can also be used as a power combiner where signals from multiple inputs are added together to produce a composite signal. Figure 11.12 gives an example use for a splitter in a RF system. Here, a power splitter feeds two antennas facing different directions such as the "up" and "down" antennas on an aircraft.

11.4.3 Antennas

In this section, we examine the concepts and computations system designers use when selecting an antenna. The relations provide characteristic performance traits and not detailed specifications for all classes of antennas. In practice, the designer must make explicit measurements to characterize fully the exact performance of an antenna system. Here, we examine antenna types found in telemetry and telecommand links, antenna beam and gain patterns, and antenna tracking systems. For more detail on antenna theory, the interested reader is referred to standard texts such as [Bala97; Krau88; Poza01; Stut98].

11.4.3.1 Antenna Types

Data transmission uses a variety of antenna types, depending upon the needs of the overall system. Typical antennas used in telemetry and telecommand systems include the following types:

- **Dish** antennas with a Cassegrain or parabolic feed configuration
- **Dipole** antennas composed of a long wire with a feed point in the middle
- **Monopole** antennas resembling a spring
- **Helical** antennas resembling a corkscrew
- **Horn** antennas with a rectangular cross-section and open at the radiating end
- **Strip** or **patch** antennas resembling a circuit board
- **Yagi** antennas with a driven element and parallel, parasitic elements
- **Earth coverage** antennas with higher gain off axis than the gain on axis (see Problem 8)

Figure 11.13 illustrates examples of several of these antenna types.

11.4.3.2 Antenna Radiation Pattern

The parameter engineers most often use to rate an antenna is the antenna gain and we will see how that parameter relates to the overall antenna characteristics. Before we look at the gain definition, we should first look at the radiation pattern generated by an antenna, as Figure 11.14 illustrates. One basic assumption engineers make about antennas is that they both radiate and receive energy equally well in both directions, so the radiation pattern should be the same for both processes. Keep in mind that Figure 11.14 really shows two-dimensional slices through a three-dimensional figure. Figure 11.14(a) shows the power pattern radiated by an *isotropic antenna*; that is, an antenna that radiates equally in all directions. Figure 11.14(b) shows the pattern radiated by an *aperture antenna* of some type such as that from a dish antenna, horn antenna, or a microwave patch antenna. It has a main lobe along the main axis and lower-gain side lobes. Figure 11.14(c) depicts the pattern emitted

(a) A 1-m Ka-Band parabolic dish antenna with an offset feed.

(b) Two Yagi antennas: one for VHF frequencies and one for UHF frequencies.

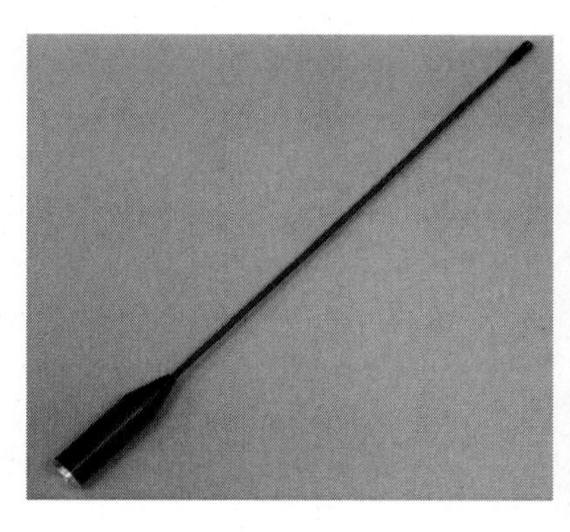

(c) A monopole antenna for UHF frequencies.

(d) A S-Band patch antenna.

Figure 11.13: Pictures of representative antenna types.

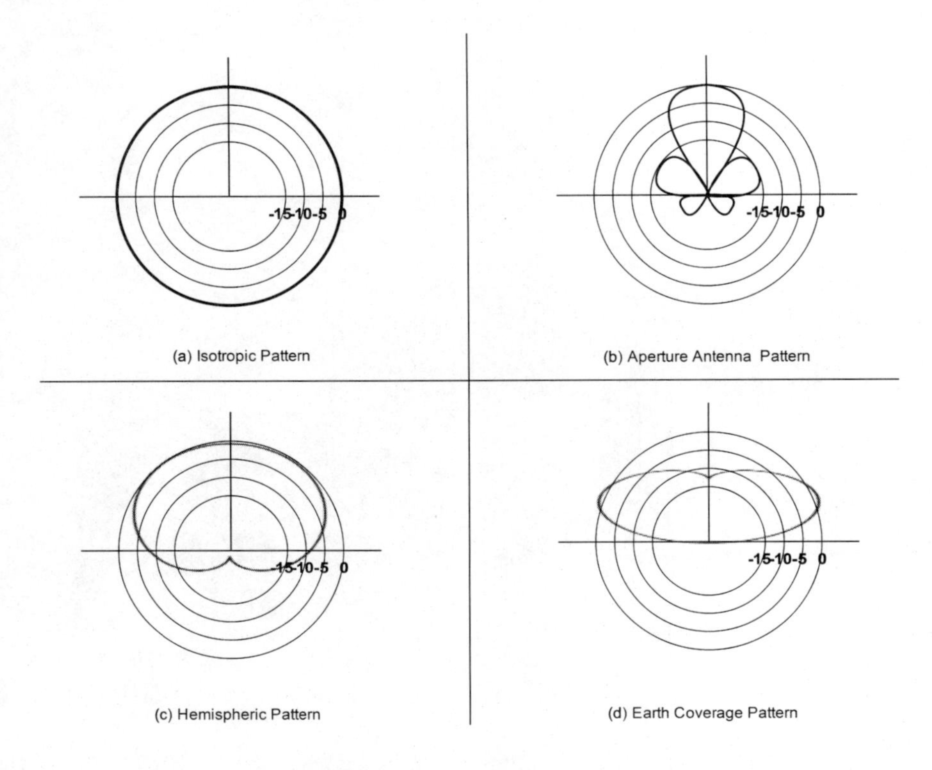

Figure 11.14: Example antenna radiation patterns.

by *hemispheric* antennas that radiates most of its signal throughout a hemisphere with a small emission to the "back" of the antenna. Figure 11.14(d) shows a pattern emitted by an Earth coverage antenna having its strongest gain off to the side and lower gain in the forward direction.

Other than the isotropic antenna, the examples in Figure 11.14 show the antenna radiation pattern is not always uniform. Ways engineers quantify the non-uniformity are via the beam solid angle, the Half-Power Beam Width (HPBW), and the beamwidth between first nulls. Figure 11.15 illustrates the three metrics. The main-beam solid angle is proportional to the radiated pattern with the constraint that all the transmitted power goes through the main lobe only. This is the large area outlined in Figure 11.15. The inner angle in Figure 11.15, the HPBW, is the angle subtended by the points where the radiated power is down to one-half ($-3\,\text{dB}$) its maximum power level. If the radiation pattern is not symmetric, a HPBW angle appears in each direction of the pattern.

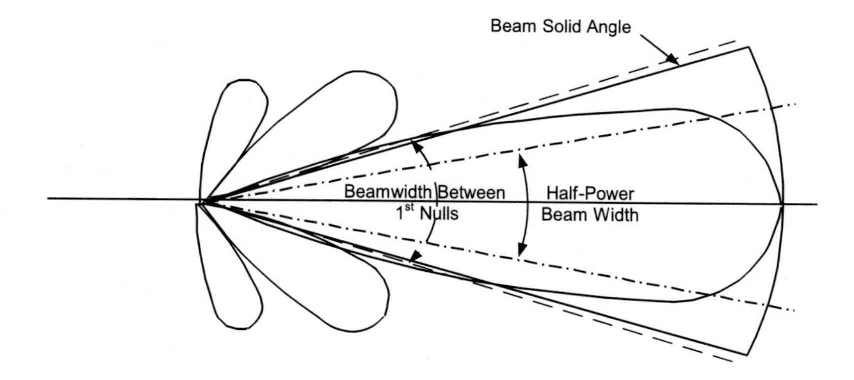

Figure 11.15: Antenna solid angle and half-power beamwidth.

The beamwidth between the first nulls is the angle subtended by the first null in the radiation pattern.

Engineers use antenna *directivity*, D, to specify how concentrated the radiation pattern is compared with an isotropic antenna. Directivity is the ratio of the maximum radiation intensity to the average radiation intensity using [Krau88]

$$D = \frac{4\pi}{\int_0^{2\pi}\int_0^{\pi} P_n(\theta,\phi)\sin\phi d\theta d\phi} = \frac{4\pi}{\Omega_A} \tag{11.1}$$

where Ω_A is the *beam solid angle* and $P_n(\theta,\phi)$ is the *normalized power pattern*. That is, if $P(\theta,\phi)$ is the measured power pattern, then it is normalized relative to its maximum value using

$$P_n(\theta,\phi) = \frac{P(\theta,\phi)}{P(\theta,\phi)_{max}} \tag{11.2}$$

Table 11.4 gives typical values for the directivity.

In addition to the radiation pattern's directivity, the maximum relative size of the side lobes is also important. The side lobe distribution determines the amount of radiation spread outside of the direction of interest and this determines the spillover of the energy to or from the ground or other noise sources.

11.4.3.3 Antenna Gain

The antenna *gain*, G, is related to the directivity, D, and is defined as the ratio of the maximum radiation intensity to the maximum radiation

Table 11.4: Typical Values for Antenna Directivity

Antenna Type	**Directivity**
Isotropic antenna	1
Short-wire dipole	1.5
Half-wave dipole	1.64
Horn	100
Medium-sized dish	1000
210-foot dish	100,000

intensity from a reference antenna with the same power output. The reference antenna is usually an isotropic antenna. The gain, relative to the directivity, is

$$G = \eta D \tag{11.3}$$

where η represents the ratio of the antenna total radiated power to the input power, that is, it is an *efficiency* measure. Another way to relate the gain to the directivity is through the antenna *insertion loss*, L

$$[L] = [D] - [G] \tag{11.4}$$

In this chapter, we use [*] to indicate that dB units are being used.

Aperture Antennas For aperture-type antennas, such as the dish or the patch antenna, engineers estimate the gain, G, in terms of the operating wavelength, λ, in meters or the operating frequency, f, in Hertz, and the speed of light, c, in meters per second by using the *universal gain equation*

$$G = \frac{4\pi A_e}{\lambda^2} = \frac{4\pi A_e f^2}{c^2} \tag{11.5}$$

A_e, is the effective aperture of the antenna in square meters which is related to the physical aperture (area), A_p, via the *aperture efficiency*, η_{ap}, by the equation

$$\eta_{ap} = \frac{A_e}{A_p} \tag{11.6}$$

Typically, aperture-type antennas have an efficiency in the range of 45% to 70%.

Normally, dish antenna designers modify the feed radiation distribution to control the side lobe levels. Figure 11.16 illustrates a *cosine tapir*

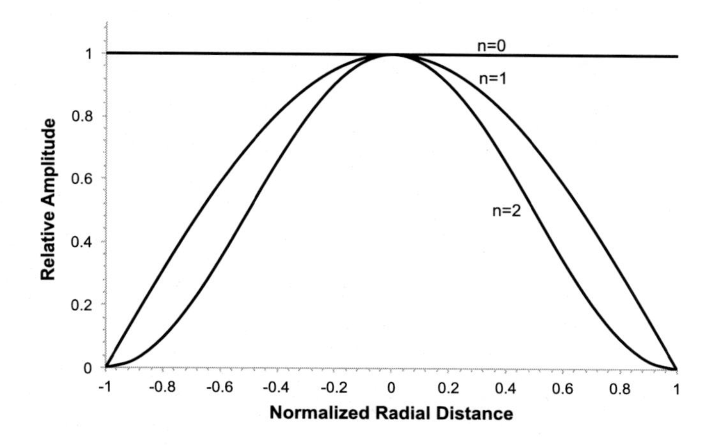

Figure 11.16: Cosine tapir function for a dish antenna.

Table 11.5: Effect of a Cosine Tapir on a Dish Antenna [Stut98]

N	HPBW (deg)	Maximum Side Lobe (dB)
0	$58.4\lambda/d$	-13.3
1	$72.8\lambda/d$	-23.0
2	$84.2\lambda/d$	-31.7

shaping function for this purpose where n specifies the order of the cosine tapir function. If $n = 0$, then there is no tapir (the radiation distribution is uniform). When $n = 1$, the tapir is proportional to the cosine. When $n = 2$, the tapir function is proportional to cosine squared. Table 11.5 gives the corresponding HPBW as a function of the antenna diameter, d, and the associated maximum side lobe level.

For example, suppose we have a 1 m dish with 50% efficiency. If we operate the link at 2 GHz, then from using Equations (11.5) and (11.6), $f = 2\,\text{GHz}$, $\lambda = c/f = 0.15\,\text{m}$, and $A_p = (\pi d^2)/4$. The gain computes to $G = 219.6 = 23.42\,\text{dB}$. If we operate the antenna at 20 GHz, then $\lambda = 0.015\,\text{m}$, and $G = 21960 = 43.42\,\text{dB}$. For this same dish with a cosine aperture tapir ($n = 1$) operating at a frequency of 2 GHz, we obtain a HPBW of $HPBW = 72.77°(0.15)/(1) = 10.9°$. At a frequency of 20 GHz and the same feed configuration, the $HPBW = 1.09°$.

Helical Antennas Helical antennas are another antenna configuration designers use in telemetry and telecommand systems. Design engineers use the following parameters to specify helical antennas [Stut98]:

- N — the number of turns in the helical antenna ($N > 3$)
- λ — the operating wavelength (meters)
- d — the diameter of the helices (meters)
- C — the circumference ($C = \pi d$) of the helices (meters)
- $S = C \tan \alpha$ where α is the helix pitch angle ($12° < \alpha < 15°$ and $0.75 < \frac{C}{\lambda} < 1.3$

With these definitions, we estimate the helical antenna HPBW using

$$HPBW = \frac{65°}{(C/\lambda)\sqrt{N\frac{S}{\lambda}}} \tag{11.7}$$

For well-designed helical antennas, the gain and directivity are

$$G \approx D = 6.2\left(\frac{C}{\lambda}\right)^2 N\frac{S}{\lambda} \tag{11.8}$$

For example, a 10-turn helical antenna operating at 1 GHz with a circumference of $C = 0.92\lambda$ or $C = 3.06$ cm and a pitch angle of $\alpha = 13°$ gives a HPBW of 48.5° and a gain of 10.5 dB. Note, these design equations are most accurate for $N \approx 10$ but they are still useful as a starting point for other designs.

11.4.3.4 Target Tracking

The antenna beam pattern relates to two important parameters for data communications: the beam footprint and the receiving station tracking accuracy. The footprint is a measure, usually performed at a given power density level, of the shape (area and physical shape) of the antenna beam as it passes the receiving antenna. The tracking accuracy is a measure of how well the receiving antenna follows the transmitting antenna's beam and stays within a specified portion of the beam. Both of these are functions of the beam pattern. In this section, we examine tracking methods. Typically, the receiving antenna must track the transmitting

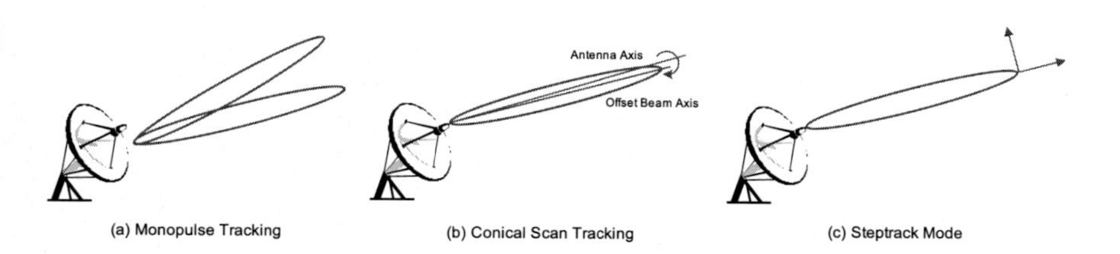

Figure 11.17: Tracking methods for antenna systems.

antenna to within a fraction of the HPBW to minimize signal loss. Figure 11.17 illustrates strategies to perform the tracking for the following types:

Conical scan tracking — the pattern is rotated and if the received wavefront intersects the antenna power pattern through its maximum point, there is no change in the received power level as the beam is rotated; if the received wavefront does not go through the maximum point, there is a change in the received power level as the beam is rotated, which is converted to a pointing error vector to move the antenna.

Monopulse tracking — uses a separate receiver system to search out the nulls in the pattern and then derive an error vector based on the pattern and the received energy; good to 1/20 of the HPBW.

Open loop — points the antenna based upon an assumed path, or other pre-computed pointing information; this is adequate for small, nondirectional dishes.

Step tracking — points the antenna based upon an elevation and azimuth error vector derived from four measurements of the received power pattern and is good to 1/5 to 1/3 of the HPBW.

The system designer selects the tracking mode based on the required pointing accuracy, the system cost, and the added power and weight to support the tracking method.

11.4.4 Software Defined Radio

So far in this overall section, we have been discussing RF components with the assumption that they are constructed using discrete electronic components such as resistors, capacitors, and inductors. Modern radio

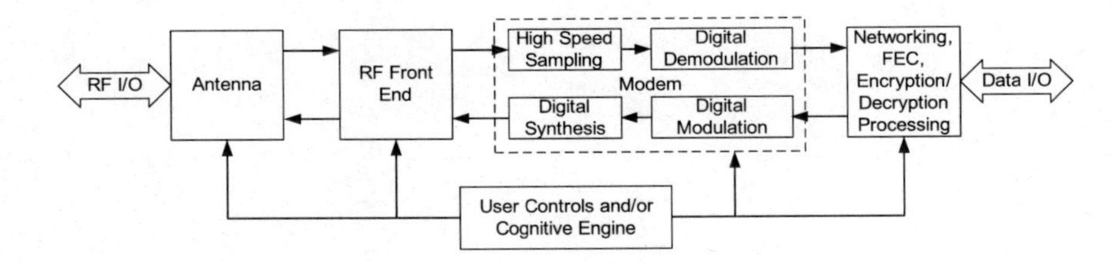

Figure 11.18: Block diagram for a potential software defined radio architecture.

design techniques are moving further from these discrete components to more signal processing intense architectures that present designers with options for flexibility and revision that discrete componets do not permit. This approach is found in the design philosophy of the Software Defined Radio (SDR) [Marg98; Mito93; Mito00; Reed02]. Figure 11.18 illustrates the possible design blocks in a full SDR radio. These blocks have the following functions:

Antenna The software-controlled antenna permits selection of both the carrier frequency and the antenna beam pattern, e.g., spot beam, hemispherical, etc.

RF Front End The carrier frequency, amplifier characteristics, and signal filtering are optimized for signal quality and channel performance.

Modem On the receiver side, high-speed sampling and signal processing recover the data from the carrier. With sufficiently rapid sampling, the demodulation signal processing components will sample the carrier directly and not need to downconvert it to a lower frequency thereby saving components. On the transmitter side, high-speed signal processing and digital-to-analog conversion will allow the modulator to structure the waveform ready for amplification and transmission.

Networking and Other Processing The baseband data is processed for forward error correction protocols, perform any necessary encryption functions, and perform networking functions that are appropriate for the communications link.

Each of these functions can be configured and changed in real time via control signals. The system designer can also revise the modules through a software upgrade to bring in new options or communications standards.

The next extension to the SDR is the *cognitive radio* design. With a cognitive radio, the radio itself attempts to optimize the communications performance based on how the system designer determines what is the optimal configuration based on metrics such as

- Desired communications quality goals
- Desired communications throughput goals
- Current channel propagation conditions
- Available channel bandwidth based on other users and regulations or allocation

In principle, the user programs the radio pairs with these metrics, and any other metrics relevant to the communications needs. The programming in the radios will adaptively configure the radios for optimal performance based on the metrics. The radios continue to monitor their environment and continue to optimize the performance as conditions evolve. In a fully cognitive system, the transmitter will choose the modulation and filtering format as well as the transmission frequency and power levels in real time. In a similar manner, the receiver will track the signal and fully determine the modulation format and data rate from the transmitted signal alone. Figure 11.18 allows for this cognitive engine to control the radio configuration in addition to direct user control signals. In practice, operational radios may need guide channels embedded with the transmitted signal to assist the receiver in fully decoding the transmitter's signal.

11.5 FREE SPACE PROPAGATION

Transmitting a radio signal between two points requires additional considerations than the ones we have discussed so far. The remainder of this chapter discusses the basics of radio propagation and the effects that interfere with signal transmission. First, we examine the free space case devoid of problems. Using this as a baseline, we examine how the atmosphere, rain, and path effects degrade the signal. At the end, the system designer has the design support tools to evaluate if the link and RF components will meet the data transmission requirements.

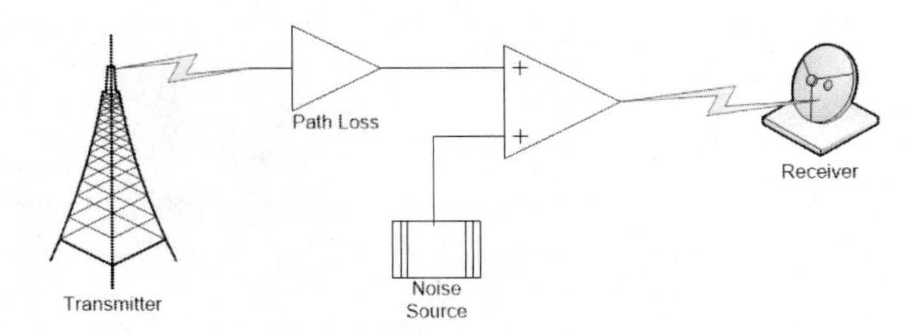

Figure 11.19: The simple transmission link model between a transmitter and a receiver with path loss and noise.

11.5.1 Friis Transmission Relationship

The link designer performs a link analysis with the goal of determining if the transmitter and receiver systems are matched and provide sufficient signal energy at the receiver to achieve the desired Signal-to-Noise Ratio (SNR) (analog) or Energy-per-Bit-to-Noise-Spectral-Density (E_b/N_o) (digital) performance. Figure 11.19 illustrates the starting channel model between the source and sink points with the channel's power loss and noise effects; see also [P.341; P.525]. The parameters that affect this analysis are

- Transmitter and receiver antenna gains
- Separation distance between the transmitter and the receiver
- Transmitter output power
- Receiver noise temperature
- Attenuation sources

Link designers use a *link budget* to account for each of these effects in the link analysis. See references such as [Boku11; Sing11; Skla79; Skla01; Yuen82a] for further details on this process.

Link analysts use the *Friis transmission relationship* as the basis for the transmission-link power budget [Frii46]. The relationship uses the physics of the transmitter and receiver systems, the $1/R^2$ power attenuation due to distance, and attenuation due to system losses between the transmitter and the receiver. We will cover attenuation sources other than distance later. Because we are working in dB units, we merely subtract all of these further attenuation sources from the margin we derive between the link specification and the available power.

If we ignore all component losses for the moment and start with an omnidirectional antenna (uniformly radiating over 4π steradians) radiating with a transmitted power, P_t, then the received signal *flux density*, F, in W/m^2, crossing a plane at some distance, R, from the transmitter is

$$F = \frac{P_t}{4\pi R^2} \tag{11.9}$$

Most system designers do not use isotropic antennas; rather, they use antennas that radiate or receive in a preferred direction with some gain, G. The analyst then replaces the transmitted power in Equation (11.9) with the transmitter's Effective Isotropic Radiated Power (EIRP)

$$EIRP = GP_t \tag{11.10}$$

When the receiving antenna aligns with the transmitting antenna, the total signal power gathered by the receiving antenna is the product of the flux density at the receiving antenna and that antenna's effective aperture. The analyst computes the *received carrier power*, C, in *Watts*, from

$$C = FA_e = \frac{EIRP\ A_e}{4\pi R^2} \tag{11.11}$$

It is more convenient for the link analyst to replace the effective area in Equation (11.11) with the universal gain equation to rewrite the equation in terms of the carrier's wavelength, λ, and the gain of the receiver's antenna, G_R, as

$$C = \frac{EIRP\ G_R\ \lambda^2}{(4\pi R)^2} \tag{11.12}$$

Ultimately, the link analyst wishes to estimate the received signal's quality as a signal-to-noise ratio by estimating the received carrier power relative to the system noise. To do this, the analyst assumes, for the moment, that all of the noise arises in the receiving equipment and is due to thermal effects. The thermal noise power, N, in *Watts*, is a function of the receiver bandwidth, B, the receiver System Noise Temperature (T_{sys}), and Boltzmann's constant, k, ($1.380\,648\,52 \times 10^{-23}$ J/K or $-228.6\,\text{dBW/K} - \text{Hz}$). The analyst computes the noise power using [Dave87; Frii44]

$$N = kBT_{sys} \tag{11.13}$$

Using Equations (11.12) and (11.13), the analyst computes the Carrier-to-Noise Power Ratio (C/N) ratio as

$$\frac{C}{N} = \frac{EIRP\, G_R\, \lambda^2}{(4\pi R)^2} \frac{1}{kBT_{sys}} \tag{11.14}$$

11.5.2 Space Loss

As we know from the physics of radio propagation, the radiated energy has a R^{-2} fall off due to the spreading of the radiation through space. Equation (11.14) results in a unitless quantity. Within Equation (11.14), we can combine the link path length, R, with the radiation wavelength, λ, to form another dimensionless quantity. Link analysts define a dimensionless loss term called the space loss, L_S, to express this ratio. The analyst computes the *space loss* using

$$L_S = (4\pi R/\lambda)^2 \tag{11.15}$$

In dB units, the space loss is

$$[L_S] = 20 \log (4\pi R/\lambda) \tag{11.16}$$

For example, if the transmission distance is 1000 m and the transmission frequency is 2 GHz, $L_S = 20log(4\pi 1000/0.15) = 98.5\,\text{dB}$. Similarly, if the frequency is 20 GHz, then $L_S = 20log(4\pi 1000/0.015) = 118.5\,\text{dB}$. We use the definition for space loss to rewrite Equation (11.14) as

$$\frac{C}{N} = \frac{EIRP}{L_S} \frac{G_R}{T_{sys}} \frac{1}{kB} \tag{11.17}$$

In Equation (11.17), we isolate the Gain-to-System-Temperature ratio or Figure of Merit (G/T) for the receiver system, by convention.

11.5.3 Noise Temperature

The *system noise temperature* is a measure of the noise in the receiver system parameterized in a single representative quantity [Dave87]. The primary noise sources are usually the receiver's electronics. However, as Figure 11.20 illustrates, contributions to the total system noise temperature come from the sun, the ground, and the atmosphere in addition to the electronics. Engineers usually express this parameterized measure of the receiver noise power in temperature units although it is not a true physical temperature. In this section, we will see how to compute

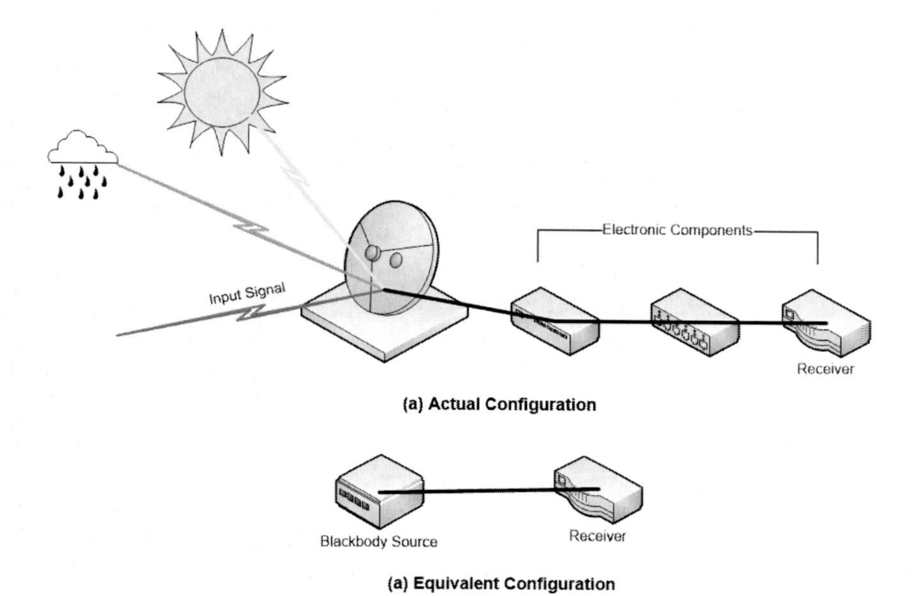

Figure 11.20: System elements contributing to the overall system noise temperature in the receiver and the equivalent model for the system noise.

the effects of the electronics and the outside influences. In all cases, we assume that the noise is random, is not a function of the transmitted data, and has no long-term correlation characteristics.

Designers compute the system noise temperature from the noise generated by each device in the system. The designer assumes that white Gaussian noise dominates so the effective noise temperature of a device, T_e, in terms of the noise power, N, that it produces, Boltzmann's constant, k, and the device bandwidth, B, is [Frii44]

$$T_e = \frac{N}{kB} \tag{11.18}$$

This effective noise temperature is the thermal temperature that a black body radiator would possess if it produced the same noise power as the device under test; it is not the operating temperature of the device. This is why it is possible to have low-noise amplifiers operating at room temperature (nominally defined as 290 K) even though the manufacturer rates as a temperature of only ≈100 K. Conversely, many receivers have

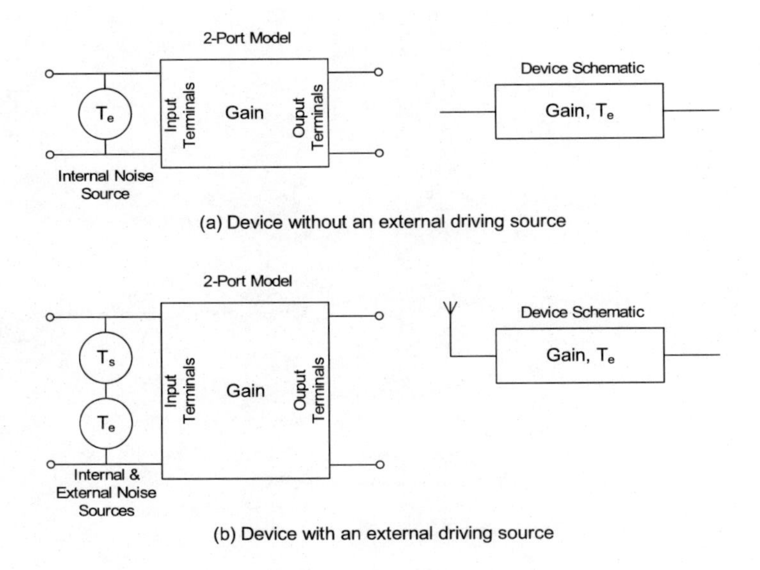

Figure 11.21: Two-port model for internal noise source modeling.

noise temperatures of 1000 K or more and show no indication of meltdown.

11.5.3.1 Two-port model

This section shows how we use the effective temperature to generate a system temperature based on a block diagram model of the system [Frii44]. Figure 11.21 illustrates the *two-port component model* designers use to produce a noise model for any component in the telemetry link. The model has two forms: one for devices without attached noise sources and one for devices with attached noise sources. Each device model consists of a temperature source acting as a driver to a two-port bulk equivalent model of the device. The device model has a forward gain, G, and a reverse gain of $1/G$. When the system is composed of multiple devices, the two-port models for each device are cascaded to form the complete system.

Figure 11.21(a) represents a device without an external noise source. The temperature source driving the two-port model is due only to the internal device noise generation, which is parameterized by the device effective temperature, T_e. At the input terminals to the two-port model, the measured noise power, N_{in}, due to the effective temperature of the device is

$$N_{in} = kT_eB \tag{11.19}$$

The measured noise, N_{out}, at the output of the two-port device is

$$N_{out} = GkT_eB \tag{11.20}$$

Figure 11.21(b) represents a device with an external noise source attached in addition to the internal noise source. A typical external noise source is an antenna, which is receiving broadband microwave radiation from the atmosphere, the surface of the Earth, and other sources in addition to the received signal it is trying to recover. We characterize this noise by a source temperature, T_S, just as we did the effective temperature for the device. In this case, the external source is a driver in series with the device's own effective temperature to make a combined driver to the two-port model. In this configuration, the measured noise power at the input terminals is

$$N_{in} = k\left(T_e + T_S\right)B \tag{11.21}$$

Similar to Equation (11.20), the measured noise power at the output terminals is

$$N_{out} = Gk\left(T_e + T_S\right)B \tag{11.22}$$

11.5.3.2 Effective Temperature Computation

Not all devices have a gain greater than unity. Some devices, for example a co-axial cable, produce a signal loss. For lossy devices, we represent the loss, L, either as a number greater than unity or as a gain, G, less than unity. The loss is related to the gain by $L = G^{-1}$ and lossy devices are modeled with two-port models as increasing the total noise contribution of the system because they decrease the signal power. The *effective noise temperature* for a lossy component is given in terms of the physical temperature, T_p, by

$$T_e = \left(L - 1\right)T_p = \frac{1 - G}{G}T_p \tag{11.23}$$

Normally, the physical temperature is taken to be the standard *reference temperature*, $T_0 = 290\,\mathrm{K}$ [Frii44].

Another way in which manufacturers specify a device is according to its noise figure or noise factor. This parameter measures noise contribution by making a relative measurement of the signal-to-noise ratio at the input and output of the device, usually referenced to a standard source at 290 K.

The effective temperature for a device with a standard *noise figure*, NF, is [Frii44]

$$T_e = (NF - 1)\,T_0 \tag{11.24}$$

Usually, the *noise factor* is in dB units so the designer converts them to a ratio before using in this form. For example, the manufacturer advertises a commercial amplifier with a noise figure of 1.2 dB (1.318 in ratio form). We compute the device effective temperature from

$$T_e = (1.318 - 1)\,290\ K = 92.3\ K \tag{11.25}$$

A co-axial cable with a 1 dB loss (or a gain of 0.794) has a noise figure of 1.259 (not in dB) and an effective temperature of

$$T_e = (1.259 - 1)\,290\ K = 75.1\ K \tag{11.26}$$

11.5.3.3 Antenna Temperature Computation

An antenna forms a discrete noise source for the receiver system. The Antenna Temperature (T_{ant}) is the noise temperature of the microwave radiation coming into the receiver from sources such as sky noise from clouds and atmospheric gasses, Earth noise, sun noise, and Milky Way galactic noise.

Notice that the received signal is not part of the antenna temperature. To obtain the total T_{ant}, one must integrate the contribution of each source weighted by the antenna pattern in that direction. If the antenna points mostly "up," then sky noise and solar noise dominate. If the antenna points mostly "down," then Earth noise at 300 K dominates.

Figure 11.22, based on equations in Flock [Floc87], shows the sky temperature as a function of the frequency and the elevation angle θ from the horizontal. The sky noise is least for directions towards the zenith because the propagation path has the shortest distance through the atmosphere. The sky noise is also least between 1 GHz and 10 GHz. Below 1 GHz, microwave radiation coming in from outside the Earth dominates the sky noise, while the noise above 10 GHz mostly comes from the molecules in the Earth's atmosphere.

A reasonable approximation for T_{ant} is found by knowing the antenna pattern solid angle seeing the sky, Ω_{sky}, the ground, Ω_g, and the sun, Ω_{sun}, the local ground voltage reflection factor, ρ, and polarization factor, p [Prit93]

$$T_{ant} = \frac{\Omega_{sky}}{4\pi\,(1 - \rho^2)} + \frac{\Omega_g}{4\pi}\left(1 - \rho^2\right) + \frac{p\Omega_{sun}}{4\pi}\left(\frac{G_{sun}}{L_R}\right) \tag{11.27}$$

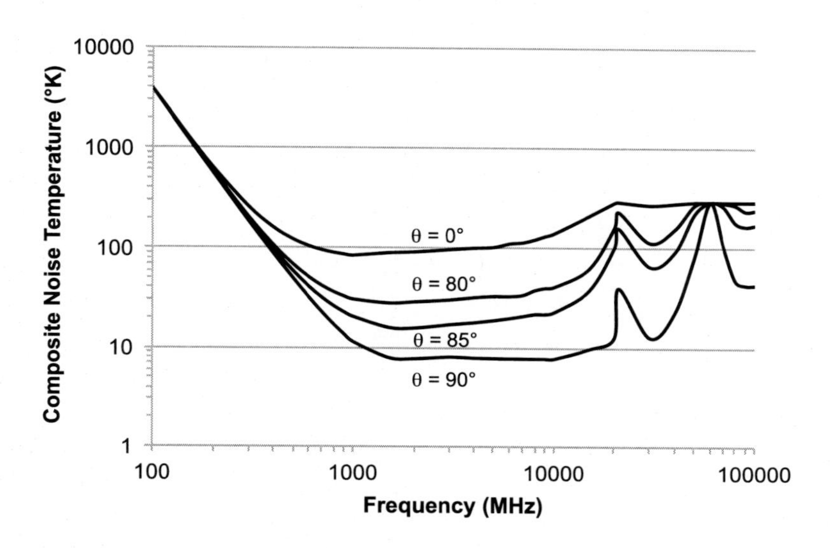

Figure 11.22: Sky temperature as a function of frequency and elevation angle θ from the horizontal [Floc87].

G_{sun} is the antenna gain in the direction of the sun, while L_R is the rain loss on the link. If the antenna has a HPBW $<0.5°$ and the antenna is pointed at the sun, the antenna temperature approaches T_{sun}. Generally, a system outage occurs under those conditions. If the antenna is low-gain, the sun does not greatly affect the antenna noise temperature. We will see an example of this rise in noise temperature due to the sun later in this chapter.

11.5.3.4 System Temperature Computation

The usual case for a receiver system is to have several components cascaded together. This section explains how designers compute the noise power for a number of components. Figure 11.23(a) illustrates the case of a single source driving two devices. We can compute the expected noise power level at several reference planes in this system: at the input to the first device, between the first and second devices, and at the output of the third device [Ippo08; Skla01]. In each case, we characterize the total noise power by a *system temperature*, T_{sys}, computed from the total noise power measurement, N, by using the basic relationship $N = kT_{sys}B$.

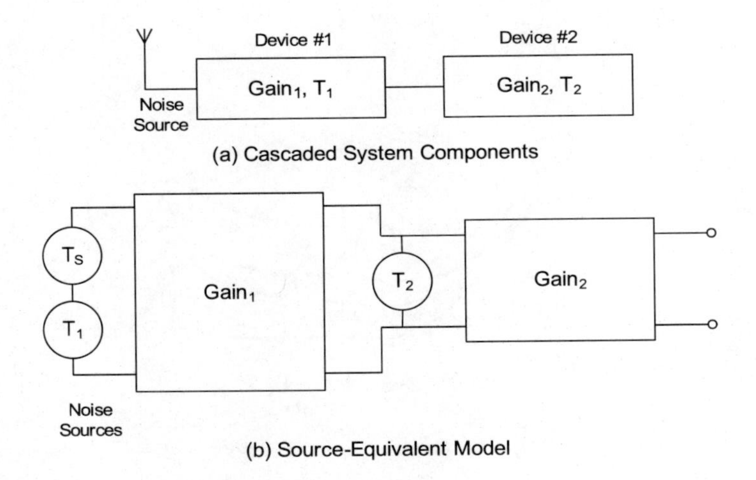

Figure 11.23: Cascaded devices for system temperature computation. Each device has its own gain and effective temperature. The antenna provides a source temperature, T_S.

The system temperature makes that equation hold true in the overall configuration.

For Figure 11.23(b), the noise power we measure at the input to the cascaded system is

$$N = k\left[(T_S + T_1) + \frac{T_2}{G_1}\right] B \tag{11.28}$$

from which we identify the system temperature as

$$T_{sys} = (T_S + T_1) + \frac{T_2}{G_1} \tag{11.29}$$

If we measure the noise power at the reference plane between Device 1 and Device 2, the total noise power is

$$N = k\left[(T_S + T_1)\, G_1 + T_2\right] B \tag{11.30}$$

Equation (11.30) gives a system temperature of

$$T_{sys} = (T_S + T_1)\, G_1 + T_2 \tag{11.31}$$

Finally, if we measure the noise power at the output of the second device, the noise power is

$$N = k\left[(T_S + T_1)\, G_1 G_2 + T_2 G_2\right] B \tag{11.32}$$

Equation (11.32) gives a system temperature of

$$T_{sys} = (T_S + T_1)\, G_1 G_2 + T_2 G_2 \tag{11.33}$$

From this example, we see several trends concerning computing the system temperature. The effective temperature of the device at the reference plane enters the computation with no gain modification. The devices to the left of the reference plane have their effective temperatures multiplied by the gains of the devices to the left of the reference plane. The devices to the right of the reference plane, except for the one to the immediate right of the reference plane, have their effective temperatures divided by the gains of all of the devices to the right of the reference plane. Therefore, for the general case of M devices in a component chain that are cascaded together in series, each with an associated gain (which may be greater than unity or less than unity), G_i, and equivalent noise temperature, T_i, the output noise power, N, measured at the output port of the Mth device is

$$N = kB\left[(T_S + T_1)\,(G_1 G_2 \ldots G_M) + T_2\,(G_2 \ldots G_M) + \cdots + T_M G_M\right] \tag{11.34}$$

As we expect, based upon the previous example, the output system temperature is

$$T_{sys} = (T_S + T_1)\,(G_1 G_2 \ldots G_M) + T_2\,(G_2 \ldots G_M) + \cdots + T_M G_M \tag{11.35}$$

The more typical case for analysis is the computation of the system temperature at a reference plane somewhere in the middle, for example at the input to the Low Noise Amplifier (LNA) (see Figure 11.24). For the same M cascaded devices, the computation of the system temperature at reference plane L, that is between device L and device $L+1 < M$, the measured noise power is

$$\begin{aligned} N = kB\Big[&(T_S + T_1)\, G_1 G_2 \ldots G_L + T_2 G_2 \ldots G_L + \cdots + T_L G_L + T_{L+1} \\ &+ \frac{T_{L+2}}{G_{L+1}} + \cdots + \frac{T_M}{G_{L+1} \ldots G_{M-1}}\Big] \end{aligned} \tag{11.36}$$

The system temperature associated with Equation (11.36) is

$$\begin{aligned} T_{sys} = &(T_S + T_1)\, G_1 G_2 \ldots G_L + T_2 G_2 \ldots G_L + \cdots + T_L G_L + T_{L+1} \\ &+ \frac{T_{L+2}}{G_{L+1}} + \cdots + \frac{T_M}{G_{L+1} \ldots G_{M-1}} \end{aligned} \tag{11.37}$$

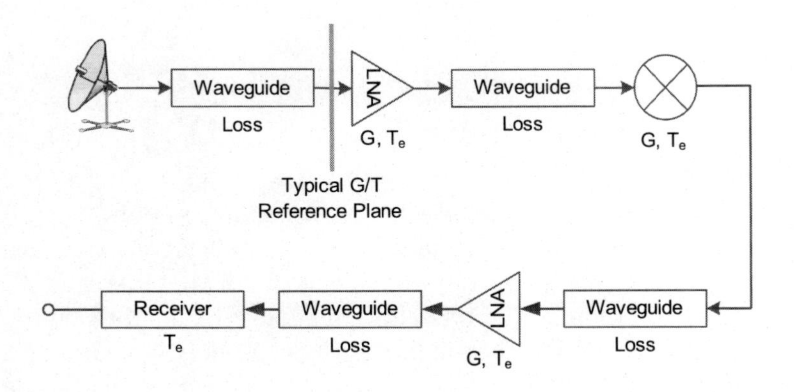

Figure 11.24: An example receiver system from the antenna to the actual receiver element.

Another parameter of interest at reference plane L is the system *figure of merit* or G/T ratio where the T is the system temperature, T_{sys}, and the G is the total gain supplied by the system both through device L in the receiver chain. We use these results in computing the link budget as in Equation (11.17).

Figure 11.24 illustrates a receiver chain that consists of an antenna, microwave components, and waveguides. We compute the system noise temperature and G/T ratio at the input to the first amplifier and use the parameters given in Table 11.6 for the components. These values are illustrative of system values found in many configurations. Taking the input to the first LNA after the antenna as the reference point, we compute the system temperature and the G/T for the receiver chain. For example, to compute the effective temperature for the 0.5-dB loss waveguide, we apply the relationship $T_e = (100.5/10 - 1)290$ K to arrive at an effective temperature of 35.4 K for the device. Similarly, for the LNA with a noise figure of 0.94 dB, we find the effective temperature of 70 K by applying the relationship $T_e = (100.94/10 - 1)290$ K. The other computations proceeded using the same relationships. From the values in Table 11.6 for the components in Figure 11.24, we compute the system temperature at the input to the LNA from

Table 11.6: Example Receiver Chain Computations

Receiver Chain Element	**Gain** (dB)	**Given T_e or NF**	**Computed T_e (K)**
Antenna	49.5	70 K	
Feed Loss	0.5		35.4
LNA	20	0.94 dB	70
LNA/Downconverter	−1		75.1
Downconverter	0	273 K	
Downconveter/Amplifier	−1		75.1
Second Amplifier	10	1.29 dB	100
Amplifier/Receiver	−1		75.1
Receiver			1000

$$T_{sys} = (T_{ant} + T_{wg1})\,G_{wg1} + T_{LNA} + \frac{T_{wg2}}{G_{LNA}} + \frac{T_{DC}}{G_{wg2}G_{LNA}} + \frac{T_{wg3}}{G_{DC}G_{wg2}G_{LNA}} + \frac{T_{AMP}}{G_{wg3}G_{DC}G_{wg2}G_{LNA}} + \frac{T_{wg4}}{G_{AMP}G_{wg3}G_{DC}G_{wg2}G_{LNA}} + \frac{T_{RCV}}{G_{wg4}G_{AMP}G_{wg3}G_{DC}G_{wg2}G_{LNA}} \tag{11.38}$$

The result of applying Equation (11.38) is a T_{sys} of 22.4 dB K. The total gain in the system as measured at the input to the LNA is

$$G = 49.5\ dB - 0.5\ dB = 49\ dB \tag{11.39}$$

At the reference plane, the figure of merit is

$$\frac{G}{T} = 49\ dB - 22.4\ dBK = 26.6\ dB/K \tag{11.40}$$

The system temperature changes if we change the reference plane to a different location along the receiver chain. While this is true, the total G/T for the receiver chain remains constant as long as the component values stay fixed.

11.5.4 Signal Margin

In addition to the electronic and antenna noise, the link analyst must also consider sources of signal loss along the entire path from where the signal is generated until it is processed in the receiver, as Figure 11.25 illustrates. Generally, the analyst characterizes sources of signal gain because they are composed of components such as amplifiers and antennas. The loss factors the link designer uses in link analysis are typically straightforward to measure; however, the designer must estimate some of them in the statistical sense. For example, attenuation sources, such as waveguides, have their loss levels determined empirically.

Link designers model the estimated parameters, such as atmospheric attenuation, on a worst-case basis based on the probability that the link must be available. Figure 11.25 shows the mix of potential attenuation and interference sources. Certain losses are point locations, while others, such as atmospheric attenuation, are bulk quantities measured over the whole path length.

Because these bulk loss terms are generally not specifically measured or are time variable, the link designer may be uncertain about the exact value to use. To cover the uncertainties in the analysis, the link designer factors in extra system performance, or *link margin*, over and above the performance level the designer deems necessary to achieve the desired link quality of service. The margin, measured in dB, is the difference between the required performance and the expected performance. The more of these individual attenuation sources the link designer identifies and evaluates, the less link margin the designer needs to cover the link uncertainties. The margin should be at least 3 dB when the link is well known and the designer accounts for all losses. The margin should be even more (>10 dB) when link is not well described or if the link designer is just beginning to formulate the estimates for the parameters.

Figure 11.25 also shows interference sources. These affect reception just as noise does and they prevent the receiver from achieving the designer's desired performance level. Link margin also helps the link designer overcome any uncertainties in the interference sources. In the next section, we will look at the basic power equation for the link and then look at an example computation for a digital link. After that, we will look at the details of estimating the losses in the link.

11.5.5 Link Analysis

We start with the basic carrier-to-noise ratio shown in Equation (11.17) and then add the bulk link losses to arrive at an end-to-end *link equation*

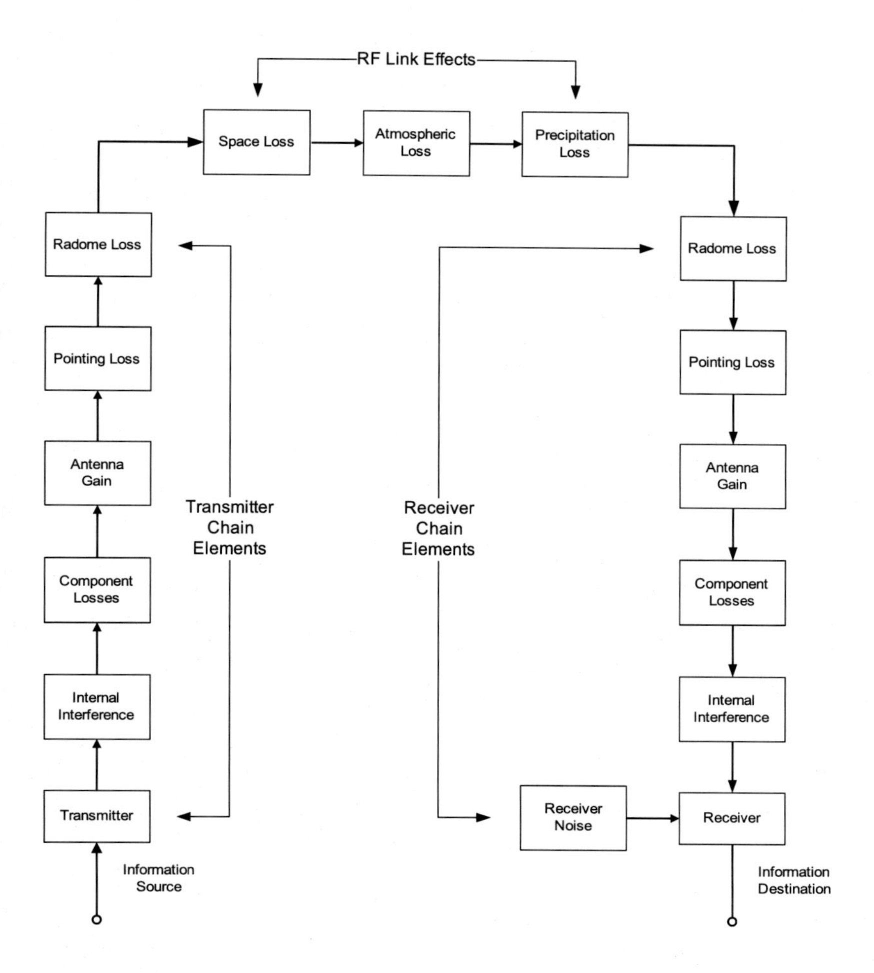

Figure 11.25: Sources of loss and interference along a link from signal generation to signal reception.

of the form

$$\left[\frac{C}{N}\right] = [EIRP] - [L_S] + \left[\frac{G}{T_{sys}}\right] - [k] - [B] - [L_R] - [L_A] - \cdots \quad (11.41)$$

L_R is the loss due to rain and L_A is the loss in the atmosphere. The link designer adds other known losses, in dB units, to Equation (11.41) until the designer accounts for all of the loss effects. For digital links, the link designer requires the E_b/N_o estimate. The energy per bit, E_b, is related to the carrier power, C, and either the bit rate, R_b, or bit period, T_b, via

$$E_b = \frac{C}{R_b} = CT_b \quad (11.42)$$

The designer computes the noise spectral density, N_o, in terms of the noise power, N, and the either the bandwidth, B, or the system temperature, T_{sys}, and Boltzmann's constant, k, using

$$N_o = \frac{N}{B} = kT_{sys} \quad (11.43)$$

The designer computes the E_b/N_o from Equations (11.42) and (11.43) using

$$\frac{E_b}{N_o} = \frac{C}{N}\frac{B}{R_b} \quad (11.44)$$

When the propagation path changes from free space to multipath propagation, the space loss term in Equation (11.15) is often modified by a factor, n, to account for the additional attenuation on the link [Rapp96]. For free space, $n = 1$, while in multipath environments, $2 < n < 3$ is typical. The space loss term now is

$$[L_S] = 20n \log(4\pi R/\lambda) \quad (11.45)$$

Figure 11.26 illustrates a *link design spreadsheet* for the link designer to use in organizing the information for computing the overall *link budget.* This link analysis spreadsheet is broken into subsections for user input as described below.

System Description The parameters necessary to describe the desired link performance

- The baseband data rate before applying any error correcting coding

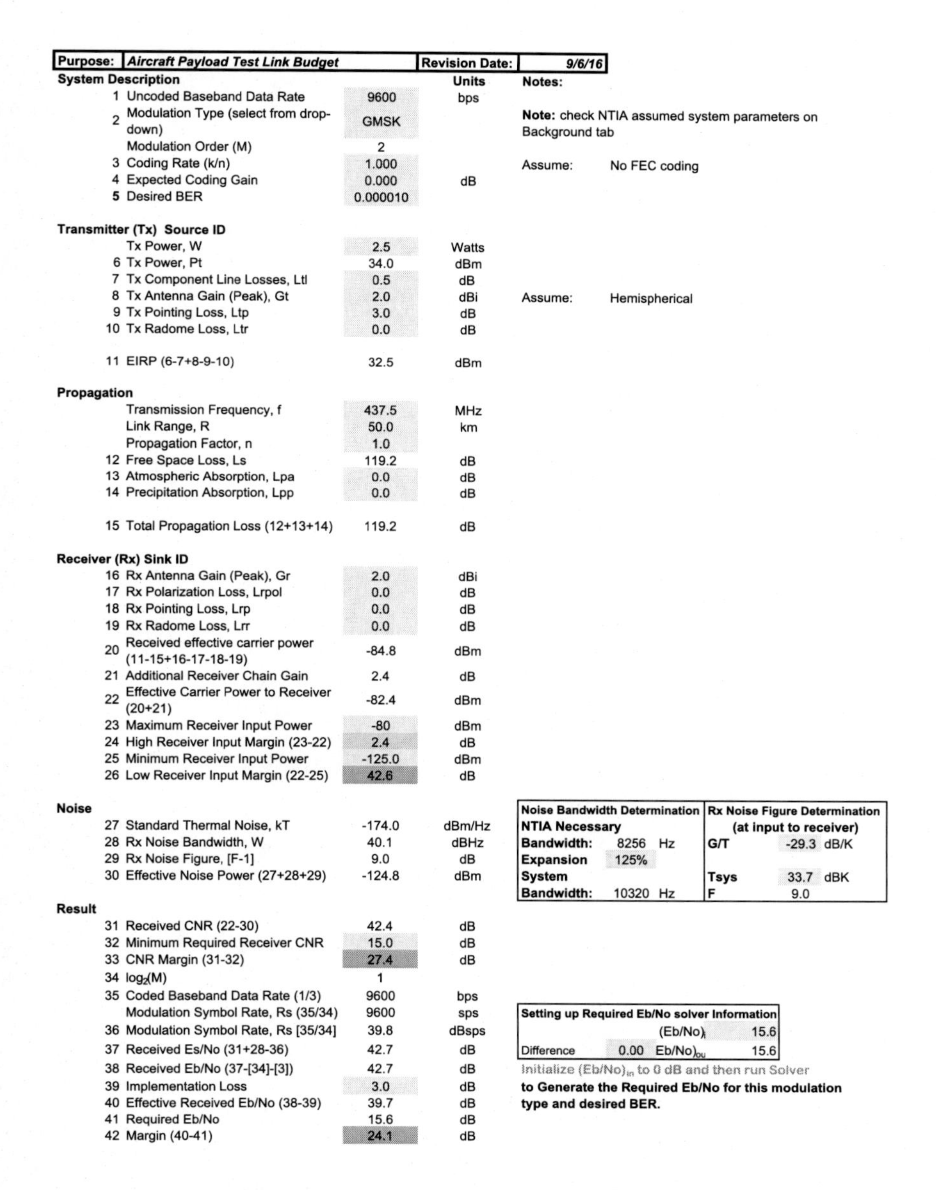

Purpose:	Aircraft Payload Test Link Budget		Revision Date:	9/6/16
System Description			**Units**	**Notes:**
1	Uncoded Baseband Data Rate	9600	bps	
2	Modulation Type (select from drop-down)	GMSK		**Note:** check NTIA assumed system parameters on Background tab
	Modulation Order (M)	2		
3	Coding Rate (k/n)	1.000		Assume: No FEC coding
4	Expected Coding Gain	0.000	dB	
5	Desired BER	0.000010		
Transmitter (Tx) Source ID				
	Tx Power, W	2.5	Watts	
6	Tx Power, Pt	34.0	dBm	
7	Tx Component Line Losses, Ltl	0.5	dB	
8	Tx Antenna Gain (Peak), Gt	2.0	dBi	Assume: Hemispherical
9	Tx Pointing Loss, Ltp	3.0	dB	
10	Tx Radome Loss, Ltr	0.0	dB	
11	EIRP (6-7+8-9-10)	32.5	dBm	
Propagation				
	Transmission Frequency, f	437.5	MHz	
	Link Range, R	50.0	km	
	Propagation Factor, n	1.0		
12	Free Space Loss, Ls	119.2	dB	
13	Atmospheric Absorption, Lpa	0.0	dB	
14	Precipitation Absorption, Lpp	0.0	dB	
15	Total Propagation Loss (12+13+14)	119.2	dB	
Receiver (Rx) Sink ID				
16	Rx Antenna Gain (Peak), Gr	2.0	dBi	
17	Rx Polarization Loss, Lrpol	0.0	dB	
18	Rx Pointing Loss, Lrp	0.0	dB	
19	Rx Radome Loss, Lrr	0.0	dB	
20	Received effective carrier power (11-15+16-17-18-19)	-84.8	dBm	
21	Additional Receiver Chain Gain	2.4	dB	
22	Effective Carrier Power to Receiver (20+21)	-82.4	dBm	
23	Maximum Receiver Input Power	-80	dBm	
24	High Receiver Input Margin (23-22)	2.4	dB	
25	Minimum Receiver Input Power	-125.0	dBm	
26	Low Receiver Input Margin (22-25)	42.6	dB	
Noise				
27	Standard Thermal Noise, kT	-174.0	dBm/Hz	
28	Rx Noise Bandwidth, W	40.1	dBHz	
29	Rx Noise Figure, [F-1]	9.0	dB	
30	Effective Noise Power (27+28+29)	-124.8	dBm	
Result				
31	Received CNR (22-30)	42.4	dB	
32	Minimum Required Receiver CNR	15.0	dB	
33	CNR Margin (31-32)	27.4	dB	
34	$log_2(M)$	1		
35	Coded Baseband Data Rate (1/3)	9600	bps	
	Modulation Symbol Rate, Rs (35/34)	9600	sps	
36	Modulation Symbol Rate, Rs [35/34]	39.8	dBsps	
37	Received Es/No (31+28-36)	42.7	dB	
38	Received Eb/No (37-[34]-[3])	42.7	dB	
39	Implementation Loss	3.0	dB	
40	Effective Received Eb/No (38-39)	39.7	dB	
41	Required Eb/No	15.6	dB	
42	Margin (40-41)	24.1	dB	

Noise Bandwidth Determination			Rx Noise Figure Determination (at input to receiver)		
NTIA Necessary Bandwidth:	8256	Hz	**G/T**	-29.3	dB/K
Expansion	125%				
System Bandwidth:	10320	Hz	**Tsys**	33.7	dBK
			F	9.0	

Setting up Required Eb/No solver Information			
		$(Eb/No)_{in}$	15.6
Difference	0.00	$(Eb/No)_{out}$	15.6

Initialize $(Eb/No)_{in}$ to 0 dB and then run Solver **to Generate the Required Eb/No for this modulation type and desired BER.**

Figure 11.26: Example link budget spreadsheet.

- The modulation format
- The error correcting code rate (enter 1.0 if no coding is applied) and the associated coding gain in dB (enter 0 dB if no coding is applied)
- the required bit error rate the link should support

Transmitter The parameters necessary to estimate the transmitter's EIRP

- The transmitter's output power; the user enters the value in Watts and the spreadsheet converts the power to dBm for computations
- The cabling and other link losses, in dB, between the transmitter and the antenna
- The transmitter antenna's peak gain, expected worse case pointing loss, and any losses associated with a radome that may cover the antenna all expressed in dB

Propagation The parameters necessary to estimate the link propagation losses

- The operating frequency in MHz
- The link range in km
- The propagation factor; $n = 1$ for free space propagation or $n = 2$ to 3 for multipath environments
- The expected worst case atmospheric and rain attenuation losses in dB

Receiver The parameters necessary to estimate the received carrier power

- The receiver antenna gain in dB
- The worst-case polarization, radome, and pointing losses in the receiver antenna system expressed in dB
- The minimum and maximum receiver signal power levels in dBm for correct receiver operation; if the received signal power level is within the bounds by at least 3 dB, the analysis highlights the associated margin cells in green, if the power level is less than 3 dB, the analysis highlights the cells in red or yellow

Noise The parameters necessary to estimate the system noise power

- The system's G/T and T_{sys} values with the system noise temperature entered in dB K $[10log(T_{sys})]$
- The bandwidth expansion factor beyond the modulation necessary bandwidth to estimate the total receiver noise bandwidth

Result The parameter to determine if there is link closure and estimate the link margin. This is the minimum carrier-to-noise ratio for receiver operation to produce the desired Bit Error Rate (BER) at a worst case. If the received C/N level is at least 3 dB, the margin cell turns green, or it turns red or yellow if below 3 dB.

To help automate the process, the spreadsheet uses the solver function in Excel® to solve the equations from Section 10.4.6 for the required E_b/N_o needed for the link to achieve the desired BER. The solver function manipulates the E_b/N_o value in the probability of error equation for the user-specified modulation type until it matches the desired BER. This determines the system's required E_b/N_o. After determining the required E_b/N_o, the spreadsheet determines the margin between it and the available E_b/N_o value computed from the inputs. If the margin exceeds 3 dB, the spreadsheet's margin cell turns green, or it turns yellow or red if the margin is less than 3 dB. If the link designer knows the statistical variations in the parameters, the designer can modify the spreadsheet to account for those variations and give an estimate of the confidence in the result.

Notice that the margin computation is a function of the designer's choice of modulation type, error correction coding, system transmitter and receiver parameters, carrier frequency, etc. The designer can use a tool such as this spreadsheet to explore different options for each of the system parameters. The designer can then optimize system performance, system cost, and frequency allocation parameters. The designer can also determine if aspects of the system are over designed yielding a margin that is too large for design efficiency.

The spreadsheet example link in Figure 11.26 describes a 437.5-MHz transmission at 9600 bps over a 50-km link. The transmission and reception antennas have a 2-dB gain. Minimal hardware losses are included to account for inefficiencies in the system. Rain and atmospheric losses are minimal at UHF frequencies. The link budget spreadsheet uses the equations developed in this section for the component computations.

Purpose:	*Aircraft Payload Test*		Revision Date:		7/17/14
Component	**Value**	**Units**	**Te**	**Units**	**Gain**
Receive Antenna Gain	2.00	dB			1.6
Maximum Antenna Temperature	300.0	K	300.0	K	
Antenna-to-Waveguide/Cable Connector Loss	0.10	dB	6.8	K	1.0
Waveguide/Cable Loss	1.00	dB	75.1	K	0.8
Waveguide/Cable-to-Diplexer Connector Loss	0.10	dB	6.8	K	1.0
Diplexer Loss	1.00	dB	75.1	K	0.8
Diplexer-to-Waveguide/Cable Connector Loss	0.10	dB	6.8	K	1.0
Waveguide/Cable Loss	2.00	dB	169.6	K	0.6
Waveguide/Cable-to-LNA Connector Loss	0.10	dB	6.8	K	1.0
LNA					
Gain	10.00	dB			10.0
Effective Temperature	100.00	K	100.0	K	
LNA-to-Waveguide/Cable Connector Loss	0.10	dB	6.8	K	1.0
Waveguide/Cable Loss	3.00	dB	288.6	K	0.5
Waveguide/Cable-to-Receiver Connector Loss	0.10	dB	6.8	K	1.0
Receiver Effective Temperature	289.00	K	289.0	K	
System Temperature @ Receiver Input	2324.23	K	33.7	dBK	
G/T @ Receiver Input	-29.26	dB/K			
Total Excess Gain to Receiver	2.40	dB			

Figure 11.27: Example G/T and T_{sys} computation for the system being analyzed in the link budget spreadsheet. The highlighted cells are the user input cells.

Figure 11.27 lists the receiver hardware used in the spreadsheet example to estimate the G/T and T_{sys} computations.

A frequent link design problem is to compute the overall link performance for a two-hop (or greater) link. The designer computes the overall C/N ratio based on the first leg's ratio, C/N_1 and the second leg's ratio, C/N_2, using the same relationship as adding resistors in parallel

$$\frac{C}{N} = \left[\left(\frac{C}{N_1}\right)^{-1} + \left(\frac{C}{N_2}\right)^{-1}\right]^{-1} \tag{11.46}$$

Analysts can extend Equation (11.46) for more hops in the overall link.

11.6 ATMOSPHERIC, SUN, AND GROUND PROPAGATION EFFECTS

So far, we have examined free-space propagation. As Figure 11.25 illustrates, the RF link suffers from attenuation sources in addition to space loss. The link designer needs additional margin to overcome loss effects from equipment problems, atmospheric attenuation sources, sun, and ground effects [P.530]. In this section, we examine atmospheric effects on RF signals due to

- Attenuation by atmospheric gas components
- Atmospheric refraction
- Diffraction by obstacles
- Atmospheric scintillation

We will also examine nonatmospheric effects due to solar microwave energy intruding on the receiver antenna and terrain propagation effects. Later in this chapter, we will also examine the effects of rain on the link propagation.

11.6.1 Gaseous Attenuation

Figure 11.22 showed that the atmosphere does not absorb the RF signal in a uniform manner above 10 GHz due to absorption by the atmosphere's molecular components. Figure 11.26 has an entry for this in the "Atmospheric Absorption" line of the link budget spreadsheet. For link designers needing to predict gaseous attenuation for line-of-sight links relatively close to the ground, the simple approximation given in [P.676] is sufficiently accurate. The approximation estimates the specific attenuation in dB/km based on temperature, atmospheric pressure, and water vapor density. The model uses two bulk elements: a dry air component and a water vapor component. Contributions from both are added together to generate the total gaseous absorption estimate. Figure 11.28 illustrates this approximation for frequencies from 1 to 66 GHz at sea level for the Earth's atmosphere at standard temperature and pressure. In this plot, we can see absorption features for water vapor near 22 GHz and oxygen near 50 GHz. The dry air component is generally smooth over this range without the specific features like water vapor.

The link analyst computes the gaseous attenuation, A_g, in dB as a

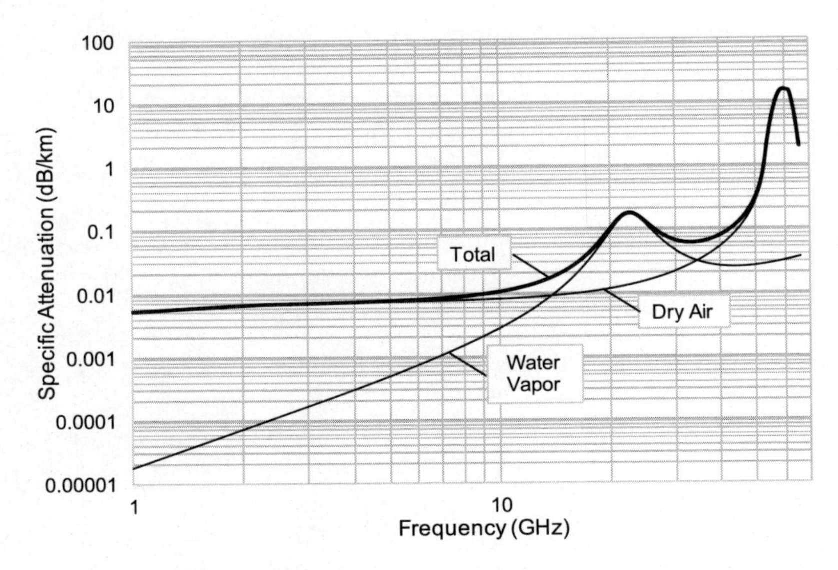

Figure 11.28: Specific attenuation for atmospheric gaseous absorption from 1 to 66 GHz based on the approximations given in [P.676].

function of the total gaseous specific attenuation, γ_{total} in dB/km, and the link range, R, in km from

$$A_g = \gamma_{total} R \tag{11.47}$$

For higher frequencies, frequencies near atmospheric absorption lines, or long slant paths through the atmosphere, the link designer should use the more exact algorithm given in [P.676]. Generally, for line-of-sight RF links for frequencies below 20 GHz, the link designer can assume a gaseous absorption below 1 dB for typical telemetry and telecommand system applications.

11.6.2 Refraction

Refraction is the bending of an electromagnetic wave due to changes in the index of refraction of the atmosphere [Ippo08; P.453; P.834; Reud74]. Figure 11.29 illustrates this schematically. The atmospheric index of refraction ranges from 1.000 301 at sea level to 1.0000 in free space. Because of refraction, the RF signal may encounter effects such as

- The radio line of sight is not the same as the optical line of sight
- The actual radio path exceeds the geometric path length

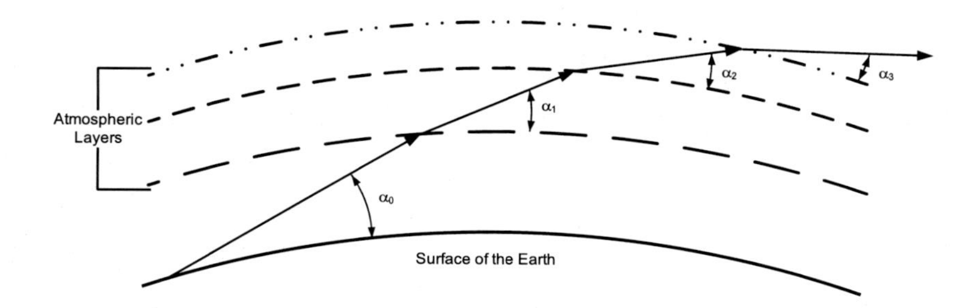

Figure 11.29: A schematic for atmospheric refraction of radio waves.

- The apparent pointing angle to a target is not the same as the geometric pointing angle, especially for paths close to the horizon
- The RF signal may appear to be spread out (defocused) due to variations in the atmosphere, especially for paths close to the horizon

The refraction of the radio waves by the atmosphere adds about 15% to the radio horizon distance relative to the optical horizon distance. Link planners model this effect by imagining that the radio waves travel along a straight ray path and vary the Earth's curvature to match the bending of the radio waves. This effectively makes the Earth look "bigger." One of the interesting consequences of refraction is that if an object appears to be on the radio horizon, it is really below the optical horizon but we still maintain communications. This may cause pointing problems if not properly accounted for.

Link planners use the K factor to describe the change in the atmosphere's index of refraction with height. The exact value for K is a function of pressure and humidity and requires knowledge of these factors as a function of height. Normally, link planners use a typical value of $K = {}^4/_3$. There is, in fact, specially lined graph paper used to plan links using the $K = {}^4/_3$ value. The designer plots the location of the transmitter and receiver and their heights above sea level on the graph paper. The designer does this for the terrain between both endpoints and for any intermediate obstructions like mountains, forests, etc. Once the designer makes the plot, the designer estimates the correct pointing angles and determines the potential interference source.

For the times in which K is not ${}^4/_3$ or the requisite graph paper is not available, then the link designer uses a procedure with normal graph paper to plot the link features [Crai96; Free87]. In this case, the designer plots apparent curvature of the Earth from the vantage of the radio wave, and the heights of the transmitter, receiver, obstructions, and terrain using the bending due to the K factor. To do this, the link designer first computes the effective radius of the Earth, a, from the K factor and the true radius, $r_0 = 6378\,\text{km}$, using

$$a = Kr_0 \tag{11.48}$$

The link designer then draws a curve of the relative surface of the Earth, $y(x)$, in terms of the relative distance from the transmitter, x, starting at the transmitter. The designer uses the effective radius of the Earth from Equation (11.48) and the following equation for the surface:

$$y(x) = \frac{-x^2}{2a} \tag{11.49}$$

The designer next plots the heights of individual terrain features, h_i, the receiver, and any intermediate obstructions as a function of their relative distance from the transmitter, x_i, on this curve using the modified elevation for each point

$$y\left(x_i, h_i\right) = h_i - \frac{-x_i^2}{2a} \tag{11.50}$$

Figure 11.30 illustrates a path profile with $K = 4/3$. Each feature is described by its height above sea level and distance from the transmitter $\{h_i, x_i\}$. The link planner modifies the heights using Equation (11.50) to give the relative terrain profile between the transmitter and the receiver. Once the link planner plots the heights of the objects with the curvature profile, the planner determines link obstructions from the point of view of the radio wave.

Figure 11.31 illustrates an actual terrain profile for a 900-MHz communications link between a ground antenna and an antenna on a mountain top. The radio path is curved due to the atmospheric refraction effects.

11.6.3 Diffraction

Electromagnetic radiation does not need to touch an object directly to interact with it. *Diffraction* is the scattering of electromagnetic energy by objects, including the Earth's surface, along the propagation path [Bert00; P.526; P.530; Smit98; Viga75; Viga81]. Propagation analysts

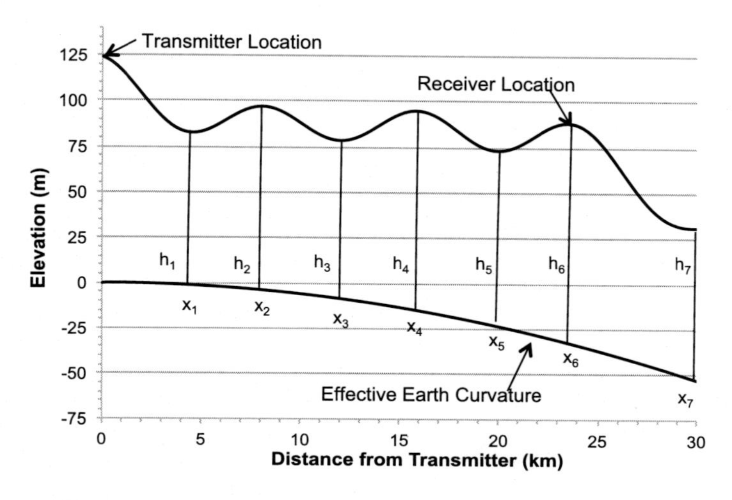

Figure 11.30: Path profile modified for refraction.

consider diffraction fading effects to start occurring when an object or the surface of the Earth are within 60% of the diameter of the first Fresnel zone of the wavefront. The Fresnel zones are functions of the radiation wavelength, λ, the distance between the transmitter and the obstruction, d_t, the receiver and the obstruction, d_r, and the height of the object into the path, h (the height is negative if the object rises above the path line and positive if it is below the path line). Figure 11.32 illustrates the relative Fresnel zone geometry for an obstruction in the path. The obstruction contributes a significant fade if the object is within the first Fresnel zone, which has a radius, r_1, computed from

$$r_1 = \sqrt{\frac{\lambda d_t d_r}{d_t + d_r}} \tag{11.51}$$

Figure 11.31 also shows the envelope of the 0.6 and 1.0 Fresnel zones for that path. At the left-hand side of the figure, we can see that there is little ground clearance for the Fresnel zones. Because of this low clearance, there may be a diffraction fading issue in this case.

When the wavefront encounters the obstruction, the obstruction scatters the signal's energy and reduces the signal power at the receiving antenna. This diffraction fading attenuates the received signal power by A_d dB that the designer can estimate using [P.530; Viga81]

$$A_d = \frac{-20h}{F_1} + 10 \tag{11.52}$$

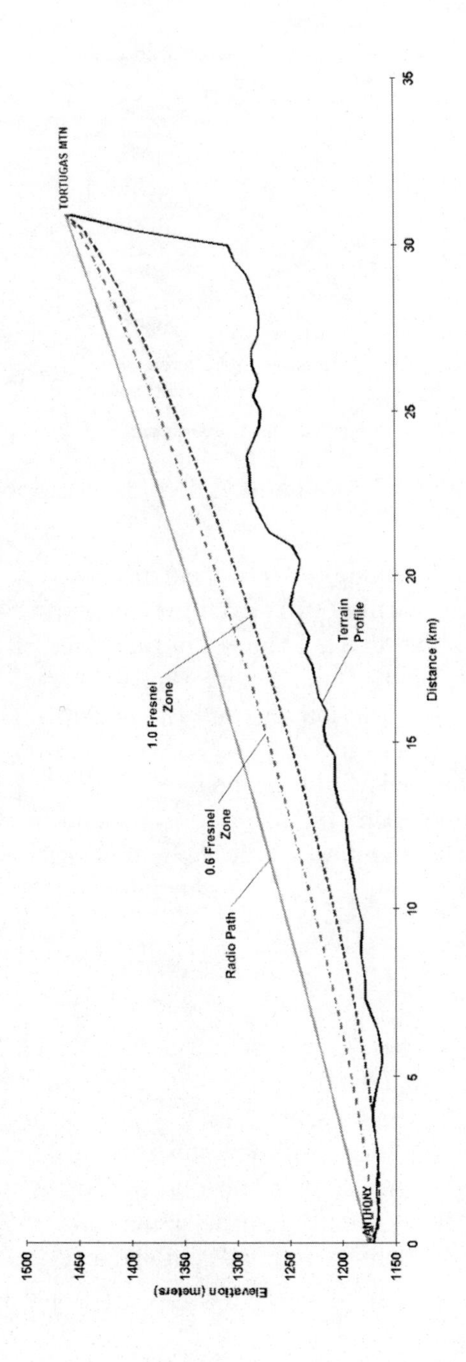

Figure 11.31: Example path profile showing the effects of refraction of the radio wave and the Fresnel zones for the propagation path.

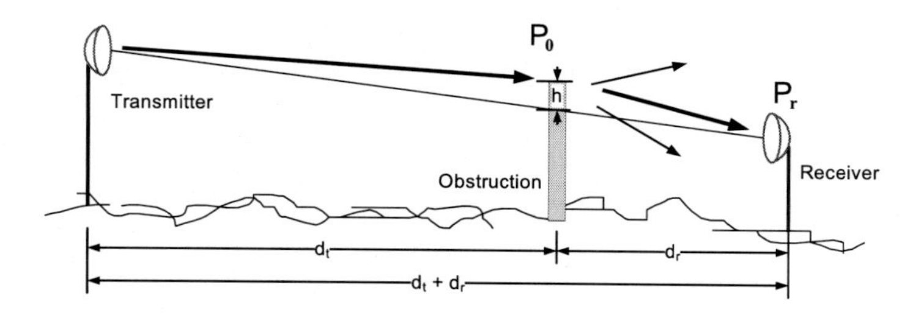

Figure 11.32: A radio wave diffraction due to interaction with an obstruction such as a building.

Here, F_1 is computed from the transmission frequency f in GHz and the total distance $d = d_t + d_r$ using

$$F_1 = 17.3\sqrt{\frac{d_t d_r}{fd}} \tag{11.53}$$

The link analyst adds the diffraction fading attenuation, in dB units, to the atmospheric loss terms in Equation (11.41).

11.6.4 General Terrain Fade Margin

If a propagation path is close to the ground (often within the first Fresnel zone) and it encounters a number of generic obstructions including terrain features such as multiple hills, trees, and other objects that may be time variable (not specific immovable objects like those considered earlier), then the link designer adds a statistical estimate to the overall fade in the link budget with Equation (11.41). The designer can use the Barnett-Vigants relationship for this estimate [Barn72; Viga75]. If the link designer knows the path length, d, in kilometers, the operating frequency, f, in GHz, a terrain factor, C, and fade attenuation level, A, that the link is designed to withstand, then the designer can estimate the probability, p, that the fades exceed that attenuation level by using

$$p = 6 \times 10^{-7} C f d^3 10^{-0.1A} \tag{11.54}$$

The terrain factor, C, ranges from 0.2 for desert terrain, to 1 for normal terrain, to 4 to 6 for smooth, damp paths. A different model is available from the ITU *Radio Regulations* in [P.530], but it requires that the link designer have a more detailed knowledge of the overall terrain characteristics.

A more useful relationship than the probability estimate found in Equation (11.54) is to compute a Fade Margin (FaM) at a given reliability level, also called a link availability level, R. This FaM is relative to the normal received signal power level without the terrain fades. To compute the FaM, we recast Equation (11.54) as

$$FaM = 30\log(d) + 10\log(6fC) - 10\log(1 - R) - 70\ dB \qquad (11.55)$$

For example, suppose we need a 10-km path at 4 GHz with a desired reliability of 0.999 999. Using Equation (11.55), we need a FaM = 27 dB to account for potential terrain fades over desert terrain.

11.6.5 Scintillation

Most of us are familiar scintillation effects on electromagnetic radiation. Scintillation is frequently observable as a twinkling when we look at stars at night. Radio waves are also subject to this effect [Ippo08]. Variations within the ionosphere, the atmosphere, and rain and clouds along the path cause scintillation. Scintillation is observable as a rapid fluctuation in the amplitude and phase of the carrier, which appears to be a form of random modulation. Figure 11.33 illustrates this effect for the Ka-Band beacon signal from the National Aeronautics and Space Administration's Advanced Communications Technology Satellite (ACTS) satellite.

Ionospheric scintillation may occur when a radio wave passes through the ionosphere and it is a potential problem on satellite communication links but not with line-of-sight links. The standard ITU *Radio Regulations*model is a function of frequency, antenna elevation angle, time of year, phase of the solar sunspot cycle, and geographical location [Ippo08]. This model is rather inexact but may be a good starting point to estimate the magnitude of the effect and the reference describes a procedure for preparing the estimate.

Rain and cloud scintillation increases the loss due to the fades caused by the storm activity. In Figure 11.34, the rain and clouds cause scintillation in addition to the signal attenuation. This scintillation is the added jaggedness to the received signal levels in addition to the pure fade such as around time 00:45:00.

With atmospheric scintillation, the scintillation level is a function of temperature, humidity, terrain factors, and, importantly, antenna elevation angle [Ippo08]. It is also a function of the time of day. For example, the ACTS beacon measurement shown in Figure 11.33 was made over mountainous terrain and the level increases as the late afternoon

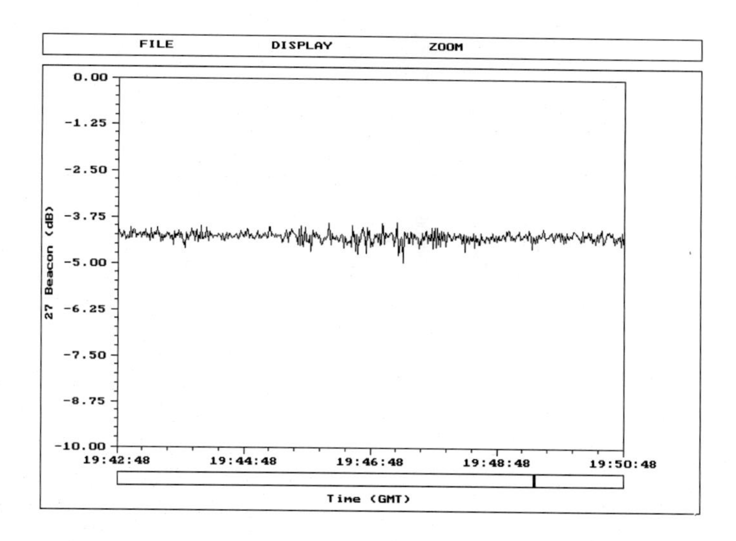

Figure 11.33: Atmospheric scintillation of a Ka-Band radio beacon.

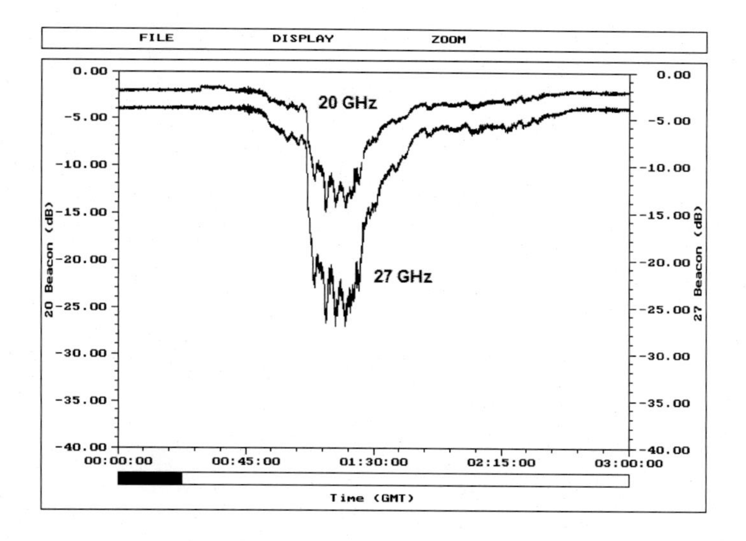

Figure 11.34: Rainfall attenuation at Ka-band frequencies as recorded with the ACTS propagation experiment.

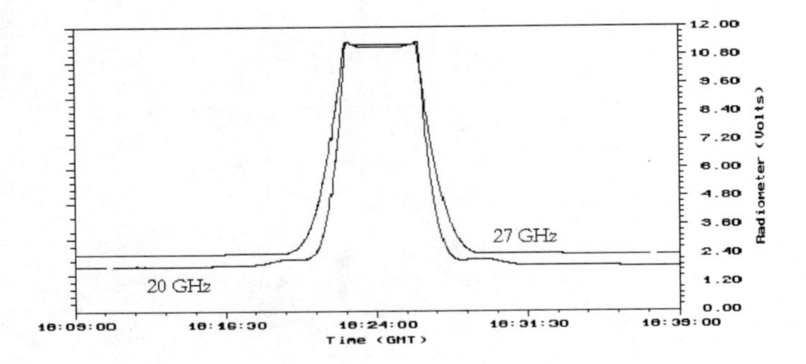

Figure 11.35: Increase in the Ka-Band antenna noise temperature due to the sun passing through the beam.

convection increases over the mountains. Generally, link analysts model scintillation statistically as a random process. The path traveled by the beacon signal to the receiver in that example is at a high elevation angle and the signal variance is less than 1 dB so there is not much effect on the link. At these frequencies at locations where the path angle is lower, the variations can be much higher — over 5 dB in variance. When this occurs, the signal amplitude has significant variations and cause data loss over short time periods.

11.6.6 Sun Intrusions

As we saw in Equation (11.27), the sun causes a rise in antenna temperature when it fills the antenna beam pattern. The antenna shown in Figure 11.13(a) has a HPBW approximately the size of the solar disk when used at Ka-Band. A *radiometer* is a special receiver for measuring the apparent sky temperature at radio frequencies. Figure 11.35 shows the rise in radiometer output at 20 and 27 GHz as the sun passes through the dish's antenna pattern. At the center of the figure, the sun has saturated the receiver electronics giving rise to a *sun intrusion*. The intrusion event lasts from 10:19:30 to 10:29:30. If this were a data receiver, the rise in the associated antenna temperature could reduce the E_b/N_o sufficiently enough to degrade system performance. The normal mitigation for this effect is to avoid pointing the antenna at the sun. Interestingly, the moon also causes this same effect but at a much lower increase in noise temperature.

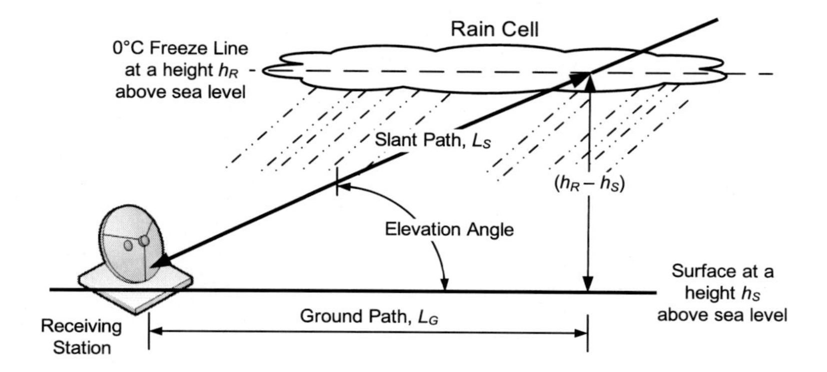

Figure 11.36: Standard geometric relationship for system elements in rainfall modeling for links with elevation angles > 5° based on [P.618].

11.7 RAIN EFFECTS MODELING

Rainfall provides a common attenuation source to a transmitted RF signal with two forms of change occurring to the signal being present: scattering of the energy and absorption of the energy. The magnitude of these effects is a function of both the carrier frequency and the rainfall rate. The raindrop shape is a function of the fall rate. Raindrops are spherical at low fall rates but as the rate increases, they become oblate spheroids. At high rates, the oblate spheroids deform, with dimples in the downward side. The size varies with the shape. To analyze fully the rainfall effects on the radiation, the link planner must consult electromagnetic scattering theory. For most practical purposes, the link planner uses a bulk approximation. This section discusses rainfall effects, the statistical representation designers use to characterize the rain loss, and the standard ITU model for rain attenuation*Radio Regulations*. We also examine how rainfall can affect antenna surfaces to cause additional attenuation.

11.7.1 Rain Effects

Rainfall on a radio propagation path has two major effects: signal attenuation that is proportional to the path length through the rainfall, and increased system temperature due to a rise in the antenna temperature. Additionally, scintillation of the radio wavefront caused by the rainfall is also present. Figure 11.36 illustrates the ITU standard geometry for Earth-space link geometries that we will use here [P.618].

The rain attenuation is of most concern to link planners designing links for operation above 10 GHz. A very intense storm can be a factor even at lower frequencies. The attenuation along a path is seen in Figure 11.34 where the ACTS Ka-Band beacon is measured during a thunderstorm at White Sands, New Mexico. The rainstorm begins at approximately 00:45 and ends around 02:55. Note the following items related to Figure 11.34:

- Attenuation at 27 GHz exceeds the attenuation at 20 GHz; this is typical for rain attenuation.
- Thunderstorms typically have well-defined edges and 30 min to 2 h durations; they do not affect the link throughout the day as a general rain storm might.
- The attenuation can be as small as 1 to 2 dB to the entire receiver link margin.

The rain attenuation level is a function of the point rain rate measured in mm/h. Figure 11.37 plots the point rain rate measured at the antenna along with the attenuation for the same storm. From the lower trace in Figure 11.37, we see that the rain rate is highly variable even during the storm. This produces variable attenuation levels during the storm and the receiver may go into and out of lock multiple times throughout the storm if it does not have adequate reception margin.

As we just saw, the link undergoes a propagation loss due to the rainstorm's attenuation. The antenna picks up additional microwave energy from the storm because the storm appears warmer than the clear sky temperature. As we saw in Figure 11.22, a typical clear sky temperature is approximately 100 K. During an intense rainstorm, the sky temperature can rise to approximately 275 K. Figure 11.38 illustrates this effect by plotting the radiometer output along with the beacon attenuation observed during the storm in Figure 11.37. The radiometer output is proportional to the apparent sky temperature. The sky appears to be warmer during the storm than the clear sky temperature.

The link designer can estimate the effective sky temperature, T_{sky}, that the receiving antenna sees due to the attenuation in dB, A, by using [P.618]

$$T_{sky} = T_{mr}\left(1 - 10^{-0.1A}\right) + (2.7)\left(10^{-0.1A}\right) \tag{11.56}$$

Here, T_{mr} is the atmospheric mean radiating temperature, in K. The designer can estimate T_{mr} from local surface temperature, T_S, also in K,

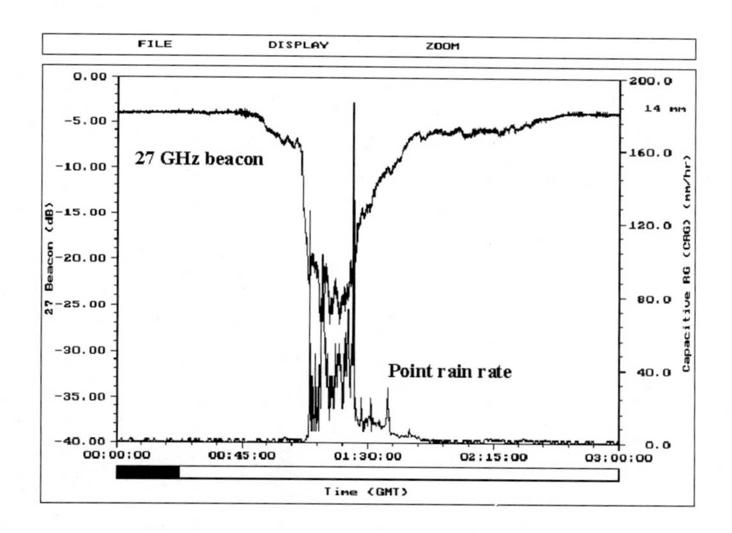

Figure 11.37: Rain attenuation (upper trace) as a function of rainfall rate (lower trace) as recorded with the ACTS propagation experiment.

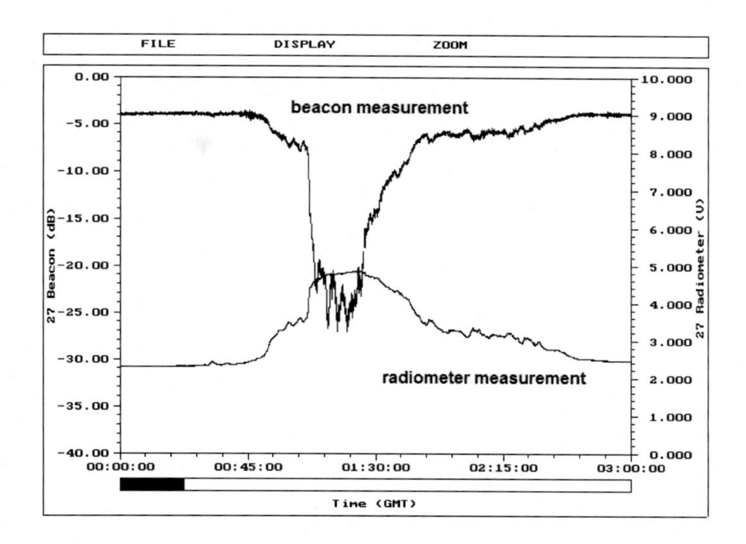

Figure 11.38: Sky temperature increase as measured by the radiometer (lower trace) with beam attenuation (upper trace) as recorded with the ACTS propagation experiment.

by using [P.618]

$$T_{mr} = 37.34\,\mathrm{K} + 0.81T_S \tag{11.57}$$

Note: although these relationships are designed for satellite links, the sky temperature increase is also valid for point-to-point terrestrial links as well. The receiver detects the sky radiation based on the pointing angle to the sky and not on the link data usage.

11.7.2 International Telecommunications Union Model

Link designers can choose from a number of predictive models for estimating the rain attenuation on a link. The estimates generated by the models do not predict specific, moment-by-moment attenuation on the link. Rather, the prediction is a statistical one that forecasts the percentage of the year that the attenuation reaches or exceeds a specified level. That is, how often does the rain attenuation exceed the value the link designer has allocated in the overall link margin estimate. Modelers have developed the estimates discussed here for predicting rain attenuation through the entire atmosphere to satellites. The link designer adapts these models for terrestrial links. In order to use them, the link designer must know the transmission frequency, link location, and link path length through the rain. While these models are easy to compute, the results may vary greatly and the models work best with actual local rain data rather than using regional point rain rates averages tabulated for the model. The common method for estimating the rainfall attenuation level along the link is for the link designer to use the following equation for the *specific attenuation*, γ_R, in dB/km [Cran96; Godd96; Ippo08; P.837]:

$$\gamma_R = kR_{\%}^{\alpha} \tag{11.58}$$

The coefficients k and α are frequency dependent and they are functions of the propagation model. The *point rain rate*, in mm/h, at a given percent of the year, $R_{\%}$, depends upon the link location. Here, we examine the ITU model to predict rain attenuation. The reader should consult the references for other models.

The ITU model attempts to predict the rain attenuation at the 0.01% level when averaged over one year. This means that the actual link attenuation is less than this level 99.99% of the year and exceeds it only 0.01% of the year. The attenuation level, in dB, at this percentage, $A_{0.01}$, has an associated point rain rate level, $R_{0.01}$.

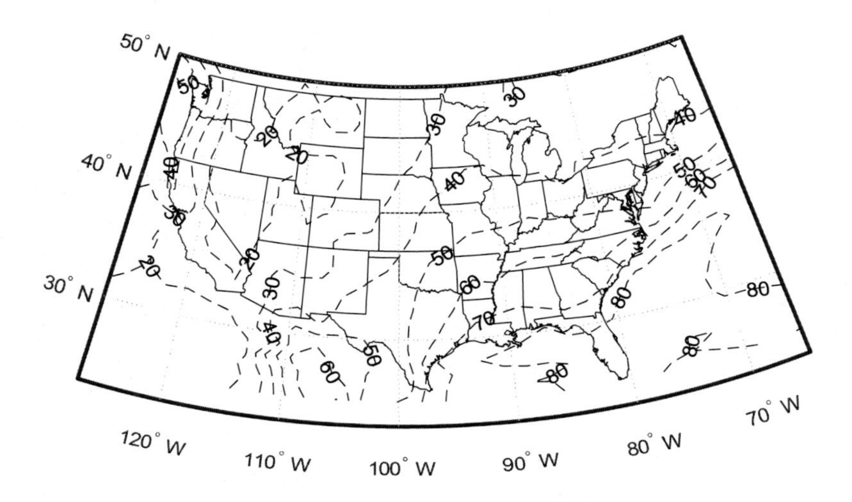

Figure 11.39: ITU 0.01% rain rate, $R_{0.01}$ in mm/h contours for the continental United States based on [P.837].

11.7.2.1 Satellite Link Path Attenuation

To compute the satellite link attenuation estimate, link designers work through a series of intermediate computations for the specific parameters to finally arrive at the expected attenuation. The evaluation steps are as follows:

Point Rain Rate The first step is to determine the appropriate 0.01% point-rain rate level. If the local climate database has the value, the link designer should use it. If not, the designer can use the ITU database program and the procedure given in [P.837] to estimate the value. As a quick start, the link designer can consult the contour maps the ITU provides to gain an order-of-magnitude estimate of the 0.01% point-rain rate level. Figure 11.39 gives an example map based on the contours for the United States. For example, the Cape Hatteras region has a $R_{0.01}$ of approximately 65 mm/h, while El Paso, Texas has a $R_{0.01}$ of approximately 40 mm/h.

Specific Attenuation Coefficients The designer's next step is to determine the coefficients used in Equation (11.58). These coefficients are both carrier frequency-dependent and polarization dependent. First, the designer computes the horizontal and vertical polarization pairs (k_H, k_V)

and (α_H, α_V) for the carrier frequency, f, in GHz. The computation uses the same series expansion for both the horizontal and vertical components but with the corresponding coefficients from Table 11.7 for the (k_H, k_V) and (α_H, α_V). The k_x and α_x values are computed with the equations [P.838]

$$\log_{10} k = \sum_{j=1}^{4} a_j \exp\left[-\left(\frac{\log_{10} f - b_j}{c_j}\right)^2\right] + m_k \log_{10} f + c_k \quad (11.59a)$$

$$\alpha = \sum_{j=1}^{5} a_j \exp\left[-\left(\frac{\log_{10} f - b_j}{c_j}\right)^2\right] + m_\alpha \log_{10} f + c_\alpha \quad (11.59b)$$

These pairs are used to compute the desired k and α values from [P.838]

$$k = \left[k_H + k_V + (k_H - k_V)\cos^2\theta \cos 2\tau\right]/2 \quad (11.60a)$$

$$\alpha = \left[k_H\alpha_H + k_V\alpha_V + (k_H\alpha_H - k_V\alpha_V)\cos^2\theta \cos 2\tau\right]/2k \quad (11.60b)$$

Here, θ is the propagation path elevation angle through the rain from Figure 11.36 and τ is the electromagnetic polarization tilt angle relative to the horizontal plane. The analyst is to use a τ of 45° with circular polarization.

With this information, the designer computes the specific rain attenuation at the 0.01% level. With this information, the designer next moves to computing the path length to convert specific attenuation to total attenuation in the next steps.

Vertical and Horizontal Path Lengths To compute the vertical and horizontal path lengths for a satellite link, the link designer must know the rain height as we saw in Figure 11.36. The designer starts with the 0 °C *Freeze Line* height, h_0, for the receiving station location derived from either local data or from the average data available in conjunction with [P.839]. The designer computes the rain height, h_R, in km, from

$$h_R = h_0 + 0.36\,\text{km} \quad (11.61)$$

For propagation paths with $\theta > 5°$, we can see from Figure 11.36 that the Slant Path (L_S) through the rain is given in terms of the receiving station height above sea level, h_S, and the rain height by

$$L_S = \frac{h_R - h_S}{\sin\theta} \quad (11.62)$$

Table 11.7: Series Expansion Coefficients for Computing k and α for the Specific Attenuation [P.838]

Series coefficients for computing k_H					
j	a_j	b_j	c_j	m_k	c_k
1	-5.33980	-0.10008	1.13098		
2	-0.35351	1.26970	0.45400	-0.18961	0.71147
3	-0.23789	0.86036	0.15354		
4	-0.94158	0.64552	0.16817		
Series coefficients for computing k_V					
j	a_j	b_j	c_j	m_k	c_k
1	-3.80595	0.56934	0.81061		
2	-3.44965	-0.22911	0.51059	-0.16398	0.63297
3	-0.39902	0.73042	0.11899		
4	0.50167	1.07319	0.27195		
Series coefficients for computing α_H					
j	a_j	b_j	c_j	m_α	c_α
1	-0.14318	1.82442	-0.55187		
2	0.29591	0.77564	0.19822		
3	0.32177	0.63773	0.13164	0.67849	-1.95537
4	-5.37610	-0.96230	1.47828		
5	16.1721	-3.29980	3.43990		
Series coefficients for computing α_V					
j	a_j	b_j	c_j	m_α	c_α
1	-0.07771	2.33840	-0.76284		
2	0.56727	0.95545	0.54039		
3	-0.20238	1.14520	0.26809	-0.053739	0.83433
4	-48.2991	0.791669	0.116226		
5	48.5833	0.791459	0.116479		

When the θ is below 5°, the designer needs to consider the atmospheric curvature and the slant path is computed using the Earth radius (R_e) with the equation [P.618]

$$L_S = \frac{2\left(h_R - h_S\right)}{\left(\sin^2\theta + \dfrac{2\left(h_R - h_S\right)}{R_e}\right)^{1/2} + \sin\theta} \tag{11.63}$$

The designer next computes the horizontal Ground Path (L_G) using

$$L_G = L_S \cos\theta \tag{11.64}$$

Rain Adjustment Factors The specific attenuation found in Equation (11.58) overestimates the attenuation level because a continuous rain cell string precipitating at uniform rates does not cover most paths. The model compensates for this by supplying a horizontal path reduction factor, $r_{0.01}$, which the designer computes using [P.618]

$$r_{0.01} = \frac{1}{1 + 0.78\sqrt{\dfrac{L_G\gamma_R}{f}} - 0.38\left(1 - \mathrm{e}^{-2L_G}\right)} \tag{11.65}$$

Next the designer computes the vertical path adjustment factor, $\nu_{0.01}$, through a series of intermediate calculations [P.618]. The first of these intermediate factors to find is

$$\zeta = \tan^{-1}\left(\frac{h_R - h_S}{L_G r_{0.01}}\right)$$

If $\zeta > \theta$, the path elevation angle, then

$$L_R = \frac{L_G r_{0.01}}{\cos\theta}$$

Otherwise,

$$L_R = \frac{L_G r_{0.01}}{\sin\theta}$$

If the receiving antenna is at a latitude, ϕ, where $|\phi < 36°|$, then the next intermediate factor, χ, is given by $\chi = 36° - |\phi|$; if not, $\chi = 0°$. The designer computes the vertical adjustment factor, $\nu_{0.01}$, using

$$\nu_{0.01} = \frac{1}{1 + \sqrt{\sin\theta}\left(31\left(1 - \mathrm{e}^{(\theta/(1+\chi))}\right)\dfrac{\sqrt{L_R\gamma_R}}{f^2} - 0.45\right)} \tag{11.66}$$

Effective Path Length The designer computes the Effective Path Length (L_E) using

$$L_E = L_R \nu_{0.01} \tag{11.67}$$

Total Attenuation Finally, the designer computes the predicted total attenuation, in dB, at the 0.01% outage level from the specific attenuation in Equation (11.58) and the effective path length using [P.618]

$$A_{0.01} = \gamma_R L_E \tag{11.68}$$

Link planners use this attenuation to compute the available C/N for the system. The link designer needs to remember that the rain statistics are highly variable on a yearly basis so this result is really a long-term average attenuation and may not represent any specific year.

11.7.2.2 Terrestrial Link Path Attenuation

The preceeding procedure is for a link between an Earth station and an orbiting satellite. The link designer needs to modify this procedure for point-to-point terrestrial links using the following steps [P.530]:

Specific Attenuation Estimate the link specific attenuation at the 0.01% level and operating frequency using Equation (11.58).

Effective Path Length For a physical path length of d km, compute the effective path length, d_{eff}, for the frequency, f, in GHz, and α from Equation (11.58) using

$$d_{eff} = \frac{d}{0.477 d^{0.633} R_{0.01}^{0.073\alpha} f^{0.123} - 10.579\left(1 - \exp\left(-0.024 d\right)\right)} \tag{11.69}$$

Path Length Check If the effective path length, d_{eff} is more than 2.5 times the physical path length, d, the model caps effective path length at 2.5 times the physical path length. Otherwise, the link designer uses the effective path length as computed.

Total Attenuation Estimate With the specific attenuation and the effective path length for the 0.01% rain rate level, the link designer estimates the total rain attenuation, in dB, using

$$A_{0.01} = \gamma_r d_{eff} \tag{11.70}$$

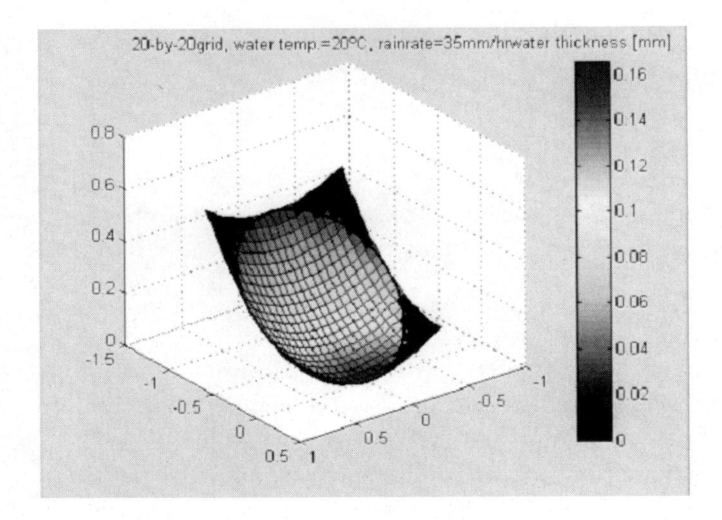

Figure 11.40: Predicted water depth on a composite antenna for a moderate rain rate [Bors99a].

11.7.3 Antenna Wetting

One interesting phenomenon that becomes important to the system designer for antennas operating at Ka-Band frequencies and higher is the effect water on the antenna surface has on the received signal strength. This effect is most pronounced for composite dishes, such as the one shown in Figure 11.13(a), and not the bare metal dishes. This is caused by the composite material placing a gap between the water on the surface and the metal reflector in the dish. Figure 11.40 shows that as the rain falls on the dish, a layer of water builds up on the surface. By solving the propagation equations at the wave level, we discover that the suspended water layer on the dish can cause destructive interference between the incoming wavefront and the reflected wave [Bors99a; Bors99b]. The magnitude of this effect is a function of the point rain rate. Moderate rain rates can cause attenuation at the antenna surface of approximately 3 to 5 dB in addition to the path attenuation. The system designer can predict the attenuation level by knowing the construction details of the dish, the operating frequency, and the rain rate.

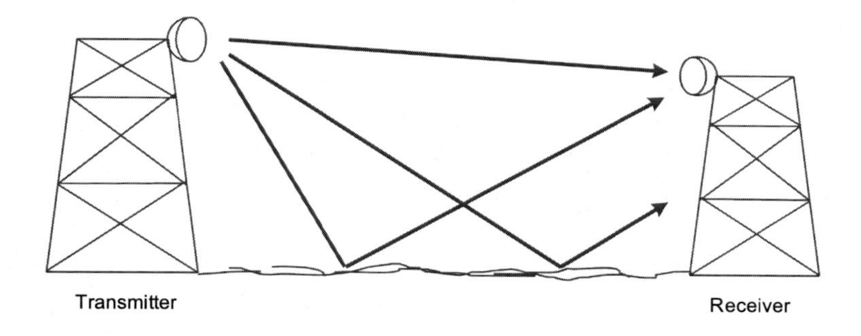

Figure 11.41: Snell's Law for electromagnetic waves reflection.

11.8 MOBILE PROPAGATION

The general free space propagation model is an over-simplification for most terrestrial channels of interest. In addition to the atmospheric effects seen earlier, there are channel complications due to multiple reflections, motion of the transmitter or the receiver, and effects due to foliage. The reference literature covers these effects in more detail [Ande99; Bert00; Gold98; Jake74; Rapp96; Skla01]. We now look at the mobile propagation concepts and how they affect the data links.

11.8.1 Channel Geometry

The mobile propagation environment is very complicated and difficult for the link designer to model. For telemetry and telecommand systems, this is especially true for systems using omnidirectional antennas. By using high-gain antennas, system designers reduce problems because the antenna confines the signal's direction to a narrow field of view.

When the system designer plans the link, the first principle the designer must understand is Snell's Law, which states that electromagnetic waves reflect from a surface at the same angle as the incident wave angle. Figure 11.41 illustrates this concept. Snell's Law has two main ramifications for the designer's consideration: (a) the reflections are symmetric about the vector normal to the surface where the radiation contacts the surface, and (b) because surfaces such as the ground are not smooth, the reflections are at a variety of angles. This implies that as the relative positions of the transmitter and receiver change, the link has a time-variable channel with contributions from the direct ray between the

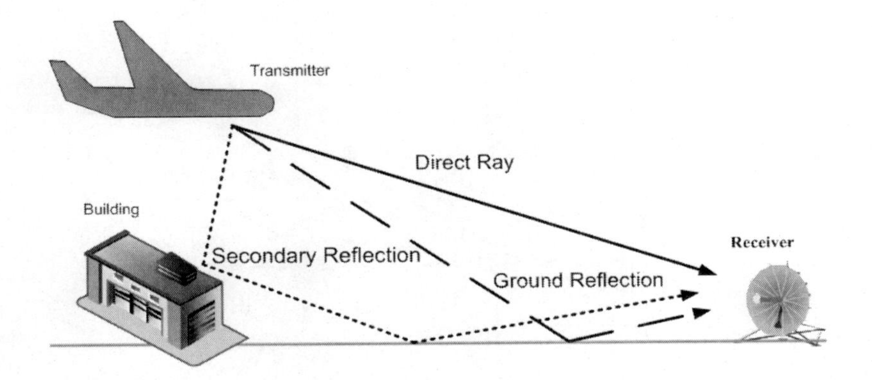

Figure 11.42: Wireless multipath environment.

transmitter and the receiver plus reflected rays coming from the ground and structures in the environment. Figure 11.42 illustrates this concept.

Naturally, the reflections do not encompass 100% of the incident energy so the first few reflections are the main ones to consider. The main variables that influence this process are (a) the reflection coefficient (ground or structural), (b) the transmitter and receiver heights, and (c) the distances involved between the objects. The resulting signal at the receiver undergoes interference from these various reflections and the resulting amplitude fades due to the interference and the motion of the transmitter and/or receiver. This leads to lower data quality and/or lower data throughput than we expect for free-space propagation.

11.8.2 Two-ray Model

The easiest multipath case for the link designer to model is the *two-ray model* in Figure 11.43. This channel type is a *frequency-selective fading* channel [Rapp96; Rice00]. The two-ray condition causes a self-interference at the receiver because the direct wave and reflected wave components traverse slightly different paths from the transmitter to the receiver. This difference causes a relative phase difference in the two copies of the carrier. The analyst can compute the path difference, Δ, betwen the direct path and the reflected path from [Reud74]

$$\Delta = d'' - d' = \sqrt{(h_t + h_r)^2 + d^2} - \sqrt{(h_t - h_r)^2 + d^2} \qquad (11.71)$$

This produces a phase difference, θ_Δ, that is a function of the carrier

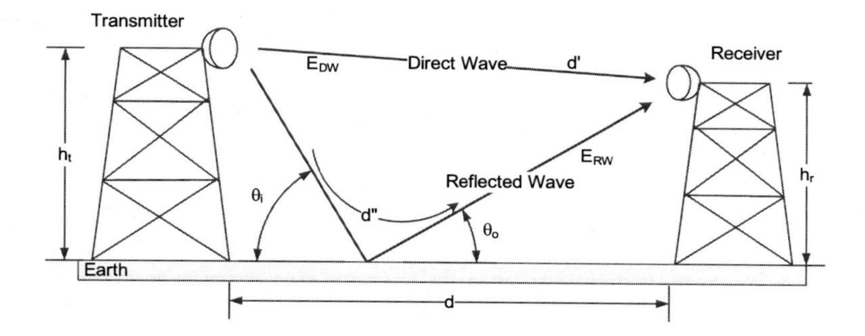

Figure 11.43: Two-ray multipath transmission model.

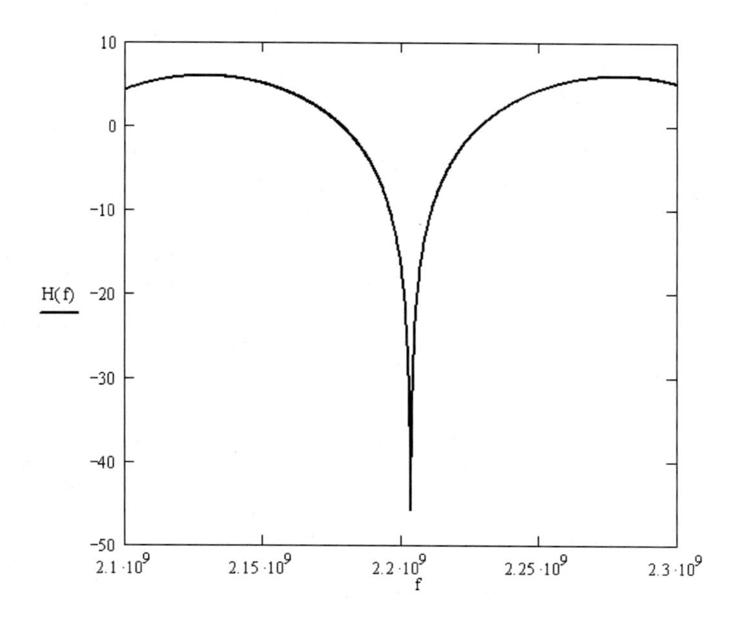

Figure 11.44: Transfer function for a frequency-selective fading channel.

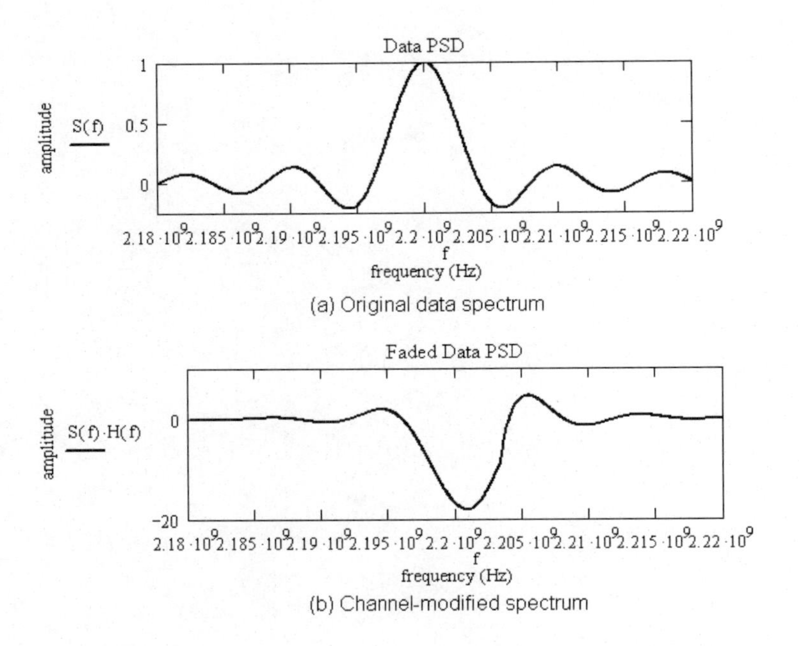

Figure 11.45: Effect of a frequency-selective fade on a PSK signal spectrum.

wavelength, λ, and is computed from

$$\theta_\Delta = \frac{2\pi\Delta}{\lambda} \tag{11.72}$$

The frequency-selective fade acts like a notch filter on the channel. Figure 11.44 shows an example of the resulting channel transfer function where an S-Band fade is simulated. Figure 11.45 demonstrates how this affects the modulated carrier for a Phase Shift Keying (PSK) link. Figure 11.45(a) is the nominal spectrum for a Non-Return to Zero-Level (NRZ-L) PSK transmission. Figure 11.45(b) shows the channel notch filter acting on this spectrum. It is not hard to imagine that the channel quality of service is hard to maintain in this environment.

In a typical case where the link geometry indicates that the incidence angle, θ_i, is small and the distance between the transmitter is large compared with the height of the transmitter and the receiver, it can be shown that the received power, P_r, is expressed as [Reud74]

$$P_r = P_t G_t G_r \left(\frac{h_t h_r}{d^2}\right)^2 \tag{11.73}$$

In Equation (11.73), P_t is the transmitted power, G_t is the transmitting antenna's gain, G_r is the receiver antenna's gain, h_t is the transmitter height, h_r is the receiver height, and d is the distance between the transmitter and the receiver. The important factor for the link designer is that the loss due to the link distance is no longer the distance squared; it is the fourth power of the distance. We anticipated this increased attenuation in Equation (11.45) where we incorporated a link propagation factor n. For free space propagation, $n = 1$, while for the two-ray multipath model, $n = 2$.

11.8.3 Multiple-ray Model

The two-ray channel is a good start but geometries that are more complicated require a more complicated fading model. A more generalized fading model adds the relative contributions from each of the reflected components. The model for the received signal, $r(t)$, in terms of the transmitted carrier, $s(t)$, is

$$r(t) = \sum_{k=1}^{N} \alpha_k s(t - \tau_k) e^{j\theta_k} \tag{11.74}$$

In Equation (11.74), α_k is the relative amplitude of the $\mathrm{k^{th}}$ signal component, τ_k is the corresponding delay for the component, and θ_k is the relative phase shift due to reflections for that component. Figure 11.46 shows an example of simulating a carrier with a four-component model. While there does not appear to be much difference during the bits, there are large transients at the bit edges. This is the time when receivers frequently make bit decisions so these transients can cause bit errors and reduce the quality of service on the link.

11.8.4 Doppler Shifts

The mobile channel has to contend with reflections and disruption of the signal due to *Doppler shifts* in the carrier due to the motion of the transmitter and/or the receiver. The link designer estimates the magnitude of the frequency shift, f_d, due to the Doppler effect from a knowledge of the carrier wavelength, λc, or the carrier frequency, f_c, the vehicle's speed, ν, and the relative angle between the wavefront and the vehicle's motion, θ, by using

$$f_d = \frac{\nu}{\lambda_c} \cos\theta = \frac{\nu f_c}{c} \cos\theta \tag{11.75}$$

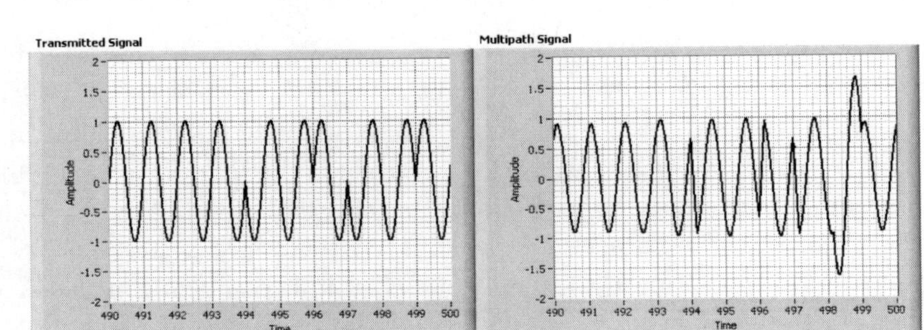

Figure 11.46: The results for a simulated four-component multipath model showing the transmitted and received signals.

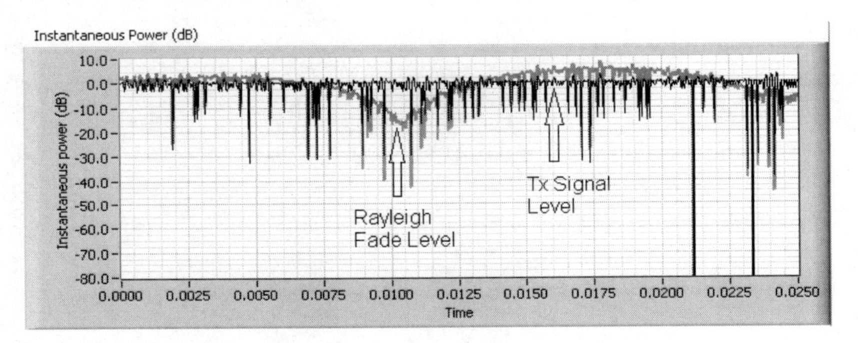

Figure 11.47: The Rayleigh mobile channel LabVIEW simulation results.

For terrestrial vehicles, the Doppler shift is typically below 100 Hz. Figure 11.47 illustrates that the Doppler shift can be combined with the multipath to create a complicated amplitude fade for the signal. The figure shows a LabVIEW® simulation of a Quadrature Phase Shift Keying (QPSK) transmission through a mobile link. The figure illustrates that the Rayleigh fading profile causes a time-varying change to the received QPSK carrier with significant amplitude modulation that affects the recovered data set.

Link analysts use two general statistical models to describe small-scale fading caused by Doppler effects [Ande99; Jake74; Rapp96; Reud74]. The first is a *Rayleigh Model* when there is no identifiable direct component and the link has many reflected components. The second is a *Rician Model* where a direct component and reflected components are present. Figure 11.48 illustrates these two modes. Rayleigh fading is modeled by considering the arriving signal amplitude for each component, B_n, is relatively constant. Each component arrives at a random angle ϕ_n with

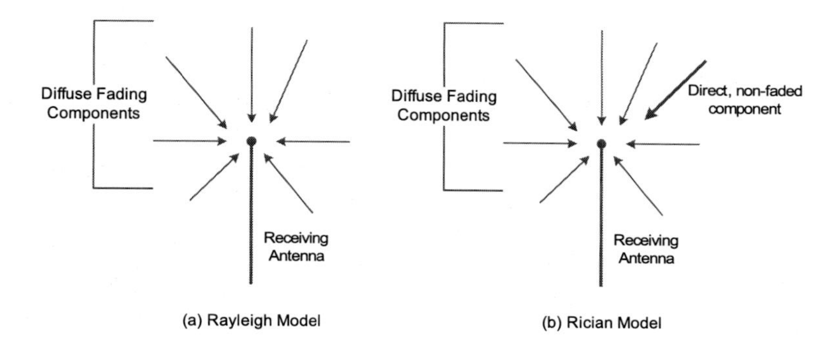

Figure 11.48: Rician and Rayleigh fading components.

a random delay variation ψ_n, and has a frequency shifted by $\Delta f = \nu/\lambda$ Hz. The electric field for the received signal is

$$E_r = \sum n = 1^N B_n \cos(\omega_0 t + \theta_n)$$
$$\theta_n = 2\pi \left(\frac{\nu}{\lambda}\right) t \cos\phi_n + \psi_n \tag{11.76}$$

In I/Q-channel form, this is

$$I(t) = \sum_{n=1}^{N} B_n \cos\theta_n$$
$$Q(t) = \sum_{n=1}^{N} B_n \sin\theta_n \tag{11.77}$$

For a "large" number of carriers ($N > 6$), we can apply statistical means to see that the envelope formed by the components in Equation (11.77) is described by a Rayleigh statistical distribution [Ande99; Jake74]. Figure 11.49 shows the variation in the received signal's envelope as a function of the Doppler frequency for the cases of 10, 100 and 500 Hz. Fades up to 20 dB are possible due to the Rayleigh fading.

If there is a strong direct component, the received signal envelope becomes a Rician distribution

$$e = \sqrt{(I(t) + A)^2 + Q(t)^2} \tag{11.78}$$

This is equivalent to the Rayleigh envelope from Equation (11.77) with the constant amplitude term, A, added from the direct ray. The

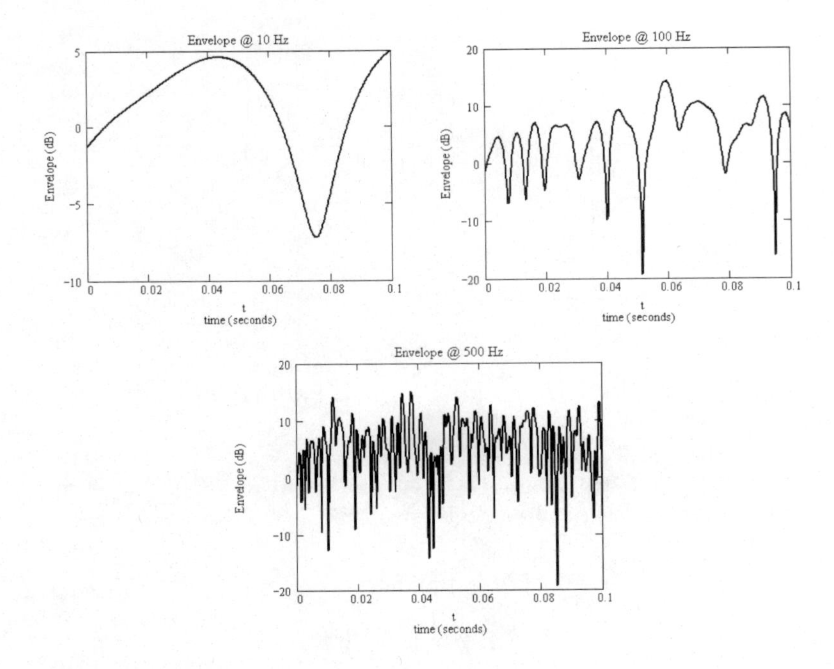

Figure 11.49: Examples of Rayleigh fading envelopes as a function of Doppler shift.

Rician channel is usually characterized by the K factor that yields the relative ratio between the direct component and the diffuse component. The K factor estimate is

$$K = \frac{A^2}{2\sigma^2}$$
$$\sigma^2 = \frac{1}{2} E\left\{|B_n|^2\right\} \tag{11.79}$$

The Rician fading channel has similar characteristics to the Rayleigh fading channel and therefore has approximately similar amplitude fades.

11.8.5 Link Planning

As we have seen, the mobile channel can have significant amplitude fades that exceed 10 dB. This implies that the link designer needs to include an additional fade term in Equation (11.41) to have sufficient margin to ensure that the C/N maintains the channel's desired quality of service. As with the rain fades, mobile amplitude fades are statistical in nature so the link designer does not need the fading margin at all times. Because of the random nature of the fades, there is a finite probability that the link exceeds the assumed fade levels and then the link BER is not maintained. The link designer must determine the margin required to ride out all of these potential amplitude fades.

11.9 REFERENCES

[Ande99] J. B. Anderson. *Digital Transmission Engineering.* Piscataway, NJ: IEEE Press, 1999.

[ARRL05] "Propagation of RF Signals." In: *The ARRL Handbook.* Ed. by D. G. Reed. 82nd Edition. Newington, CT: ARRL, 2005. Chap. 20, pp. 1 –25. ISBN: 0-87259-929-9.

[Bala97] C. A. Balanis. *Antenna Theory.* 2nd Edition. New York, NY: John Wiley & Sons, 1997. ISBN: 0-471-59268-4.

[Barn72] W. T. Barnett. "Multipath Propagation at 4, 6, and 11 GHz." In: *Bell System Technical Journal* 51.2 (Feb. 1972), pp. 321 –361.

[Bert00] H. L. Bertoni. *Radio Propagation for Wireless Systems.* Upper Saddle River, NJ: Prentice Hall PTR, 2000. ISBN: 0-13-026373-7.

[Boku11] R. S. Bokulic and C. C. DeBoy. "Telemetry, Tracking, and Command (TT&C) Subsystem." In: *Space Mission Engineering: The New SMAD.* Ed. by J. R. Wertz, D. F. Everett, and J. J. Puschell. Hawthorne: Microcosm Press, 2011. Chap. 21.1, pp. 627–641. ISBN: 978-1-881-883-15-9. URL: www.sme-smad.com. CA.

[Bors99a] A. Borsholm and S. Horan. "A MATLAB Program for Predicting the Antenna Wetting Effects of the ACTS 1.2-meter Antenna." In: *Proc. 23rd NASA Propagation Experimenters Meeting (NAPEX XXIII) and the Advanced Communications Technology Satellite (ACTS) Propagation Studies Workshop.* JPL Publication 99-16. Falls Church, VA, 1999.

[Bors99b] A. Borsholm. "Modeling and Experimental Validation of the Surface Attenuation Effects of Rain on Composite Antenna Structures at Ka-Band." MS Thesis. Las Cruces, NM: New Mexico State University, May 1999.

[COSPAR] COSPAR International Reference Atmosphere Working Group. *COSPAR International Reference Atmosphere: 1986 (0 km to 120 km).* National Aeronautics and Space Administration, Goddard Space Flight Center. 1986. URL: http://ccmc.gsfc.nasa.gov/modelweb/atmos/cospar1.html.

[Crai96] K. H. Craig. "Clear-air characteristics of the troposphere." In: *Propagation of Radiowaves*. Ed. by M. P. M. Hall, L. W. Barclay, and M. T. Hewitt. London, U. K.: The Institution of Electrical Engineers, 1996. Chap. 6, pp. 105 –130. ISBN: 0-85296-819-1.

[Cran96] R. K. Crane. *Electromagnetic Wave Propagation Through Rain*. New York, NY: John Wiley & Sons, 1996. ISBN: 0-471-61376-2.

[Dave87] W. B. Davenport and W. L. Root. *An Introduction to the Theory of Random Signals and Noise*. New York, NY: IEEE Press, 1987. ISBN: 0-87942-235-1.

[Floc87] W. L. Flock. *Propagation Effects on Satellite Systems at Frequencies Below 10 GHz. A Handbook for Satellite Systems Design*. NASA Reference Publication 1102. Version 02. National Aeronautics and Space Administration. Washington, D.C., Dec. 1987.

[Free87] R. L. Freeman. *Radio System Design for Telecommunications (1-100 GHz)*. New York, NY: John Wiley & Sons, 1987. ISBN: 0-471-81236-6.

[Frii44] H.T. Friis. "Noise Figures of Radio Receivers." In: *Proceedings of the IRE* 32 (7 July 1944), pp. 419 –422. DOI: 10.1109/JRPROC.1944.232049.

[Frii46] H.T. Friis. "A Note on a Simple Transmission Formula." In: *Proceedings of the IRE* 34 (5 May 1946), pp. 254 –256. DOI: 10.1109/JRPROC.1946.234568.

[Godd96] J. W. F. Goddard. "Prediction of reliability when degraded by precipitation and cloud." In: *Propagation of Radiowaves*. Ed. by M. P. M. Hall, L. W. Barclay, and M. T. Hewitt. London, U. K.: The Institution of Electrical Engineers, 1996. Chap. 9, pp. 179 –195. ISBN: 0-85296-819-1.

[Gold98] J. Goldhirsh and W. J. Vogel. *Handbook of Propagation Effects for Vehicular and Personal Mobile Satellite Systems. Overview of Experimental and Modeling Results*. The Johns Hopkins University Applied Physics Laboratory and The University of Texas at Austin. Laurel, MD and Austin, TX, Dec. 1998.

[Hall96] M. P. M. Hall. "Overview of radio propagation." In: *Propagation of Radiowaves*. Ed. by M. P. M. Hall, L. W. Barclay, and M. T. Hewitt. London, U. K.: The Institution of Electrical Engineers, 1996. Chap. 1, pp. 1 –23. ISBN: 0-85296-819-1.

[Hick07] I. Hickman. *Practical RF Handbook*. 4th Edition. Oxford, U.K.: Newnes, 2007. ISBN: 0-7506-8039-3.

[IEEE521] *IEEE Standard for Letter Designations for Radar-Frequency Bands.* IEEE Std 521-2002. New York, NY: Institute of Electrical and Electronics Engineers, Jan. 2003. ISBN: ISBN 0-7381-3356-6. DOI: 10.1109/IEEESTD.2003.94224.

[Ippo08] L. J. Ippolito. *Satellite Communications Systems Engineering: Atmospheric Effects, Satellite Link Design and System Performance.* New York, NY: John Wiley & Sons, 2008. ISBN: ISBN 978-0-470-72527-6.

[Jake74] W. C. Jakes. "Multipath interference." In: *Microwave Mobile Communications.* Ed. by W. C. Jakes. New York, NY: IEEE Press, 1974. Chap. 1, pp. 11 –78. ISBN: 0-7803-1069-1.

[Krau88] J. D. Kraus. *Antennas.* 2nd Edition. New York, NY: McGraw-Hill, 1988. ISBN: 0-07-035422-7.

[Marg98] A.S. Margulies and J. Mitola. "Software Defined Radios: A Technical Challenge and a Mitigation Strategy." In: *Proceed. IEEE Fifth International Symposium on Spread Spectrum Techniques and Applications.* 3 vols. Sun City, South Africa, Sept. 1998. ISBN: 0-7803-4281-X. DOI: 10.1109/ISSSTA.1998.723845.

[Mito00] J. Mitola. *Software Radio Architecture Object-Oriented Approaches to Wireless Systems Engineering.* New York: John Wiley & Sons, 2000. ISBN: 0-471-38492-5.

[Mito93] J. Mitola. "Software Radios Survey, Critical Evaluation and Future Directions." In: *IEEE Aerospace and Electronic Systems Magazine* 8 (4 Apr. 1993), pp. 25 –36. DOI: 10.1109/62.210638.

[NTIA13] *Manual of Regulations and Procedures for Federal Radio Frequency Management.* 2013th ed. Dept. of Commerce, National Telecommunications and Information Administration. Washington, D.C., May 2013. URL: http://www.ntia.doc.gov/files/ntia/publications/redbook/2013/May_2013_Edition_of_the_NTIA_Manual.pdf.

[P.341] *Recommendation ITU-R P.341-5. The Concept of Transmission Loss for Radio Links.* International Telecommunications Union, 1999. URL: https://www.itu.int/rec/R-REC-P.341/en.

[P.453] *Recommendation ITU-R P.453-11. The radio refractive index: its formula and refractivity data.* International Telecommunications Union, July 2015. URL: https://www.itu.int/rec/R-REC-P.453/en.

[P.525] *Recommendation ITU-R P.525-2. Calculation of Free-Space Attenuation.* International Telecommunications Union, 1994. URL: https://www.itu.int/rec/R-REC-P.525/en.

[P.526] *Recommendation ITU-R P.526-13. Propagation by diffraction.* International Telecommunications Union, Nov. 2013. URL: https://www.itu.int/rec/R-REC-P.526/en.

[P.530] *Recommendation ITU-R P.530-16. Propagation data and prediction methods required for the design of terrestrial line-of-sigth systems.* International Telecommunications Union, July 2015. URL: https://www.itu.int/rec/R-REC-P.530/en.

[P.618] *Recommendation ITU-R P.618-12. Propagation data and prediction methods required for the design of Earth-space telecommunications systems.* International Telecommunications Union, July 2015. URL: https://www.itu.int/rec/R-REC-P.618/en.

[P.676] *Recommendation ITU-R P.676-10. Attenuation by atmospheric gases.* International Telecommunications Union, Sept. 2013. URL: https://www.itu.int/rec/R-REC-P.676/en.

[P.834] *Recommendation ITU-R P.834-7. Effects of tropospheric refraction on radiowave propagation.* International Telecommunications Union, Oct. 2015. URL: https://www.itu.int/rec/R-REC-P.834/en.

[P.837] *Recommendation ITU-R P.837-6. Characteristics of precipitation for propagation modelling.* International Telecommunications Union, Feb. 2012. URL: https://www.itu.int/rec/R-REC-P.837/en.

[P.838] *Recommendation ITU-R P.838-3. Specific attenuation model for rain for use in prediction methods.* International Telecommunications Union, 2005. URL: https://www.itu.int/rec/R-REC-P.838/en.

[P.839] *Recommendation ITU-R P.839-4. Rain height model for prediction methods.* International Telecommunications Union, Sept. 2013. URL: https://www.itu.int/rec/R-REC-P.839/en.

[Poza01] D. M. Pozar. *Microwave and RF Wireless Systems.* New York, NY: John Wiley & Sons, 2001. ISBN: 0-471-32282-2.

[Poza05] D. M. Pozar. *Microwave Engineering.* 3rd Edition. New York, NY: John Wiley & Sons, 2005. ISBN: 978-0-471-44878-5.

[Prit93] W. L. Pritchard, H. G. Suyderhoud, and R. A. Nelson. *Satellite Communication Systems Engineering.* Englewood Cliffs, NJ: Prentice Hall PTR, 1993. ISBN: 0-13-791468-7.

[Rapp96] T. S. Rappaport. *Wireless Communications Principles and Practice.* Upper Saddle River, NJ: Prentice Hall PTR, 1996. ISBN: 0-13-461088-1.

[Reed02] J.H. Reed. *Software Radio: A Modern Approach to Radio Engineering.* Upper Saddle River, NJ: Prentice Hall PTR, 2002. ISBN: 0-13-081158-0.

[Reud74] D. O. Reudink. "Large-scale variations of the average signal." In: *Microwave Mobile Communications.* Ed. by W. C. Jakes. New York, NY: IEEE Press, 1974. Chap. 2, pp. 79 –132. ISBN: 0-7803-1069-1.

[Rice00] M. Rice. "PCM/FM aeronautical telemetry in frequency selective multipath interference." In: *IEEE Transactions on Aerospace and Electronic Systems* 36.4 (Oct. 2000), pp. 1090 –1098. DOI: 10.1109/7.892660.

[Sing11] R. Singh, E. Pentaleri, and K. M. Price. "Communications Payloads." In: *Space Mission Engineering: The New SMAD.* Ed. by J. R. Wertz, D. F. Everett, and J. J. Puschell. Hawthorne: Microcosm Press, 2011. Chap. 16, pp. 455–492. ISBN: 978-1-881-883-15-9. URL: www.sme-smad.com. CA.

[Skla01] B. Sklar. *Digital Communications: Fundamentals and Applications.* 2nd Edition. Upper Saddle River, NJ: Prentice Hall PTR, 2001.

[Skla79] B. Sklar. "What the System Link Budget Tells the System Engineer or How I Learned to Count in Decibels." In: *Proceedings of the International Telemetering Conference.* Vol. XV. San Diego, CA, Nov. 1979, pp. 331 –344.

[Smit98] A. A. Smith. *Radio Frequency Principles and Applications.* New York, NY: IEEE Press, 1998. ISBN: 0-7803-3431-0.

[Stut98] W. L. Stutzman and G. A. Thiele. *Antenna Theory and Design, 2nd Edition.* New York, NY: Wiley, 1998.

[Viga75] A. Vigants. "Space-Diversity Engineering." In: *Bell System Technical Journal* 54.1 (Jan. 1975), pp. 103 –142.

[Viga81] A. Vigants. "Microwave Radio Obstruction Fading." In: *Bell System Technical Journal* 60.6 (July-August 1981), pp. 785 –801.

[Yuen82a] J. H. Yuen. "Telecommunications System Design." In: *Deep Space Telecommunications Systems Engineering.* Ed. by J. H. Yuen. JPL Publication 82-76. Pasadena, CA: Jet Propulsion Laboratory, California Institute of Technology, National Aeronautics, and Space Administration, 1982. Chap. 1, pp. 1 – 22.

11.10 PROBLEMS

1. Complete the following table for a dish antenna with 50% efficiency assumed:

Diameter	Operating Frequency	Antenna Gain
	1 GHz	40 dB
1 m	10 MHz	
1 m	10 GHz	

2. Using the receiver configuration of Figure 11.24, show that the G/T remains constant by evaluating G, T_{sys}, and G/T at the input to the downconverter and the input to the receiver.

3. For the antennas in Problem 1, what are the HPBW and side lobe levels when a $n = 2$ tapir is used?

4. Determine the gain and HPBW for a 21-turn helical antenna operating at 2.2 GHz. Assume a pitch angle of 13° and a circumference of 0.92λ. If this were mounted on a satellite in geostationary orbit (approximate path length of 36 000 km), how much of the Earth is within the HPBW?

5. Suppose that you replace the LNA in Figure 11.24 having a gain of 20 dB with two 10 dB LNAs. What is the new G/T?

6. For the components in the following figure, determine the G/T at the LNA input

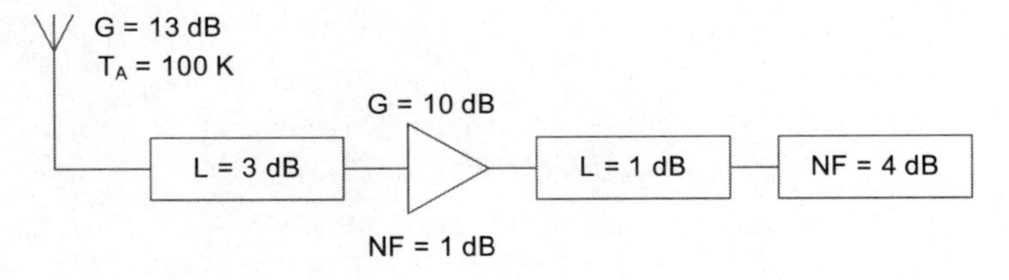

7. Is there link closure for the following Binary Phase Shift Keying (BPSK) link:

Component	Specification
Transmit EIRP	1 dB W
Receiver Dish	10-foot; 45% efficient
Receiver T_{sys}	300 K
Link Distance	100 km
Receiver Bandwidth	250 MHz
Operating Frequency	2200 MHz
Minimum BER	0.0001
Minimum Fade/Attenuation Margin	10 dB
Data Rate	100 kbps

8. Using the knowledge from the link analysis, why does the Earth coverage antenna pattern shown in Figure 11.14 make sense. Consider that an orbiting spacecraft may have a link range of 300 km near the zenith and approximately 1000 km near the horizon. Justify your answer.

9. Consider antenna patterns that have a_1 values of 0.75, 0.85, and 0.90, a_3 values that can be taken to be negligible, and elevation angles of 5° and 55°. Estimate the sky noise temperature at 1 GHz, 10 GHz and 20 GHz.

10. Determine the necessary gain that the receiver antenna needs to close the link operating at 139 MHz when using Frequency Shift Keying (FSK) modulation. This link needs a 20-dB margin to account for multipath fading. The link is a satellite link with a data rate of 9600 bps, a minimum BER of 10^{-5}, a transmitter power of 1 W, a transmitter antenna gain 2 dB, a transmitter pointing loss margin of 3 dB, and a maximum path length of 500 km. The receiver has a NF of 5 dB and a noise bandwidth of 15 kHz. Make any other assumptions about the link you think reasonable. What antenna type would you choose for this task?

11. A commercial encryption algorithm is used on a BPSK telecommand link. The algorithm encodes each command data packet into a fixed-length packet 256 bits long. If there is an error on the link, the decryption algorithm effectively scrambles the command packet so that the resulting error rate is 50%. If the channel has a base BER of 10^{-5}, then

(a) Compute the "coding loss" for this encryption on the BPSK transmission

(b) How much larger would the EIRP need to be to make the packet error rate less than 10^{-6} given that all of the other system parameters stay the same

(c) Alternatively, suggest a coding technique to achieve the same result

Provide the analysis to support your answers.

12. Estimate the gaseous absorption for the path profile illustrated in Figure 11.31. For a worst-case result, assume the atmosphere is at standard temperature and pressure. Estimate the attenuation at 1, 5, 10, 15 and 20 GHz.

13. Show the path profile for a transmitter, receiver and obstructions given that $K = 4/3$.

Path Profile

Object	**Height Above Sea Level** (m)	**Distance from Transmitter** (km)
Transmitter	1000	0
Mountain ridge	4500	20
Mountain ridge	4500	30
Receiver	1000	40

14. For the path profile in Problem 13, compute the size of the Fresnel zones given that the transmitting frequency is (a) 460 MHz and (b) 4 GHz.

15. For the terrain profile given in Figure 11.31, suggest at least two mitigation approaches for the interference at the "ANTHONY" antenna site?

16. Compute the expected rain-fade margin required to support a 50-km path through rain at an operational frequency of 30 GHz and for rain rates of 10 mm/h and 90 mm/h. Compute the expected attenuation by using the ITU model.

17. Repeat Problem 16 using the ITU rain contours for your geographic region.

18. Derive the path length difference in Equation (11.71).

19. Sketch the expected BPSK carrier with a carrier frequency of 10 Hz, a data rate of 1 bps, and a four-component multipath channel. Assume that you model the components with the following parameters:

Component	**Path Delay (sec)**	**Relative Amplitude**	**Relative Phase Shift**
1	0.0	0.9	0°
2	0.05123456	0.5	180°
3	0.0198765544322	0.2	0°
4	0.15647387278	0.1	180°

20. Use a simulation package such as LabVIEW® or MATLAB® to investigate Rician fading profiles. Consider a QPSK signal; use K factors of 1, 10, and 100, and Doppler spreads of 10 Hz and 100 Hz. What do you notice about the amplitude changes as the parameters change?

APPENDIX A

Acronyms, Abbreviations, and Symbols

A.1 ACRONYMS AND ABBREVIATIONS

AC	Alternating Current
ACTS	Advanced Communications Technology Satellite
ADC	Analog-to-Digital Converter
AES	Advanced Encryption Standard
AM	Amplitude Modulation
ANOVA	Analysis of Variance
ARQ	Automatic Repeat Request
ASCII	American Standard Code for Information Interchange
AWGN	Additive White Gaussian Noise
BC	Bus Controller
BCD	Binary Coded Decimal
BCH	Bose Ray-Chaudhuri Hocquenghem
BER	Bit Error Rate

BFSK	Binary Frequency Shift Keying
Biϕ	Bi-Phase
Biϕ-L	Bi-Phase-Level
Biϕ-M	Bi-Phase-Mark
Biϕ-S	Bi-Phase-Space
BIST	Built-In Self-Test
BM	Bus Monitor
BPF	Band Pass Filter
BPSK	Binary Phase Shift Keying
BT	Bandwidth-Bit-Period
BUFR	Binary Universal Form for data Representation
CCSDS	Consultative Committee for Space Data Systems
CDF	Cumulative Distribution Function
CDMA	Code Division Multiple Access
CF	Control Functions
CMD	Command
CPFSK	Continuous Phase Frequency Shift Keying
CPU	Central Processing Unit
CRC	Cyclic Redundancy Check
CRG	Capacitive Rain Gauge
CTS	Clear To Send
DAC	Digital-to-Analog Converter
DC	Direct Current
DCE	Data Communications Equipment

DST	Daylight Savings Time
DTE	Data Terminal Equipment
DTN	Delay Tolerant Networking
ECG	Electrocardiograph
EHF	Extremely High Frequency
EIRP	Effective Isotropic Radiated Power
ELF	Extremely Low Frequency
ENOB	Effective Number of Bits
EOS	Earth Observing System
FaM	Fade Margin
FCC	Federal Communications Commission
FDM	Frequency Division Multiplexing
FFID	Frame Format Identifier
FM	Frequency Modulation
FQPSK	Feher's Quadrature Phase Shift Keying
FSK	Frequency Shift Keying
ftp	File Transfer Protocol
GMSK	Gaussian Minimum Shift Keying
GMT	Greenwich Mean Time
GPS	Global Positioning System
GUI	Graphical User Interface
GUM	Guide to the expression of Uncertainty in Measurement
HDLC	High-level Data Link Control
HF	High Frequency

HMI	Human-Machine Interface
HOW	Hand Over Word
HPA	High Power Amplifier
HPBW	Half-Power Beam Width
HPF	High Pass Filter
I^2C	Inter-Integrated Circuit
I/F	Interface
IF	Intermediate Frequency
IM	Intermodulation
I/O	Input/Output
IoT	Internet of Things
IP	Internet Protocol
IRIG	Inter-Range Instrumentation Group
ISM	Industrial, Scientific, and Medical
ISO	International Organization for Standardization
ITU	International Telecommunications Union
JD	Julian Date
JPEG	Joint Photographic Experts Group
LAN	Local Area Network
LDPC	Low Density Parity Check
LF	Low Frequency
LNA	Low Noise Amplifier
LPF	Low Pass Filter
LR-WPAN	Low-Rate Wireless Personal Area Network

LVDS Low Voltage Differential Signaling

M-LVDS Multipoint Low Voltage Differential Signaling

MEL Master Equipment List

MaF Major Frame

MF Medium Frequency

MiF Minor Frame

MJD Modified Julian Date

MMI Man-Machine Interface

mse Mean Square Error

MSK Minimum Shift Keying

MUF Maximum Usable Frequency

MUX multiplexer

NASA National Aeronautics and Space Administration

NBFM Narrowband Frequency Modulation

NIST National Institute of Standards and Technology

NMEA National Marine Electronics Association

NRZ Non-Return to Zero

NRZ-L Non-Return to Zero-Level

NRZ-M Non-Return to Zero-Mark

NRZ-S Non-Return to Zero-Space

NTIA National Telecommunications and Information Administration

O/S Operating System

OQPSK Offset Quadrature Phase Shift Keying

PCM Pulse Code Modulation

PDF Probability Density Function

PDU Protocol Data Unit

PLL Phase Locked Loop

PM Phase Modulation

PN Pseudorandom Noise

ppb Parts Per Billion

ppm Parts Per Million

PPS Pulse Per Second

POB Power Out-of-Band

PSD Power Spectral Density

PSK Phase Shift Keying

PWM Pulse Width Modulation

QAM Quadrature Amplitude Modulation

QPSK Quadrature Phase Shift Keying

RAM Random Access Memory

ReT Resistance Thermometer

RF Radio Frequency

RFID Radio Frequency Identification

rms Root Mean Square

ROM Read Only Memory

R-S Reed-Solomon

RT Remote Terminal

RTD Resistance Temperature Detector

RTS Request To Send

RZ	Return to Zero
$\Sigma\Delta$	Sigma-Delta
SA	Selective Availability
SBS	Straight Binary Seconds
SCL	Serial Clock
SDA	Serial Data
SDR	Software Defined Radio
SE	Standard Error
SPI	Serial Peripheral Interface
SSE	Sum Square Error
SEU	Single Event Upset
SF	Scale factor
SF	SubFrame
SFID	SubFrame Identifier
SHA	Sample-and-Hold Amplifier
SHF	Super High Frequency
SI	International System of Units
SNR	Signal-to-Noise Ratio
SOQPSK	Shaped Offset Quadrature Phase Shift Keying
SQPSK	Staggered Quadrature Phase Shift Keying
SSR	Sum Square Regression
ST	Sidereal Time
TAI	International Atomic Time
TC	Telecommand

TCP/IP	Transmission Control Protocol/Internet Protocol
TDM	Time Division Multiplexing
TM	Telemetry
TMoIP	Telemetry over Internet Protocol
TT	Terrestrial Time
TWTA	Traveling Wave Tube Amplifier
UDP/IP	User Datagram Protocol/Internet Protocol
UHF	Ultra High Frequency
USB	Universal Serial Bus
UT	Universal Time
UTC	Coordinated Universal Time
VCO	Voltage Controlled Oscillator
VHF	Very High Frequency
VI	Virtual Instrument
VIM	International Vocabulary of Metrology
VLF	Very Low Frequency
VPN	Virtual Private Network
WBFM	Wideband Frequency Modulation

A.2 FUNCTIONS, SYMBOLS, UNITS, AND VARIABLES

A	Area, m^2
A_d	Diffraction Fading Attenuation, dB
A_g	Gaseous Attenuation, dB
A_n	Accuracy of the nth measurement
B	Baseband Bandwidth, Hz

β_{FM}	FM Modulation Index
β_{PM}	PM Modulation Index
bps	Bits per second
c	Speed of Light, 299 792 458 m/s
°C	Degree Celsius
C	Capacitance, F
C/N	Carrier-to-Noise Power Ratio
D	Deviation Ratio
D	Directivity
dB	decibel
dB W	decibel Watt
dBm	decibel milliWatt
ϵ	Strain, m/m
E	Young's modulus
E_b/N_o	Energy-per-Bit-to-Noise-Spectral-Density
E_s	Energy in the signal $s(t)$
erf	Error Function
erfc	Complementary Error Function
f	Frequency, Hz
f	f statistic
$f_\alpha(m,n)$	Critical value of the F distribution
F	Farads
F	Force, Newtons
F_α	Cumulative Distribution Function of α

fph	frames per hour
fpm	frames per minute
fps	frames per second
f_S	Sampling rate, sps
g	Gravitational constant, $6.674\,08 \times 10^{-11}\ \mathrm{m^3/kg - s^2}$
G	Unit strain
G	Gain
Gbps	1 000 000 000 bps
G_C	Coding gain
G	Generator matrix
GHz	1 000 000 000 Hz
γ_R	Specific Attenuation for Rain, dB/km
γ_{total}	Specific Attenuation for Total Gaseous Attenuation, dB/km
G/T	Gain-to-System-Temperature ratio or Figure of Merit, K^{-1}
h	Strain coefficient, V/m
h	CPFSK modulation index
h	hour
H	Parity check matrix
Hz	Frequency unit
$H(\omega)$	Transfer function in the frequency domain
i	Current, A
I	Identity matrix
j	$\sqrt{-1}$

k	Boltzmann's constant, $1.380\,648\,52 \times 10^{-23}$ J/K or -228.6 dBW/K − Hz
k	Coverage Factor
K	Atmospheric index of refraction change with height factor
K	Kelvin
kbps	1000 bps
kHz	1000 Hz
km	1000 m
ksps	1000 sps
L	Length, m
L	Number of voltage levels
L_E	Effective Path Length, km
L_G	Ground Path, km
L_S	Slant Path, km
m	meter
μ	Mean value
Mbps	1 000 000 bps
MHz	1 000 000 Hz
mm	0.001 m
Msps	1 000 000 sps
mV	0.001 V
µV	0.000 001 V
N	Number of points
N	Noise power, W

N_o	Noise density, W/Hz
$p(x)$	Probability as a function of x
p_e	probability of error
P	Pressure
P	Parity check matrix
P_W^c	Coded word error probability
P_{FL}	Probability of false lock
P_k	Probability of k or fewer errors
P_M	Probability of miss
P_n	Precision of the nth measurement
P_W^{uc}	Uncoded word error probability
q	Sampling voltage increment, V
$Q(x)$	Q Function
ρ	Resistivity, $\Omega - \mathrm{m}$
ρ_i	Residual of the ith point
r_1	Radius of the first Fresnel Zone
R	Range, km
R	Resistance, Ω
R_e	Earth radius, 8500 km
$R_{\%}$	Point Rain Rate, mm/h
R_b	Bit Rate, bps
R_g	Gauge resistance, Ω
R^2	squared correlation
s	second

s	Standard error
s	Complex frequency, $s = j\omega$
$s(t)$	Time domain signal
$S(\omega)$	Frequency domain signal
σ	Standard deviation
σ^2	Variance
σ_μ	Uncertainty in the mean
σ_μ^2	Error in the mean
sps	Samples per second
τ	Time
t	Thickness, m
$t_{\alpha/2}$	Critical value of the t distribution
T	Temperature, K
T_{ant}	Antenna Temperature, K
T_b	Bit Period, s
T_s	Sample period, s
T_{sys}	System Noise Temperature, K
U	Expanded Uncertainty
$u(x)$	Unit step function
V	Volt
V_{pp}	Peak-to-peak voltage, V
ω	Radian frequency
Ω	Ohms
W	Channel Bandwidth, Hz
W	Watt
$z_{\alpha/2}$	Critical value of the normal distribution

APPENDIX B

Supporting Tables

Table B.1: Computed erf, erfc, and Q Functions

x	erf(x)	erfc(x)	Q(x)	x	erf(x)	erfc(x)	Q(x)	x	erf(x)	erfc(x)	Q(x)
0.00	0.000000	1.000000	0.500000	1.35	0.943762	0.056238	0.088508	2.70	0.999866	1.3433E-04	0.003467
0.05	0.056372	0.943628	0.480061	1.40	0.952285	0.047715	0.080757	2.75	0.999899	1.0062E-04	0.002980
0.10	0.112463	0.887537	0.460172	1.45	0.959695	0.040305	0.073529	2.80	0.999925	7.5013E-05	0.002555
0.15	0.167996	0.832004	0.440382	1.50	0.966105	0.033895	0.066807	2.85	0.999944	5.5656E-05	0.002186
0.20	0.222703	0.777297	0.420740	1.55	0.971623	0.028377	0.060571	2.90	0.999959	4.1098E-05	0.001866
0.25	0.276326	0.723674	0.401294	1.60	0.976348	0.023652	0.054799	2.95	0.999970	3.0203E-05	0.001589
0.30	0.328627	0.671373	0.382089	1.65	0.980376	0.019624	0.049471	3.00	0.999978	2.2090E-05	0.001350
0.35	0.379382	0.620618	0.363169	1.70	0.983790	0.016210	0.044565	3.05	0.999984	1.6080E-05	0.001144
0.40	0.428392	0.571608	0.344578	1.75	0.986672	0.013328	0.040059	3.10	0.999988	1.1649E-05	9.6760E-04
0.45	0.475482	0.524518	0.326355	1.80	0.989091	0.010909	0.035930	3.15	0.999992	8.3982E-06	8.1635E-04
0.50	0.520500	0.479500	0.308538	1.85	0.991111	0.008889	0.032157	3.20	0.999994	6.0258E-06	6.8714E-04
0.55	0.563323	0.436677	0.291160	1.90	0.992790	0.007210	0.028717	3.25	0.999996	4.3028E-06	5.7703E-04
0.60	0.603856	0.396144	0.274253	1.95	0.994179	0.005821	0.025588	3.30	0.999997	3.0577E-06	4.8342E-04
0.65	0.642029	0.357971	0.257846	2.00	0.995322	0.004678	0.022750	3.35	0.999998	2.1625E-06	4.0406E-04
0.70	0.677801	0.322199	0.241964	2.05	0.996258	0.003742	0.020182	3.40	0.999998	1.5220E-06	3.3693E-04
0.75	0.711156	0.288844	0.226627	2.10	0.997021	0.002979	0.017864	3.45	0.999999	1.0661E-06	2.8029E-04
0.80	0.742101	0.257899	0.211855	2.15	0.997639	0.002361	0.015778	3.50	0.999999	7.4310E-07	2.3263E-04
0.85	0.770668	0.229332	0.197663	2.20	0.998137	0.001863	0.013903	3.55	0.999999	5.1548E-07	1.9262E-04
0.90	0.796908	0.203092	0.184060	2.25	0.998537	0.001463	0.012224	3.60	1.000000	3.5586E-07	1.5911E-04
0.95	0.820891	0.179109	0.171056	2.30	0.998857	0.001143	0.010724	3.65	1.000000	2.4448E-07	1.3112E-04
1.00	0.842701	0.157299	0.158655	2.35	0.999111	8.8927E-04	0.009387	3.70	1.000000	1.6715E-07	1.0780E-04
1.05	0.862436	0.137564	0.146859	2.40	0.999311	6.8851E-04	0.008198	3.75	1.000000	1.1373E-07	8.8417E-05
1.10	0.880205	0.119795	0.135666	2.45	0.999469	5.3058E-04	0.007143	3.80	1.000000	7.7004E-08	7.2348E-05
1.15	0.896124	0.103876	0.125072	2.50	0.999593	4.0695E-04	0.006210	3.85	1.000000	5.1886E-08	5.9059E-05
1.20	0.910314	0.089686	0.115070	2.55	0.999689	3.1066E-04	0.005386	3.90	1.000000	3.4792E-08	4.8096E-05
1.25	0.922900	0.077100	0.105650	2.60	0.999764	2.3603E-04	0.004661	3.95	1.000000	2.3217E-08	3.9076E-05
1.30	0.934008	0.065992	0.096800	2.65	0.999822	1.7849E-04	0.004025	4.00	1.000000	1.5417E-08	3.1671E-05

Note: Values computed using Excel®.

Table B.2: Critical Values of the Normal Distribution

Probability	0.8	0.9	0.95	0.98	0.99	0.995	0.998
α	0.2	0.1	0.05	0.02	0.01	0.005	0.002
$\alpha/2$	0.1	0.05	0.025	0.01	0.005	0.0025	0.001
Critical Value	1.282	1.645	1.960	2.326	2.576	2.807	3.090

Note: Values computed using Excel®.

Table B.3: Critical Values of the t Distribution

Probability	0.8	0.9	0.95	0.98	0.99	0.995	0.998
α	0.2	0.1	0.05	0.02	0.01	0.005	0.002
$\alpha/2$	0.1	0.05	0.025	0.01	0.005	0.0025	0.001
ν degrees of freedom							
1	3.078	6.314	12.706	31.821	63.657	127.321	318.309
2	1.886	2.920	4.303	6.965	9.925	14.089	22.327
3	1.638	2.353	3.182	4.541	5.841	7.453	10.215
4	1.533	2.132	2.776	3.747	4.604	5.598	7.173
5	1.476	2.015	2.571	3.365	4.032	4.773	5.893
6	1.440	1.943	2.447	3.143	3.707	4.317	5.208
7	1.415	1.895	2.365	2.998	3.499	4.029	4.785
8	1.397	1.860	2.306	2.896	3.355	3.833	4.501
9	1.383	1.833	2.262	2.821	3.250	3.690	4.297
10	1.372	1.812	2.228	2.764	3.169	3.581	4.144
11	1.363	1.796	2.201	2.718	3.106	3.497	4.025
12	1.356	1.782	2.179	2.681	3.055	3.428	3.930
13	1.350	1.771	2.160	2.650	3.012	3.372	3.852
14	1.345	1.761	2.145	2.624	2.977	3.326	3.787
15	1.341	1.753	2.131	2.602	2.947	3.286	3.733
16	1.337	1.746	2.120	2.583	2.921	3.252	3.686
17	1.333	1.740	2.110	2.567	2.898	3.222	3.646
18	1.330	1.734	2.101	2.552	2.878	3.197	3.610
19	1.328	1.729	2.093	2.539	2.861	3.174	3.579
20	1.325	1.725	2.086	2.528	2.845	3.153	3.552
25	1.316	1.708	2.060	2.485	2.787	3.078	3.450
30	1.310	1.697	2.042	2.457	2.750	3.030	3.385
40	1.303	1.684	2.021	2.423	2.704	2.971	3.307
50	1.299	1.676	2.009	2.403	2.678	2.937	3.261
100	1.290	1.660	1.984	2.364	2.626	2.871	3.174

Note:: Values computed using Excel®.

Table B.4: Critical Values of the $f_{0.1}(m,n)$ Distribution

n/m	1	2	3	4	5	6	7	8	9	10	15	20
1	39.86	49.50	53.59	55.83	57.24	58.20	58.91	59.44	59.86	60.19	61.22	61.74
2	8.526	9.000	9.162	9.243	9.293	9.326	9.349	9.367	9.381	9.392	9.425	9.441
3	5.538	5.462	5.391	5.343	5.309	5.285	5.266	5.252	5.240	5.230	5.200	5.184
4	4.545	4.325	4.191	4.107	4.051	4.010	3.979	3.955	3.936	3.920	3.870	3.844
5	4.060	3.780	3.619	3.520	3.453	3.405	3.368	3.339	3.316	3.297	3.238	3.207
6	3.776	3.463	3.289	3.181	3.108	3.055	3.014	2.983	2.958	2.937	2.871	2.836
7	3.589	3.257	3.074	2.961	2.883	2.827	2.785	2.752	2.725	2.703	2.632	2.595
8	3.458	3.113	2.924	2.806	2.726	2.668	2.624	2.589	2.561	2.538	2.464	2.425
9	3.360	3.006	2.813	2.693	2.611	2.551	2.505	2.469	2.440	2.416	2.340	2.298
10	3.285	2.924	2.728	2.605	2.522	2.461	2.414	2.377	2.347	2.323	2.244	2.201
11	3.225	2.860	2.660	2.536	2.451	2.389	2.342	2.304	2.274	2.248	2.167	2.123
12	3.177	2.807	2.606	2.480	2.394	2.331	2.283	2.245	2.214	2.188	2.105	2.060
13	3.136	2.763	2.560	2.434	2.347	2.283	2.234	2.195	2.164	2.138	2.053	2.007
14	3.102	2.726	2.522	2.395	2.307	2.243	2.193	2.154	2.122	2.095	2.010	1.962
15	3.073	2.695	2.490	2.361	2.273	2.208	2.158	2.119	2.086	2.059	1.972	1.924
16	3.048	2.668	2.462	2.333	2.244	2.178	2.128	2.088	2.055	2.028	1.940	1.891
17	3.026	2.645	2.437	2.308	2.218	2.152	2.102	2.061	2.028	2.001	1.912	1.862
18	3.007	2.624	2.416	2.286	2.196	2.130	2.079	2.038	2.005	1.977	1.887	1.837
19	2.990	2.606	2.397	2.266	2.176	2.109	2.058	2.017	1.984	1.956	1.865	1.814
20	2.975	2.589	2.380	2.249	2.158	2.091	2.040	1.999	1.965	1.937	1.845	1.794
25	2.918	2.528	2.317	2.184	2.092	2.024	1.971	1.929	1.895	1.866	1.771	1.718
30	2.881	2.489	2.276	2.142	2.049	1.980	1.927	1.884	1.849	1.819	1.722	1.667
40	2.835	2.440	2.226	2.091	1.997	1.927	1.873	1.829	1.793	1.763	1.662	1.605
50	2.809	2.412	2.197	2.061	1.966	1.895	1.840	1.796	1.760	1.729	1.627	1.568
100	2.756	2.356	2.139	2.002	1.906	1.834	1.778	1.732	1.695	1.663	1.557	1.494

Note: Values computed using Excel®.

Table B.5: Critical Values of the $f_{0.05}(m,n)$ Distribution

n/m	1	2	3	4	5	6	7	8	9	10	15	20
1	161.4	199.5	215.7	224.6	230.2	234.0	236.8	238.9	240.5	241.9	245.9	248.0
2	18.51	19.00	19.16	19.25	19.30	19.33	19.35	19.37	19.38	19.40	19.43	19.45
3	10.13	9.552	9.277	9.117	9.013	8.941	8.887	8.845	8.812	8.786	8.703	8.660
4	7.709	6.944	6.591	6.388	6.256	6.163	6.094	6.041	5.999	5.964	5.858	5.803
5	6.608	5.786	5.409	5.192	5.050	4.950	4.876	4.818	4.772	4.735	4.619	4.558
6	5.987	5.143	4.757	4.534	4.387	4.284	4.207	4.147	4.099	4.060	3.938	3.874
7	5.591	4.737	4.347	4.120	3.972	3.866	3.787	3.726	3.677	3.637	3.511	3.445
8	5.318	4.459	4.066	3.838	3.687	3.581	3.500	3.438	3.388	3.347	3.218	3.150
9	5.117	4.256	3.863	3.633	3.482	3.374	3.293	3.230	3.179	3.137	3.006	2.936
10	4.965	4.103	3.708	3.478	3.326	3.217	3.135	3.072	3.020	2.978	2.845	2.774
11	4.844	3.982	3.587	3.357	3.204	3.095	3.012	2.948	2.896	2.854	2.719	2.646
12	4.747	3.885	3.490	3.259	3.106	2.996	2.913	2.849	2.796	2.753	2.617	2.544
13	4.667	3.806	3.411	3.179	3.025	2.915	2.832	2.767	2.714	2.671	2.533	2.459
14	4.600	3.739	3.344	3.112	2.958	2.848	2.764	2.699	2.646	2.602	2.463	2.388
15	4.543	3.682	3.287	3.056	2.901	2.790	2.707	2.641	2.588	2.544	2.403	2.328
16	4.494	3.634	3.239	3.007	2.852	2.741	2.657	2.591	2.538	2.494	2.352	2.276
17	4.451	3.592	3.197	2.965	2.810	2.699	2.614	2.548	2.494	2.450	2.308	2.230
18	4.414	3.555	3.160	2.928	2.773	2.661	2.577	2.510	2.456	2.412	2.269	2.191
19	4.381	3.522	3.127	2.895	2.740	2.628	2.544	2.477	2.423	2.378	2.234	2.155
20	4.351	3.493	3.098	2.866	2.711	2.599	2.514	2.447	2.393	2.348	2.203	2.124
25	4.242	3.385	2.991	2.759	2.603	2.490	2.405	2.337	2.282	2.236	2.089	2.007
30	4.171	3.316	2.922	2.690	2.534	2.421	2.334	2.266	2.211	2.165	2.015	1.932
40	4.085	3.232	2.839	2.606	2.449	2.336	2.249	2.180	2.124	2.077	1.924	1.839
50	4.034	3.183	2.790	2.557	2.400	2.286	2.199	2.130	2.073	2.026	1.871	1.784
100	3.936	3.087	2.696	2.463	2.305	2.191	2.103	2.032	1.975	1.927	1.768	1.676

Note: Values computed using Excel®.

Table B.6: Critical Values of the $f_{0.01}(m,n)$ Distribution

n/m	**1**	**2**	**3**	**4**	**5**	**6**	**7**	**8**	**9**	**10**	**15**	**20**
1	4052	5000	5403	5625	5764	5859	5928	5981	6022	6056	6157	6209
2	98.50	99.00	99.17	99.25	99.30	99.33	99.36	99.37	99.39	99.40	99.43	99.45
3	34.12	30.82	29.46	28.71	28.24	27.91	27.67	27.49	27.35	27.23	26.87	26.69
4	21.20	18.00	16.69	15.98	15.52	15.21	14.98	14.80	14.66	14.55	14.20	14.02
5	16.26	13.27	12.06	11.39	10.97	10.67	10.46	10.29	10.16	10.05	9.722	9.553
6	13.75	10.92	9.780	9.148	8.746	8.466	8.260	8.102	7.976	7.874	7.559	7.396
7	12.25	9.547	8.451	7.847	7.460	7.191	6.993	6.840	6.719	6.620	6.314	6.155
8	11.26	8.649	7.591	7.006	6.632	6.371	6.178	6.029	5.911	5.814	5.515	5.359
9	10.56	8.022	6.992	6.422	6.057	5.802	5.613	5.467	5.351	5.257	4.962	4.808
10	10.04	7.559	6.552	5.994	5.636	5.386	5.200	5.057	4.942	4.849	4.558	4.405
11	9.646	7.206	6.217	5.668	5.316	5.069	4.886	4.744	4.632	4.539	4.251	4.099
12	9.330	6.927	5.953	5.412	5.064	4.821	4.640	4.499	4.388	4.296	4.010	3.858
13	9.074	6.701	5.739	5.205	4.862	4.620	4.441	4.302	4.191	4.100	3.815	3.665
14	8.862	6.515	5.564	5.035	4.695	4.456	4.278	4.140	4.030	3.939	3.656	3.505
15	8.683	6.359	5.417	4.893	4.556	4.318	4.142	4.004	3.895	3.805	3.522	3.372
16	8.531	6.226	5.292	4.773	4.437	4.202	4.026	3.890	3.780	3.691	3.409	3.259
17	8.400	6.112	5.185	4.669	4.336	4.102	3.927	3.791	3.682	3.593	3.312	3.162
18	8.285	6.013	5.092	4.579	4.248	4.015	3.841	3.705	3.597	3.508	3.227	3.077
19	8.185	5.926	5.010	4.500	4.171	3.939	3.765	3.631	3.523	3.434	3.153	3.003
20	8.096	5.849	4.938	4.431	4.103	3.871	3.699	3.564	3.457	3.368	3.088	2.938
25	7.770	5.568	4.675	4.177	3.855	3.627	3.457	3.324	3.217	3.129	2.850	2.699
30	7.562	5.390	4.510	4.018	3.699	3.473	3.304	3.173	3.067	2.979	2.700	2.549
40	7.314	5.179	4.313	3.828	3.514	3.291	3.124	2.993	2.888	2.801	2.522	2.369
50	7.171	5.057	4.199	3.720	3.408	3.186	3.020	2.890	2.785	2.698	2.419	2.265
100	6.895	4.824	3.984	3.513	3.206	2.988	2.823	2.694	2.590	2.503	2.223	2.067

Note: Values computed using Excel®.

Index